W9-DIU-148

F V

WITHDRAWN

St. Louis Community
College

Library

5801 Wilson Avenue
St. Louis, Missouri 63110

ELECTRONIC
DEVICES

ELECTRONIC DEVICES

Second Edition

Thomas L. Floyd

Merrill Publishing Company
A Bell & Howell Information Company
Columbus   Toronto   London   Melbourne

Cover Photo: Larry Hamill

Published by Merrill Publishing Company
A Bell & Howell Information Company
Columbus, Ohio 43216

This book was set in Times Roman.

Administrative Editor: Steve Helba
Developmental Editor: Don Thompson
Production Coordinator: Molly Kyle
Art Coordinator: Peter Robison
Cover Designer: Cathy Watterson
Text Designer: Cynthia Brunk

Library of Congress Catalog Card Number: 87–61393
International Standard Book Number: 0–675–20883–1
Printed in the United States of America
1 2 3 4 5 6 7 8 9—93 92 91 90 89 88

To Sheila, with love

MERRILL'S INTERNATIONAL SERIES IN ELECTRICAL AND ELECTRONICS TECHNOLOGY

Preface

The second edition of *Electronic Devices* provides the most thorough, comprehensive, and practical coverage of modern electronic devices, circuits, and applications that is possible within a single volume. This book is well balanced with respect to discrete devices and linear integrated circuits. Chapters 1 and 2 are introductory in nature; Chapters 3–11 are devoted to discrete semiconductor devices and circuits, including diodes, bipolar junction transistors, field-effect transistors, and amplifiers; and Chapters 12–17 cover operational amplifiers and other linear ICs with applications. Coverage of oscillators has been expanded to a separate full chapter (Chapter 15), and Chapters 18 and 19 are devoted to thyristors, unijunction transistors, and optoelectronic devices.

All of the features that were so well received in the first edition have been retained and enhanced in this second edition. On the basis of comments and suggestions from instructors in the course of daily use and from interviews, surveys, and extensive reviews, we have incorporated certain changes in pedagogy and organization, as well as additional coverage.

The major new features in this edition are:

- [ ] *Functional* two-color format
- [ ] Chapter objectives
- [ ] Devices applied in a system
- [ ] Greater emphasis on applications
- [ ] Improved troubleshooting coverage
- [ ] A chapter on oscillators
- [ ] Completely revised coverage of active filters
- [ ] Practice exercises related to selected examples
- [ ] Computer programs integrated into example problems
- [ ] Placement of the thyristor and optoelectronic chapters at the end of the book

This edition retains and improves the popular format of the first edition. Each chapter begins with a list of objectives, a brief introduction, and a section listing. A section review follows each chapter section to reinforce key concepts and to provide frequent feedback on the student's comprehension. Answers to these section reviews appear at the ends of the chapters. Self-tests are found at the end of each chapter, with solutions at the end of the book. Beginning with Chapter 3, all chapters have problems, grouped by section numbers for easier assignment and reference. Answers to odd-numbered problems appear at the end of the book, and the Instructor's Manual provides full solutions to all problems.

New in this edition are the practice exercises associated with selected examples throughout the book. These practice exercises require the student to go through the procedure demonstrated in the example under conditions, or with a set of parameters, different from those used in the example itself. This encourages the student to work independently through the selected examples, thus providing additional reinforcement of the relevant concept. Computer programs are now integrated into example problems so that the student can see how helpful a computer can be in solving specific kinds of problems. Answers to the practice exercises are also found at the end of each chapter, along with a list of formulas and a summary.

A unique feature of this second edition is the inclusion in each chapter of a system application. Beginning in Chapter 1, a specific representative system to which the student can relate (a superheterodyne AM/FM receiver and power supply) is introduced at a level compatible with the student's background at that point. Throughout the book, each chapter then opens with a brief introduction to a selected portion of the representative system or a related system that is highlighted in that chapter. The last section of each chapter discusses the typical use of a selected device or devices in the highlighted portion of the system. This approach familiarizes the student with the system concept and how individual devices and circuits fit into a typical system application.

Suggestions for Use

Electronic Devices, Second Edition, can be used in several ways to accommodate a variety of scheduling needs:

☐ A two-term electronic devices and circuits course can cover most of the book by breaking after the material on diodes, BJTs, FETs, and small-signal amplifiers in Chapters 3–9, or after the complete body of amplifier material in Chapters 8–11, with the second term covering operational amplifiers and linear ICs, oscillators, thyristors, UJTs, and optoelectronic devices.

☐ Chapters 12–16 provide sufficient material for a one-term course in op-amps and linear integrated circuits, either after a discrete devices sequence (Chapters 1–13 and Section 15–1, 15–2, and 15–4) or as a completely independent course for students with other backgrounds. Chapters 17–19 can then be added as time allows.

☐ With selective presentation, this text can be used for a one-term devices course by omitting certain topics and by maintaining a rigorous schedule. An example of this approach for a course covering only discrete devices (basically, Chapters 1–11) is to omit Chapter 1, cover Chapter 2 lightly, cover Chapter 4 selectively, and omit or cover Chapter 11 selectively. For a one-term course that needs some linear IC coverage, additional cuts in Chapters 2–11 may be necessary. However, this text's organization, writing style, and price (only a dollar or two more than most shorter, so-called one-semester books) should make one-term use reasonable when necessary, with the subsequent advantage that the student then has an excellent reference tool for further study.

Acknowledgments

This second edition of *Electronic Devices* is the result of many people's efforts. In particular, I want to express my appreciation to Don Thompson, Tim McEwen, Steve Helba, Molly Kyle, Bruce Johnson, and Peter Robison of Merrill Publishing Company for their enthusiasm and dedication to producing the best product possible. My thanks also to the many users of the first edition who offered constructive suggestions, especially William Campas (John Tyler Community College), Edward Kerley (Delaware Technical and Community College—Terry Campus), Sam Hilburn (University of South Carolina), Helmut Seemann (Algonquin College); to Gary Pfeiffer (DeVry Institute of Technology—Columbus) and Jim Morgan (ITT Technical Institute—Dayton), for diligently checking the revised manuscript for accuracy; and the following instructors who reviewed the manuscript and provided many valuable suggestions: Paul Svatik (Owens Technical College); Floyd Martin (Rancho Santiago College); Greg Dixon (Microcomputer Technical Institute); Ulrich Zeisler (Utah Technical College at Salt Lake City); Jill Harlamert (DeVry Institute of Technology—Columbus); Joseph Labok (Los Angeles Valley College); and Duane Henninger (Lincoln Technical Institute—Allentown, Pa.) A special note of appreciation again goes to my wife, Sheila, for her support.

Tom Floyd

Contents

CHAPTER 12
OPERATIONAL AMPLIFIERS   446

CHAPTER 13
OP-AMP FREQUENCY RESPONSE, STABILITY, AND COMPENSATION   496

CHAPTER 14
BASIC OP-AMP APPLICATIONS   534

ELECTRONIC DEVICES

Introduction and Review

In this chapter you will learn

- ☐ A brief history of the development of electronics
- ☐ Who invented the transistor
- ☐ Basic discrete electronic devices and their symbols
- ☐ Basic linear integrated circuits and their symbols
- ☐ Some basic device applications
- ☐ About the system concept
- ☐ How a specific electronic system is organized as an introduction to the use of the devices in system applications

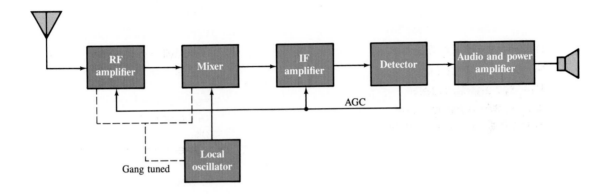

The field of linear electronics encompasses those circuit functions designed to process *analog* quantities as opposed to *digital* quantities. Analog quantities are those that vary continuously over a specified range.

This chapter provides an overview of the various semiconductor devices, both discrete components and integrated circuits, that find applications in linear electronics. We will also introduce a specific system to familiarize you with the system concept and to show how the various devices and circuits we will cover throughout the text can be used in a systems application. We have selected the superheterodyne receiver for this purpose because of its widespread use. The purpose of the system presentation in this chapter is to give you a basic idea of what the system does; complete, detailed coverage is beyond the scope and purpose of this book. After the introduction in this chapter, we will use the system throughout the book to further explore its operation and to illustrate applications of the devices covered in various chapters.

HISTORICAL BACKGROUND

1-1 The Beginning of Electronics

The early experiments in electronics involved electric currents in glass vacuum tubes. One of the first to conduct such experiments was a German named Heinrich Geissler (1814–1879). Geissler removed most of the air from a glass tube and found that the tube glowed when there was an electric current through it. Around 1878, British scientist Sir William Crookes (1832–1919) experimented with tubes similar to those of Geissler. In his experiments, Crookes found that the current in the vacuum tubes seemed to consist of particles.

Thomas Edison (1847–1931), experimenting with the carbon-filament light bulb he had invented, made another important finding. He inserted a small metal plate in the bulb. When the plate was positively charged, there was a current from the filament to the plate. This device was the first thermionic diode. Edison patented it but never used it.

The electron was discovered in the 1890s. The French physicist Jean Baptiste Perrin (1870–1942) demonstrated that the current in a vacuum tube consists of the movement of negatively charged particles in a given direction. Some of the properties of these particles were measured by Sir Joseph Thomson (1856–1940), a British physicist, in experiments he performed between 1895 and 1897. These negatively charged particles later became known as electrons. The charge on the electron was accurately measured by an American physicist, Robert A. Millikan (1868–1953), in 1909. As a result of these discoveries, electrons could be controlled, and the electronic age was ushered in.

Putting the Electron to Work

A vacuum tube that allowed electrical current in only one direction was constructed in 1904 by British scientist John A. Fleming. The tube was used to detect electromagnetic waves. Called the Fleming valve, it was the forerunner of the more recent vacuum diode tubes. Major progress in electronics, however, awaited the development of a device that could boost, or amplify, a weak electromagnetic wave or radio signal. This device was the audion, patented in 1907 by Lee deForest, an American. It was a triode vacuum tube capable of amplifying small electrical ac signals.

Two other Americans, Harold Arnold and Irving Langmuir, made great improvements in the triode vacuum tube between 1912 and 1914. About the same time, deForest and Edwin Armstrong, an electrical engineer, used the triode tube in an oscillator circuit. In 1914, the triode was incorporated in the telephone system and made the transcontinental telephone network possible. The tetrode tube was invented in 1916 by Walter Schottky, a German. The tetrode, along with the pentode (invented in 1926 by Dutch engineer Tellegen), greatly improved the triode. The first television picture tube, called the kinescope, was developed in the 1920s by Vladimir Sworykin, an American researcher.

During World War II, several types of microwave tubes were developed that made possible modern microwave radar and other communications systems. In 1939, the magnetron was invented in Britain by Henry Boot and John Randall. In the same year, the klystron microwave tube was developed by two Americans, Russell Varian and his brother Sigurd Varian. The traveling-wave tube (TWT) was invented in 1943 by Rudolf Komphner, an Austrian-American.

Solid State Electronics

The crystal detectors used in early radios were the forerunners of modern solid state devices. However, the era of solid state electronics began with the invention of the transistor in 1947 at Bell Labs. The inventors were Walter Brattain, John Bardeen, and William Shockley. Figure 1–1 shows these three men.

In the early 1960s, the integrated circuit (IC) was developed. It incorporated many transistors and other components on a single small *chip* of semiconductor material. Integrated circuit technology has been continuously developed and improved, allowing increasingly more complex circuits to be built on smaller chips.

Around 1965, the first *integrated general-purpose operational amplifier* was introduced. This low-cost, highly versatile device incorporated nine transistors and twelve resistors in a small package. It proved to have many advantages over comparable discrete component circuits in terms of reliability and performance. Since this introduction the IC

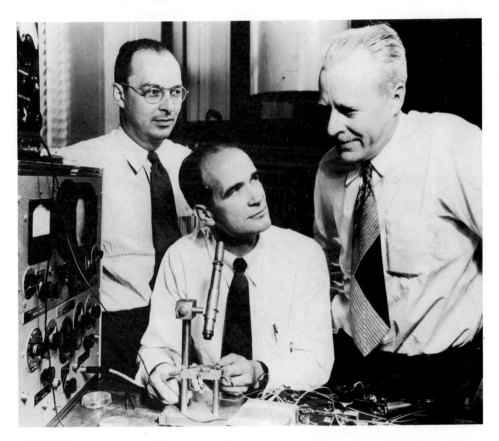

FIGURE 1–1
Nobel Prize winners Drs. John Bardeen, William Shockley, and Walter Brattain, shown left to right, with apparatus used in their first investigations that led to the invention of the transistor. The trio received the 1956 Nobel Physics award for their invention of the transistor, which was announced by Bell Laboratories in 1948. (Courtesy of Bell Laboratories)

operational amplifier has become a *basic building block* for a great variety of linear systems.

---

1. Invention of the transistor occurred where and when?

2. What was the first type of linear integrated circuit to be introduced?

SEMICONDUCTOR DEVICES

1–2

In this text, you will study many different electronic devices, their characteristics, and their applications. These devices—including diodes, transistors, thyristors, and linear integrated circuits—come in a wide range of shapes, sizes, and operational characteristics. They vary from relatively simple, single-component devices to complex circuits housed in a single package. Figure 1–2 shows representative types of package configurations for many of these device categories. Approximately two-thirds of this book deals with these *discrete* semiconductor devices, and one-third of the book covers the principles of linear integrated circuits (ICs).

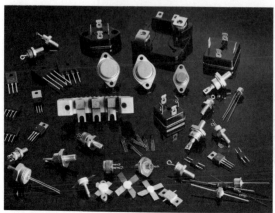

(a) Discrete devices

(b) Integrated circuits

FIGURE 1–2
A grouping of typical semiconductor devices.

Basically, the discrete devices are *individual* components that must be interconnected with other components to form a complete functional circuit. Examples of discrete components that you may already be familiar with are resistors, capacitors, and inductors. In contrast, integrated circuits are basically complete functional circuits constructed on a single piece of semiconductor material and housed in small packages. Discrete semiconductor devices include various types of *diodes, transistors,* and *thyristors.* Linear integrated circuits include *operational amplifiers, comparators, audio amplifiers, radio frequency amplifiers, voltage regulators,* and *interface circuits.*

Discrete Semiconductor Devices

This section is a survey of many of the semiconductor devices that you will learn about in the following chapters.

Diodes.   Diodes are two-terminal devices that allow current in only one direction. They can be classified according to the type of semiconductor material—germanium diodes or silicon diodes. They can also be classified according to the intended application or particular characteristics—rectifier diodes, zener diodes, high-frequency diodes, light-emitting diodes (LEDs), photodiodes, and so on. Figure 1–3 shows the schematic symbols for several common diode types.

FIGURE 1–3
Some diode schematic symbols.

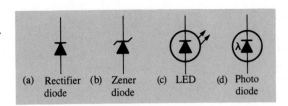

(a) Rectifier   (b) Zener   (c) LED   (d) Photo
     diode   diode   diode

Rectifier Diodes.   Diodes in this category are intended for use in *rectifier* circuits. Rectifier circuits are used to convert an ac voltage into a pulsating dc voltage, as illustrated in Figure 1–4.

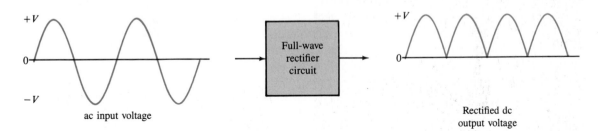

FIGURE 1–4
One type of rectifier function.

Zener Diodes.   Zener diodes are used in voltage regulator circuits or for voltage references. Regulation is the process of maintaining a dc voltage at a constant value. This is illustrated in Figure 1–5.

Other Diodes.   *High-frequency diodes* are specifically designed for applications in communications circuits and other areas where high-frequency voltages and currents are used. *Light-emitting diodes* (LEDs) emit light only when conducting current. They are used in displays, indicators, and optical coupling applications. *Photodiodes* conduct current when a certain amount of light falls on the light-sensitive semiconductor material. They are used in applications requiring detection of specified levels of light. Another type of diode is the

FIGURE 1–5
FIGURE 1–5
Basic voltage regulator function.

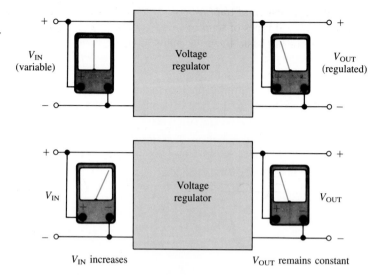

tuning diode or *varactor*. These diodes are used for their variable capacitance characteristic in circuits such as the tuner in TV receivers.

Transistors.   Transistors are semiconductor devices that exhibit the property of *amplification*, whereby a small time-varying voltage (or current) can be increased (amplified) to a larger voltage (or current) that is a replica of the smaller.

There are two basic categories of transistor: the *bipolar junction transistor* (BJT) and the *field-effect transistor* (FET). These two types of devices differ greatly in construction and theory of operation, but their broad applications are similar. Figure 1–6 shows the schematic symbols for the two types of bipolar transistor (pnp and npn) and several varieties of field-effect transistors.

FIGURE 1–6
Transistor symbols.

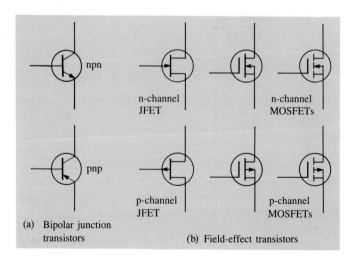

A major application of transistors is the *amplification* of electrical signals, as previously mentioned. All types of transistors are used in a wide variety of amplifier circuits. Figure 1–7 shows the basic linear amplifier concept.

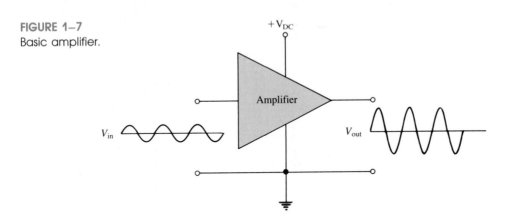

FIGURE 1–7
Basic amplifier.

Oscillators are another type of circuit in which transistors (or integrated circuits) are used. Basically, an oscillator produces an ac output voltage with only a dc input by virtue of its *regenerative* property. Oscillators are used as sources of electrical signals (signal generators), as illustrated in Figure 1–8.

Transistors are also used in voltage regulator circuits to improve the performance by regulating variations of both input voltage (line regulation) and load current (load regulation). Figure 1–9 illustrates these two aspects of regulator operations.

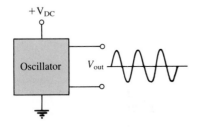

FIGURE 1–8
Basic oscillator.

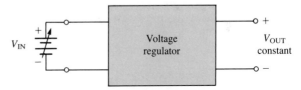

(a) Line regulation (V_{OUT} constant for variable V_{IN})

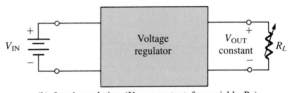

(b) Load regulation (V_{OUT} constant for variable R_L)

FIGURE 1–9
Voltage regulation.

Thyristors.   Thyristors are in a group of semiconductor devices that act as open or closed switches. They are switched into the conducting state with various control mechanisms. This group includes the *Shockley diode,* the *silicon-controlled rectifier* (SCR), the *silicon-controlled switch* (SCS), the *diac,* and the *triac.* Figure 1–10 shows the schematic symbols for each of these devices.

FIGURE 1–10
Thyristor symbols.

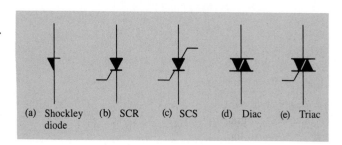

(a) Shockley   (b) SCR   (c) SCS   (d) Diac   (e) Triac
      diode

Shockley Diode.   This device conducts current only when a certain voltage level is reached across its two terminals. Once in conduction, it will continue to conduct until the current drops below a specified minimum value.

Silicon-Controlled Rectifier (SCR).   An SCR is a three-terminal device that can be made to conduct current from its *anode* terminal to its *cathode* terminal by application of a short-duration positive voltage (trigger) to its *gate* terminal. Once conducting, it continues to conduct until the anode current drops below a specified minimum value, as in the Shockley diode.

Silicon-Controlled Switch (SCS).   This device is similar to the SCR except that it has four terminals, two of which are control terminals that can be used to turn on and turn off the device.

Diac.   The diac is a two-terminal thyristor that permits current in either direction. Current will be in a direction determined by the polarity of the voltage across the terminals when the breakover level is reached.

Triac.   The triac is basically a diac with a third terminal for controlling turn-on of the device in either direction.

Unijunction Transistor (UJT).   The UJT is not in the thyristor family, but it is a trigger-controlled device. It is turned on and conducts current between its two base terminals when a certain threshold voltage level is reached on the control (emitter) terminal. The UJT is often used as a trigger device for the SCR or other thyristors. The schematic symbol is shown in Figure 1–11.

FIGURE 1–11
Unijunction transistor symbol.

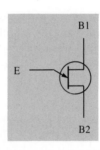

Linear Integrated Circuits

Figure 1–12 shows an integrated circuit chip mounted in a typical package. There are two general categories of integrated circuit: linear and digital. The linear IC is an analog type of circuit, as opposed to the digital type. As mentioned before, an analog function has *continuous* values within a specified range, whereas a digital function has discrete values or steps. Linear integrated circuits can be broken down into several main categories.

FIGURE 1–12
Cutaway view showing an IC chip mounted in a dual-in-line package.

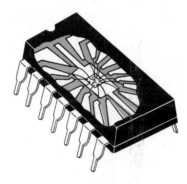

Operational Amplifiers.   The operational amplifier is an extremely versatile device possessing several unique characteristics. Among these are *high voltage gain, high input impedance,* and *low output impedance.* The schematic symbol for an *op-amp* is shown in Figure 1–13. Op-amp applications are diverse and will be covered in detail in later chapters.

FIGURE 1–13
Schematic symbol for an operational amplifier. (dc power supply and ground connections are omitted.)

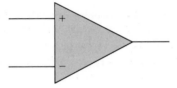

Voltage Regulators.   As mentioned earlier, the purpose of a voltage regulator is to provide a constant output voltage independent of input supply voltage, output load current, and temperature. One basic type of integrated circuit regulator is known as the three-terminal regulator. It has an input, an output, and a ground connection, as shown in Figure 1–14.

FIGURE 1–14
Three-terminal voltage regulator symbol.

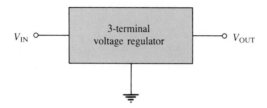

Communications Circuits.   This category of linear ICs includes RF/IF amplifiers, modulators and demodulators, video amplifiers, audio amplifiers, oscillators, phase-locked loops, and others that fulfill more specialized functions.

Interface Circuits.   Large electronic systems typically consist of many different subsystems, all of which must work together (interface) properly to make functional systems. Linear and digital circuits in both discrete and integrated form are used in many systems. Interface circuits "tie" the parts of a system together and usually contain both linear and digital elements. Some important types of interface circuits follow.

Comparators.   These interface devices produce a two-level (digital) output that indicates whether the two input voltages are approximately equal or not equal. The basic schematic symbol is the same as that for an op-amp.

Sense Amplifiers.   The sense amplifier is closely related to the comparator. The basic application is for *sensing* and converting very small voltages or currents normally produced by transducers and some types of memories into voltage levels compatible with other circuits.

Line Drivers and Receivers.   These circuits interface transmission lines with transmitting and receiving systems.

Analog-to-Digital and Digital-to-Analog Converters.   The analog-to-digital (A/D) converter takes values of a continuous voltage input at closely spaced intervals and converts each value to a digital code. The voltage input over a period of time is then represented by a series of digital codes. The digital-to-analog (D/A) converter performs the reverse process, converting the digital codes into an approximate replica of the original analog voltage.

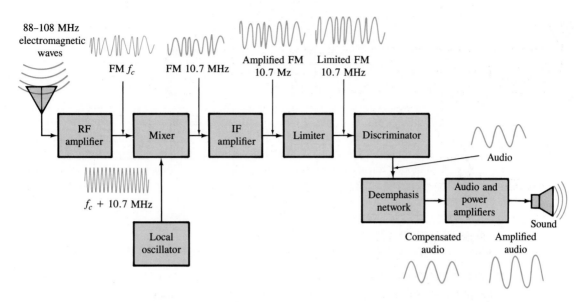

FIGURE 1–21
Example of signal flow through an FM receiver.

The Power Supply

All systems require power to operate. Any AM/FM system requires a power supply to provide dc voltage and current, and the power supply can be considered an integral part of the system. Figure 1–22 shows a block diagram of a typical dc power supply. The input is usually the 120 V, 60 Hz ac from a wall outlet. A transformer couples this ac voltage to a *rectifier,* which converts the ac to pulsating dc. The *filter* smooths out the pulsating dc and produces a relatively constant dc voltage level. The *regulator* serves to keep the dc voltage constant over a range of input or load fluctuations. Some systems may operate without a regulator.

The system application sections at the end of the chapters will each highlight a certain block in the receiver system. They explain the basic operation of the blocks, and you will see how a specific device or circuit fits into the block and what its function is.

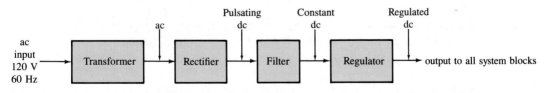

FIGURE 1–22
A typical power supply block diagram.

SECTION REVIEW 1–4

1. What do AM and FM mean?
2. How do AM and FM differ?
3. What are the standard broadcast frequency bands for AM and FM?

SUMMARY

☐ A discrete device is a single component, whereas an integrated circuit consists of many components interconnected on a single small chip of semiconductor material.

☐ Diodes are semiconductor devices that permit current in only one direction.

☐ Zener diodes are used for voltage regulation and voltage references.

☐ Thyristors are devices that conduct current when triggered and continue to conduct until the current is sufficiently reduced.

☐ Transistors are semiconductor devices that exhibit the property of amplification, or they can be used as electronic switches.

☐ The two basic types of transistor are the bipolar junction transistor (BJT) and the field-effect transistor (FET).

☐ Two important classes of linear integrated circuits are the operational amplifier and the voltage regulator.

☐ Basically, a linear circuit is one whose output is linearly related to the input.

☐ A system is a composition of many types of devices and circuits that work together to perform a specified function.

SELF-TEST

1. What is the purpose of the rectifier diode?
2. What are zener diodes used for?
3. What does an amplifier do?
4. What is the purpose of a rectifier?
5. Name the two basic types of transistor.
6. Name three types of thyristors.
7. Name an important type of linear integrated circuit.
8. Sketch the schematic symbol(s) for each of these devices: **(a)** Rectifier diode; **(b)** Zener diode; **(c)** Bipolar transistor; **(d)** SCR; **(e)** JFET; **(f)** MOSFET.
9. Draw the block diagrams for an AM and an FM superheterodyne receiver and a power supply.
10. What is the basic function of a voltage regulator?

Section 1–1
1. Bell Labs, 1947.

2. Operational amplifier.

Section 1–2
1. Integrated circuits consist of a large number of electronic devices formed on a single tiny chip of silicon material and interconnected to form a complete circuit that performs a specified function. Discrete devices are individual components that must be connected with other devices on a circuit board to form a complete circuit.

Section 1–3
1. Not usually.

2. Devices are individual components; circuits are several devices connected together to produce a desired result; and systems consist of many circuits that operate together to produce a desired result.

Section 1–4
1. AM—amplitude modulation, FM—frequency modulation.

2. In AM, the amplitude of a carrier signal varies according to a prescribed pattern corresponding to a modulating signal. In FM, the frequency of a carrier signal varies according to a pattern corresponding to a modulating signal.

3. AM—540 kHz to 1640 kHz, FM—88 MHz to 108 MHz.

Semiconductor Materials and pn Junctions

In this chapter you will learn
- [ ] The definition of an atom
- [ ] What an electron is and how it fits into the atomic structure
- [ ] What ionization is
- [ ] The basic structure of silicon and germanium atoms
- [ ] How atoms bond together to form crystals
- [ ] How the energy levels within an atom's structure relate to current in a material
- [ ] The definition of p-type and n-type semiconductors
- [ ] How a pn junction is formed
- [ ] The practical use of pn junctions in electronic devices
- [ ] What a aiode is
- [ ] How to forward-bias and reverse-bias a diode
- [ ] How diodes are used in a specific system application

To acquire a basic understanding of semiconductors, some knowledge of atomic theory and the structure of materials is necessary. In this chapter you will learn about the basic semiconductor materials used in the manufacture of the diodes, transistors, and other semiconductor devices studied in the following chapters. We will also cover the important concepts of pn junction theory, which are necessary for understanding diode and transistor operation.

The diode, introduced in this chapter, is important in many applications. One application in which the diode is always found is in power supply rectifiers (indicated by the shaded block in the system diagram). The basic property of the diode that permits current to flow in only one direction is directly applicable in converting ac voltage to dc voltage, as you will see in the system application section at the end of this chapter.

ATOMS

2–1

An atom is the smallest particle of an element that retains the characteristics of that element. Each known element has atoms that are different from the atoms of all other elements. This gives each element a unique atomic structure. According to the classical Bohr model, atoms have a *planetary* type of structure which consists of a central *nucleus* surrounded by orbiting *electrons*. The nucleus consists of positively charged particles called *protons* and uncharged particles called *neutrons*. Electrons are the basic particles of *negative charge*.

Each type of atom has a certain number of electrons and protons that distinguishes it from the atoms of all other elements. For example, the simplest atom is that of hydrogen. It has one proton and one electron, as shown in Figure 2–1(a). The helium atom, shown in Figure 2–1(b), has two protons and two neutrons in the nucleus orbited by two electrons.

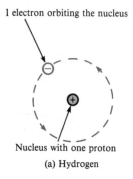

1 electron orbiting the nucleus

Nucleus with one proton

(a) Hydrogen

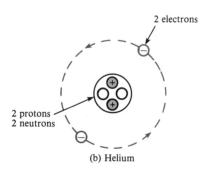

2 electrons

2 protons
2 neutrons

(b) Helium

FIGURE 2–1
Hydrogen and helium atoms.

Atomic Weight and Number

All elements are arranged in the *periodic table of the elements* in order according to their *atomic number,* which equals the number of electrons in an electrically balanced atom. The elements can also be arranged by their *atomic weight,* which is approximately the number of protons and neutrons in the nucleus. For example, hydrogen has an atomic number of one and an atomic weight of one. The atomic number of helium is two, and its atomic weight is four. In their normal, or neutral, state, all atoms of a given element have the same number of electrons as protons; the positive charges cancel the negative charges, and the atom has a net charge of 0.

Electron Shells and Orbits

Electrons orbit the nucleus at certain distances from the nucleus. Electrons near the nucleus have less energy than those in more distant orbits. It is known that only *discrete* values of electron energies exist within atomic structures. Therefore, electrons must orbit only at discrete distances from the nucleus.

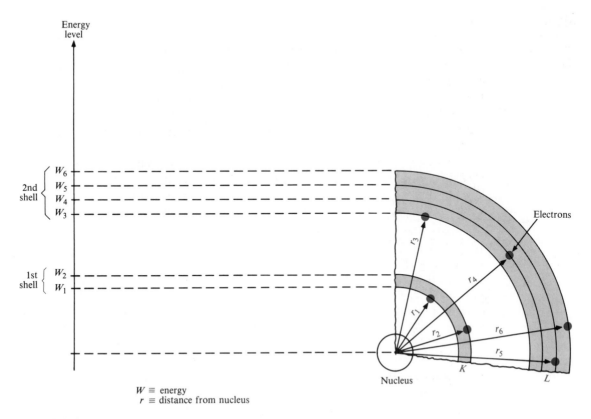

FIGURE 2–2
Energy levels increase as distance from nucleus increases.

Each discrete distance (orbit) from the nucleus corresponds to a certain energy level. In an atom, orbits are grouped into energy bands known as *shells*. A given atom has a fixed number of shells. Each shell has a fixed maximum number of electrons at permissible energy levels (orbits). The differences in energy levels within a shell are much smaller than the difference in energy between shells. The shells are designated *K*, *L*, *M*, *N*, and so on, with *K* being closest to the nucleus. This concept is illustrated in Figure 2–2.

Valence Electrons

Electrons in orbits farther from the nucleus are less tightly bound to the atom than those closer to the nucleus. This is because the force of attraction between the positively charged nucleus and the negatively charged electron increases with decreasing distance. Electrons with the highest energy levels exist in the outermost shell of an atom and are relatively loosely bound to the atom. These *valence electrons* contribute to chemical reactions and bonding within the structure of a material. The *valence* of an atom is the number of electrons in its outermost shell.

Ionization

When an atom absorbs energy from a heat source or from light, for example, the energy levels of the electrons are raised. When an electron gains energy, it moves to an orbit farther from the nucleus. Since the valence electrons possess more energy and are more loosely bound to the atom than inner electrons, they can jump to higher orbits more easily when external energy is absorbed.

If a valence electron acquires a sufficient amount of energy, it can be completely removed from the outer shell and the atom's influence. The departure of a valence electron leaves a previously neutral atom with an excess of positive charge (more protons than electrons). The process of losing a valence electron is known as *ionization* and the resulting positively charged atom is called a *positive ion*. For example, the chemical symbol for hydrogen is H. When it loses its valence electron and becomes a positive ion, it is designated H^+. The escaped valence electron is called a *free electron*. When a free electron falls into the outer shell of a neutral hydrogen atom, the atom becomes negatively charged (more electrons than protons) and is called a *negative ion,* designated H^-.

Silicon and Germanium Atoms

Two widely used types of semiconductor materials are *silicon* and *germanium*. Both the silicon and the germanium atoms have four valence electrons. They differ in that silicon has fourteen protons in its nucleus and germanium has 32. Figure 2–3 shows the atomic structure for both materials.

FIGURE 2–3
Silicon and germanium atoms.

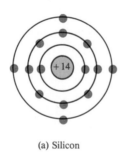 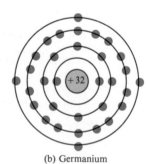

(a) Silicon   (b) Germanium

SECTION REVIEW 2–1

1. Define *atom*.
2. Name the components of an atom.
3. What is a valence electron?
4. What is a free electron?

ATOMIC BONDING

When silicon atoms combine into molecules to form a solid material, they arrange themselves in a fixed pattern called a *crystal*. The atoms within the crystal structure are held together by *covalent bonds,* which are created by interaction of the valence electrons of each atom.

Figure 2–4 shows how each silicon atom positions itself with four adjacent atoms. Since an atom can have up to eight electrons in its outer shell, a silicon atom with its four valence electrons shares an electron with each of its four neighbors. This sharing of valence electrons produces the covalent bonds that hold the atoms together, because each shared electron is attracted equally by the two adjacent atoms which share it. Covalent bonding of a pure (intrinsic) silicon crystal is shown in Figure 2–5. Bonding for germanium is similar because it also has four valence electrons.

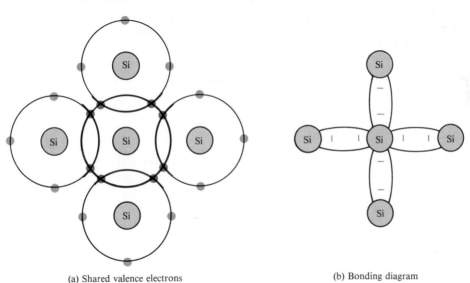

(a) Shared valence electrons
form covalent bonds.

(b) Bonding diagram

FIGURE 2–4
Covalent bonds in silicon.

FIGURE 2–5
Covalent bonds in a pure silicon crystal.

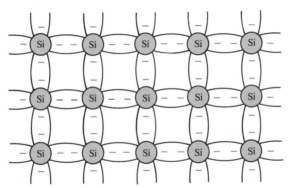

1. Name two semiconductor materials.

2. What is a covalent bond?

CONDUCTION IN SEMICONDUCTOR CRYSTALS

2–3   As you have seen, the electrons of an atom can exist only within prescribed energy bands. Each shell around the nucleus corresponds to a certain energy band and is separated from adjacent shells by energy gaps, in which no electrons can exist. This is shown in Figure 2–6 for an unexcited silicon atom (no external energy). This condition occurs only at absolute 0° temperature.

FIGURE 2–6
Energy band diagram for *unexcited* silicon atom.

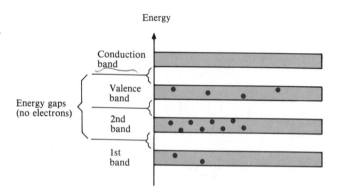

Conduction Electrons and Holes

A pure silicon crystal at room temperature derives heat (thermal) energy from the surrounding air, causing some valence electrons to gain sufficient energy to jump the gap from the *valence band* into the *conduction band,* becoming free electrons. This is illustrated in the energy diagram of Figure 2–7(a) and in the bonding diagram of Figure 2–7(b).

When an electron jumps to the conduction band, a vacancy is left in the valence band. This vacancy is called a *hole.* For every electron raised to the conduction band by thermal or light energy, there is one hole left in the valence band, creating what is called an *electron-hole pair. Recombination* occurs when a conduction band electron loses energy and falls back into a hole in the valence band.

To summarize, a piece of pure silicon at room temperature has, at any instant, a number of conduction band (free) electrons that are unattached to any atom and are essentially drifting randomly throughout the material. There is also an equal number of holes in the valence band created when these electrons jump into the conduction band. This is illustrated in Figure 2–8.

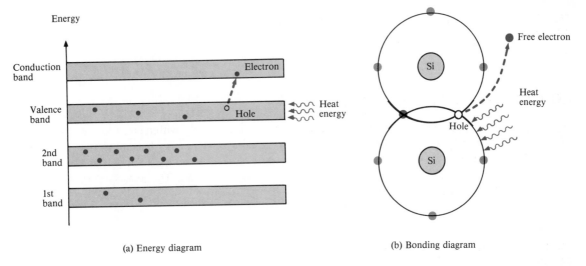

(a) Energy diagram

(b) Bonding diagram

FIGURE 2–7
Creation of an electron-hole pair in an excited silicon atom.

FIGURE 2–8
Electron-hole pairs in a silicon crystal.

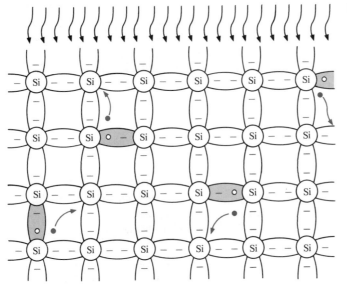

Germanium versus Silicon

The situation in a germanium crystal is similar to that in silicon except that, because of its atomic structure, pure germanium has more free electrons than silicon and therefore a higher conductivity. Silicon, however, is the favored semiconductor material and is far more widely used than germanium. One reason for this is that silicon can be used at a much higher temperature than germanium.

Electron and Hole Current

When a voltage is applied across a piece of silicon, as shown in Figure 2–9, the free electrons in the conduction band are easily attracted toward the positive end. This movement of free electrons is one type of current in a semiconductor material, called *electron current*.

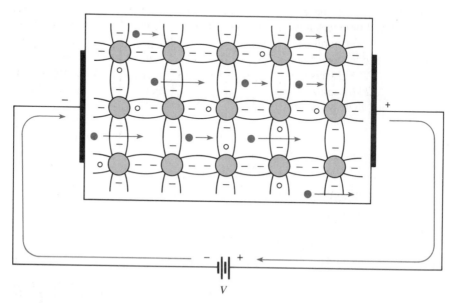

FIGURE 2–9
Free electron current in silicon.

Another current mechanism occurs at the valence level, where the holes created by the free electrons exist. Electrons remaining in the valence band are still attached to their atoms and are not free to move randomly in the crystal structure. However, a valence electron can ''fall'' into a nearby hole, with little change in its energy level, thus leaving another hole where it came from. Effectively the hole has moved from one place to another in the crystal structure, as illustrated in Figure 2–10. This is called *hole current*.

FIGURE 2–10
Hole current in silicon.

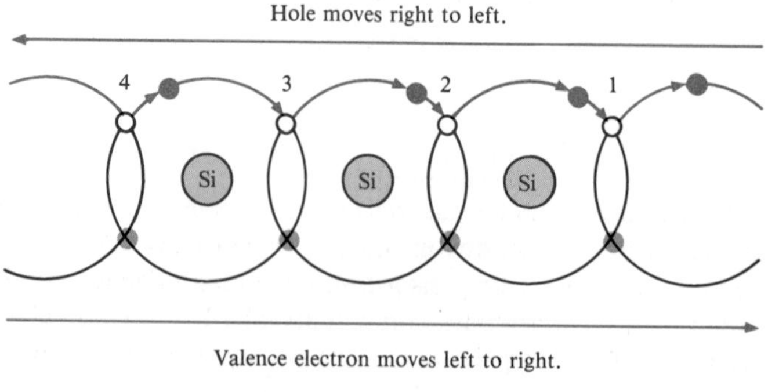

Hole moves right to left.

4   3   2   1

Si   Si   Si

Valence electron moves left to right.

Semiconductors, Conductors, and Insulators

In a pure (intrinsic) semiconductor, there are relatively few free electrons; so neither silicon nor germanium is very useful in its intrinsic state. They are neither insulators nor good conductors because current in a material depends directly on the number of free electrons.

A comparison of the energy bands in Figure 2–11 for the three types of materials shows the essential differences among them regarding conduction. The energy gap for an insulator is so wide that hardly any electrons acquire enough energy to jump into the conduction band. The valence band and the conduction band in a conductor (like copper) overlap so that there are always many conduction electrons, even without the application of external energy. The semiconductor, as the figure shows, has an energy gap that is much narrower than that in an insulator.

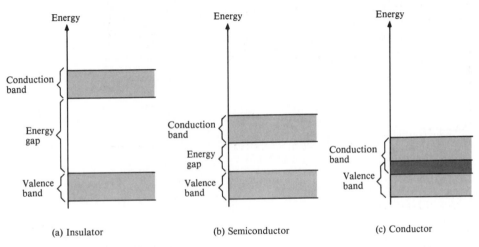

FIGURE 2–11
Energy diagrams for three categories of materials.

---

1. In the atomic structure of a semiconductor, within which energy band do free electrons exist? Valence electrons?
2. How are holes created in an intrinsic semiconductor?
3. Why is current more easily established in a semiconductor than in an insulator?

SECTION REVIEW 2–3

n-TYPE AND p-TYPE SEMICONDUCTORS

2–4

Intrinsic semiconductor materials do not conduct current very well because of the limited number of free electrons in the conduction band. This means that the resistivity of a semiconductor is much greater than that of a conductor. For example, a one-cubic-centimeter sample of silver has a resistivity of 10^{-6} $\Omega \cdot$ cm, whereas the resistivity is about 45 $\Omega \cdot$ cm for pure germanium and several thousand ohms $\cdot$ cm for pure silicon.

Doping

The resistivities of silicon and germanium can be drastically reduced and controlled by the addition of *impurities* to the pure semiconductor material. This process, called *doping*, *increases* the number of current carriers (electrons or holes), thus increasing the conductivity and decreasing the resistivity. The two categories of impurities are *n-type* and *p-type*.

n-Type Semiconductor

To increase the number of conduction-band electrons in pure silicon, *pentavalent* impurity atoms are added. These are atoms with five valence electrons such as *arsenic, phosphorus,* and *antimony*.

As illustrated in Figure 2–12, each pentavalent atom (antimony, in this case) forms covalent bonds with four adjacent silicon atoms. Four of the antimony atom's valence electrons are used to form the covalent bonds, leaving *one extra electron*. This extra electron becomes a conduction electron because it is not attached to any atom. The number of conduction electrons can be controlled by the amount of impurity added to the silicon.

FIGURE 2–12
Pentavalent impurity atom in a silicon crystal. An antimony (Sb) impurity atom is shown in the center.

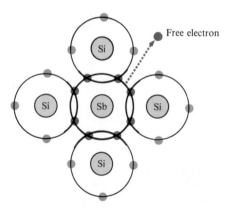

Free electron

Since most of the current carriers are *electrons*, silicon (or germanium) doped in this way is an *n-type* semiconductor material where the *n* stands for the *negative* charge on an electron. The *electrons* are called the *majority carriers* in n-type material. Although the great majority of current carriers in n-type material are electrons, there are some holes. *Holes* in an n-type material are called *minority carriers*.

p-Type Semiconductor

To increase the number of holes in pure silicon, *trivalent* impurity atoms are added. These are atoms with three valence electrons such as *aluminum, boron,* and *gallium*. As illustrated in Figure 2–13, each trivalent atom (boron, in this case) forms covalent bonds with four adjacent silicon atoms. All three of the boron atom's valence electrons are used in the covalent bonds; and, since four electrons are required, a *hole* is formed with each trivalent

FIGURE 2–13
Trivalent impurity atom in a silicon crystal.
A boron (B) impurity atom is shown in the
center.

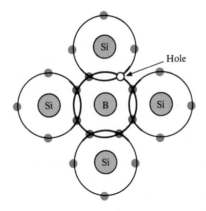

atom. The number of holes can be controlled by the amount of trivalent impurity added to
the silicon.

Since most of the current carriers are *holes*, silicon (or germanium) doped in this
way is called a *p-type* semiconductor material because holes can be thought of as positive
charges. The *holes* are the *majority carriers* in p-type material. Although the great majority of current carriers in p-type material are holes, there are some electrons. *Electrons* in
p-type material are called *minority carriers*.

1. How is an n-type semiconductor formed?

2. How is a p-type semiconductor formed?

3. What is a majority carrier?

SECTION
REVIEW
2–4

pn JUNCTIONS

2–5

When a piece of silicon is doped so that half is n-type and the other half is p-type, a *pn
junction* is formed between the two regions, as shown in Figure 2–14(a). This device is
known as a *semiconductor diode*. The n region has many conduction electrons and the p
region has many holes, as shown in Figure 2–14(b). The pn junction is fundamental to the
operation not only of diodes but also of transistors and other solid state devices.

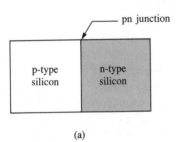

— pn junction

p-type
silicon

n-type
silicon

(a)

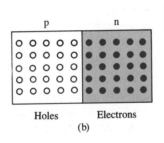

p   n

Holes   Electrons

(b)

FIGURE 2–14
Basic pn structure at the instant of junction
formation.

The Depletion Layer

With no external voltage, the conduction electrons in the n region are aimlessly drifting in all directions. At the instant of junction formation, some of the electrons near the junction *diffuse* across into the p region and recombine with holes near the junction. For each electron that crosses the junction and recombines with a hole, a pentavalent atom is left with a net positive charge in the n region near the junction, making it a *positive ion*. Also, when the electron recombines with a hole in the p region, a trivalent atom acquires net negative charge, making it a *negative ion*.

As a result of this recombination process, a large number of positive and negative ions build up near the pn junction. As this build-up occurs, the electrons in the n region must overcome both the attraction of the positive ions and the repulsion of the negative ions in order to migrate into the p region. Thus, as the ion layers build up, the area on both sides of the junction becomes essentially depleted of any conduction electrons or holes and is known as the *depletion layer*. This condition is illustrated in Figure 2–15. When an equilibrium condition is reached, the depletion layer has widened to a point where no further electrons can cross the pn junction.

FIGURE 2–15
pn junction equilibrium condition.

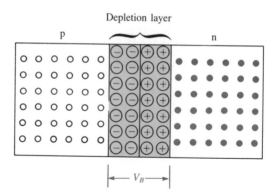

The existence of the positive and negative ions on opposite sides of the junction creates a *barrier potential* (V_B) across the depletion layer, as indicated in the figure. At 25°C, the barrier potential is approximately 0.7 V for silicon and 0.3 V for germanium. As the junction temperature *increases*, the barrier potential *decreases*, and vice versa.

Energy Diagram of the pn Junction

Now, we will look at the operation of the pn junction in terms of its energy level. First consider the pn junction at the instant of its formation. The energy bands of the trivalent impurity atoms in the p-type material are at a slightly higher level than those of the pentavalent impurity atom in the n-type material, as shown in Figure 2–16. This is because the core attraction for the valence electrons (+3) in the trivalent atom is less than the core attraction for the valence electrons (+5) in the pentavalent atom. So, the trivalent valence electrons are in a slightly higher orbit and, thus, at a higher energy level.

Notice in Figure 2–16 that there is some overlap of the conduction bands in the p and n regions and also some overlap of the valence bands in the p and n regions. This permits the electrons of higher energy near the top of the n-region conduction band to begin diffusing across the junction into the lower part of the p-region conduction band. As soon as an electron diffuses across the junction, it recombines with a hole in the valence band. As diffusion continues, the depletion layer begins to form. Also, the energy bands in the n region "shift" down as the electrons of higher energy are lost to diffusion. When the top of the n-region conduction band reaches the same level as the bottom of the p-region conduction band, diffusion ceases and the equilibrium condition is reached. This is shown in terms of energy levels in Figure 2–17. There is an *energy gradient* across the depletion layer rather than an abrupt change in energy level.

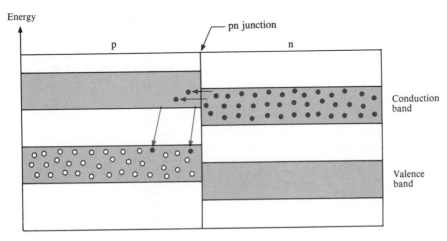

FIGURE 2–16
pn junction energy diagram as diffusion begins.

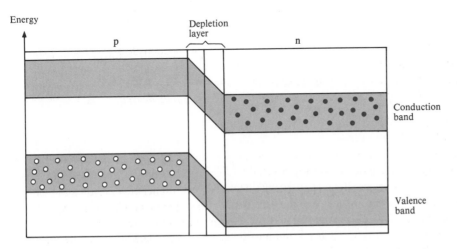

FIGURE 2–17
Energy diagram at equilibrium. The n-region bands are shifted down from their original positions when diffusion begins.

1. What is a pn junction?

2. When p and n regions are joined, a depletion layer forms. Describe the depletion layer characteristics.

3. The barrier potential for silicon is greater than for germanium (T or F).

4. What is the barrier potential for silicon at 25°C?

BIASING THE pn JUNCTION

2–6

As you have seen, there is no current across a pn junction at equilibrium. The primary usefulness of the pn junction diode is its ability to allow current in only *one* direction and to prevent current in the other direction as determined by the bias. There are two bias conditions for a pn junction: *forward* and *reverse*. Either of these conditions is created by application of an external voltage of the proper polarity.

Forward Bias

The term *bias* in electronics normally refers to a fixed voltage that sets the operating conditions for a semiconductor device. *Forward bias* is the condition that permits current across a pn junction. Figure 2–18 shows a dc voltage connected in a direction to forward-bias the diode. Notice that the negative terminal of the battery is connected to the n region (called the cathode), and the positive terminal is connected to the p region (called the anode).

FIGURE 2–18

Forward-bias connection. The resistor limits the forward current to prevent damage to the diode.

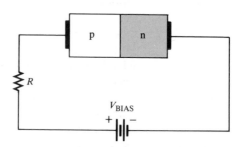

This is the basic operation of forward bias: The negative terminal of the battery pushes the conduction-band electrons in the n region toward the junction, while the positive terminal pushes the holes in the p region also toward the junction. (Recall that like charges repel each other.) When it overcomes the barrier potential, the external voltage source provides the n-region electrons with enough energy to penetrate the depletion layer and cross the junction, where they combine with the p-region holes. As electrons leave the n region, more flow in from the negative terminal of the battery. So, current through the n region is the movement of conduction electrons (majority carriers) toward the junction.

Once the conduction electrons enter the p region and combine with holes, they become valence electrons. They then move as valence electrons from hole to hole toward the positive connection of the battery. The movement of these valence electrons is the

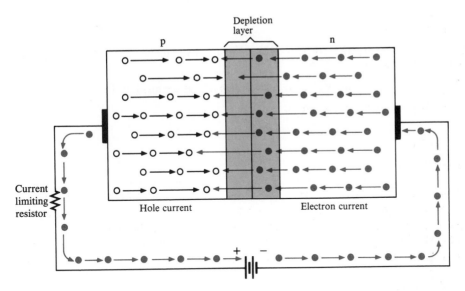

FIGURE 2–19
Forward current in a diode causes the depletion layer to narrow.

same as the movement of holes in the opposite direction. So, current in the p region is the movement of holes (majority carriers) toward the junction. Figure 2–19 illustrates current in a forward-biased diode.

Effect of Barrier Potential on Forward Bias.   The barrier potential of the depletion layer can be envisioned as acting as a small battery that *opposes* bias, as illustrated in Figure 2–20(a). The resistances R_p and R_n represent the *bulk resistances* of the p and n materials. The external bias voltage must overcome the barrier potential before the diode conducts, as illustrated in Figure 2–20(b). Conduction occurs at approximately 0.7 V for silicon and 0.3 V for germanium. Once the diode is conducting in the forward direction, the voltage drop across it remains at approximately the barrier potential and changes very little with changes in forward current (I_F) except for bulk resistance effects, as illustrated in Figure 2–20(c). The bulk resistances are usually only a few ohms and result in only a small voltage drop when the diode conducts. Often this drop can be neglected.

Energy Diagram for Forward Bias.   Forward bias raises the energy levels of the conduction electrons in the n region, allowing them to move into the p region and combine with holes in the valence band. This condition is shown in Figure 2–21.

Reverse Bias

Reverse bias is the condition that prevents current across the pn junction. Figure 2–22 shows a dc voltage source connected to reverse-bias the diode. Notice that the negative terminal of the battery is connected to the p region, and the positive terminal to the n region. The negative terminal of the battery attracts holes in the p region away from the pn junction, while the positive terminal also attracts electrons away from the junction. As

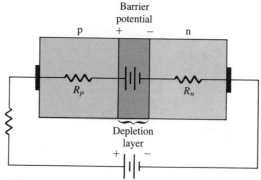

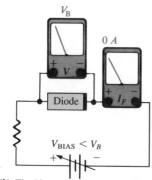

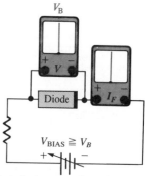

(a) Barrier potential and bulk resistance equivalent.

(b) The bias voltage is too small to overcome the barrier potential.

(c) The diode begins to conduct when forward-bias voltage is slightly greater than barrier potential. (Banded end represents the n-region.)

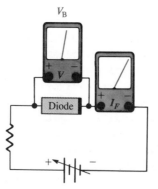

(d) The forward diode drop increases very little with a large increase in forward current.

FIGURE 2–20
Effects of diode barrier potential and bulk resistance.

Energy

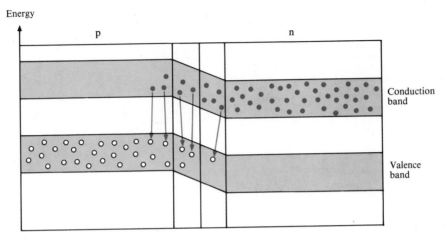

FIGURE 2–21
Energy diagram for forward bias showing recombination in depletion layer and in p region as conduction electrons move across junction.

FIGURE 2–22
Reverse-bias connection.

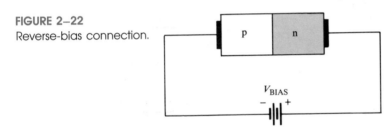

electrons and holes move away from the junction, the depletion layer widens; more positive ions are created in the n region, and more negative ions are created in the p region, as shown in Figure 2–23(a).

The depletion layer widens until the potential difference across it equals the external bias voltage. At this point, the holes and electrons stop moving away from the junction and majority current ceases, as indicated in Figure 2–23(b). The initial movement of majority carriers away from the junction is called *transient current* and lasts only for a very short time upon application of reverse bias.

When the diode is reverse-biased, the depletion layer effectively acts as an insulator between the layers of oppositely charged ions. This forms an effective capacitance, as illustrated in Figure 2–23(c). Since the depletion layer widens with increased reverse-biased voltage, the capacitance decreases and vice versa. This internal capacitance is called the *depletion-layer capacitance* and can be represented by an equivalent circuit as shown in Figure 2–23(d).

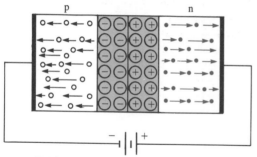

(a) Transient current as depletion layer widens

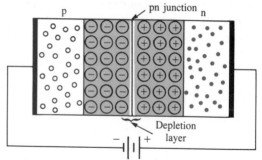

(b) Current ceases when barrier potential equals bias voltage

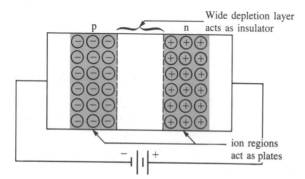

(c) Depletion layer widens as reverse bias increases

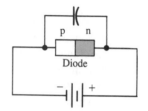

(d) Diode circuit with depletion-layer capacitance

FIGURE 2–23
Reverse bias.

Reverse Leakage Current. As you have learned, *majority current* very quickly becomes 0 when reverse bias is applied. There is, however, a very small *leakage current* produced by minority carriers during reverse bias. Germanium, as a rule, has a greater leakage current than silicon. This current is typically in the μA or nA range. A relatively small number of thermally produced electron-hole pairs exist in the depletion layer. Under the influence of the external voltage, some electrons manage to diffuse across the pn junction before recombination. This process establishes a small minority carrier current throughout the material. The reverse leakage current is dependent primarily on the junction temperature and not on the amount of reverse-biased voltage. A temperature increase causes an increase in leakage current.

Reverse Breakdown. If the external reverse-biased voltage is increased to a large enough value, *avalanche breakdown* occurs. Here is what happens: Assume that one minority conduction-band electron acquires enough energy from the external source to accelerate it toward the positive end of the diode. During its travel, it collides with an atom and imparts enough energy to knock a valence electron into the conduction band.

There are now two conduction-band electrons. Each will collide with an atom, knocking two more valence electrons into the conduction band. There are now four conduction-band electrons which, in turn, knock four more into the conduction band. This rapid multiplication of conduction-band electrons, known as an *avalanche effect*, results in a rapid build-up of reverse current. Most diodes are normally not operated in reverse breakdown and can be damaged by the resulting excessive power if they are. However, a particular type of diode (to be studied later), known as a zener diode, is optimized for reverse-breakdown operation.

---

1. Name the two bias conditions.
2. Which bias condition produces majority carrier current?
3. Which bias condition produces a widening of the depletion region?
4. Minority carriers produce the current during avalanche breakdown (T or F).

SECTION REVIEW 2–6

A SYSTEM APPLICATION

2–7

The characteristic of a pn junction diode that permits current to flow when forward-biased and prevents current from flowing when reverse-biased is extremely useful in the conversion of an ac voltage to a dc voltage. This is what the *rectifier* portion of a dc power supply does. It converts the 60 Hz sine wave voltage from an electrical outlet to a dc voltage that pulsates on each half-cycle or on alternate half-cycles (depending on the type of rectifier) of the input sine wave. The output of a rectifier is dc because it does not change polarity.

Let's see how the diode is used to rectify an ac voltage. We will use a very basic circuit in this section to illustrate the application and then expand the concept in the next chapter. Figure 2–24 illustrates a simple rectifier circuit in the power supply block diagram. On the positive alternations of the input sine wave voltage, the diode is forward-

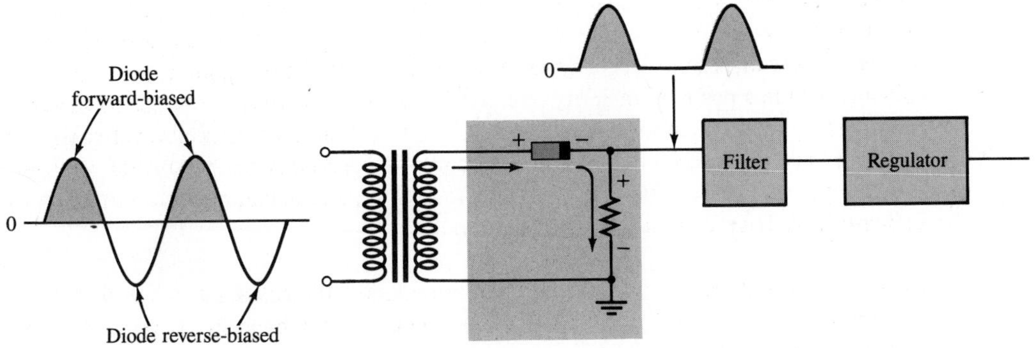

FIGURE 2–24
A simple diode rectifier in a dc power supply system.

biased and current flows through the diode and resistor as indicated. The current creates a voltage across R with the shape of half a sine wave with a polarity as indicated. On the negative alternations of the sine wave input, the diode is reverse-biased and no current can flow in the rectifier, so the voltage across R is zero, as indicated. As you can see, the result is a pulsating voltage with a single polarity. This pulsating dc voltage is then applied to the filter block to be smoothed out.

SUMMARY

- ☐ An atom is the smallest particle of an element that retains the characteristics of that element.
- ☐ An atom is described as a nucleus containing protons and neutrons orbited by electrons.
- ☐ Protons are positive, neutrons are neutral, and electrons are negative.
- ☐ The atomic number is the quantity of electrons in orbit within the atom.
- ☐ The atomic weight is approximately the number of protons and neutrons in the nucleus.
- ☐ Atomic shells are energy bands.
- ☐ The outermost shell containing electrons is the valence shell.
- ☐ A positive ion is an atom that has lost a valence electron.
- ☐ A negative ion is an atom that has gained an extra valence electron.
- ☐ A conduction band electron is called a *free electron*.
- ☐ Silicon and germanium are the predominate semiconductor materials.
- ☐ Atoms within a crystal structure are held together with covalent bonds.
- ☐ Electron-hole pairs are thermally produced.
- ☐ An intrinsic semiconductor is a pure material with relatively few free electrons.
- ☐ The process of adding impurities to an intrinsic semiconductor to increase and control conductivity is called *doping*.
- ☐ A p-type semiconductor is doped with trivalent impurity atoms.
- ☐ An n-type semiconductor is doped with pentavalent impurity atoms.
- ☐ A pn junction is the boundary between n-type and p-type materials.
- ☐ The depletion layer is a region adjacent to the pn junction containing no majority carriers.
- ☐ The barrier potential is the inherent voltage across the depletion layer.
- ☐ Forward bias permits majority-carrier current through the pn junction.
- ☐ Reverse bias prevents majority-carrier current.
- ☐ A pn structure is called a *diode*.
- ☐ Reverse leakage current is due to thermally produced electron-hole pairs.
- ☐ Reverse breakdown occurs when the reverse-biased voltage exceeds a specified value.

1. Define *atom*.

2. What is the Bohr model?

3. Which particles in an atom have a negative charge? Positive charge? No charge?

4. What is the atomic number of silicon?

5. What is the atomic weight of germanium?

6. Both silicon and germanium have a valence of _____.

7. What is the valence shell designation in a silicon atom?

8. Distinguish between a neutral atom, a positive ion, and a negative ion.

9. Define *covalent bond*.

10. In what energy band do free electrons exist?

11. Describe what happens when an electron-hole pair is produced.

12. Define *recombination*.

13. What are the two types of current in a semiconductor?

14. What is the essential difference between a semiconductor and an insulator?

15. How does a trivalent impurity modify an intrinsic semiconductor?

16. How does a pentavalent impurity modify an intrinsic semiconductor?

17. What are the majority carriers in an n-type semiconductor?

18. Approximately what value does a bias voltage have to be in order to forward-bias a silicon pn junction?

19. What is the current in the circuit of Figure 2–25 if the total bulk resistance of the silicon diode is 10 Ω?

FIGURE 2–25

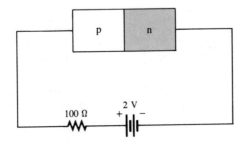

**ANSWERS
TO
SECTION
REVIEWS**

Section 2–1

1. The smallest particle that retains the characteristics of its element.

2. Electrons, protons, neutrons.

3. An electron in the outermost shell (valence band).

4. An electron in the conduction band.

Section 2–2

1. Silicon, germanium.

2. The sharing of electrons with neighboring atoms.

Section 2–3

1. Conduction band; valence band.

2. When an electron is thermally raised to the conduction band, leaving a hole in the valence band.

3. The energy gap between the valence band and the conduction band is less for a semiconductor.

Section 2–4

1. By the addition of pentavalent atoms to the intrinsic semiconductor material.

2. By the addition of trivalent atoms to the intrinsic semiconductor material.

3. The particle in greatest abundance: electrons in n-type material and holes in p-type material.

Section 2–5

1. The boundary between n-type and p-type materials.

2. Devoid of majority carriers, contains only positive and negative ions.

3. True.

4. 0.7 V.

Section 2–6

1. Forward, reverse.

2. Forward.

3. Reverse.

4. True.

Diodes and Applications

In this chapter you will learn
- [ ] What a diode characteristic curve shows
- [ ] What the terms *anode* and *cathode* mean
- [ ] How to identify the terminals of a diode
- [ ] What is meant by *half-wave rectification* and how a basic half-wave rectifier circuit works
- [ ] How to determine the average value of a half-wave signal
- [ ] What is meant by *full-wave rectification*
- [ ] How to determine the average value of a full-wave rectified signal
- [ ] How a center-tapped full-wave rectifier works
- [ ] How a bridge rectifier works
- [ ] How to determine the PIV (peak inverse voltage) across the diodes in a rectifier
- [ ] How a capacitor-input filter smooths out a rectified voltage
- [ ] What ripple voltage is
- [ ] How the value of the filter capacitor affects the amount of ripple
- [ ] How an inductor-input filter improves the dc output voltage
- [ ] How to troubleshoot rectifier circuits
- [ ] How diode limiters work
- [ ] How diode clamping circuits work
- [ ] How voltage doublers and triplers work
- [ ] How to use a diode data sheet
- [ ] How rectifiers are used in a specific system application

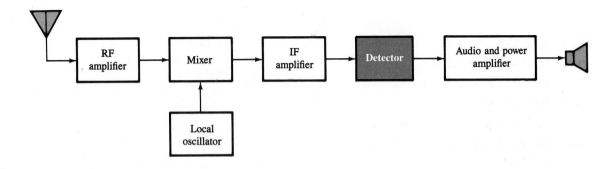

In the last chapter you learned that a single pn junction device is known as a *semiconductor diode*. The importance of this diode as a practical electronic device lies in its ability to conduct current in one direction and block current in the other. Many applications stem from this characteristic, particularly in the area of *ac rectification,* which is covered in this chapter. Other diode characteristics—including reverse breakdown, pn junction capacitance, and energy emission—are optimized for certain applications.

This chapter describes various types of diodes and their applications and discusses important performance parameters and ratings. As you already know, the diode is an important electronic device that can be used in numerous applications. A specific application is in the detector circuit of an AM superheterodyne receiver, as indicated by the shaded block in the system diagram. Recall that the function of the detector is to extract the audio signal from the IF carrier signal when it comes out of the IF amplifier. The system application section at the end of this chapter presents the basic operation of a typical detector and shows how the diode fits into the operation of the circuit.

RECTIFIER DIODES

3–1

Rectifier diodes form a major group of semiconductor diodes. Besides rectification, there are other uses to which this type of diode can be applied. In fact, many devices in this category are referred to as *general-purpose diodes*.

Diode Characteristic Curve

As you learned in the last chapter, a diode conducts current when it is forward-biased if the bias voltage exceeds the barrier potential, and the diode prevents current when it is reverse-biased at less than the breakdown voltage. Figure 3–1 is a graph of diode current versus voltage. The upper right quadrant of the graph represents the *forward-biased condition*. As you can see, there is very little forward current (I_F) for forward voltages (V_F) below the barrier potential. As the forward voltage approaches the value of the barrier potential (0.7 V for silicon and 0.3 V for germanium), the current begins to increase. Once the forward voltage reaches the barrier potential, the current increases drastically and must be limited by a series resistor. *The voltage across the forward-biased diode remains approximately equal to the barrier potential.*

FIGURE 3–1
Diode characteristic curve.

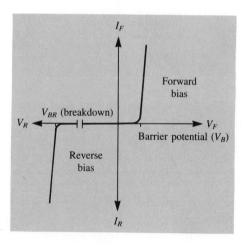

The lower left quadrant of the graph represents the *reverse-biased condition*. As the reverse voltage (V_R) increases to the left, the current remains near 0 until the breakdown voltage (V_{BR}) is reached. When breakdown occurs, there is a large reverse current which, if not limited, can destroy the diode. Typically, the breakdown voltage is greater than 50 V for most rectifier diodes. Most diodes should not be operated in reverse breakdown.

Diode Symbol

Figure 3–2(a) is the standard schematic symbol for a general-purpose diode. The arrow points in the direction of *conventional current*. The two terminals of the diode are the *anode* and *cathode*. When the anode is positive with respect to the cathode, the diode is

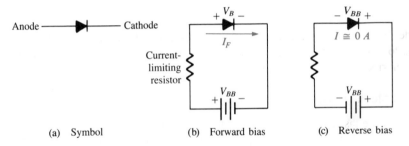

(a) Symbol   (b) Forward bias   (c) Reverse bias

FIGURE 3–2
General-purpose diode and conditions of forward bias and reverse bias.

forward-biased and current is from anode to cathode, as shown in Figure 3–2(b). Remember that when the diode is forward-biased, the barrier potential always appears between anode and cathode, as indicated in the figure. When the anode is negative with respect to the cathode, the diode is *reverse-biased,* as shown in Figure 3–2(c).

Some typical diodes are shown in Figure 3–3(a) to illustrate the variety of physical structures. Part (b) illustrates terminal identification.

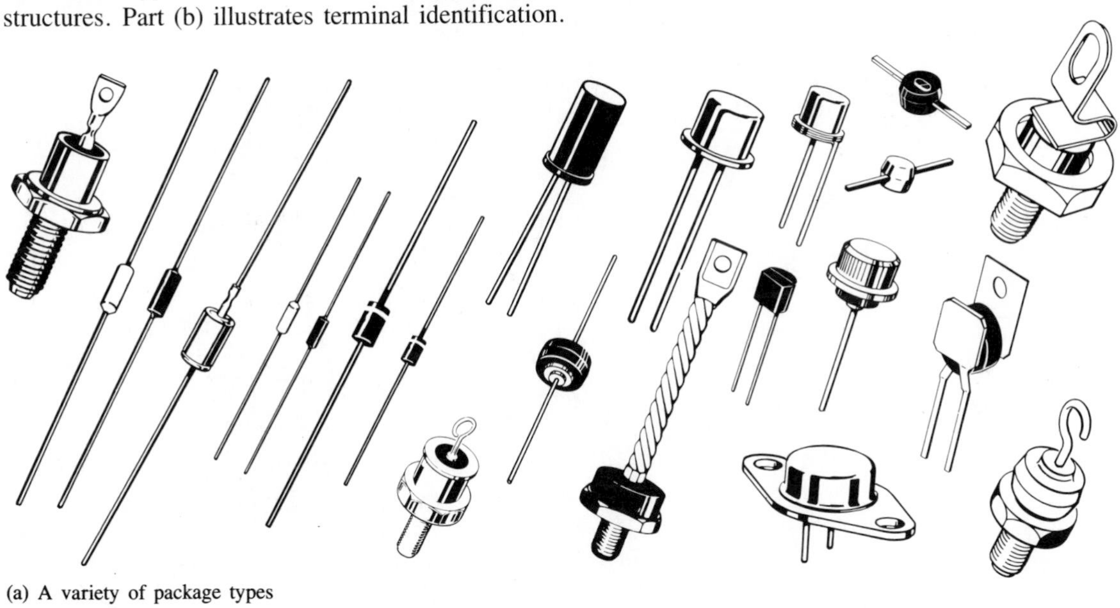

(a) A variety of package types

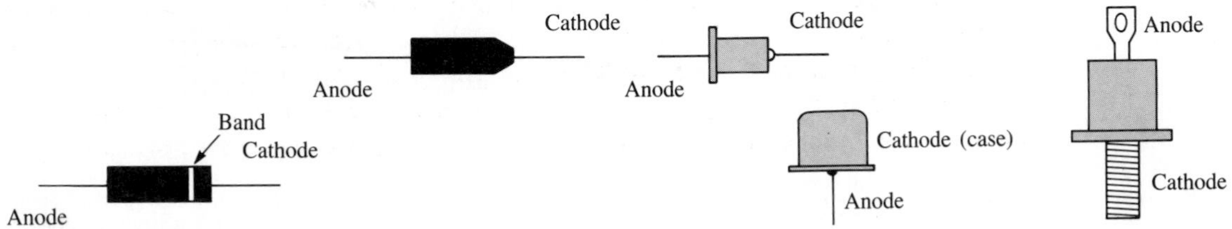

(b) Examples of terminal identification

FIGURE 3–3
Typical diodes.

Testing the Diode with an Ohmmeter

The internal battery in certain ohmmeters will forward-bias or reverse-bias a diode, permitting a quick and simple check for proper functioning. This works for 1.5 V ohmmeters such as VOMs, but not for some of the digital multimeters. Many digital multimeters have a diode test position.

To check the diode in the forward direction, the positive meter lead is connected to the anode and the negative lead to the cathode, as shown in Figure 3–4(a). When a diode is forward-biased, its internal resistance is low (typically around 100 Ω, more or less). When the meter leads are reversed, as shown in Figure 3–4(b), the internal ohmmeter battery reverse-biases the diode, and a very large resistance value (ideally infinite) is indicated. The pn junction is shorted if a low resistance is indicated in both bias conditions; it is open if a very high resistance is read for both checks.

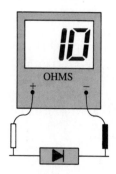

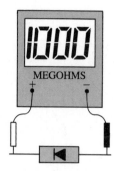

(a)   Typical forward check   (b) Typical reverse check

FIGURE 3–4
Checking a semiconductor diode with an ohmmeter.

Diode Approximations

The simplest way to visualize diode operation is to think of it as a *switch*. When forward-biased, the diode acts as a closed (on) switch, and when reverse-biased it acts as an open (off) switch, as in Figure 3–5. The characteristic curve for this approximation is also shown. Note that the forward voltage and the reverse current are always 0. This ideal model, of course, neglects the effect of the barrier potential, the internal resistances, and other parameters. But in many cases it is accurate enough.

The next higher level of accuracy is the barrier potential model. In this approximation, the forward-biased diode is represented as a closed switch in series with a small "battery" equal to the barrier potential V_B (0.7 V for Si and 0.3 V for Ge), as shown in Figure 3–6(a). The positive end of the battery is toward the anode. Keep in mind that the barrier potential cannot actually be measured with a voltmeter, but only has the *effect* of a battery when forward bias is applied. The reverse-biased diode is represented by an open switch, as in the ideal case, because the barrier potential does not affect reverse bias. This is shown in Figure 3–6(b). The characteristic curve for this model is shown in part (c). This book uses the barrier potential for analysis unless otherwise stated.

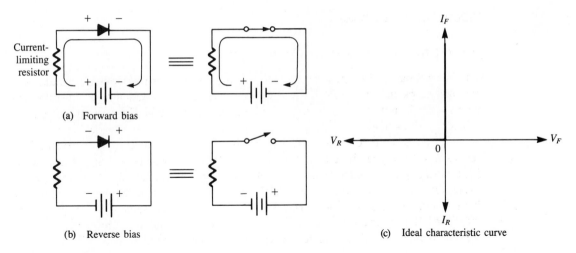

(a) Forward bias

(b) Reverse bias

(c) Ideal characteristic curve

FIGURE 3–5
Diode/switch equivalent circuits.

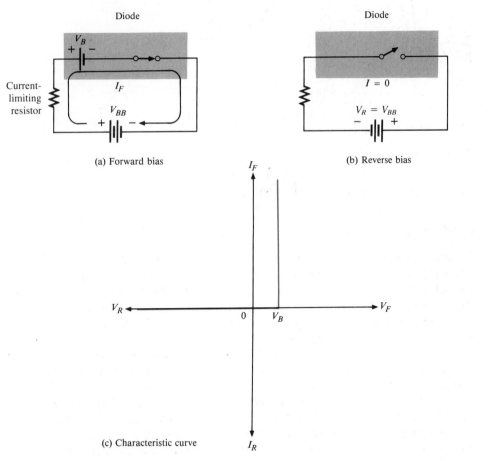

(a) Forward bias

(b) Reverse bias

(c) Characteristic curve

FIGURE 3–6
Diode approximation including barrier potential.

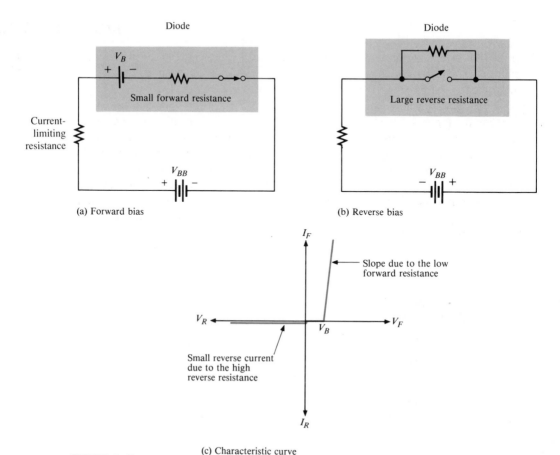

(c) Characteristic curve

FIGURE 3–7

Diode approximation including barrier potential, forward resistance, and reverse leakage resistance.

One more level of accuracy will be considered at this point. Figure 3–7(a) shows the forward-biased diode model with both the *barrier potential* and the low *forward (bulk) resistance.* Figure 3–7(b) shows how the high *reverse resistance* (leakage resistance) affects the reverse-biased model. The characteristic curve is shown in part (c). Other parameters such as junction capacitance and breakdown voltage become important only under certain operating conditions and will be considered where appropriate.

SECTION REVIEW 3–1

1. What are the two conditions for normal rectifier diode operation?

2. Sketch a rectifier diode symbol and label the terminals.

3. For a normal diode, the forward resistance is quite low and the reverse resistance is very high (T or F).

4. An open switch ideally represents a _____-biased diode. A closed switch ideally represents a _____-biased diode.

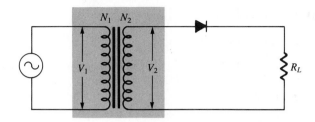

FIGURE 3–15
Half-wave rectifier with transformer-coupled input.

If $N_2 > N_1$, the primary voltage is less than the secondary voltage. If $N_2 < N_1$, the primary voltage is greater than the secondary voltage. If $N_2 = N_1$, then $V_2 = V_1$.

EXAMPLE 3–3

Determine the peak value of the output voltage for Figure 3–16.

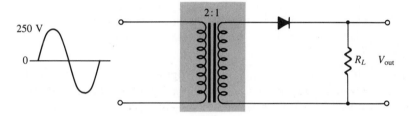

FIGURE 3–16

Solution
The turns ratio is

$$\frac{N_2}{N_1} = \frac{1}{2} = 0.5$$

The secondary peak voltage is

$$V_2 = \left(\frac{N_2}{N_1}\right)V_1$$
$$= 0.5(250 \text{ V})$$
$$= 125 \text{ V}$$

The peak rectified output voltage is

$$V_{p(\text{out})} = 125 \text{ V} - 0.7 \text{ V}$$
$$= 124.3 \text{ V}$$

Practice Exercise 3–3
Determine the peak value of the output voltage for Figure 3–16 if the turns ratio is 1:2 and $V_{p(\text{in})} = 50$ V.

**SECTION
REVIEW
3–2**

1. At what point on the input cycle does the PIV occur?
2. For a half-wave rectifier, there is current through the load for approximately what percentage of the input cycle?
3. What is the average value of the voltage shown in Figure 3–17?

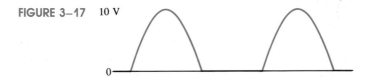

FIGURE 3–17   10 V

FULL-WAVE RECTIFIERS

3–3

The difference between full-wave and half-wave rectification is that a full-wave rectifier allows unidirectional current to the load during the entire input cycle, and the half-wave rectifier allows this only during one half-cycle. The result of full-wave rectification is a dc output voltage that pulsates every half-cycle of the input, as shown in Figure 3–18.

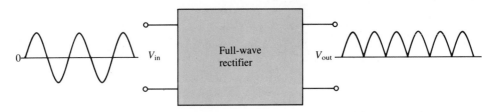

FIGURE 3–18
Full-wave rectification.

The average value for a full-wave rectified voltage is twice that of the half-wave, as expressed in equation (3–5).

$$V_{AVG} = \frac{2\,V_p}{\pi}$$ (3–5)

EXAMPLE 3–4

Find the average value of the full-wave rectified voltage in Figure 3–19.

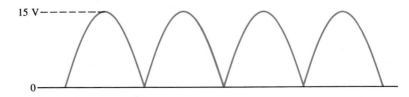

15 V

0

FIGURE 3–19

Solution

$$V_{AVG} = \frac{2 V_p}{\pi} = \frac{2(15 \text{ V})}{\pi} = 9.55 \text{ V}$$

Center-Tapped Full-Wave Rectifier

This type of full-wave rectifier circuit uses two diodes connected to the secondary of a center-tapped transformer, as shown in Figure 3–20. The input signal is coupled through the transformer to the center-tapped secondary. Half of the secondary voltage appears between the center tap and each end of the secondary winding as shown.

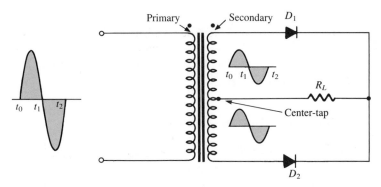

FIGURE 3–20
Center-tapped full-wave rectifier.

For a positive half-cycle of the input voltage (t_0 to t_1), the polarities of the secondary voltages are shown in Figure 3–21(a). This condition *forward-biases the upper diode D_1* and *reverse-biases the lower diode D_2*. The current path is through D_1 and the load resistor, as indicated. For a negative half-cycle of the input voltage (t_1 to t_2), the voltage polarities on the secondary are shown in Figure 3–21(b). This condition *reverse-biases D_1* and *forward-biases D_2*. The current path is through D_2 and the load resistor, as indicated. Because the output current during both the positive and negative portions of the input cycle is in the same direction through the load, the output voltage developed across the load is a full-wave rectified dc voltage, as shown in Figure 3–21(c).

Effect of the Turns Ratio on Full-Wave Output Voltage

If the transformer's turns ratio is one, the peak value of the rectified output voltage equals half the peak value of the primary input voltage less the barrier potential (diode drop), as illustrated in Figure 3–22. This is because half of the input voltage appears across each half of the secondary winding.

In order to obtain an output voltage equal to the input (less the barrier potential), a step-up transformer with a turns ratio of one-to-two must be used, as shown in Figure

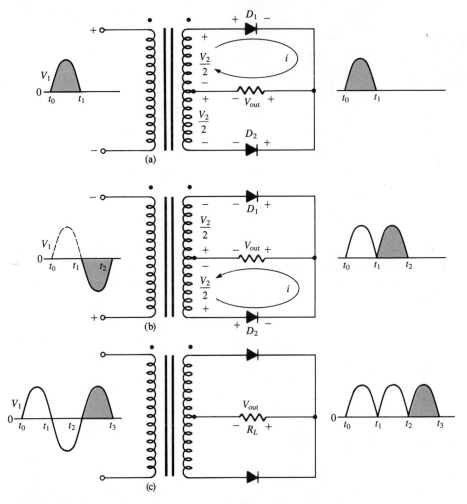

FIGURE 3–21
Operation of center-tapped full-wave rectifier.

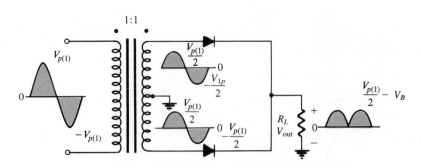

FIGURE 3–22
Center-tapped full-wave rectifier with a transformer turns ratio of 1.

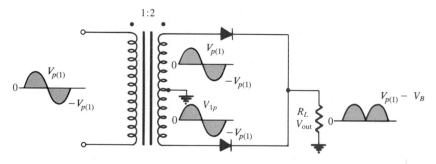

FIGURE 3–23
Center-tapped full-wave rectifier with a transformer turns ratio of 2.

3–23. In this case, the total secondary voltage V_2 is twice the primary voltage ($2\,V_1$), so the voltage across each half of the secondary is equal to V_1.

In any case, the output voltage of a center-tapped full-wave rectifier is always *one-half of the total secondary voltage,* no matter what the turns ratio is.

$$V_{out} = \frac{V_2}{2} \qquad (3\text{--}6)$$

To include the diode drop V_B, subtract it from $V_2/2$.

$$V_{out} = \frac{V_2}{2} - V_B \qquad (3\text{--}7)$$

Peak Inverse Voltage

Each diode in the full-wave rectifier is alternately forward-biased and then reverse-biased. The maximum reverse voltage that each diode must withstand is the peak secondary voltage (V_2). This can be seen by examining Figure 3–24. When the total secondary voltage V_2 has the polarity shown, the anode of D_1 is $+V_2/2$ and the anode of D_2 is $-V_2/2$. Since D_1 is forward-biased, its cathode is at the same voltage as its anode ($+V_2/2$, neglecting the barrier potential); this is also the voltage on the cathode of D_2.

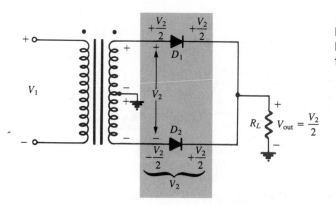

FIGURE 3–24
Diode reverse voltage (D_2 shown reverse-biased). The PIV is twice the peak value of the output voltage.

The total reverse voltage across D_2 is therefore $+V_2/2 - (-V_2/2) = V_2/2 + V_2/2 = V_2$. Since $V_2 = 2 V_{out}$, the *peak inverse voltage* across either diode in the center-tapped full-wave rectifier is

$$\text{PIV} = 2 V_{p(out)} \qquad\qquad\qquad \textbf{(3–8)}$$

EXAMPLE 3–5

Show the voltage waveforms across the secondary winding and across R_L when a 25 V peak sine wave is applied to the primary winding in Figure 3–25. Also, what minimum PIV rating must the diodes have?

FIGURE 3–25

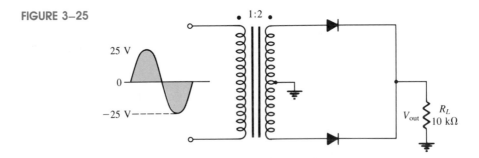

Solution
The total peak secondary voltage is

$$V_2 = \left(\frac{N_2}{N_1}\right)V_1 = (2)25 \text{ V} = 50 \text{ V}$$

There is a 25 V peak across each half of the secondary. The output load voltage has a peak value of 25 V, less the 0.7 V drop across the diode. Each diode must have a minimum PIV rating of 50 V (neglecting diode drop). The waveforms are shown in Figure 3–26.

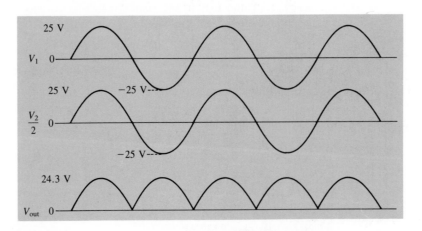

FIGURE 3–26

Full-Wave Bridge Rectifier

This type of full-wave rectifier uses four diodes, as shown in Figure 3–27. When the input cycle is positive as in part (a), diodes D_1 and D_2 are forward-biased and conduct current in the direction shown. A voltage is developed across R_L which looks like the positive half of the input cycle. During this time, diodes D_3 and D_4 are reverse-biased. When the input cycle is negative as in part (b), diodes D_3 and D_4 are forward-biased and conduct current in the same direction through R_L as during the positive half-cycle. During the negative half-cycle, D_1 and D_2 are reverse-biased. A full-wave rectified output voltage appears across R_L as a result of this action.

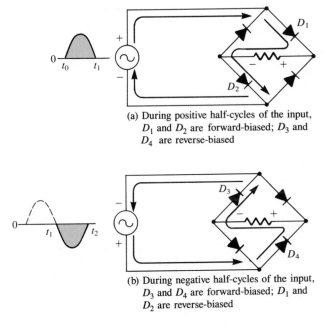

FIGURE 3–27
Full-wave bridge rectifier operation.

(a) During positive half-cycles of the input, D_1 and D_2 are forward-biased; D_3 and D_4 are reverse-biased

(b) During negative half-cycles of the input, D_3 and D_4 are forward-biased; D_1 and D_2 are reverse-biased

Bridge Output Voltage

A bridge rectifier with a transformer-coupled input is shown in Figure 3–28(a). During the positive half-cycle of the total secondary voltage, diodes D_1 and D_2 are forward-biased. Neglecting the diode drops, the secondary voltage V_2 appears across the load resistor. The same is true when D_3 and D_4 are forward-biased during the negative half-cycle.

$$V_{out} = V_2 \qquad (3–9)$$

As you can see in Figure 3–28(b), *two* diodes are always in series with the load resistor during both the positive and negative half-cycles. If these diode drops are taken into account, the output voltage is

$$V_{out} = V_2 - 2\,V_B \qquad (3–10)$$

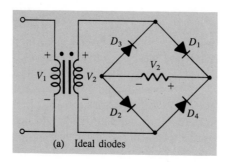
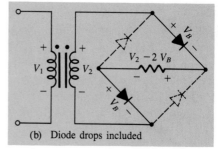

(a)   Ideal diodes   (b)   Diode drops included

FIGURE 3–28
Bridge output voltage.

Peak Inverse Voltage

Let us assume that D_1 and D_2 are forward-biased and examine the reverse voltage across D_3 and D_4. Visualizing D_1 and D_2 as shorts (ideally), as in Figure 3–29, you can see that D_3 and D_4 have a peak inverse voltage equal to the peak secondary voltage. Since the output voltage is *ideally* equal to the secondary voltage, we have

$$\text{PIV} \cong V_{p(\text{out})} \qquad (3\text{--}11)$$

The PIV rating of the bridge diodes is half that required for the center-tapped configuration.

FIGURE 3–29
Peak inverse voltage in a bridge rectifier during the positive half-cycle of the input voltage.

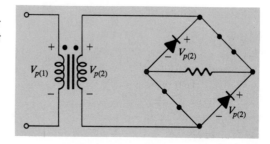

EXAMPLE 3–6

Determine the output voltage for the bridge rectifier in Figure 3–30. What minimum PIV rating is required for the silicon diodes?

FIGURE 3–30

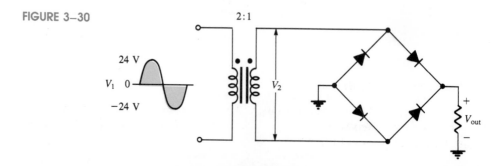

Solution

The peak output voltage is (taking into account the two diode drops)

$$V_{p(\text{out})} = V_2 - 2 V_B$$

$$= \left(\frac{N_2}{N_1}\right) V_{p(\text{in})} - 2 V_B$$

$$= \frac{1}{2} V_{p(\text{in})} - 2 V_B$$

$$= 12 \text{ V} - 1.4 \text{ V}$$

$$= 10.6 \text{ V}$$

The PIV for each diode is

$$\text{PIV} \cong V_{p(\text{out})} = 10.6 \text{ V}$$

SECTION REVIEW 3–3

1. What is the average value of a full-wave rectified voltage with a peak value of 60 V?
2. Which type of full-wave rectifier has the greater output voltage for the same input voltage and transformer turns ratio?
3. For a given output voltage, the PIV for bridge rectifier diodes is less than for center-tapped rectifier diodes (T or F).

RECTIFIER FILTERS

3–4

In most power supply applications, the standard 60 Hz ac power line voltage must be converted to a sufficiently constant dc voltage. The 60 Hz pulsating dc output of a half-wave rectifier or the 120 Hz pulsating output of a full-wave rectifier must be *filtered* to virtually eliminate the large voltage variations. Figure 3–31 illustrates the filtering concept showing an ideally smooth dc output voltage. A full-wave rectifier voltage is applied to the filter's input and, ideally, a constant dc level appears on the output.

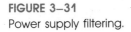

FIGURE 3–31
Power supply filtering.

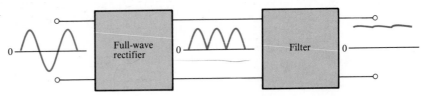

Capacitor-Input Filter

A half-wave rectifier with a capacitor-input filter is shown in Figure 3–32. R_L represents the load resistance. We will use the half-wave rectifier to illustrate the principle, and then expand the concept to full-wave.

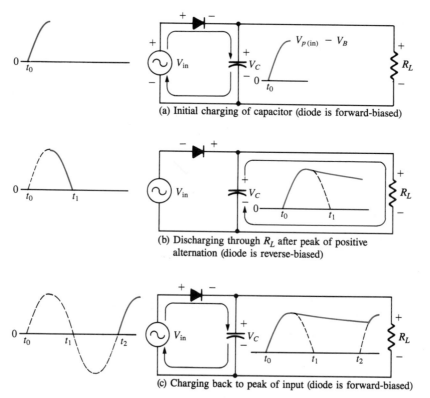

FIGURE 3–32
Operation of a half-wave rectifier with a capacitor-input filter.

During the positive first quarter-cycle of the input, the diode is forward-biased, allowing the capacitor to charge to within a diode drop of the input peak, as illustrated in Figure 3–32(a). When the input begins to decrease below its peak, as shown in part (b), the capacitor retains its charge and the diode becomes reverse-biased. During the remaining part of the cycle, the capacitor can discharge only through the load resistance at a rate determined by the $R_L C$ time constant. The larger the time constant, the less the capacitor will discharge. During the first quarter of the next cycle, the diode will again become forward-biased when the input voltage exceeds the capacitor voltage by approximately a diode drop. This is illustrated in part (c).

Ripple

As you have seen, the capacitor quickly charges at the beginning of a cycle and slowly discharges after the positive peak (when the diode is reverse-biased). The variation in the output voltage due to the charging and discharging is called the *ripple*. The smaller the ripple, the better the filtering action, as illustrated in Figure 3–33.

For a given input frequency, the output frequency of a full-wave rectifier is twice that of a half-wave rectifier, as illustrated in Figure 3–34. This makes a full-wave rectifier

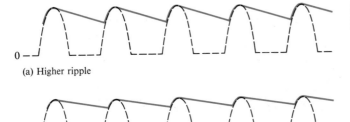

FIGURE 3–33
Half-wave ripple voltage (solid line).

(a) Higher ripple

(b) Lower ripple

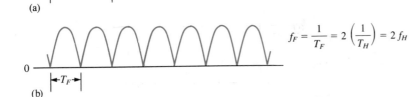

$$f_H = \frac{1}{T_H}$$

$$T_H = 2\,T_F$$

(a)

$$f_F = \frac{1}{T_F} = 2\left(\frac{1}{T_H}\right) = 2\,f_H$$

(b)

FIGURE 3–34
Frequencies of half-wave and full-wave signals.

easier to filter. When filtered, the full-wave rectified voltage has less ripple than does a half-wave signal for the same load resistance and capacitor values. This is because the capacitor discharges less during the shorter interval between full-wave pulses, as shown in Figure 3–35.

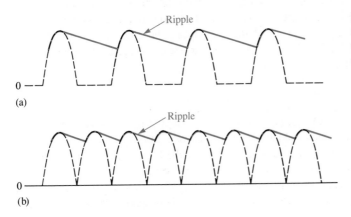

Ripple

(a)

Ripple

(b)

FIGURE 3–35
Comparison of ripple voltages for half-wave and full-wave signals with same filter and same input frequency.

The *ripple factor* is an indication of the effectiveness of the filter and is defined as

$$r = \frac{V_r}{V_{dc}} \tag{3-12}$$

where V_r is the rms ripple voltage and V_{dc} is the dc (average) value of the filter's output voltage, as illustrated in Figure 3–36. The lower the ripple factor, the better the filter. The ripple factor can be lowered by increasing the value of the filter capacitor.

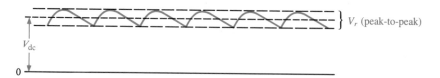

FIGURE 3–36
V_r and V_{dc} determine the ripple factor.

For a full-wave rectifier with a sufficiently high capacitance-input filter, if V_{dc} is very near in value to the peak rectified input voltage, then the expressions for V_r and V_{dc} are as follows. (The detailed derivations are in Appendix B.)

$$V_{dc} = \left(1 - \frac{0.00417}{R_L C}\right)V_{p(in)} \tag{3-13}$$

$$V_r = \frac{0.0024}{R_L C}V_{p(in)} \tag{3-14}$$

where $V_{p(in)}$ is the peak rectified voltage applied to the filter.

EXAMPLE 3–7

Determine the ripple factor for the filtered bridge rectifier in Figure 3–37.

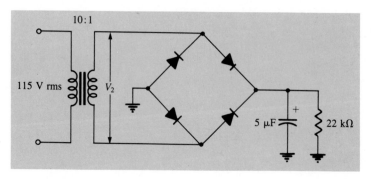

FIGURE 3–37

Solution

The peak primary voltage is

$$V_{p(1)} = (1.414)115 \text{ V}$$
$$= 162.6 \text{ V}$$

The peak secondary voltage is

$$V_{p(2)} = \left(\frac{1}{10}\right)162.6 \text{ V}$$
$$= 16.26 \text{ V}$$

The peak full-wave rectified voltage at the filter input is

$$V_{p(\text{in})} = V_{p(2)} - 2\,V_B$$
$$= 16.26 \text{ V} - 1.4 \text{ V}$$
$$= 14.86 \text{ V}$$

The filtered dc output voltage is

$$V_{\text{dc}} = \left(1 - \frac{0.00417}{R_L C}\right)V_{p(\text{in})}$$
$$= \left(1 - \frac{0.00417}{(22 \text{ k}\Omega)(5 \text{ }\mu\text{F})}\right)14.86 \text{ V}$$
$$= (1 - 0.0379)14.86 \text{ V}$$
$$= 14.3 \text{ V}$$

The rms ripple is

$$V_r = \frac{0.0024\,V_{p(\text{in})}}{R_L C}$$
$$= \frac{0.0024(14.86 \text{ V})}{(22 \text{ k}\Omega)(5 \text{ }\mu\text{F})}$$
$$= 0.324 \text{ V}$$

The ripple factor is

$$r = \frac{V_r}{V_{\text{dc}}}$$
$$= \frac{0.324 \text{ V}}{14.3 \text{ V}}$$
$$= 0.0227$$

The percent ripple is 2.27 percent.

The following BASIC program computes the percent ripple for a bridge rectifier when you input the turns ratio, the load resistance, and the filter capacitance.

```
10  CLS
20  PRINT "THIS PROGRAM COMPUTES THE RIPPLE"
30  PRINT "FACTOR FOR THE BRIDGE RECTIFIER"
40  PRINT "IN FIGURE 3-37 GIVEN N2/N1,"
50  PRINT "RL, AND C,"
60  PRINT:PRINT:PRINT
70  INPUT "TO CONTINUE PRESS 'ENTER'";X
80  CLS
90  INPUT "TURNS RATIO (N2/N1)";N
100 INPUT "LOAD RESISTANCE RL IN OHMS";RL
110 INPUT "FILTER CAPACITANCE C IN FARADS";C
120 CLS
130 V2P=N*162.6
140 VIP=V2P-1.4
150 VDC=(1-.00417/(RL*C))*VIP
160 VR=.0024*VIP/(RL*C)
170 PRINT "THE PERCENT RIPPLE IS";(VR/VDC)*100
```

Choosing the Filter Capacitance

A formula for calculating the capacitance needed to produce a specified ripple factor for a given load resistance in a full-wave rectifier is determined as follows: Let $V_{dc} = V_{p(in)}$. Then

$$r = \frac{V_r}{V_{dc}}$$

$$= \frac{0.0024\,V_{dc}}{R_L C\,V_{dc}}$$

$$= \frac{0.0024}{R_L C}$$

and

$$C = \frac{0.0024}{R_L r} \tag{3-15}$$

EXAMPLE 3–8

Determine the capacitance required to produce a ripple factor of no greater than 0.05 for a full-wave bridge rectifier having a load resistance of 10 kΩ.

Solution

$$C = \frac{0.0024}{(10\ \text{k}\Omega)(0.05)} = 4.8\ \mu F$$

This is a *minimum* value for C. Using a larger value will result in less than a 5 percent ripple.

Surge Current in the Capacitor Filter

Before the switch in Figure 3–38(a) is closed, the filter capacitor is uncharged. At the instant the switch is closed, voltage is connected to the bridge and the capacitor appears as a short, as shown. This produces an initial surge of current $I_{s(max)}$ through the two forward-biased diodes. The worst-case situation occurs when the switch is closed at a peak of the secondary voltage and a maximum surge current is produced, as illustrated in part (b) of the figure.

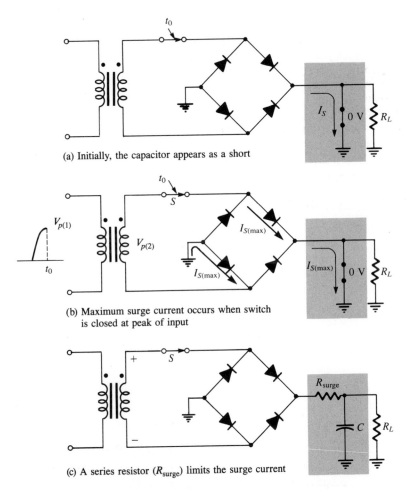

(a) Initially, the capacitor appears as a short

(b) Maximum surge current occurs when switch is closed at peak of input

(c) A series resistor (R_{surge}) limits the surge current

FIGURE 3–38
Surge current in a capacitor-input filter.

It is possible that the surge current could destroy the diodes, and for this reason a surge-limiting resistor is sometimes connected, as shown in Figure 3–38(c). The value of this resistor must be small compared to R_L. Also, the diodes must have a forward current rating such that they can withstand the momentary surge of current.

Inductor Input Filter

When a choke is added to the filter input, as in Figure 3–39, a reduction in the ripple voltage is achieved. The choke has a high reactance at the ripple frequency, and the capacitive reactance is low compared to both X_L and R_L. The two reactances form an ac voltage divider that tends to significantly reduce the ripple from that of a straight capacitor-input filter, as shown in Figure 3–40.

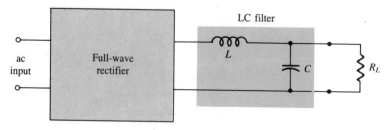

FIGURE 3–39
Rectifier with an LC filter.

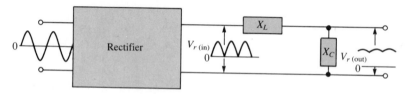

FIGURE 3–40
The LC filter as it looks to the ac component.

The magnitude of the ripple voltage out of the filter is determined with the voltage divider equation.

$$V_{r(out)} = \left(\frac{X_C}{|X_L - X_C|} \right) V_{r(in)} \qquad (3-16)$$

To the dc (average) value of the rectified input, the choke presents a winding resistance (R_W) in series with the load resistance, as shown in Figure 3–41. This resis-

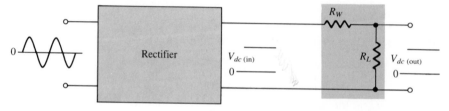

FIGURE 3–41
The LC filter as it looks to the dc component.

tance produces an undesirable reduction of the dc value, and therefore R_W must be small compared to R_L. The dc output voltage is determined as follows.

$$V_{dc(out)} = \left(\frac{R_L}{R_W + R_L}\right)V_{dc(in)} \qquad\qquad (3\text{--}17)$$

EXAMPLE 3–9

A 120 Hz full-wave rectified voltage with a peak value of 162.6 V is applied to the LC filter in Figure 3–42. Determine the filter output in terms of its dc value and the rms ripple voltage. What is the ripple factor? Compare this to the capacitor filter in Example 3–7.

FIGURE 3–42

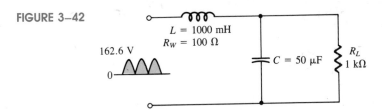

$L = 1000$ mH
$R_W = 100\ \Omega$

162.6 V

$C = 50\ \mu F$

R_L
$1\ k\Omega$

Solution
First, we determine the dc value of the full-wave rectified input using equation (3–5).

$$V_{dc(in)} = V_{AVG} = \frac{2\ V_p}{\pi}$$

$$= \frac{2(162.6\text{ V})}{\pi}$$

$$= 103.5\text{ V}$$

Next, we find the amount of rms input ripple using a formula derived in Appendix B for the rms ripple of an unfiltered full-wave rectified signal.

$$V_{r(in)} = 0.308\ V_p$$
$$= 0.308(162.6\text{ V})$$
$$= 50.1\text{ V}$$

Now that the input values are known, the output values can be calculated.

$$V_{dc(out)} = \left(\frac{R_L}{R_W + R_L}\right)V_{dc(in)}$$

$$= \left(\frac{1\text{ k}\Omega}{1.1\text{ k}\Omega}\right)103.5\text{ V}$$

$$= 94.1\text{ V}$$

For the ripple calculation we need X_L and X_C.

$$X_L = 2\pi f L$$
$$= 2\pi(120 \text{ Hz})(1000 \text{ mH})$$
$$= 754 \text{ } \Omega$$

$$X_C = \frac{1}{2\pi f C}$$
$$= \frac{1}{2\pi(120 \text{ Hz})(50 \text{ } \mu F)}$$
$$= 26.5 \text{ } \Omega$$

$$V_{r(out)} = \left(\frac{X_C}{|X_L - X_C|}\right) V_{r(in)}$$
$$= \left(\frac{26.5 \text{ } \Omega}{|754 \text{ } \Omega - 26.5 \text{ } \Omega|}\right) 50.1 \text{ V}$$
$$= 1.82 \text{ V rms}$$

The ripple factor is

$$r = \frac{V_{r(out)}}{V_{dc}}$$
$$= \frac{1.82 \text{ V}}{94.1 \text{ V}}$$
$$= 0.0193$$

The percent ripple is 1.93, which is less than that for the capacitor filter in Example 3–7.

Practice Exercise 3–9
A 120 Hz full-wave rectified voltage with a peak value of 50 V is applied to the LC filter in Figure 3–42 with $L = 300$ mH, $R_W = 50$ Ω, $C = 100$ μF, and $R_L = 10$ kΩ. Determine the filter output in terms of its dc value and the rms ripple voltage. What is the ripple factor?

It should be noted at this time that an inductor-input filter produces an output with a dc value approximately equal to the *average* value of the rectified input. The capacitor-input filter, however, produces an output with a dc value approximately equal to the *peak* value of the input. Another point of comparison is that the amount of ripple voltage in the capacitor-input filter varies inversely with the load resistance. Ripple voltage in the LC filter is essentially independent of the load resistance and depends only on X_L and X_C, as long as X_C is sufficiently less than R_L.

π-Type Filter

A one-section π-type filter is shown in Figure 3–43. It can be thought of as a capacitor filter followed by an LC filter. It combines the peak filtering action of the single-capacitor filter with the reduced ripple and load independence of the LC filter.

FIGURE 3–43
π-type LC filter.

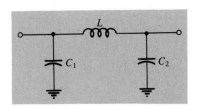

---

1. A 60 Hz sine wave is applied to a half-wave rectifier. What is the output frequency? What is the output frequency for a full-wave rectifier?
2. What causes the ripple voltage on the output of a capacitor filter?
3. The load resistance of a capacitor-filtered full-wave rectifier is reduced. What effect does this have on the ripple voltage?
4. Calculate the ripple factor for $V_{dc} = 30$ V and for $V_r = 0.2$ V rms.
5. Name one advantage of an LC filter over a capacitor filter. Name one disadvantage.
6. What is the *ideal* dc output voltage for a capacitor filter with a full-wave rectified input having a peak value of 220 V? For an LC filter?

TROUBLESHOOTING RECTIFIER CIRCUITS

3–5

Several types of failures can occur in power-supply rectifiers. In this section we will examine some possible failures and the effects they would have on a circuit's operation.

Open Diode

A half-wave rectifier with a diode that has opened (a common failure mode) is shown in Figure 3–44. In this case, you would measure 0 V dc across the load resistor, as depicted.

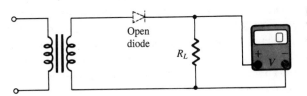

FIGURE 3–44
Test for an open diode in a half-wave rectifier.

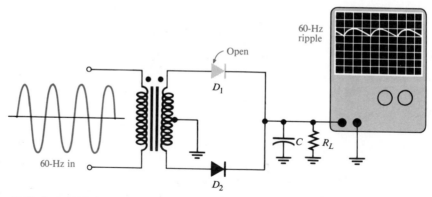

(a) Ripple should be less and have a frequency of 120 Hz.

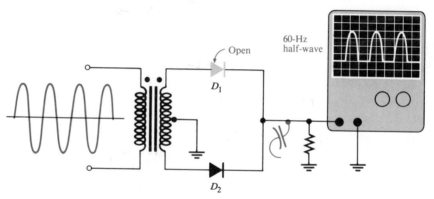

(b) With C removed, output should be a full-wave 120-Hz signal.

FIGURE 3–45
Test for an open diode in a full-wave center-tapped rectifier.

Now consider the center-tapped full-wave rectifier in Figure 3–45. Assume that diode D_1 has failed open. Here is what you would observe with an oscilloscope connected to the output, as shown in part (a): You would see a larger-than-normal ripple voltage at a frequency of 60 Hz rather than 120 Hz. Disconnecting the filter capacitor, you would observe a *half-wave* rectified voltage, as in part (b). Now let's examine the reason for these observations. If diode D_1 is open, there will be current through R_L only during the negative half-cycle of the input signal. During the positive half-cycle, an open path prevents current through R_L. The result is a half-wave voltage, as illustrated in Figure 3–46(a). With the filter capacitor in the circuit, the half-wave signal will allow it to discharge more than it would with a normal full-wave signal, resulting in a larger ripple voltage, as shown in part (b). The same observations would be made for an open failure of diode D_2.

An open diode in a bridge rectifier would create symptoms identical to those just discussed for the center-tapped rectifier. As illustrated in Figure 3–47, the open diode would prevent current through R_L during half of the input cycle (in this case, the negative

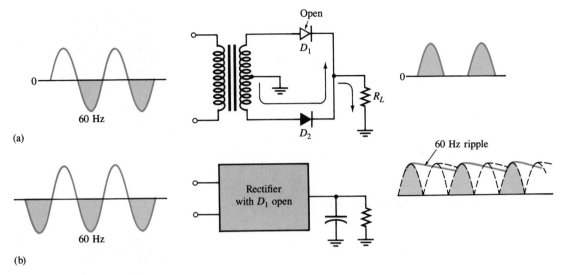

(a)

(b)

FIGURE 3–46
Effects of an open diode in a center-tapped rectifier.

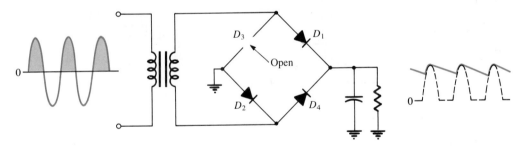

FIGURE 3–47
Effects of an open diode in a bridge rectifier.

half). This would result in a half-wave output and an increased ripple voltage at 60 Hz, as discussed before.

Shorted Diode

A shorted diode is one that has failed such that it has a very low resistance in both directions. If a diode suddenly became shorted in a bridge rectifier, it is likely that a sufficiently high current would exist during one half of the input cycle such that the shorted diode itself would burn open or the other diode in series with it would open. The transformer could also be damaged. This is illustrated in Figure 3–48 with D_1 shorted.

In part (a) of the figure, current is supplied to the load through the shorted diode during the first positive half-cycle, just as though it were forward-biased. During the negative half-cycle, the current is shorted through D_1 and D_4, as shown in part (b). Again, damage to the transformer is possible. It is likely that this excessive current would burn either or both of the diodes open. If only one of the diodes opened, you would still observe

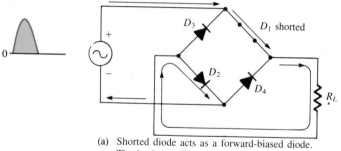

(a) Shorted diode acts as a forward-biased diode.
     The load current is normal.

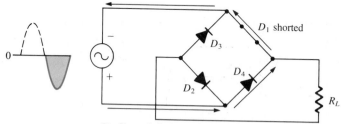

(b) Shorted diode produces short across source,
     causing a high short-circuit current during
     one alternation.

FIGURE 3–48
Effects of a shorted diode in a bridge rectifier circuit.

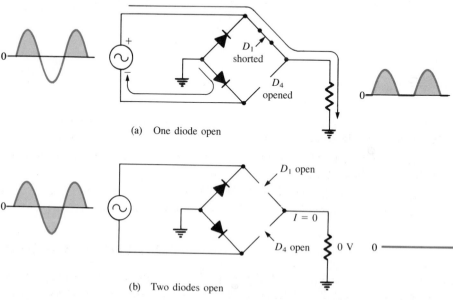

(a)   One diode open

(b)   Two diodes open

FIGURE 3–49
Effects of open diodes in a bridge rectifier circuit.

a half-wave voltage on the output. If both diodes opened, there would be no voltage developed across the load. These conditions are illustrated in Figure 3–49.

Shorted or Leaky Filter Capacitor

A *shorted capacitor* would most likely cause some or all of the diodes in a full-wave rectifier to open due to excessive current. In any event, there would be no dc voltage on the output. Figure 3–50 illustrates the condition. A *leaky capacitor* can be represented by a leakage resistance in parallel with the capacitor, as shown in Figure 3–51(a). The effect of the leakage resistance is to reduce the discharging time constant. This would cause an increase in ripple voltage on the output, as shown in Figure 3–51(b).

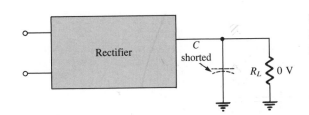

FIGURE 3–50
Shorted filter capacitor produces an output of 0 V.

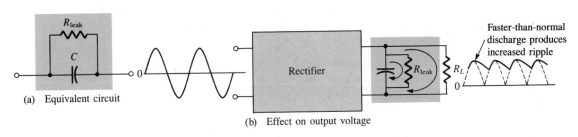

(a) Equivalent circuit

(b) Effect on output voltage

FIGURE 3–51
Leaky filter capacitor.

---

1. What effect would an open D_2 produce in the rectifier of Figure 3–45?

2. You are checking a 60 Hz full-wave bridge rectifier and observe that the output has a 60 Hz ripple. What failure(s) would you suspect?

3. If you observe that the output ripple of a full-wave rectifier is much greater than normal but its frequency is still 120 Hz, what component would you suspect?

**SECTION
REVIEW
3–5**

DIODE LIMITING AND CLAMPING CIRCUITS

3–6

Diode circuits are sometimes used to clip off portions of signal voltages above or below certain levels; these circuits are called *limiters* or *clippers*. Another type of diode circuit is used to restore a dc level to an electrical signal; these are called *clampers*. Both limiter and clamper diode circuits will be examined in this section.

DIODES AND APPLICATIONS

Diode Limiters

Figure 3–52(a) shows a diode circuit that limits the positive part of the input voltage. As the input signal goes positive, the diode becomes forward-biased. Since the cathode is at ground potential (0 V), the anode cannot exceed 0.7 V (assuming silicon). So point A is limited to +0.7 V when the input exceeds this value. When the input goes back below 0.7 V, the diode reverse-biases and appears as an open. The output voltage looks like the negative part of the input, but with a magnitude determined by the R_s and R_L voltage divider as follows:

$$V_{out} = \left(\frac{R_L}{R_s + R_L} \right) V_{in}$$

If R_s is small compared to R_L, then $V_{out} \cong V_{in}$.

Turn the diode around, as in Figure 3–52(b), and the *negative* part of the input is clipped off. When the diode is forward-biased during the negative part of the input, point A is held at −0.7 V by the diode drop. When the input goes above −0.7 V, the diode is no longer forward-biased and a voltage appears across R_L proportional to the input.

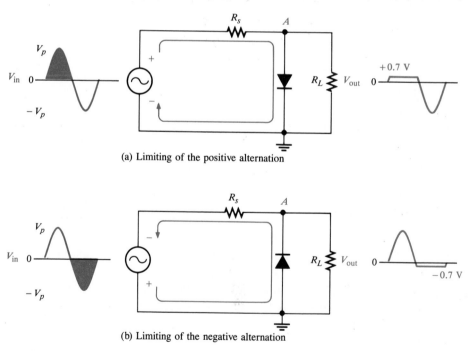

(a) Limiting of the positive alternation

(b) Limiting of the negative alternation

FIGURE 3–52
Diode limiters.

EXAMPLE 3–10

What would you expect to see displayed on an oscilloscope connected as shown in Figure 3–53?

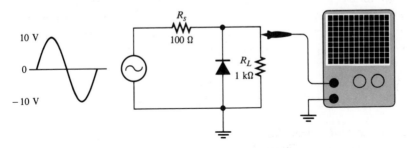

FIGURE 3–53

Solution

The diode conducts when the input voltage goes below −0.7 V. So, we have a negative limiter with a peak output voltage determined by the following equation.

$$V_{p(out)} = \left(\frac{R_L}{R_s + R_L}\right)V_{p(in)}$$

$$= \left(\frac{1\ k\Omega}{1.1\ k\Omega}\right)10\ V$$

$$= 9.09\ V$$

The scope will display an output waveform as shown in Figure 3–54.

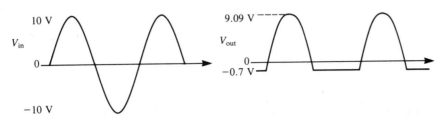

FIGURE 3–54
Waveforms for Figure 3–53.

Adjustment of the Limiting Level

The level to which a signal voltage is limited can be adjusted by adding a *bias* voltage in series with the diode, as shown in Figure 3–55. The voltage at point *A* must equal

DIODES AND APPLICATIONS

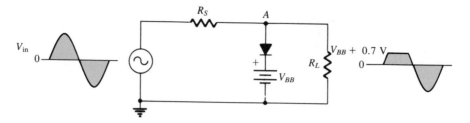

FIGURE 3–55
Positively biased limiter.

$V_{BB} + 0.7$ V before the diode will conduct. Once the diode begins to conduct, the voltage at point A is limited to $V_{BB} + 0.7$ V so that all input voltage above this level is clipped off.

If the bias voltage is varied up or down, the limiting level changes correspondingly, as shown in Figure 3–56. If the polarity of the bias voltage is reversed, as in Figure 3–57, voltages above $-V_{BB} + 0.7$ V are clipped, resulting in an output waveform as shown. The diode is reverse-biased only when the voltage at point A goes below $-V_{BB} + 0.7$ V.

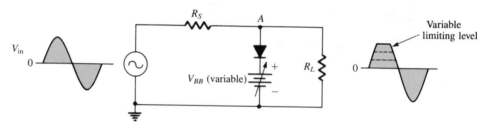

FIGURE 3–56
Positive limiter with variable bias.

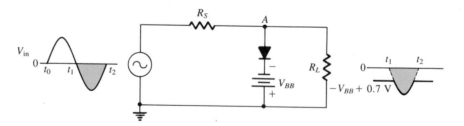

FIGURE 3–57

If it is necessary to limit a voltage to a specified negative level, then the diode and bias battery must be connected as in Figure 3–58. In this case, the voltage at point A must go below $-V_{BB} - 0.7$ V to forward-bias the diode and initiate limiting action, as shown.

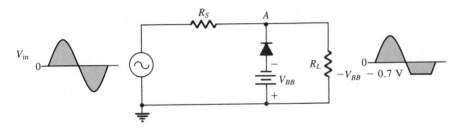

FIGURE 3–58

EXAMPLE 3–11

Figure 3–59 shows a circuit combining a positive-biased limiter with a negative-biased limiter. Determine the output waveform.

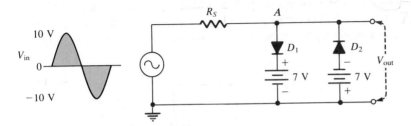

FIGURE 3–59

Solution

When the voltage at point A reaches $+7.7$ V, diode D_1 conducts and limits the waveform to $+7.7$ V. Diode D_2 does not conduct until the voltage reaches -7.7 V. Therefore, positive voltages above $+7.7$ V and negative voltages below -7.7 V are clipped off. The resulting output waveform is shown in Figure 3–60.

FIGURE 3–60
Output waveform for Figure 3–59.

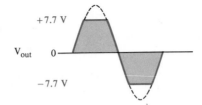

Practice Exercise 3–11

Determine the output waveform in Figure 3–59 if both dc sources are 10 V and the input has a peak value of 20 V.

DIODES AND APPLICATIONS

Diode Clampers

The purpose of a clamper is to add a dc level to an ac signal. Clampers are sometimes known as *dc restorers*. Figure 3–61 shows a diode clamper that inserts a *positive* dc level. The operation of this circuit can be seen by considering the first negative half-cycle of the input voltage. When the input initially goes negative, the diode is forward-biased, allowing the capacitor to charge to near the peak of the input ($V_{p(in)} - 0.7$ V), as shown in Figure 3–61(a). Just above the negative peak, the diode is reverse-biased. This is because the cathode is held near ($V_{p(in)} - 0.7$ V) by the charge on the capacitor. The capacitor can only discharge through the high resistance of R_L. So, from the peak of one negative half-cycle to the next, the capacitor discharges very little. The amount that is discharged, of course, depends on the value of R_L. For good clamping action, the RC time constant should be at least ten times the period of the input frequency.

The net effect of the clamping action is that the capacitor retains a charge approximately equal to the peak value of the input less the diode drop. The capacitor voltage acts essentially as a battery in series with the input signal, as shown in Figure 3–61(b). The dc voltage of the capacitor adds to the input voltage by superposition, as in Figure 3–61(c). If the diode is turned around, a *negative* dc voltage is added to the input signal, as shown in Figure 3–62.

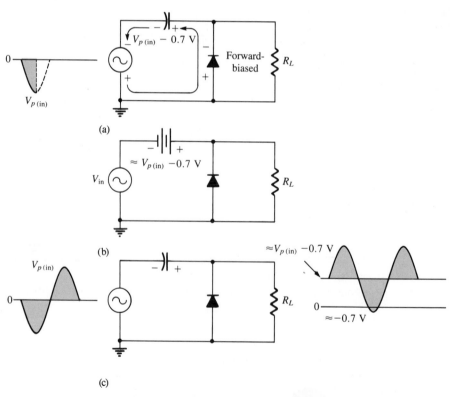

FIGURE 3–61
Positive clamper operation.

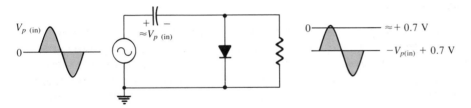

FIGURE 3–62
Negative clamper.

A Clamper Application

A clamping circuit is often used in television receivers as a dc restorer. The incoming composite video signal is normally processed through capacitively coupled amplifiers which eliminate the dc component, thus losing the black and white reference levels and the blanking level. Before being applied to the picture tube, these reference levels must be restored. Figure 3–63 illustrates this process in a general way.

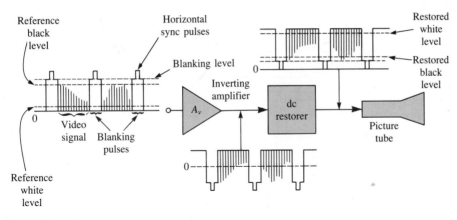

FIGURE 3–63
Clamper (dc restorer) in a TV receiver.

EXAMPLE 3–12

What is the output voltage that you would expect to observe across R_L in the clamper circuit of Figure 3–64? Assume that RC is large enough to prevent significant capacitor discharge.

FIGURE 3–64

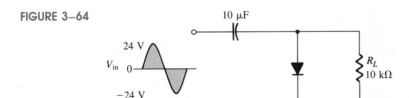

Solution

Ideally, a negative dc value equal to the input peak less the diode drop is inserted by the clamping circuit:

$$V_{dc} \cong V_{p(in)} - 0.7 \text{ V}$$
$$= 24 \text{ V} - 0.7 \text{ V}$$
$$= 23.3 \text{ V}$$

Actually, the capacitor will discharge slightly between peaks, and, as a result, the output voltage will have an average value of slightly less than that calculated above. The output waveform goes to approximately 0.7 V above ground, as shown in Figure 3–65.

FIGURE 3–65

Output waveform for Figure 3–64.

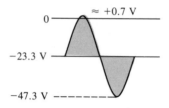

Practice Exercise 3–12

What is the output voltage that you would observe across R_L in Figure 3–64 if $C = 22 \ \mu\text{F}$ and $R_L = 18 \text{ k}\Omega$?

**SECTION
REVIEW
3–6**

1. Determine the output waveform for the circuit of Figure 3–66.

FIGURE 3–66

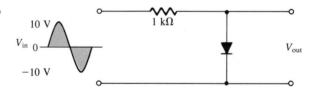

2. What is the output voltage in Figure 3–67?

FIGURE 3–67

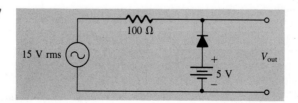

3. Sketch the approximate output waveform for a positive clamper having a sine wave input with a peak of 50 V.

VOLTAGE MULTIPLIERS

Voltage multipliers utilize clamping action to increase peak rectified voltages without the necessity of increasing the input transformer's voltage rating. Multiplication factors of two, three, and four are commonly used.

Voltage Doublers

A half-wave voltage doubler is shown in Figure 3–68. During the positive half-cycle of the secondary voltage, diode D_1 is forward-biased and D_2 is reverse-biased. Capacitor C_1 is charged to the peak of the secondary voltage (V_p) less the diode drop with the polarity shown in part (a). During the negative half-cycle, diode D_2 is forward-biased and D_1 reverse-biased, as shown in part (b). The peak voltage on C_1 adds to the secondary voltage to charge C_2 to $2 V_p$. Applying Kirchhoff's law around the loop,

$$V_{C_1} - V_{C_2} + V_p = 0$$
$$V_{C_2} = V_p + V_{C_1}$$

since $V_{C_1} \cong V_p$,

$$V_{C_2} = 2 V_p$$

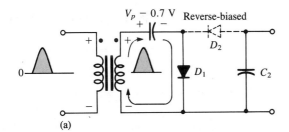

(a)

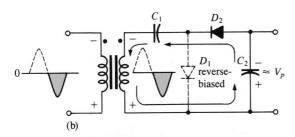

(b)

FIGURE 3–68
Half-wave voltage doubler operation.

    Under a no-load condition, C_2 remains charged to approximately $2 V_p$. If a load resistance is connected across the output, C_2 discharges through the load on the next positive half-cycle and is again recharged to $2 V_p$ on the following negative half-cycle. The resulting output is a half-wave, capacitor-filtered voltage. The peak inverse voltage across each diode is $2 V_p$.

    A full-wave doubler is shown in Figure 3–69. When the secondary voltage is positive, diode D_1 is forward-biased and C_1 charges to approximately V_p, as shown in part

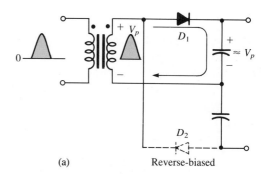

(a)   Reverse-biased

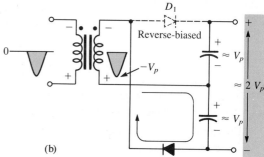

FIGURE 3–69
Full-wave voltage doubler operation.   (b)

(a). During the negative half-cycle, D_2 is forward-biased and C_2 charges to approximately V_p, as shown in part (b). The output voltage, $2\ V_p$, is taken across the two capacitors in series.

Voltage Tripler

The addition of another diode-capacitor section to the half-wave voltage doubler creates a voltage tripler, as shown in Figure 3–70. The operation is as follows: On the positive half-cycle of the secondary voltage, C_1 charges to V_p through D_1. During the negative half-cycle, C_2 charges to $2\ V_p$ through D_2, as described for the doubler. During the next positive half-cycle, C_3 charges to $2\ V_p$ through D_3. The tripler output is taken across C_1 and C_3, as shown in the figure.

FIGURE 3–70
Voltage tripler.

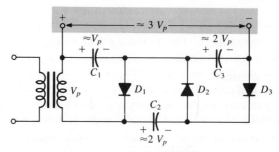

Voltage Quadrupler

The addition of still another diode-capacitor section, as shown in Figure 3–71, produces an output four times the peak secondary voltage. C_4 charges to $2 V_p$ through D_4 on a negative half-cycle. The $4 V_p$ output is taken across C_2 and C_4, as shown. In both the tripler and quadrupler circuits, the PIV of each diode is $2 V_p$.

FIGURE 3–71
Voltage quadrupler.

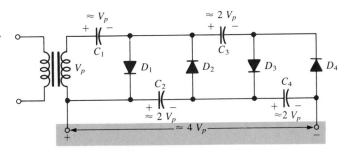

SECTION REVIEW 3–7

1. What must be the peak voltage rating of the transformer secondary for a voltage doubler that produces an output of 200 V?

2. The output voltage of a quadrupler is 240 V. What minimum PIV rating must each diode have?

INTERPRETING DIODE DATA SHEETS

3–8

A manufacturer's data sheet gives detailed information on a device so that it can be used properly in a given application. A typical data sheet provides *maximum ratings, electrical characteristics, mechanical data,* and *graphs of various parameters*. We use a specific example to illustrate a typical data sheet. Table 3–1 shows the *maximum ratings* for a certain series of rectifier diodes (1N4001 through 1N4007). These are the absolute maximum values under which the diode can be operated without damage to the device. For greatest reliability and longer life, the diode should always be operated well under these maximums. Generally, the maximum ratings are specified at 25°C and must be adjusted downward for greater temperatures.

An explanation of some of the parameters from Table 3–1 is as follows:

V_{RRM} The maximum reverse peak voltage that can be applied repetitively across the diode. Notice that in this case, it is 50 V for the 1N4001 and 1 kV for the 1N4007.

V_R The maximum reverse dc voltage that can be applied across the diode.

V_{RSM} The maximum reverse peak value of nonrepetitive voltage (60 Hz minimum) that can be applied across the diode.

I_0 The maximum average value of a 60 Hz full-wave rectified forward current.

I_{FSM} The maximum peak value of nonrepetitive (one cycle) forward current. The graph in Figure 3–72 expands on this parameter to show values for more than one cycle at temperatures of 25°C and 175°C. The dashed lines represent values where typical failures occur.

TABLE 3–1
Maximum ratings

Rating	Symbol	1N4001	1N4002	1N4003	1N4004	1N4005	1N4006	1N4007	Unit
Peak repetitive reverse voltage Working peak reverse voltage dc blocking voltage	V_{RRM} V_{RWM} V_R	50	100	200	400	600	800	1000	V
Nonrepetitive peak reverse voltage (half-wave, single-phase, 60 Hz)	V_{RSM}	60	120	240	480	720	1000	1200	V
rms reverse voltage	$V_{R(rms)}$	35	70	140	280	420	560	700	V
Average rectified forward current (single-phase, resistive load, 60 Hz, $T_A = 75°C$)	I_0	1.0							A
Nonrepetitive peak surge current (surge applied at rated load conditions)	I_{FSM}	30 (for 1 cycle)							A
Operating and storage junction temperature range	T_J, T_{stg}	−65 to +175							°C

FIGURE 3–72
Nonrepetitive surge capability.

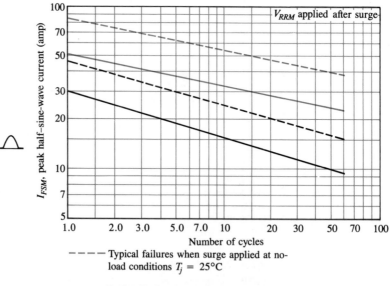

- – – – Typical failures when surge applied at no-load conditions $T_j = 25°C$
- ——— Design limits when surge applied at no-load conditions $T_j = 25°C$
- ▬▬▬ Typical failures when surge applied at rated-load conditions $T_j = 175°C$
- ▬ ▬ ▬ Design limits when surge applied at rated-load conditions $T_j = 175°C$

Notice what happens on the lower solid line when ten cycles of I_{FSM} are applied. The limit is 15 A rather than the one-cycle value of 30 A.

Table 3–2 lists typical and maximum values of certain electrical characteristics. These items differ from the maximum ratings in that they are not selected by design but are the result of operating the diode under specified conditions. A brief explanation of these parameters is as follows:

v_F The instantaneous voltage across the forward-biased diode when the forward current is 1 A at 25°C. Figure 3–73 shows how the forward voltages vary with forward current.

$V_{F(AV)}$ The maximum forward voltage drop averaged over a full cycle.

I_R The maximum current when the diode is reverse-biased with a dc voltage.

$I_{R(AV)}$ The maximum reverse current averaged over one cycle (when reverse-biased with an ac voltage).

The mechanical data for these particular diodes as they appear on a typical data sheet are shown in Figure 3–74.

FIGURE 3–73
Forward voltage.

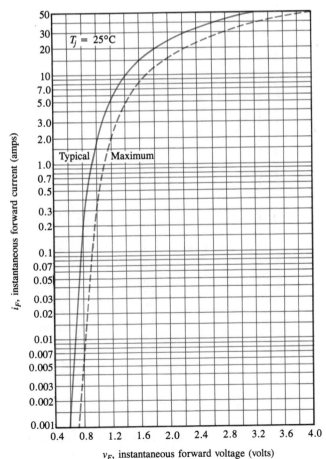

TABLE 3–2
Electrical characteristics

Characteristic and conditions	Symbol	Typical	Maximum	Unit
Maximum instantaneous forward voltage drop ($i_F = 1$ A, $T_J = 25°$C)	v_F	0.93	1.1	V
Maximum full-cycle average forward voltage drop ($I_0 = 1$ A, $T_L = 75°$C, 1 inch leads)	$V_{F(AV)}$	–	0.8	V
Maximum reverse current (rated dc voltage) $T_J = 25°$C $T_J = 100°$C	I_R	0.05 1.0	10.0 50.0	μA
Maximum full-cycle average reverse current ($I_0 = 1$ A, $T_L = 75°$C, 1 inch leads)	$I_{R(AV)}$	–	30.0	μA

FIGURE 3–74
Mechanical data

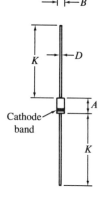

	Millimeters		Inches	
DIM	Min	Max	Min	Max
A	5.97	6.60	0.235	0.260
B	2.79	3.05	0.110	0.120
D	0.76	0.86	0.030	0.034
K	27.94	–	1.100	–

Mechanical characteristics

Case: Transfer Molded Plastic
Maximum lead temperature for soldering purposes: 350°C, ⅜″ from case for 10 seconds at 5 lbs. tension
Finish: All external surfaces are corrosion-resistant, leads are readily solderable
Polarity: Cathode indicated by color band
Weight: 0.40 Grams (approximately)

SECTION REVIEW 3–8

1. List the three diode ratings categories.
2. Identify each of the following parameters: **(a)** V_F; **(b)** I_R; **(c)** I_0

A SYSTEM APPLICATION

3–9

The detector circuit in an AM superhet receiver removes the 455 kHz carrier signal from the AM signal coming out of the IF amplifier, leaving only the audio signal that originally modulated the transmitted RF carrier. (If necessary, review the coverage of the AM receiver in Chapter 1.) Figure 3–75(a) shows a typical detector circuit, which is one of the simplest types found in receiver systems. The 455 kHz amplitude modulated signal is transformer-coupled from the output of the IF amplifier. To illustrate the operation, the amplitude modulated signal shown in Figure 3–75(b) is assumed to be the input to the detector. As you know, the diode will rectify an alternating input resulting in removal of

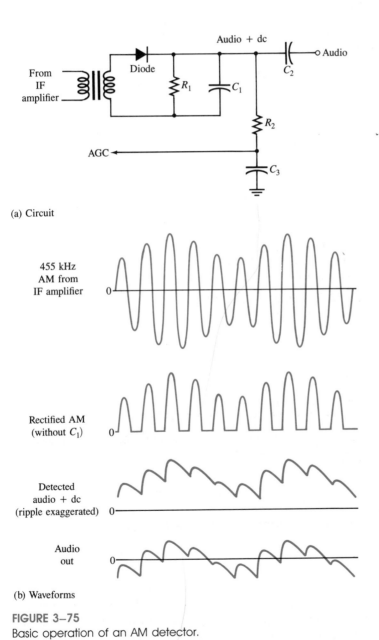

(a) Circuit

(b) Waveforms

FIGURE 3–75
Basic operation of an AM detector.

the negative portion of the AM signal, as indicated. The charging and discharging action of the capacitor C_1 removes (filters out) the 455 kHz carrier, leaving the original audio signal with only a very small amount of 455 kHz ripple that will be removed by the audio amplifier because its response is limited to around 20 kHz.

To take a more detailed look at how capacitor C_1 removes the carrier, assume that, initially, the capacitor charges to a peak of one of the carrier cycles. When the carrier sine wave begins to decrease below the peak, the capacitor will discharge slowly through the high resistance of R_1 before the next peak of the carrier. When the next peak occurs, the capacitor will charge to its value. The resulting output is a varying voltage that follows the peak variation of the modulated carrier, which has a nonzero average value or dc component. Capacitor C_2 blocks the dc component and allows only the detected audio signal to pass to the audio amplifier.

Automatic gain control (AGC) can be obtained from the detector by extracting the dc component of the audio signal with the low-pass filter composed of R_2 and C_3 as shown in Figure 3–75(a). The dc component fluctuates with signal strength and is fed back to the IF amplifier and/or RF amplifier to raise or lower the gain accordingly and keep the volume constant.

FORMULAS

(3–1)	$V_{\text{AVG}} = \dfrac{V_p}{\pi}$	Half-wave average value
(3–2)	$V_{p(\text{out})} = V_{p(\text{in})} - 0.7 \text{ V}$	Peak half-wave rectifier output (silicon)
(3–3)	$V_{p(\text{out})} = V_{p(\text{in})} - 0.3 \text{ V}$	Peak half-wave rectifier output (germanium)
(3–4)	$V_2 = \left(\dfrac{N_2}{N_1}\right) V_1$	Secondary voltage
(3–5)	$V_{\text{AVG}} = \dfrac{2\,V_p}{\pi}$	Full-wave average value
(3–6)	$V_{\text{out}} = \dfrac{V_2}{2}$	Center-tapped full-wave output
(3–7)	$V_{\text{out}} = \dfrac{V_2}{2} - V_B$	Center-tapped full-wave output (including diode drop, V_B)
(3–8)	$\text{PIV} = 2\,V_{p(\text{out})}$	Diode peak inverse voltage, center-tapped rectifier (neglecting V_B)
(3–9)	$V_{\text{out}} = V_2$	Bridge full-wave output (ideal)
(3–10)	$V_{\text{out}} = V_2 - 2\,V_B$	Bridge full-wave output (including diode drops)
(3–11)	$\text{PIV} \cong V_{p(\text{out})}$	Diode peak inverse voltage, bridge rectifier
(3–12)	$r = \dfrac{V_r}{V_{\text{dc}}}$	Ripple factor

$$\textbf{(3–13)} \quad V_{dc} = \left(1 - \frac{0.00417}{R_L C}\right) V_{p(in)}$$ dc output voltage, capacitor-input filter

$$\textbf{(3–14)} \quad V_r = \frac{0.0024}{R_L C} V_{p(in)}$$ Ripple voltage, capacitor-input filter

$$\textbf{(3–15)} \quad C = \frac{0.0024}{R_L r}$$ Filter capacitor value

$$\textbf{(3–16)} \quad \overline{V_{r(out)}} = \left(\frac{X_C}{|X_L - X_C|}\right) V_{r(in)}$$ Ripple output, LC filter

$$\textbf{(3–17)} \quad V_{dc(out)} = \left(\frac{R_L}{R_W + R_L}\right) V_{dc(in)}$$ dc output, LC filter

$$V_{r(?)} = 0.304 \, V_p$$

SUMMARY

- ☐ A diode conducts current when forward-biased and blocks current when reverse-biased.
- ☐ The forward-biased barrier potential is 0.7 V for a silicon diode and 0.3 V for a germanium diode.
- ☐ Reverse breakdown voltage for a rectifier diode is typically greater than 50 V.
- ☐ A functioning diode presents an open circuit when reverse-biased and a very low resistance when forward-biased.
- ☐ The single diode in a half-wave rectifier conducts for 180° of the input cycle.
- ☐ The output frequency of a half-wave rectifier equals the input frequency.
- ☐ The average (dc) value of a half-wave rectified signal is 0.318 ($1/\pi$) times its peak value.
- ☐ PIV (peak inverse voltage) is the maximum voltage appearing across the diode in reverse bias.
- ☐ Each diode in a full-wave rectifier conducts for 180° of the input cycle.
- ☐ The output frequency of a full-wave rectifier is twice the input frequency.
- ☐ The basic types of full-wave rectifier are center-tapped and bridge.
- ☐ The output voltage of a center-tapped full-wave rectifier is approximately one-half of the total secondary voltage.
- ☐ The PIV for each diode in a center-tapped full-wave rectifier is twice the output voltage.
- ☐ The output voltage of a bridge rectifier equals the total secondary voltage.
- ☐ The PIV for each diode in a bridge rectifier is half that required for the center-tapped configuration and is approximately equal to the peak output voltage.
- ☐ A capacitor-input filter provides a dc output approximately equal to the peak of the input.
- ☐ Ripple voltage is caused by the charging and discharging of the filter capacitor.
- ☐ The smaller the ripple, the better the filter.
- ☐ An LC filter provides improved ripple reduction over the capacitor-input filter.
- ☐ An inductor-input filter produces a dc output voltage approximately equal to the average value of the rectified input.
- ☐ Diode limiters cut off voltage above or below specified levels.
- ☐ Diode clampers add a dc level to an ac signal.

SELF-TEST

1. What voltage is indicated by each of the meters in Figure 3–76?

FIGURE 3–76

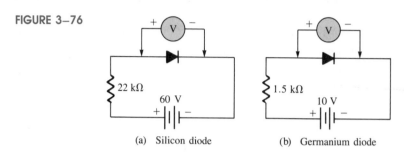

(a) Silicon diode   (b) Germanium diode

2. A silicon diode is in series with a 1 kΩ resistor and a 5 V battery. If the anode is connected to the positive source terminal, what is the cathode voltage with respect to the negative source terminal?

3. For the ohmmeter measurements in Figure 3–77, determine whether or not the diode in part (a) is good. Repeat for part (b).

FIGURE 3–77

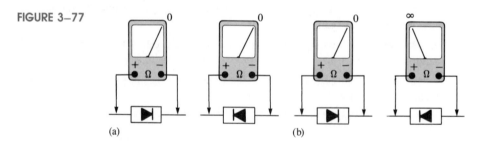

(a)   (b)

4. How can an ideal forward-biased diode be represented? An ideal reverse-biased diode?

5. Determine the current for each diode circuit in Figure 3–78. Assume that the forward resistance for each silicon diode is 25 Ω and the reverse leakage resistance is 100 MΩ.

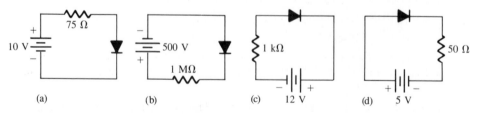

(a)   (b)   (c)   (d)

FIGURE 3–78

6. Define *rectification*.

7. Calculate the average value of a half-wave rectified voltage with a peak value of 200 V.

8. A 60 Hz sine wave is applied to the input of a half-wave rectifier. What is the frequency of the output?

9. Sketch the waveforms for the load current and voltage for Figure 3–79. Show the peak values. (The diode is silicon.)

FIGURE 3–79

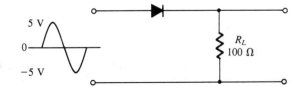

10. Can a diode with a PIV rating of 50 V be used in the circuit of Figure 3–79?

11. Calculate the average value of a full-wave rectified voltage with peak value of 75 V.

12. A 60 Hz sine wave is applied to the input of a full-wave rectifier. What is the frequency of the output?

13. The total secondary voltage in a center-tapped full-wave rectifier is 125 V rms. Neglecting the diode drop, what is the rms output voltage?

14. What is the PIV for each diode in a center-tapped full-wave rectifier if the peak output voltage is 100 V?

15. The rms output voltage of a bridge rectifier is 20 V. What is the peak inverse voltage across the diodes?

16. The *ideal* dc output voltage of a capacitor-input filter is the (peak, average) value of the rectified input.

17. A certain power-supply filter produces an output with a ripple of 100 mV and a dc value of 20 V. Calculate the ripple factor.

18. A 60 V peak full-wave rectified voltage is applied to a capacitor-input filter. Assuming that the dc output is near in value to the peak input, determine the ripple voltage and the actual dc output if $R_L = 10$ kΩ and $C = 10$ μF.

19. Find the minimum capacitance necessary for a ripple factor of 0.001 if the load resistance of a bridge rectifier is 22 kΩ.

20. What is the dc output of an LC filter for a full-wave rectified input having an average value of 15 V? The load resistance is 1.5 kΩ and the winding resistance of the coil is 30 Ω.

21. If one of the diodes in a bridge rectifier opens, what happens to the output?

22. Determine the voltage across R_L for each limiter circuit in Figure 3–80.

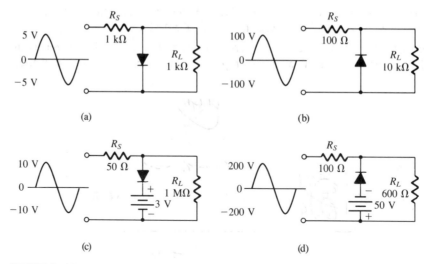

(a)   (b)

(c)   (d)

FIGURE 3–80

23. What is the approximate dc component in the output of a positive diode clamper with a peak input of 15 V?

24. The input signal to a voltage tripler has an rms value of 12 V. What is the dc output voltage?

PROBLEMS **Section 3–1**

3–1 Determine whether each diode in Figure 3–81 is forward- or reverse-biased.

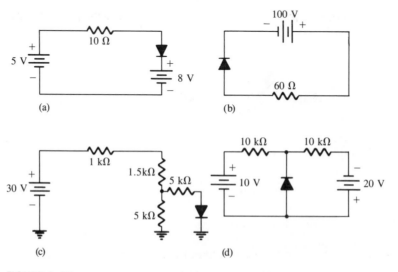

(a)   (b)

(c)   (d)

FIGURE 3–81

3–2 Determine the voltage across each diode in Figure 3–81, assuming that they are germanium diodes with a reverse resistance of 50 MΩ.

3–3 Consider the meter indications in each circuit of Figure 3–82, and determine whether the diode is functioning properly, or whether it is open or shorted.

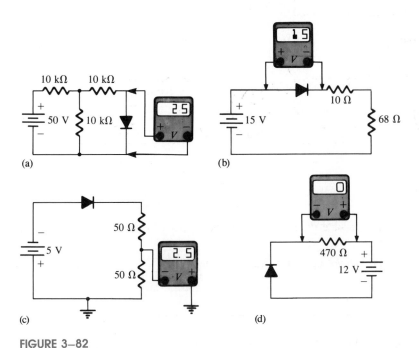

FIGURE 3–82

3–4 Determine the voltage with respect to ground at each point in Figure 3–83. (The diodes are silicon.)

FIGURE 3–83

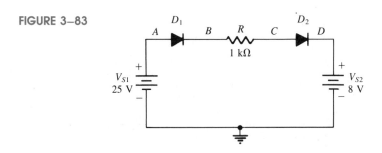

Section 3–2
3–5 Draw the output waveform for each circuit in Figure 3–84.

3–6 What is the peak forward current through each diode in Figure 3–84?

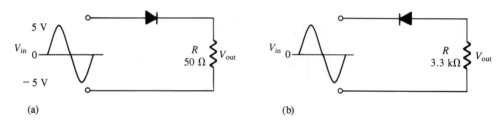

FIGURE 3–84

3–7 A power-supply transformer has a turns ratio of 5:1. What is the secondary voltage if the primary is connected to a 115 V rms source?

3–8 Determine the peak and average power delivered to R_L in Figure 3–85.

FIGURE 3–85

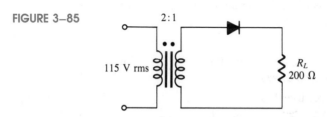

Section 3–3

3–9 Find the average value of each voltage in Figure 3–86.

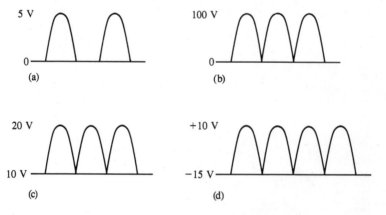

FIGURE 3–86

3–10 Consider the circuit in Figure 3–87.
   (a) What type of circuit is this?
   (b) What is the total peak secondary voltage?
   (c) Find the peak voltage across each half of the secondary.
   (d) Sketch the voltage waveform across R_L.
   (e) What is the peak current through each diode?
   (f) What is the PIV for each diode?

FIGURE 3–87

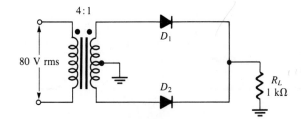

3–11 Calculate the peak voltage rating of each half of a center-tapped transformer used in a full-wave rectifier which has an average output voltage of 110 V.

3–12 Show how to connect the diodes in a center-tapped rectifier in order to produce a *negative-going* full-wave voltage across the load resistor.

3–13 What PIV rating is required for the diodes in a bridge rectifier that produces an average output voltage of 50 V?

Section 3–4

3–14 A certain rectifier filter produces a dc output voltage of 75 V with an rms ripple of 0.5 V. Calculate the ripple factor.

3–15 A certain full-wave rectifier has a peak output voltage of 30 V. A 50 μF capacitor-input filter is connected to the rectifier. Calculate the rms ripple and the dc output voltage developed across a 600 Ω load resistor.

3–16 What is the percentage of ripple in problem 3–15?

3–17 What value of filter capacitor is required to produce a 1 percent ripple for a full-wave rectifier having a load resistance of 1.5 kΩ?

3–18 A full-wave rectifier produces an 80 V peak rectified voltage from a 60 Hz ac source. If a 10 μF filter capacitor is used, determine the ripple factor for a 100 mA peak load current.

3–19 Determine the ripple and dc output voltages in Figure 3–88. The line voltage has a frequency of 60 Hz, and the winding resistance of the coil is 100 Ω.

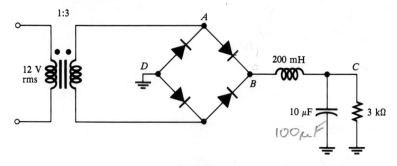

FIGURE 3–88

3–20 Refer to Figure 3–88 and sketch the following voltage waveforms in relationship to the input waveforms: V_{AB}, V_{AD}, V_{BD}, and V_{CD}.

Section 3–5

3–21 From the meter readings in Figure 3–89, determine if the rectifier circuit is functioning properly. If it is not, determine the most likely failure(s).

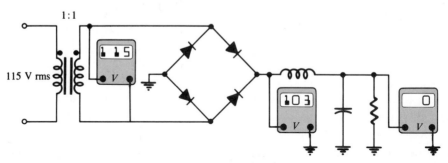

FIGURE 3–89

3–22 Each part of Figure 3–90 shows oscilloscope displays of rectifier output voltages. In each case, determine whether or not the rectifier is functioning properly and if it is not, the most likely failure(s).

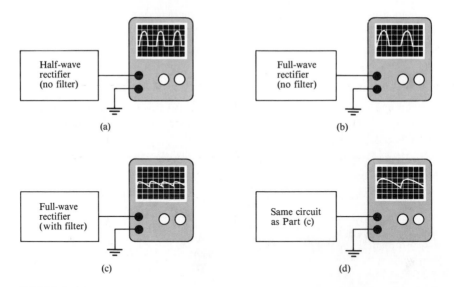

FIGURE 3–90

3–23 Based on the values given, would you expect the circuit in Figure 3–91 to fail? Why?

Section 3–6

3–24 Determine the output of the circuit in Figure 3–92(a) for each input in (b), (c), and (d).

3–25 Determine the output waveform for each circuit in Figure 3–93.

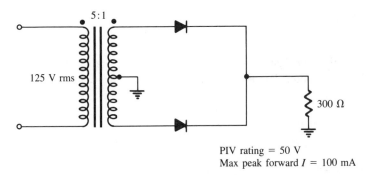

$$PIV \text{ rating} = 50 \text{ V}$$
$$\text{Max peak forward } I = 100 \text{ mA}$$

FIGURE 3–91

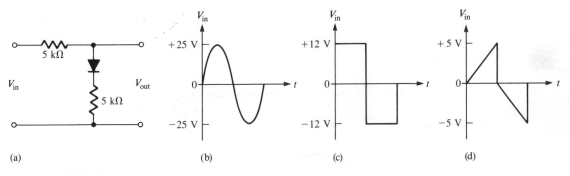

(a) (b) (c) (d)

FIGURE 3–92

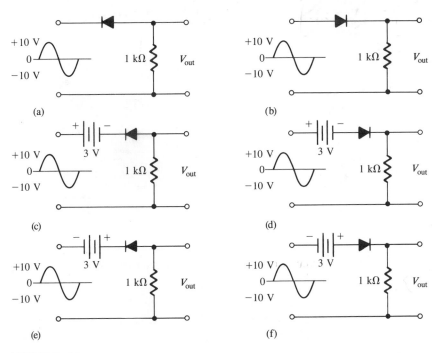

(a) (b)

(c) (d)

(e) (f)

FIGURE 3–93

3–26 Sketch the output waveforms for each circuit in Figure 3–94.

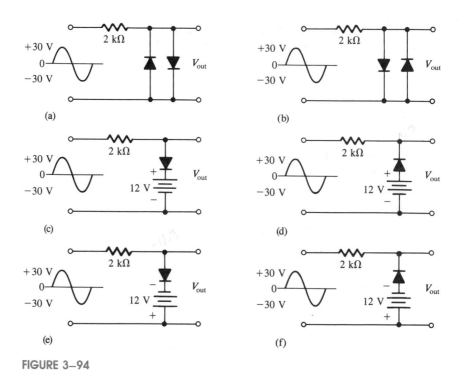

(a)   (b)

(c)   (d)

(e)   (f)

FIGURE 3–94

3–27 Describe the output waveform of each circuit in Figure 3–95. Assume the RC time constant is much greater than the period of the input.

FIGURE 3–95

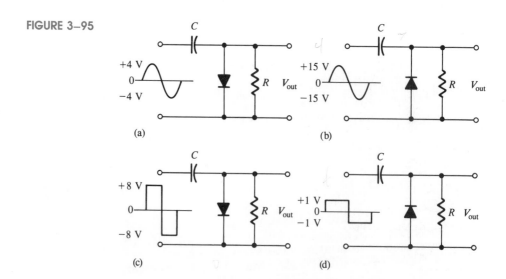

(a)   (b)

(c)   (d)

3–28 Repeat problem 3–27 with the diodes turned around.

Section 3–7

3–29 A certain voltage doubler has 20 V rms on its input. What is the output voltage? Sketch the circuit, indicating the output terminals and PIV rating for the diode.

3–30 Repeat problem 3–29 for a voltage tripler and quadrupler.

ANSWERS TO SECTION REVIEWS

Section 3–1

1. Forward and reverse bias.
2. See Figure 3–96.

FIGURE 3–96

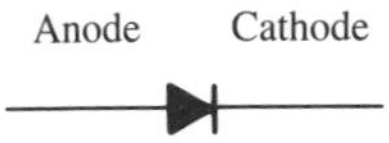
Anode
Cathode

3. True.
4. Reverse, forward.

Section 3–2

1. 270°.
2. 50 percent.
3. 3.18 V.

Section 3–3

1. 38.2 V.
2. Bridge.
3. True.

Section 3–4

1. 60 Hz, 120 Hz.
2. Capacitor charging and discharging slightly.
3. Increases ripple.
4. 0.0067.
5. Reduced ripple, less dc voltage output.
6. 220 V, 140 V.

Section 3–5

1. Half-wave output.
2. Open diode.
3. Leaky filter capacitor.

Section 3–6

1. See Figure 3–97.

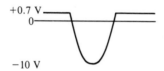

FIGURE 3–97

2. See Figure 3–98.

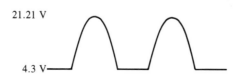

FIGURE 3–98

3. See Figure 3–99.

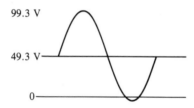

FIGURE 3–99

Section 3–7
1. 100 V.
2. 120 V.

Section 3–8
1. Forward I and V, reverse I and V, thermal.
2. **(a)** Maximum dc forward voltage drop; **(b)** Maximum dc reverse current; **(c)** Average of maximum forward current.

Example 3–3:   99.3 V

Example 3–9: $V_{dc(out)} = 31.6$ V, $V_{r(out)} = 0.96$ V, $r = 0.03$

Example 3–11:   See Figure 3–100

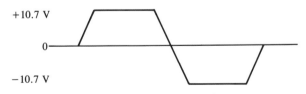

FIGURE 3–100

Example 3–12:   The same as Figure 3–65

Special Diodes

In this chapter you will learn
- [ ] The characteristics of the zener diode and how it is used as a voltage regulator
- [ ] The meaning of *line regulation* and *load regulation*
- [ ] What a varactor diode is and how it is used as a variable capacitor
- [ ] What a Schottky diode is
- [ ] What a tunnel diode is
- [ ] What an LED is
- [ ] What a photodiode is
- [ ] What a PIN diode is
- [ ] What a step-recovery diode is
- [ ] What an IMPATT diode is
- [ ] What a Gunn diode is
- [ ] How zener diodes, varactor diodes, and LEDs are used in a specific system application

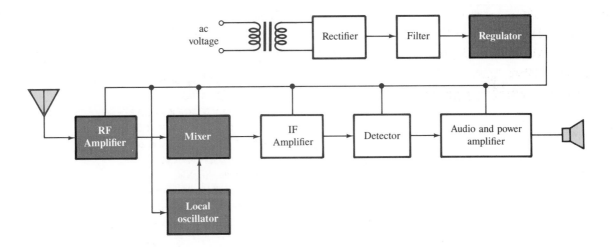

The last chapter was devoted to general-purpose and rectifier diodes, which are the most widely used types. In this chapter we will cover several other types of diodes optimized for specific applications, including zener, varactor, Schottky, pin, tunnel, step-recovery, photo-, and light-emitting diodes.

Zener diodes are sometimes used in dc power supplies to provide voltage regulation. In the superheterodyne receiver, the regulated dc voltage is used to supply each of the circuits in the system. In systems that require more than one dc voltage, multiple zener regulators can be used. Varactor diodes are important in modern receiver systems with electronic tuning. Varactor diodes can be used in the RF amplifier, the mixer, and the local oscillator to permit variation of the response over the AM (or FM) band. A third type of diode, the LED (light-emitting diode), is commonly used for displaying frequency and other information in many receiver systems.

ZENER DIODES

The zener diode is used for voltage regulation and, like the general-purpose rectifier diode, is important in many power-supply applications. The schematic symbol is shown in Figure 4–1. The zener is a silicon pn junction device that differs from the rectifier diode in that it is optimized for operation in the *reverse breakdown region*. The breakdown voltage of a zener diode is set by carefully controlling the doping level during manufacture. Recall, from the discussion of the diode characteristic curve in the last chapter, that when a diode reaches reverse breakdown, its voltage remains almost constant even though the current may change drastically. This volt-ampere characteristic is shown again in Figure 4–2.

FIGURE 4–1
Zener diode symbol.

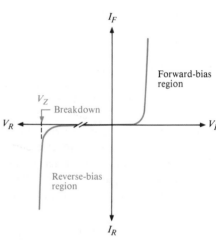

FIGURE 4–2
General diode characteristic.

Zener Breakdown

There are two types of reverse breakdown in a zener diode. One is the *avalanche* breakdown that was discussed in Chapter 2; this also occurs in rectifier diodes at a sufficiently high reverse voltage. The other type is *zener* breakdown which occurs in a zener diode at low reverse voltages. A zener diode is heavily doped to reduce the breakdown voltage. This causes a very narrow depletion layer. As a result, an intense electric field exists within the depletion layer. Near the breakdown voltage (V_Z), the field is intense enough to pull electrons from their valence bands and create current.

Zener diodes with breakdown voltages of less than approximately 5 V operate predominately in zener breakdown. Those with breakdown voltages greater than approximately 5 V operate predominately in avalanche breakdown. Both types, however, are called zener diodes. Zeners are commercially available with breakdown voltages of 1.8 V to 200 V.

Breakdown Characteristics

Figure 4–3 shows the reverse portion of a zener diode's characteristic curve. Notice that as the reverse voltage (V_R) is increased, the reverse current (I_R) remains extremely small

FIGURE 4–3
Reverse characteristic of zener diode.

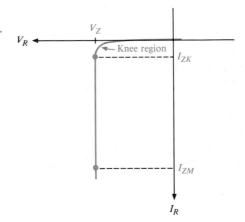

up to the "knee" of the curve. At this point, the breakdown effect begins; the zener resistance (r_Z) begins to decrease as the current (I_Z) increases rapidly. From the bottom of the knee, the breakdown voltage (V_Z) remains essentially constant. This regulating ability is the key feature of the zener diode. *It maintains an essentially constant voltage across its terminals over a specified range of reverse current values.*

A *minimum* value of reverse current, I_{ZK}, must be maintained in order to keep the diode in regulation. You can see on the curve that when the reverse current is reduced below the knee of the curve, the voltage changes drastically and regulation is lost. Also, there is a *maximum* current, I_{ZM}, above which the diode may be damaged. So, basically, the zener diode maintains a constant voltage across its terminals for values of reverse current ranging from I_{ZK} to I_{ZM}.

Zener Equivalent Circuit

Figure 4–4(a) shows the ideal approximation of a zener diode in reverse breakdown. It acts simply as a battery having a value equal to the zener voltage. Figure 4–4(b)

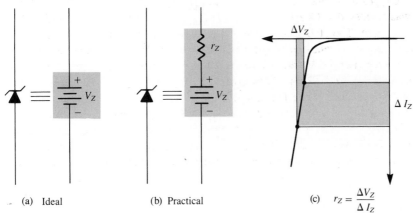

(a) Ideal   (b) Practical   (c) $r_Z = \dfrac{\Delta V_Z}{\Delta I_Z}$

FIGURE 4–4
Zener equivalent circuits.

represents the practical equivalent of a zener, where the zener resistance r_Z is included. Since the voltage curve is not ideally vertical, a small change in reverse current produces a small change in zener voltage, as illustrated in Figure 4–4(c). The ratio of ΔV_Z to ΔI_Z is the resistance, as expressed in the following equation.

$$r_Z = \frac{\Delta V_Z}{\Delta I_Z} \qquad\qquad (4\text{--}1)$$

Normally, r_Z is specified at a particular value of reverse current, I_{ZT}, called the *zener test current*. In most cases, this value of r_Z is approximately constant over the full range of reverse-current values.

EXAMPLE 4–1

A certain zener diode exhibits a 50 mV change in V_Z for a 2 mA change in I_Z. What is the zener resistance?

Solution

$$r_Z = \frac{\Delta V_Z}{\Delta I_Z} = \frac{50 \text{ mV}}{2 \text{ mA}} = 25 \text{ } \Omega$$

EXAMPLE 4–2

A 6.8 V zener diode has a resistance of 5 Ω. What is the actual voltage across its terminals when the current is 20 mA?

Solution

Figure 4–5 represents the diode. The 20 mA current causes a voltage across r_Z as follows:

$$I_Z r_Z = (20 \text{ mA})(5 \text{ } \Omega) = 100 \text{ mV}$$

FIGURE 4–5

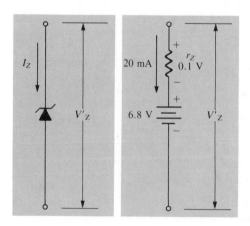

The polarity is in the same direction as the zener voltage, as shown. It therefore adds to V_Z, giving a total terminal voltage of

$$V'_Z = V_Z + I_Z r_Z = 6.8 \text{ V} + 100 \text{ mV} = 6.9 \text{ V}$$

So, for this particular current value, the terminal voltage is 0.1 V greater than the specified normal zener voltage of 6.8 V. This, of course, changes slightly with current.

Temperature Coefficient

The temperature coefficient specifies the percent change in zener voltage for each °C change in temperature. For example, a 12 V zener diode with a temperature coefficient of 0.1%/°C will exhibit a 0.012 V increase in V_Z when the junction temperature increases 1 Celsius degree. The formula for calculating the change in zener voltage over a given junction temperature change, for a specified temperature coefficient, is as follows:

$$\Delta V_Z = V_Z \times TC \times \Delta T \qquad \text{(4–2)}$$

where V_Z is the nominal zener voltage at 25°C, TC is the temperature coefficient, and ΔT is the change in temperature. A positive TC means that the zener voltage increases with an increase in temperature, and a negative TC means that the zener voltage decreases with an increase in temperature.

EXAMPLE 4–3

An 8.2 V zener diode (8.2 V at 25°C) has a positive temperature coefficient of 0.048%/°C. What is the zener voltage at 60°C?

$$V_Z = V_{ZT}$$

Solution

$$\begin{aligned}
\Delta V_Z &= V_Z \times TC \times \Delta T \\
&= (8.2 \text{ V})(0.048\%/°C)(60°C - 25°C) \\
&= (8.2 \text{ V})(0.00048/°C)(35°C) \\
&= 0.138 \text{ V}
\end{aligned}$$

The zener voltage at 60°C is

$$V_Z + \Delta V_Z = 8.2 \text{ V} + 0.138 \text{ V} = 8.338 \text{ V}$$

1. Zener diodes are normally operated in the breakdown region (T or F).

2. A certain 10 V zener diode has a resistance of 8 Ω at 30 mA. What is the terminal voltage?

3. What does a positive temperature coefficient of 0.05%/°C mean?

SECTION
REVIEW
4–1

ZENER APPLICATIONS

4–2

This section introduces two basic applications of zener diodes: *voltage regulation* and *voltage limiting*.

Output Voltage Regulation with a Varying Input Voltage

Zener diodes are widely used for voltage regulation. Figure 4–6 illustrates how a zener diode can be used to regulate a varying dc voltage. This is called *input* or *line regulation*. As the input voltage varies (within limits), the zener diode maintains an essentially constant voltage across the output terminals. However, as V_{IN} changes, I_Z will change proportionally so that the limitations on the input variation are set by the minimum and maximum current values with which the zener can operate. R is the series current-limiting resistor. For example, suppose that the zener diode in Figure 4–7 can maintain regulation over a range of current values from 4 mA to 40 mA.

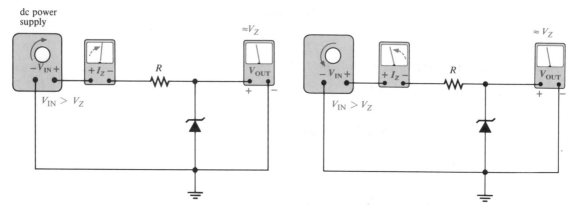

(a) As input voltage increases, V_{OUT} remains constant. (b) As input voltage decreases, V_{OUT} remains constant.

FIGURE 4–6
Zener regulation of a variable input voltage.

FIGURE 4–7

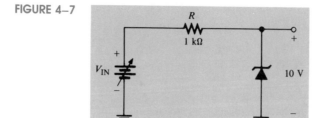

For the *minimum* current, the voltage across the 1 kΩ resistor is $V_R = $ (4 mA)(1 kΩ) = 4 V. Since $V_R = V_{IN} - V_Z$, then $V_{IN} = V_R + V_Z = 4 \text{ V} + 10 \text{ V} = $ 14 V. For the *maximum* current, the voltage across the 1 kΩ resistor is $V_R = $ (40 mA)(1 kΩ) = 40 V. Therefore $V_{IN} = 40 \text{ V} + 10 \text{ V} = 50 \text{ V}$. This shows that this zener diode can regulate an input voltage from 14 V to 50 V and maintain an approximate 10 V output. (The output will vary slightly because of the zener resistance.)

EXAMPLE 4–4

Determine the minimum and the maximum input voltages that can be regulated by the zener diode in Figure 4–8. Assume $I_{ZK} = 1$ mA, $I_{ZM} = 15$ mA, $V_Z = 5.1$ V, and $r_Z = 10$ Ω.

FIGURE 4–8

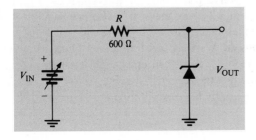

Solution

The equivalent circuit is shown in Figure 4–9. At $I_{ZK} = 1$ mA, the output voltage is

$$V_{OUT} = V_Z + I_{ZK}r_Z$$
$$= 5.1 \text{ V} + (1 \text{ mA})(10 \text{ Ω})$$
$$= 5.1 \text{ V} + 0.010 \text{ V}$$
$$= 5.11 \text{ V}$$

Therefore,

$$V_{IN(min)} = I_{ZK}R + V_{OUT}$$
$$= (1 \text{ mA})(600 \text{ Ω}) + 5.11 \text{ V}$$
$$= 5.71 \text{ V}$$

At $I_{ZM} = 15$ mA, the output voltage is

$$V_{OUT} = V_Z + I_{ZM}r_Z$$
$$= 5.1 \text{ V} + (15 \text{ mA})(10 \text{ Ω})$$
$$= 5.1 \text{ V} + 0.15 \text{ V}$$
$$= 5.25 \text{ V}$$

FIGURE 4–9
Equivalent of circuit in Figure 4–8.

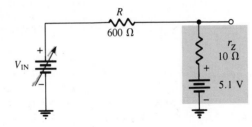

Therefore,

$$V_{IN(max)} = I_{ZM}R + V_{OUT}$$
$$= (15 \text{ mA})(600 \text{ }\Omega) + 5.25 \text{ V}$$
$$= 14.25 \text{ V}$$

Practice Exercise 4–4

Determine the minimum and maximum input voltages that can be regulated by the zener diode in Figure 4–8. $I_{ZK} = 5$ mA, $I_{ZM} = 30$ mA, $V_Z = 6.8$ V, and $r_Z = 8$ Ω.

Voltage Regulation with a Varying Load

Figure 4–10 shows a zener regulator with a variable load resistor across the terminals. The zener diode maintains a constant voltage across R_L as long as the zener current is greater than I_{ZK} and less than I_{ZM}. This is called *load regulation*.

FIGURE 4–10
Zener regulation with a variable load.

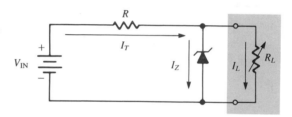

No Load to Full Load

When the output terminals are open ($R_L = \infty$), the load current is zero and all of the current is through the zener. When a load resistor is connected, part of the total current is through the zener and part through R_L. As R_L is decreased, I_L goes up and I_Z goes down. The zener diode continues to regulate until I_Z reaches its minimum value, I_{ZK}. At this point the load current is maximum. The following example will illustrate this.

EXAMPLE 4–5

Determine the minimum and the maximum load currents for which the zener diode in Figure 4–11 will maintain regulation. What is the minimum R_L that can be used? $V_Z = 12$ V, $I_{ZK} = 3$ mA, and $I_{ZM} = 90$ mA. Assume $r_Z = 0$ Ω.

FIGURE 4–11

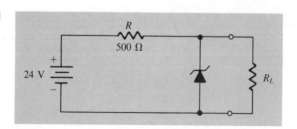

Solution

When $I_L = 0$ A, I_Z is maximum and equal to the total circuit current I_T.

$$I_Z = \frac{V_{IN} - V_Z}{R} = \frac{24 \text{ V} - 12 \text{ V}}{500 \text{ }\Omega} = 24 \text{ mA}$$

Since this is much less than I_{ZM}, 0 A is an acceptable minimum for I_L.

$$I_{L(min)} = 0 \text{ A}$$

The maximum value of I_L occurs when I_Z is minimum, so we can solve for $I_{L(max)}$ as follows:

$$
\begin{aligned}
I_{L(max)} &= I_T - I_{Z(min)} \\
&= 24 \text{ mA} - 3 \text{ mA} \\
&= 21 \text{ mA}
\end{aligned}
$$

The minimum value of R_L is

$$
\begin{aligned}
R_{L(min)} &= \frac{V_Z}{I_{L(max)}} \\
&= \frac{12 \text{ V}}{21 \text{ mA}} \\
&= 571 \text{ }\Omega
\end{aligned}
$$

Practice Exercise 4–5

Find the minimum and maximum load currents for which the circuit in Figure 4–11 will maintain regulation. Determine the minimum R_L that can be used. $V_Z = 3.3$ V, $I_{ZK} = 2$ mA, $I_{ZM} = 75$ mA, and $r_Z = 0$ Ω.

Percent Regulation

The percent regulation is a figure of merit used to specify the performance of a voltage regulator. It can be in terms of input (line) regulation or load regulation. The *percent input regulation* specifies how much change occurs in the output voltage for a given change in input voltage. It is usually expressed as a percent change in V_{OUT} for a one volt change in V_{IN} (%/V). The *percent load regulation* specifies how much change occurs in the output voltage over a certain range of load current values, usually from minimum current (no load) to maximum current (full load). It is normally expressed as a percentage and can be calculated with the following formula.

$$\text{Percent load regulation} = \frac{V_{NL} - V_{FL}}{V_{FL}} \times 100 \qquad (4\text{–}3)$$

where V_{NL} is the output voltage with no load, and V_{FL} is the output voltage with full load.

EXAMPLE 4–6

A certain regulator has a no-load output voltage of 6 V and a full-load output of 5.82 V. What is the percent load regulation?

Solution

$$\text{Percent load regulation} = \frac{V_{NL} - V_{FL}}{V_{FL}} \times 100\%$$

$$= \frac{6 \text{ V} - 5.82 \text{ V}}{5.82 \text{ V}} \times 100\%$$

$$= 3.09\%$$

Zener Limiting

Zener diodes can be used in ac applications to limit voltage swings to desired levels. Figure 4–12 shows three basic ways the limiting action of a zener diode can be used. Figure 4–12(a) shows a zener used to limit the positive peak of a signal voltage to the selected zener voltage. During the negative alternation, the zener acts as a conventional forward-biased diode and limits the negative voltage to −0.7 V. When the zener is turned around, as in part (b), the negative peak is limited by zener action and the positive voltage is limited to 0.7 V. Two back-to-back zeners limit both peaks to the zener voltage plus 0.7 V, as shown in part (c). During the positive alternation, D_2 is functioning as the zener limiter and D_1 is functioning as a conventional forward-biased diode. During the negative alternation, the roles are reversed.

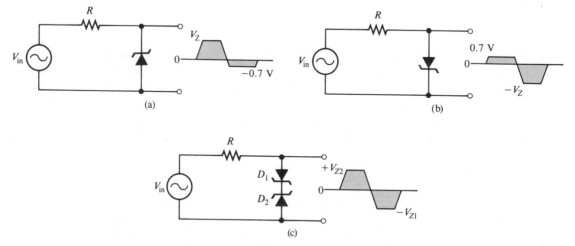

FIGURE 4–12
Basic zener limiting action.

1. Explain the difference between input (line) regulation and load regulation.
2. In a zener diode regulator, for what value of load resistance is the zener current maximum?
3. A zener regulator has an output voltage of 12 V with no load and 11.9 V with full load. What is the percent load regulation?

VARACTOR DIODES

Varactor diodes are used as *voltage-variable capacitors*. A varactor is basically a *reverse-biased* pn junction that utilizes the inherent capacitance of the depletion layer. The depletion layer, created by the reverse bias, acts as a capacitor *dielectric* because of its nonconductive characteristic. The p and n regions are conductive and act as the capacitor *plates*, as illustrated in Figure 4–13.

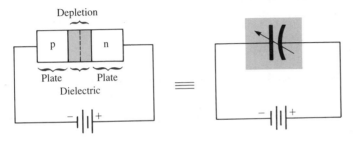

FIGURE 4–13
The reverse-biased varactor diode acts as a variable capacitor.

When the reverse-bias voltage increases, the depletion layer widens, effectively increasing the dielectric thickness and thus *decreasing* the capacitance. When the reverse-bias voltage decreases, the depletion layer narrows, thus *increasing* the capacitance. This action is shown in Figure 4–14(a) and (b). A general curve of capacitance versus voltage is shown in Figure 4–14(c). Recall that capacitance is determined by the *plate area*, *dielectric constant*, and *dielectric thickness*, as expressed in the following formula.

$$C = \frac{A\epsilon}{d} \qquad (4\text{–}4)$$

In a varactor diode, the capacitance parameters are controlled by the method of doping in the depletion layer and the size and geometry of the diode's construction. Varactor capacitances typically range from a few picofarads to a few hundred picofarads. Figure 4–15(a) shows a common symbol for a varactor, and part (b) shows a simplified equivalent circuit. R_S is the reverse series resistance, and C_V is the variable capacitance.

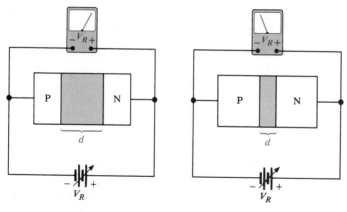

(a) Greater V_R, less capacitance   (b) Less V_R, greater capacitance

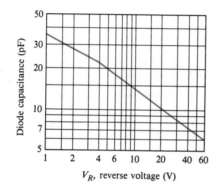

(c) Typical graph of capacitance versus reverse voltage

FIGURE 4–14
Varactor diode capacitance varies with reverse voltage.

FIGURE 4–15
Varactor diode.

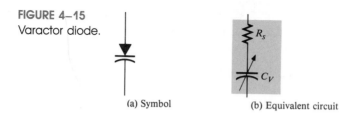

(a) Symbol   (b) Equivalent circuit

Applications

A major application of varactors is in tuning circuits. For example, electronic tuners in TV and other commercial receivers utilize varactors as one of their elements. When used in a resonant circuit, the varactor acts as a variable capacitor, thus allowing the resonant frequency to be adjusted by a variable voltage level, as illustrated in Figure 4–16 where

FIGURE 4–16
Varactors in a resonant circuit.

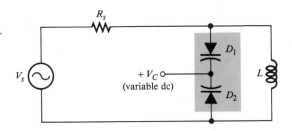

two varactor diodes provide the total variable capacitance in a parallel resonant (tank) circuit.

V_C is a variable dc voltage that controls the reverse bias and therefore the capacitance of the diodes. Recall that the resonant frequency of the tank circuit is

$$f_r \cong \frac{1}{2\pi\sqrt{LC}} \qquad \textbf{(4–5)}$$

This approximation is valid for $Q > 10$.

EXAMPLE 4–7

The capacitance of a certain varactor can be varied from 5 pF to 50 pF. The diode is used in a tuned circuit similar to that shown in Figure 4–16. Determine the tuning range for the circuit if $L = 10$ mH.

Solution
The equivalent circuit is shown in Figure 4–17. Notice that the varactor capacitances are in series.

FIGURE 4–17

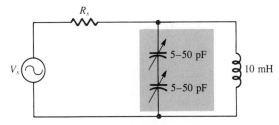

The *minimum* total capacitance is

$$C_{T(\text{min})} = \frac{C_{1(\text{min})}C_{2(\text{min})}}{C_{1(\text{min})} + C_{2(\text{min})}}$$

$$= \frac{(5 \text{ pF})(5 \text{ pF})}{10 \text{ pF}}$$

$$= 2.5 \text{ pF}$$

The *maximum* resonant frequency is therefore

$$f_{r(\text{max})} = \frac{1}{2\pi\sqrt{LC}}$$

$$= \frac{1}{2\pi\sqrt{(10 \text{ mH})(2.5 \text{ pF})}}$$

$$\cong 1 \text{ MHz}$$

The *maximum* total capacitance is

$$C_{T(\text{max})} = \frac{C_{1(\text{max})}C_{2(\text{max})}}{C_{1(\text{max})} + C_{2(\text{max})}}$$

$$= \frac{(50 \text{ pF})(50 \text{ pF})}{100 \text{ pF}}$$

$$= 25 \text{ pF}$$

The *minimum* resonant frequency is therefore

$$f_{r(\text{min})} = \frac{1}{2\pi\sqrt{LC}}$$

$$= \frac{1}{2\pi\sqrt{(10 \text{ mH})(25 \text{ pF})}}$$

$$\cong 318 \text{ kHz}$$

SECTION REVIEW 4–3

1. What is the purpose of a varactor diode?
2. Based on the general curve in Figure 4–14(c), what happens to the diode capacitance when the reverse voltage is increased?

OTHER TYPES OF DIODES

4–4   The Schottky Diode

Schottky diodes are used primarily in high-frequency and fast-switching applications. They are also known as *hot-carrier* diodes. A Schottky diode symbol is shown in Figure 4–18. A Schottky diode is formed by joining a doped semiconductor region (usually n-type) with a metal such as gold, silver, or platinum. So, rather than a pn junction, there is a metal-to-semiconductor junction, as shown in Figure 4–19.

The Schottky diode operates only with *majority* carriers. There are no minority carriers as in other types of diodes. The metal region is heavily occupied with *conduction-band* electrons, and the n-type semiconductor region is lightly doped. When forward-biased, the higher-energy electrons in the n region are injected into the metal region where

A simple photodiode application is depicted in Figure 4–30. Here a beam of light continuously passes across a conveyor belt and into a transparent window behind which is a photodiode circuit. When the light beam is interrupted by an object passing by on the conveyor belt, the sudden reduction in diode current activates a control circuit that advances a counter by one. The total count of objects that have passed that point is displayed by the counter. This basic concept can be extended and used for production control, shipping, and monitoring activity on production lines.

PIN Diode

This type of diode consists of heavily doped p and n regions separated by an intrinsic (undoped) region, as shown in Figure 4–31(a). When reverse-biased, the PIN diode acts like an almost *constant* capacitance. When forward-biased, it acts like a variable resistance, as shown in parts (b) and (c). The forward resistance of the intrinsic region decreases with increasing current.

The PIN diode is used as a dc-controlled microwave switch operated by rapid changes in bias or as a modulating device that takes advantage of the variable forward-resistance characteristic. Since no rectification occurs at the pn junction, a high-frequency signal can be modulated (varied) by a lower-frequency bias variation. A PIN diode can also be used in attenuator applications because its resistance can be controlled by the amount of current.

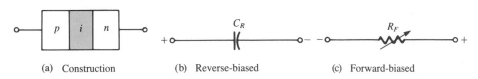

(a) Construction   (b) Reverse-biased   (c) Forward-biased

FIGURE 4–31
PIN diode.

Step-Recovery Diode

The step-recovery diode employs *graded doping* where the doping level of the semiconductor materials is reduced as the pn junction is approached. This produces an abrupt turn-off time by allowing a very fast release of stored charge when switching from forward to reverse bias. It also allows a rapid re-establishment of forward current when switching from reverse to forward bias. This diode is used in very fast switching applications.

IMPATT Diode

IMPATT stands for *impact avalanche and transit time* diode. This is a special microwave diode that utilizes the delay time required for attaining an avalanche condition plus transit time to produce a negative-resistance characteristic. It is used in microwave oscillators (10–100 GHz).

Gunn Diode

The Gunn diode is also a negative-resistance microwave device for oscillator applications. It is not constructed with a pn junction, but consists of a thin slice of n-type gallium arsenide between two metal conductors.

---

1. What bias condition produces light emission in an LED?
2. A photodiode operates in _____ bias.
3. What are the three regions in a PIN diode?

A SYSTEM APPLICATION

4–5

Zener diodes are used to provide the voltage regulation in some power supplies. The input to the zener regulator comes from the power-supply filter as shown in Figure 4–32(a). The constant output voltage of the zener regulator supplies all the blocks in a receiver system that require dc voltage. In systems that require more than one value of dc voltage, a corresponding number of zener diodes with appropriate zener voltages can be used.

Varactor diodes are used in superhet receivers to provide electronic tuning. In a typical system, the RF amplifier, the mixer, and the local oscillator may be electronically tuned with varactors. Generally, the input from the antenna is coupled through a transformer. A varactor diode across one or both transformer windings creates a resonant circuit. Similarly, the output of the RF amplifier or input to the mixer may be varactor-tuned. The frequency of the local oscillator is controlled by a resonant circuit with a varactor as the variable capacitor. A general arrangement of varactor-tuned resonant circuits in a receiver system is shown in Figure 4–32(b).

The tuning control on the receiver (knob or push button) provides a variable voltage to each varactor diode. As you adjust the tuning control, the capacitances of the varactors vary simultaneously to keep the amplifier, mixer, and local oscillator in step. For example, if you change your receiver from 92 MHz to 100 MHz on the FM band, the response of the RF amplifier and mixer is changed from 92 MHz to 100 MHz, and the local oscillator frequency is changed from 102.7 MHz to 110.7 MHz. This concept is illustrated in part (b).

Light-emitting diodes are used to display the frequency to which you are tuned as well as other information on many receiver systems. For numerical display, the LEDs are used as segments in an "eight" formation. By lighting the appropriate LEDs, a given number can be displayed. To implement a digital readout, the local oscillator frequency can be monitored and converted to digital form. By subtracting 455 kHz for AM or 10.7 MHz for FM from the local oscillator frequency, the frequency of the incoming signal to which you are tuned is obtained. This number in digital form is then used to drive the LED display. This concept is illustrated in Figure 4–32(c).

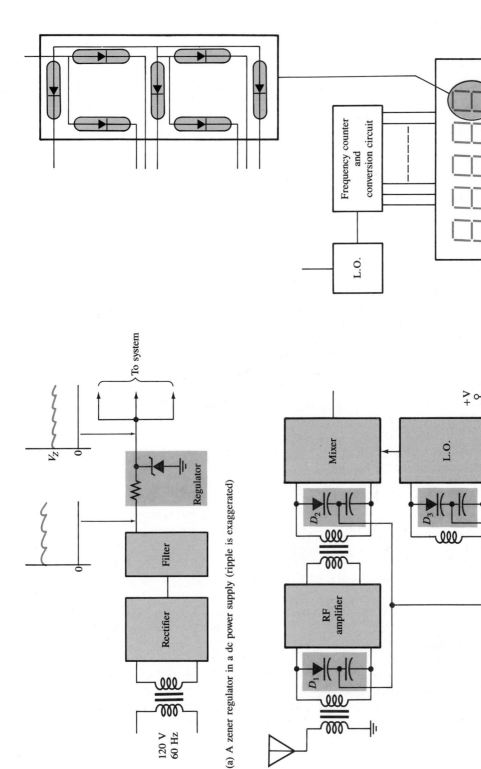

(a) A zener regulator in a dc power supply (ripple is exaggerated)

(b) The varactor diode as a tuning element in the "front end" of a receiver

(c) LEDs provide a digital readout of the received frequency

FIGURE 4–32

Diodes in system applications.

FORMULAS

$(4\text{--}1)$ $r_Z = \dfrac{\Delta V_Z}{\Delta I_Z}$ Zener resistance

$(4\text{--}2)$ $\Delta V_Z = V_Z \times TC \times \Delta T$ V_Z temperature change

$(4\text{--}3)$ $\dfrac{\text{Percent load}}{\text{regulation}} = \dfrac{V_{NL} - V_{FL}}{V_{FL}} \times 100\%$ Load regulation

$(4\text{--}4)$ $C = \dfrac{A\epsilon}{d}$ Capacitance

$(4\text{--}5)$ $f_r \cong \dfrac{1}{2\pi\sqrt{LC}}$ Resonant frequency

$(4\text{--}6)$ $R_F = -\dfrac{\Delta V_F}{\Delta I_F}$ Negative resistance

SUMMARY

☐ The zener diode operates in reverse breakdown.

☐ There are two breakdown mechanisms in a zener diode: *avalanche* breakdown and *zener* breakdown.

☐ When $V_Z < 5$ V, *zener* breakdown is predominant.

☐ When $V_Z > 5$ V, *avalanche* breakdown is predominant.

☐ A zener diode maintains an essentially constant voltage across its terminals over a specified range of zener currents.

☐ Zener diodes are used as shunt voltage regulators.

☐ Regulation of output voltage over a range of input voltages is called *input* or *line* regulation.

☐ Regulation of output voltage over a range of load currents is called *load* regulation.

☐ The *smaller* the percent regulation, the better.

☐ A *varactor* diode acts as a variable capacitor under reverse-biased conditions.

☐ The capacitance of a varactor varies inversely with reverse-biased voltage.

☐ The Schottky diode has a metal-to-semiconductor junction. It is used in fast-switching applications.

☐ The tunnel diode has a negative-resistance characteristic.

☐ An LED emits light when forward-biased.

☐ The photodiode exhibits an increase in reverse current with light intensity.

☐ The PIN diode has a p region, an n region, and an intrinsic region, and displays a variable resistance characteristic when forward-biased and a constant capacitance when reverse-biased.

SELF-TEST

1. A certain zener diode has a zener voltage of 3.6 V. Does it operate in *zener* breakdown or *avalanche* breakdown?

2. For a particular 12 V zener, a 10 mA change in I_Z produces a 0.1 V change in V_Z. Determine the zener resistance.

3. For the zener diode in problem 2, determine the minimum and maximum terminal voltages over a range of $I_Z = 0.5$ mA to $I_Z = 35$ mA.

4. In a certain zener regulator, the output voltage changes 0.2 V when the input voltage goes from 5 V to 10 V. What is the percent input regulation?

5. The output voltage of a zener regulator is 3.6 V at no load and 3.4 V at full load. Determine the percent load regulation.

6. Assume that the capacitance of a certain varactor diode decreases 5 pF for each 1 V increase in the reverse-bias control voltage. If the control voltage is varied from 2 V to 6 V, what is the capacitance at 6 V, if it is 30 pF at 2 V?

7. When operated on the negative-resistance portion of its characteristic curve, a certain tunnel diode exhibits a decrease in its forward current of 0.5 mA for a 0.1 V increase in forward voltage. What is the resistance?

8. Which of the LEDs in Figure 4–33 emit light, and which do not?

FIGURE 4–33

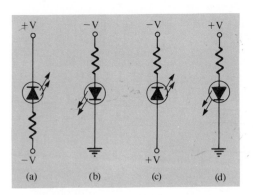

9. A photodiode is connected in a circuit so that it is reverse-biased. A microammeter is connected in series with the diode. If the intensity of the incident light on the diode is reduced, does the meter indicate an increase or a decrease?

10. What type of diode is constructed of three semiconductor regions?

Section 4–1   **PROBLEMS**

4–1 A certain zener diode has a $V_Z = 7.5$ V and an $r_Z = 0.5\ \Omega$. Sketch the equivalent circuit.

4–2 From the characteristic curve in Figure 4–34, what is the approximate minimum zener current and the approximate zener voltage?

4–3 When the reverse current in a particular zener diode increases from 20 mA to 30 mA, the zener voltage changes from 5.6 V to 5.65 V. What is the resistance of this device?

4–4 A 4.7 V zener has a resistance of 15 Ω. What is its terminal voltage when the current is 20 mA?

FIGURE 4–34

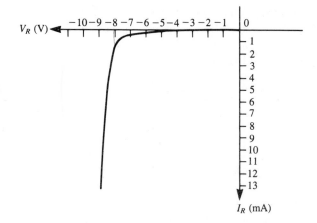

4–5 A certain zener diode has the following specifications: $V_Z = 6.8$ V at 25°C and $TC = +0.040\%/°C$. Determine the zener voltage at 70°C.

Section 4–2

4–6 Determine the *minimum* input voltage required for regulation to be established in Figure 4–35. Assume an ideal zener diode with $I_{ZK} = 1.5$ mA and $V_Z = 14$ V.

FIGURE 4–35

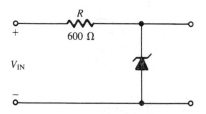

4–7 Repeat problem 4–6 with $r_Z = 20\ \Omega$.

4–8 To what value must R be adjusted in Figure 4–36 to make $I_Z = 40$ mA? Assume $V_Z = 12$ V and $r_Z = 30\ \Omega$.

FIGURE 4–36

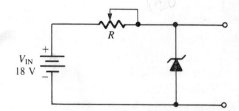

4–9 A 20 V peak sine wave voltage is applied to the circuit in Figure 4–36 in place of the dc source. Sketch the output waveform. Use the parameter values established in problem 4–8.

4–10   A loaded zener regulator is shown in Figure 4–37. For $V_Z = 5.1$ V, $I_{ZK} = 5$ mA, $I_{ZM} = 70$ mA, and $r_Z = 12$ Ω, determine the minimum and maximum load currents.

FIGURE 4–37

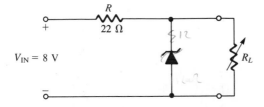

4–11   Find the percent load regulation in problem 4–10.

4–12   For the circuit of problem 4–10, assume the input voltage is varied from 6 V to 12 V. Determine the percent input regulation with no load.

4–13   The no-load output voltage of a certain zener regulator is 8.23 V, and the full-load output is 7.98 V. Calculate the percent load regulation.

Section 4–3

4–14   Figure 4–38 is a curve of reverse voltage versus capacitance for a certain varactor. Determine the change in capacitance if V_R varies from 5 V to 20 V.

FIGURE 4–38

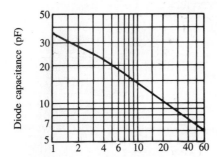

V_R, reverse voltage (V)

4–15   Refer to Figure 4–38 and determine the value of V_R that produces 25 pF.

4–16   What capacitance value is required for each of the varactors in Figure 4–39 to produce a resonant frequency of 1 MHz?

FIGURE 4–39

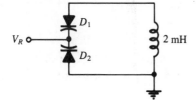

4–17 At what value must the control voltage be set in problem 4–16 if the varactors have the characteristic curve in Figure 4–38?

Section 4–5

4–18 The current in a certain tunnel diode changes from 0.25 mA to 0.15 mA when the forward voltage is adjusted from 125 mV to 200 mV. What is the resistance in this region?

4–19 In what type of circuit are tunnel diodes commonly used?

Section 4–6

4–20 When the switch in Figure 4–40 is closed, will the microammeter reading increase or decrease? Assume D_1 and D_2 are optically coupled.

FIGURE 4–40

 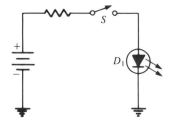 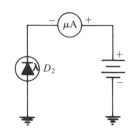

4–21 With no incident light, a certain amount of reverse current flows in a photodiode. What is this current called?

ANSWERS TO SECTION REVIEWS

Section 4–1

1. True.
2. 10.24 V.
3. V_Z increases 0.05 percent for each Celsius degree rise in temperature.

Section 4–2

1. *Input regulation:* Constant output voltage for varying input voltage. *Load regulation:* Constant output voltage for varying load current.
2. Maximum R_L.
3. 0.833 percent.

Section 4–3

1. It is a variable capacitor.
2. It decreases.

Section 4–4

1. Forward.
2. Reverse.
3. p, i, and n.

Example 4–4: $V_{\text{IN(min)}} = 9.84$ V, $V_{\text{IN(max)}} = 25$ V

Example 4–5: $I_{L(\text{min})} = 0$ A, $I_{L(\text{max})} = 39$ mA, $R_{L(\text{min})} = 85$ Ω

Bipolar Junction Transistors

In this chapter you will learn
☐ The basic construction of bipolar junction transistors
☐ The internal operation of a bipolar transistor
☐ The difference between pnp and npn transistors
☐ The transistor currents and how they are related
☐ What the characteristic curves of a transistor mean
☐ The meanings of *cutoff* and *saturation* in a transistor
☐ How a transistor produces current gain and voltage gain
☐ Why the bipolar transistor is a current-controlled device
☐ How a transistor can be used as an amplifier or as a switch
☐ The significance of transistor data sheet parameters
☐ How to test a transistor
☐ How to recognize transistors by their packaging
☐ How to identify the base, emitter, and collector terminals of a transistor
☐ How bipolar junction transistors are used in a specific system application

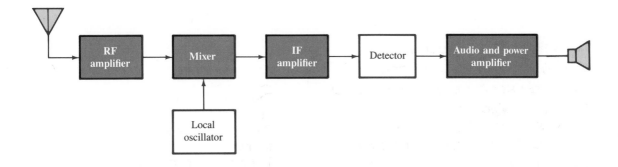

As mentioned in Chapter 1, the transistor was invented by John Bardeen, Walter Brattain, and William Shockley at Bell Laboratories in 1947. Since that time, it has revolutionized electronics by facilitating smaller, lighter, more efficient, more complex, and less expensive circuits and systems. More recent revolutionary developments, such as the integrated circuit, are an outgrowth of transistor technology.

There are two basic types of transistors—the *bipolar junction transistor* (BJT), which you will study in this chapter, and the *field-effect transistor* (FET), which we will cover later. The bipolar junction transistor is used in two broad areas of application: to boost or *amplify* an electrical signal and as an electronic *switch*. The amplification property of a transistor is applied in many types of systems, including the superheterodyne receiver. The shaded blocks in the diagram indicate where the transistor is used to amplify signals in a receiver system.

It is necessary in most receivers to amplify the extremely small signal voltages picked up by the antenna at the carrier frequency. This is the function of the transistor in the RF amplifier circuit. The mixer in most receivers provides further amplification of the modulated RF carrier. The IF amplifier typically consists of two or more bipolar amplifier stages that provide most of the amplification in the system. The audio and power amplifiers use bipolar transistors to provide the final boost in the audio signal to a level sufficient to drive the speakers.

TRANSISTOR CONSTRUCTION

5–1

The bipolar junction transistor is constructed with *three* doped semiconductor regions separated by *two* pn junctions. The three regions are called *emitter, base,* and *collector.* The two types of bipolar transistors are shown in Figure 5–1. One type consists of two n regions separated by a p region (npn), and the other consists of two p regions separated by an n region (pnp).

FIGURE 5–1
Bipolar transistor construction.

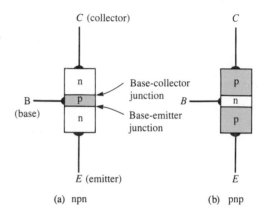

(a) npn   (b) pnp

The pn junction joining the base region and the emitter region is called the *base-emitter* junction. The junction joining the base region and the collector region is called the *base-collector* junction, as indicated. A wire lead connects to each of the three regions, as shown. These leads are labeled *E, B,* and *C* for emitter, base, and collector, respectively.

The base material is lightly doped and very narrow compared to the heavily doped emitter and collector materials. (The reason for this is discussed in the next section.) Figure 5–2 shows the schematic symbols for the npn and pnp bipolar transistors. The term *bipolar* refers to the use of both *holes* and *electrons* as carriers in the transistor structure.

FIGURE 5–2
Standard transistor symbols.

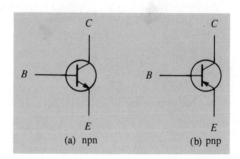

(a) npn   (b) pnp

SECTION REVIEW 5–1

1. Name the two types of bipolar transistors according to the construction.
2. The transistor is a three-terminal device. Name the three terminals.

BASIC TRANSISTOR OPERATION

In order for the transistor to operate properly as an amplifier, the two pn junctions must be correctly *biased* with external voltages. In our discussion, we use the npn transistor for illustration. The operation of the pnp is the same as for the npn except that the roles of the electrons and holes, the bias voltage polarities, and the current directions are all reversed. Figure 5–3 shows the proper bias arrangement for both npn and pnp transistors. Notice that in both cases the *base-emitter (BE) junction is forward-biased* and the *base-collector (BC) junction is reverse-biased*.

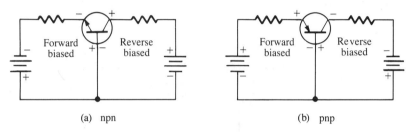

(a) npn   (b) pnp

FIGURE 5–3
Forward-reverse bias of a bipolar transistor.

Now, let's examine what happens inside the transistor when it is forward-reverse biased. The forward bias from base to emitter *narrows* the *BE* depletion layer, and the reverse bias from base to collector *widens* the *BC* depletion layer, as depicted in Figure 5–4(a). The n-type emitter region is teeming with conduction-band (free) electrons which easily diffuse across the *BE* junction into the p-type base region, just as in a forward-biased diode. The base region is lightly doped and very thin so that it has a very limited number of holes. Thus, only a small percentage of all the electrons flowing across the *BE* junction combine with the available holes. These relatively few recombined electrons flow out of the base lead as valence electrons, forming the small base current, I_B, as shown in Figure 5–4(b).

Most of the electrons flowing from the emitter into the base region diffuse into the *BC* depletion layer. Once in this layer they are pulled across the *BC* junction by the depletion layer field set up by the force of attraction between the positive and negative ions. Actually, you can think of the electrons as being pulled across the reverse-biased *BC* junction by the attraction of the positive ions on the other side. This is illustrated in part (c). The electrons now move through the collector region, out through the collector lead, and into the positive terminal of the external dc source. This forms the collector current, I_C, as shown. The amount of collector current depends directly on the amount of base current and is essentially independent of the dc collector voltage.

Transistor Currents

The directions of conventional current in an npn transistor are as shown in Figure 5–5(a), and those for a pnp are shown in Figure 5–5(b). The currents are indicated on the corresponding schematic symbols in parts (c) and (d) of the figure. Notice that the arrow on the

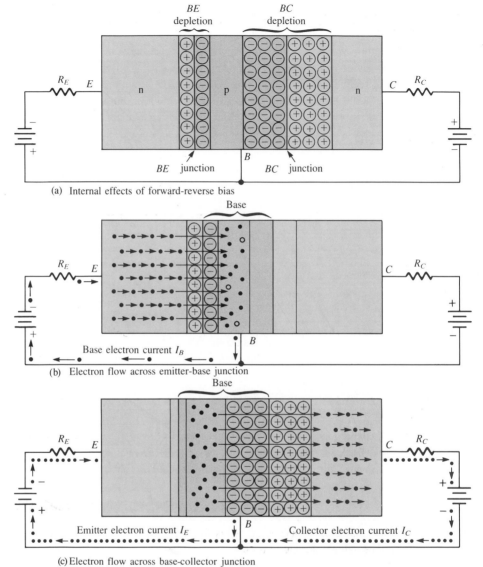

(a) Internal effects of forward-reverse bias

(b) Electron flow across emitter-base junction

(c) Electron flow across base-collector junction

FIGURE 5–4
Transistor action.

emitter of the transistor symbols points in the direction of *conventional* current. An examination of these diagrams shows that the emitter current is the *sum* of the collector and base currents, expressed as follows.

$$I_E = I_C + I_B \qquad (5-1)$$

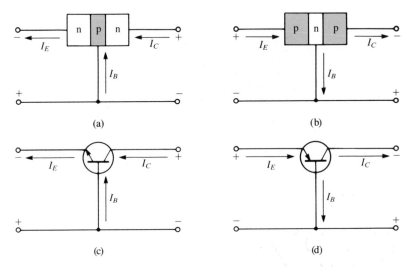

FIGURE 5–5
Transistor conventional current directions.

As mentioned before, I_B is very small compared to I_E or I_C. The capital-letter subscripts indicate dc values.

---

1. For proper amplifier operation, the *BE* junction must be _____-biased, and the *BC* junction must be _____-biased.

2. A certain transistor has a collector current of 1 mA and a base current of 1 μA. Determine the emitter current.

TRANSISTOR PARAMETERS AND RATINGS

5–3

When a transistor is connected to bias voltages, as shown in Figure 5–6 for both npn and pnp types, V_{BB} forward-biases the base-emitter junction, and V_{CC} reverse-biases the base-collector junction.

dc Beta (β_{dc}) and dc Alpha (α_{dc})

The ratio of the collector current I_C to the base current I_B is the dc current gain (β_{dc}) of the transistor.

$$\beta_{dc} = \frac{I_C}{I_B} \qquad\qquad (5\text{–}2)$$

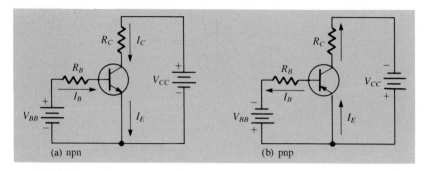

FIGURE 5–6
Transistor bias circuits.

Typical values of β_{dc} range from 20 to 200 or higher. β_{dc} is usually designated as h_{FE} on transistor data sheets. The ratio of the collector current to the emitter current I_E is the dc alpha (α_{dc}).

$$\alpha_{dc} = \frac{I_C}{I_E} \qquad \text{(5–3)}$$

Typically, values of α_{dc} range from 0.95 to 0.99 or greater.

Relationship of α_{dc} and β_{dc}

Starting with the current formula $I_E = I_C + I_B$ and dividing by I_C, we get

$$\frac{I_E}{I_C} = \frac{I_C}{I_C} + \frac{I_B}{I_C}$$

$$= 1 + \frac{I_B}{I_C}$$

Since $\beta_{dc} = I_C/I_B$ and $\alpha_{dc} = I_C/I_E$, the equation becomes

$$\frac{1}{\alpha_{dc}} = 1 + \frac{1}{\beta_{dc}}$$

Rearranging, we get

$$\frac{1}{\alpha_{dc}} = \frac{\beta_{dc} + 1}{\beta_{dc}}$$

$$\alpha_{dc} = \frac{\beta_{dc}}{\beta_{dc} + 1} \qquad \text{(5–4)}$$

Equation (5–4) allows us to calculate α_{dc} if we know β_{dc}. By simple algebra, a formula for β_{dc} in terms of α_{dc} is derived as follows from equation (5–4).

$$\alpha_{dc}(\beta_{dc} + 1) = \beta_{dc}$$

$$\alpha_{dc}\beta_{dc} + \alpha_{dc} = \beta_{dc}$$

$$\alpha_{dc} = \beta_{dc} - \alpha_{dc}\beta_{dc}$$

$$\beta_{dc}(1 - \alpha_{dc}) = \alpha_{dc}$$

$$\beta_{dc} = \frac{\alpha_{dc}}{1 - \alpha_{dc}} \qquad (5\text{–}5)$$

EXAMPLE 5–1

Determine β_{dc} and α_{dc} for a transistor where $I_B = 50\ \mu A$ and $I_C = 3.65\ mA$.

Solution

$$\beta_{dc} = \frac{I_C}{I_B} = \frac{3.65\ mA}{50\ \mu A} = 73$$

$$\alpha_{dc} = \frac{\beta_{dc}}{\beta_{dc} + 1} = \frac{73}{74} = 0.986$$

dc Analysis

Consider the circuit configuration in Figure 5–7. There are three transistor currents and three voltages: I_B, I_E, I_C, V_{BE}, V_{CB}, and V_{CE}.

FIGURE 5–7
Transistor currents and voltages.

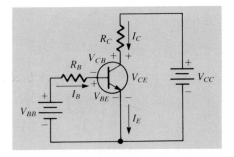

As you know, V_{BB} forward-biases the base-emitter junction and V_{CC} reverse-biases the base-collector junction. When the silicon base-emitter junction is forward-biased, it is like a diode and has a forward voltage drop of

$$V_{BE} \cong 0.7\ V \qquad (5\text{–}6)$$

The voltage across R_B is

$$V_{R_B} = V_{BB} - V_{BE}$$

and

$$V_{R_B} = I_B R_B$$

Substituting,

$$I_B R_B = V_{BB} - V_{BE}$$

and

$$I_B = \frac{V_{BB} - V_{BE}}{R_B} \qquad (5\text{--}7)$$

From equation (5–3), the expression for I_E is

$$I_E = \frac{I_C}{\alpha_{dc}}$$

From equation (5–2), the collector current is

$$I_C = \beta_{dc} I_B$$

The drop across R_C is

$$V_{R_C} = I_C R_C$$

The voltage at the collector with respect to the emitter (ground) is

$$V_{CE} = V_{CC} - I_C R_C \qquad (5\text{--}8)$$

The voltage between the base and collector is

$$V_{CB} = V_{CE} - V_{BE} \qquad (5\text{--}9)$$

EXAMPLE 5–2

Determine I_B, I_C, I_E, α_{dc}, V_{CE}, and V_{CB} in the circuit of Figure 5–8. The transistor has a $\beta_{dc} = 150$.

FIGURE 5–8

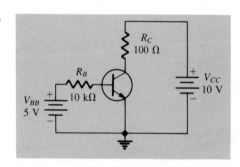

Solution

$$I_B = \frac{V_{BB} - V_{BE}}{R_B} = \frac{5 \text{ V} - 0.7 \text{ V}}{10 \text{ k}\Omega} = 430 \text{ } \mu\text{A}$$

$$I_C = \beta_{dc}I_B = (150)(430 \text{ } \mu\text{A}) = 64.5 \text{ mA}$$

$$\alpha_{dc} = \frac{\beta_{dc}}{\beta_{dc} + 1} = \frac{150}{151} = 0.993$$

$$I_E = \frac{I_C}{\alpha_{dc}} = \frac{64.5 \text{ mA}}{0.993} = 64.95 \text{ mA}$$

$$V_{CE} = V_{CC} - I_C R_C = 10 \text{ V} - (64.5 \text{ mA})(100 \text{ } \Omega)$$

$$= 10 \text{ V} - 6.45 \text{ V} = 3.55 \text{ V}$$

$$V_{CB} = V_{CE} - V_{BE} = 3.55 \text{ V} - 0.7 \text{ V} = 2.85 \text{ V}$$

Since the collector is at a higher voltage than the base, the collector-base junction is reverse-biased.

The following computer program computes the transistor dc currents and voltages when you input V_{BB}, V_{CC}, R_B, R_C, and β_{dc}.

```
10   CLS
20   PRINT "THIS PROGRAM COMPUTES THE DIRECT CURRENTS"
30   PRINT "AND DC VOLTAGES FOR THE CIRCUIT IN FIGURE"
40   PRINT "5-8 GIVEN VALUES FOR VBB, VCC, RB, RC,"
50   PRINT "AND DC BETA."
60   PRINT:PRINT:PRINT
70   INPUT "TO CONTINUE PRESS 'ENTER'";X
80   CLS
90   INPUT "VBB IN VOLTS";VBB
100  INPUT "VCC IN VOLTS";VCC
110  INPUT "RB IN OHMS";RB
120  INPUT "RC IN OHMS";RC
130  INPUT "DC BETA";B
140  CLS
150  IB=(VBB-.7)/RB
160  IC=B*IB
170  ALPHA=B/(B+1)
180  IE=IC/ALPHA
190  VCE=VCC-IC*RC
200  IF VCE<=0 THEN PRINT "TRANSISTOR IS SATURATED" ELSE 220
210  END
220  VBC=.7-VCE
230  PRINT "IB =";IB;"A"
240  PRINT "IC =";IC;"A"
250  PRINT "IE =";IE;"A"
260  PRINT "VCE =";VCE;"V"
270  PRINT "VCB =";VCB;"V"
```

Collector Curves

With a circuit like Figure 5–9(a), a set of curves can be generated showing how I_C varies with V_{CE} for various values of I_B. These are the *collector characteristic* curves. Notice that both V_{BB} and V_{CC} are adjustable. If V_{BB} is set to produce a specific value of I_B and V_{CC} is zero, then $I_C = 0$ and $V_{CE} = 0$. Now, as V_{CC} is gradually increased, V_{CE} will increase and so will I_C. This is indicated on the portion of the curve between points A and B in Figure 5–9(b).

When V_{CE} reaches approximately 0.7 V, the base-collector junction becomes reverse-biased and I_C reaches its full value determined by the relationship $I_C = \beta_{dc}I_B$. At this point the I_C levels off to an almost constant value as V_{CE} continues to increase. This action appears to the right of point B on the curve. Actually, I_C increases slightly as V_{CE} increases due to widening of the base-collector depletion layer which results in fewer holes for recombination in the base region. By using other values of I_B, additional I_C-versus-V_{CE} curves can be produced, as shown in Figure 5–9(c). These curves constitute a *family* of collector curves for a given transistor.

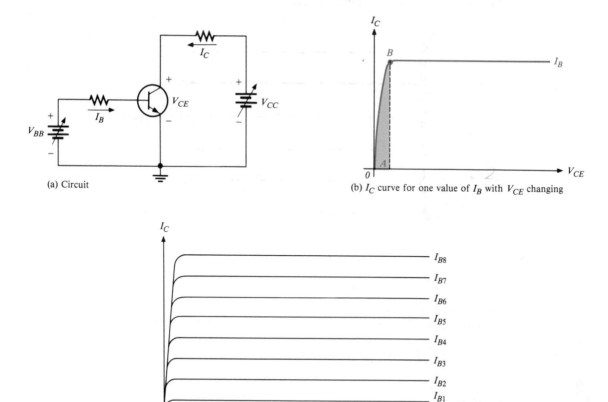

(a) Circuit

(b) I_C curve for one value of I_B with V_{CE} changing

(c) Family of collector curves ($I_{B1} < I_{B2} < I_{B3}$, etc.)

FIGURE 5–9
Collector characteristic curves.

EXAMPLE 5–3

Sketch the family of collector curves for the circuit in Figure 5–10 for $I_B = 5\ \mu A$ to $25\ \mu A$ in $5\ \mu A$ increments. Assume $\beta_{dc} = 100$.

FIGURE 5–10

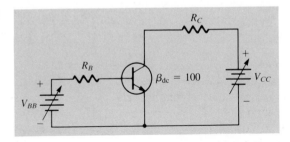

Solution

Using the relationship $I_C = \beta_{dc}I_B$, values of I_C are calculated and tabulated in Table 5–1. The resulting curves are plotted in Figure 5–11.

TABLE 5–1

I_B	I_C
$5\ \mu A$	0.5 mA
$10\ \mu A$	1.0 mA
$15\ \mu A$	1.5 mA
$20\ \mu A$	2.0 mA
$25\ \mu A$	2.5 mA

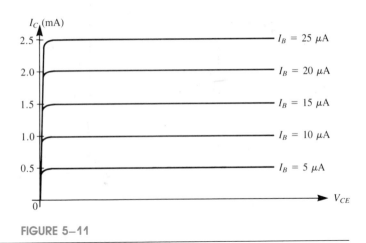

FIGURE 5–11

Cutoff and Saturation

When $I_B = 0$, the transistor is *cut off*. This is shown in Figure 5–12 with the base lead open to produce a base current of 0. Under this condition, there is a very small amount of collector leakage current, I_{CEO}, due mainly to thermally produced carriers. In cutoff, both the base-emitter and the base-collector junctions are *reverse-biased*.

FIGURE 5–12
Collector leakage current (I_{CEO}) in cutoff.

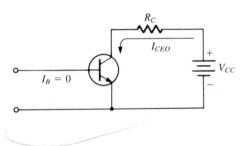

Now let's consider the condition known as *saturation*. When the base current in Figure 5–13(a) is increased, the collector current also increases and V_{CE} decreases as a result of more drop across R_C. When V_{CE} reaches a value called $V_{CE(sat)}$, the base-collector junction becomes *forward-biased* and I_C can increase no further even with a continued increase in I_B. At the point of saturation, the relation $I_C = \beta_{dc}I_B$ is no longer valid. $V_{CE(sat)}$ for a transistor occurs somewhere below the knee of the collector curves, as shown in Figure 5–13(b), and is usually only a few tenths of a volt for silicon transistors.

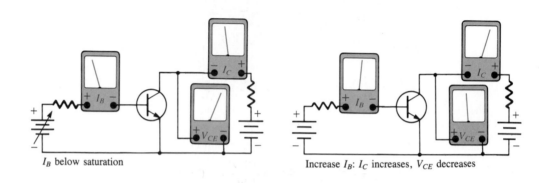

I_B below saturation   Increase I_B: I_C increases, V_{CE} decreases

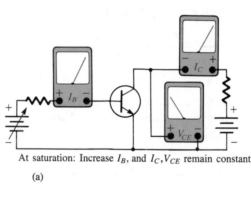

At saturation: Increase I_B, and I_C,V_{CE} remain constant

(a)

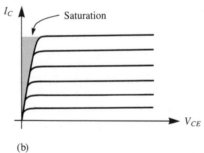

(b)

FIGURE 5–13
Saturation.

EXAMPLE 5–4

Determine whether or not the transistor in Figure 5–14 is in saturation. Assume $V_{CE(sat)}$ is small enough to neglect.

FIGURE 5–14

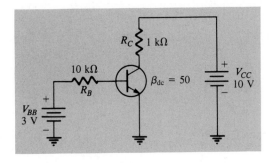

Solution

First, determine $I_{C(sat)}$.

$$I_{C(sat)} = \frac{V_{CC} - V_{CE(sat)}}{R_C} \cong \frac{10 \text{ V}}{1 \text{ k}\Omega} = 10 \text{ mA}$$

Now, let's see if I_B is large enough to produce $I_{C(sat)}$.

$$I_B = \frac{V_{BB} - 0.7 \text{ V}}{R_B} = \frac{2.3 \text{ V}}{10 \text{ k}\Omega} = 0.23 \text{ mA}$$

$$I_C = \beta_{dc}I_B = (50)(0.23 \text{ mA}) = 11.5 \text{ mA}$$

This shows that with the specified β_{dc}, this base current is capable of producing an I_C greater than $I_{C(sat)}$. Therefore, *the transistor is saturated*, and the collector current value of 11.5 mA is never reached.

Practice Exercise 5–4

Determine whether or not the transistor in Figure 5–14 is saturated for the following values: $\beta_{dc} = 25$, $V_{BB} = 1.5$ V, $R_B = 6.8$ kΩ, $R_C = 1.8$ kΩ, and $V_{CC} = 12$ V.

More about β_{dc}

The β_{dc} is a very important bipolar transistor parameter that we need to examine further. β_{dc} varies with both collector current and temperature. Keeping the junction temperature constant and increasing I_C causes β_{dc} to increase to a maximum. A further increase in I_C beyond this maximum point causes β_{dc} to decrease. If I_C is held constant and the temperature is varied, β_{dc} changes directly with the temperature. If the temperature goes up, β_{dc} goes up and vice versa. Figure 5–15 shows the variation of β_{dc} with I_C and junction temperature (T_J) for a typical transistor.

A transistor data sheet usually specifies β_{dc} (h_{FE}) at specific I_C values. Even at fixed values of I_C and temperature, β_{dc} varies from device to device for a given transistor. The

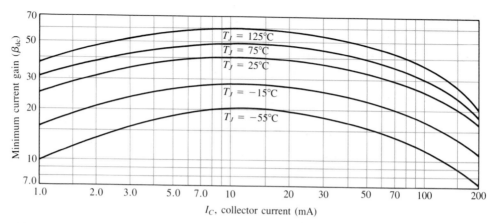

FIGURE 5–15
Variation of β_{dc} with I_C for several temperatures.

β_{dc} specified at a certain value of I_C is usually the minimum value, $\beta_{dc(min)}$, although the maximum and typical values are also sometimes specified.

Maximum Ratings

The transistor, like any other electronic device, has limitations on its operation. These limitations are stated in the form of *maximum ratings* and are normally specified on the manufacturer's data sheet. Typically, maximum ratings are given for *collector-to-base voltage, collector-to-emitter voltage, emitter-to-base voltage, collector current,* and *power dissipation.*

The product of V_{CE} and I_C must not exceed the maximum power dissipation. Both V_{CE} and I_C cannot be maximum at the same time. If V_{CE} is maximum, I_C can be calculated as

$$I_C = \frac{P_{D(max)}}{V_{CE}} \qquad\qquad (5\text{--}10)$$

If I_C is maximum, V_{CE} can be calculated by rearranging equation (5–10) as follows:

$$V_{CE} = \frac{P_{D(max)}}{I_C}. \qquad\qquad (5\text{--}11)$$

For a given transistor, a maximum power dissipation curve can be plotted on the collector curves, as shown in Figure 5–16(a). These values are tabulated in part (b). The $P_{D(max)}$ for this transistor is 0.5 W, $V_{CE(max)}$ is 20 V, and $I_{C(max)}$ is 50 mA. The curve shows that this particular transistor cannot be operated in the shaded portion of the graph. $I_{C(max)}$ is the limiting rating between points A and B, $P_{D(max)}$ is the limiting rating between points B and C, and $V_{CE(max)}$ is the limiting rating between points C and D.

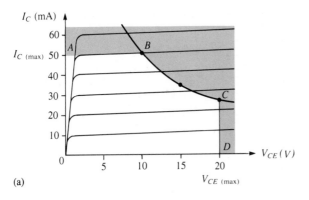

$P_{D\ (max)}$	V_{CE}	I_C
0.5 W	5 V	100 mA
0.5 W	10 V	50 mA
0.5 W	15 V	33 mA
0.5 W	20 V	25 mA

(b)

(a)

FIGURE 5–16
Maximum power dissipation curve.

EXAMPLE 5–5

A certain transistor is to be operated with $V_{CE} = 6$ V. If its maximum power rating is 0.25 W, what is the most collector current that it can withstand?

Solution

$$I_C = \frac{P_{D(max)}}{V_{CE}} = \frac{0.25\ W}{6\ V} = 41.67\ mA$$

Note that this is not necessarily the maximum I_C. The transistor can handle more collector current if V_{CE} is reduced, as long as $P_{D(max)}$ is not exceeded.

EXAMPLE 5–6

The silicon transistor in Figure 5–17 has the following maximum ratings: $P_{D(max)} = 0.8$ W, $V_{CE(max)} = 15$ V, $I_{C(max)} = 100$ mA, $V_{CB(max)} = 20$ V, and $V_{EB(max)} = 10$ V. Determine the maximum value to which V_{CC} can be adjusted without exceeding a rating. Which rating would be exceeded first?

FIGURE 5–17

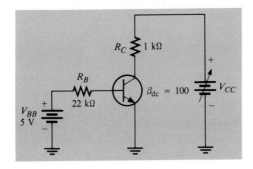

Solution

First find I_B so that I_C can be determined.

$$I_B = \frac{V_{BB} - V_{BE}}{R_B} = \frac{5 \text{ V} - 0.7 \text{ V}}{22 \text{ k}\Omega} = 195.5 \ \mu\text{A}$$

$$I_C = \beta_{dc}I_B = (100)(195.5 \ \mu\text{A}) = 19.55 \text{ mA}$$

I_C is much less than $I_{C(\text{max})}$ and will not change with V_{CC}. It is determined only by I_B and β_{dc}.

The voltage drop across R_C is

$$V_{R_C} = I_C R_C = (19.55 \text{ mA})(1 \text{ k}\Omega) = 19.55 \text{ V}$$

Now we can determine the value of V_{CC} when $V_{CE} = V_{CE(\text{max})} = 15$ V.

$$V_{R_C} = V_{CC} - V_{CE}$$

So

$$\begin{aligned} V_{CC(\text{max})} &= V_{CE(\text{max})} + V_{R_C} \\ &= 15 \text{ V} + 19.55 \text{ V} \\ &= 34.55 \text{ V} \end{aligned}$$

V_{CC} can be increased to 34.55 V, under the existing conditions, before $V_{CE(\text{max})}$ is exceeded. However, we do not know whether or not $P_{D(\text{max})}$ has been exceeded at this point. Let's find out.

$$P_D = V_{CE(\text{max})}I_C = (15 \text{ V})(19.55 \text{ mA}) = 0.293 \text{ W}$$

Since $P_{D(\text{max})}$ is 0.8 W, it is *not* exceeded when $V_{CE} = 34.55$ V. So, $V_{CE(\text{max})}$ is the limiting rating in this case. It should be noted that if the base current is removed causing the transistor to turn off, $V_{CE(\text{max})}$ will be exceeded because the entire supply voltage, V_{CC}, will be dropped across the transistor.

Practice Exercise 5–6

The transistor in Figure 5–17 has the following maximum ratings: $P_{D(\text{max})} = 0.5$ W, $V_{CE(\text{max})} = 25$ V, $I_{C(\text{max})} = 200$ mA, $V_{CB(\text{max})} = 30$ V, and $V_{EB(\text{max})} = 15$ V. Determine the maximum value to which V_{CC} can be adjusted without exceeding a rating. Which rating would be exceeded first?

Derating $P_{D(\text{max})}$

$P_{D(\text{max})}$ is usually specified at 25°C. For higher temperatures, $P_{D(\text{max})}$ is less. Data sheets often give *derating factors* for determining $P_{D(\text{max})}$ at any temperature above 25°C. For example, a derating factor of 2 mW/°C indicates that the maximum power dissipation is reduced 2 mW for each °C *increase* in temperature.

EXAMPLE 5–7

A certain transistor has a $P_{D(max)}$ of 1 W at 25°C. The derating factor is 5 mW/°C. What is the $P_{D(max)}$ at a temperature of 70°C?

Solution

The change (reduction) in $P_{D(max)}$ is

$$\Delta P_{D(max)} = (5 \text{ mW/°C})(70°C - 25°C)$$
$$= (5 \text{ mW/°C})(45°C)$$
$$= 225 \text{ mW}$$

Therefore the $P_{D(max)}$ at 70°C is

$$1 \text{ W} - 225 \text{ mW} = 775 \text{ mW} = 0.775 \text{ W}$$

Practice Exercise 5–7

A transistor has a $P_{D(max)} = 5$ W at 25°C. The derating factor is 10 mW/°C. What is the $P_{D(max)}$ at 70°C?

---

1. $I_B = 8$ μA and $I_C = 640$ μA. Determine β_{dc}.
2. If $\alpha_{dc} = 0.972$, what is β_{dc}?
3. The β_{dc} of a transistor increases with temperature (T or F).
4. Generally, what effect does an increase in I_C have on the β_{dc}?
5. What is the allowable collector current in a transistor with $P_{D(max)} = 0.32$ W when $V_{CE} = 8$ V?

THE BIPOLAR TRANSISTOR AS A VOLTAGE AMPLIFIER

5–4

Amplification is the process of linearly increasing the amplitude of an electrical signal and is one of the major properties of a transistor. As you learned, the transistor exhibits current gain (called β_{dc}). When a transistor is biased, as previously described, the *BE* junction has a *low resistance* due to forward bias and the *BC* junction has a *high resistance* due to reverse bias.

Since I_B is extremely small, I_C is approximately equal to I_E. Actually, I_C is always slightly less than I_E. Therefore, equation (5–1) can be restated as an approximation:

$$I_E \cong I_C$$

With this in mind, let's consider the transistor in Figure 5–18(a) with an ac input voltage, V_{in}, applied in series with the *BE* bias voltage and with an external resistor, R_C, connected in series with the *BC* bias voltage. Because the dc sources appear ideally as shorts to the ac voltage, the *ac equivalent* circuit is as shown in Figure 5–18(b). The forward-biased

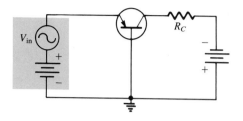 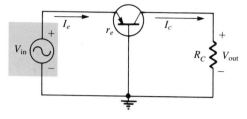

(a) ac input and bias   (b) ac equivalent circuit

FIGURE 5–18
Biased transistor with ac input signal.

base-emitter junction appears as a *low resistance* to the ac signal. This ac emitter resistance is called r_e. The ac emitter current in Figure 5–18(b) is

$$I_e = \frac{V_{in}}{r_e}$$

Since $I_c \cong I_e$, the output voltage developed across R_C is

$$V_{out} \cong I_e R_C$$

The ratio of V_{out} to V_{in} is called the ac *voltage gain* (A_v) and is expressed as follows:

$$A_v = \frac{V_{out}}{V_{in}} \qquad\qquad (5\text{–}12)$$

Substituting, we get

$$A_v = \frac{V_{out}}{V_{in}} \cong \frac{I_e R_C}{I_e r_e}$$

$$A_v \cong \frac{R_C}{r_e} \qquad\qquad (5\text{–}13)$$

This shows that the transistor circuit in Figure 5–18 provides amplification or voltage gain, dependent on the value of R_C and r_e. Remember that lowercase subscripts indicate ac quantities.

EXAMPLE 5–8

Determine the voltage gain and output voltage in Figure 5–19 if $r_e = 50 \ \Omega$.

FIGURE 5–19

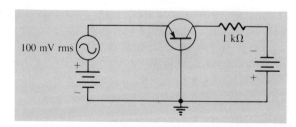

Solution
The voltage gain is

$$A_v \cong \frac{R_C}{r_e} = \frac{1 \text{ k}\Omega}{50 \text{ }\Omega} = 20$$

Therefore, the output voltage is

$$V_{\text{out}} = A_v V_{\text{in}} = (20)(100 \text{ mV}) = 2 \text{ V rms}$$

**SECTION
REVIEW
5–4**

1. Define *amplification*.
2. A certain transistor circuit has an output voltage of 5 V rms and an input of 250 mV rms. What is the voltage gain?
3. A transistor connected as in Figure 5–19 has an $r_e = 20$ Ω. If R_C is 1200 Ω, what is the voltage gain?

THE BIPOLAR TRANSISTOR AS A SWITCH

5–5

In the previous section, we discussed the transistor as an amplifier. The second major application area is switching applications. When used as an electronic switch, a transistor is normally operated alternately in cutoff and saturation. Figure 5–20 illustrates the basic operation of a transistor as a switching device. In part (a), the transistor is cut off because the base-emitter junction is *not* forward-biased. In this condition, there is, ideally, an *open* between collector and emitter. In part (b), the transistor is saturated because the base-emitter junction is forward-biased and the base current is large enough to cause the collector current to reach its saturated value. In this condition, there is, ideally, a *short* between collector and emitter. Actually, a voltage drop of a few tenths of a volt normally occurs.

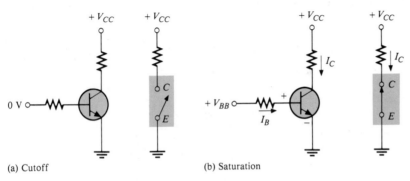

(a) Cutoff   (b) Saturation

FIGURE 5–20
Ideal switching action of a transistor.

Conditions in Cutoff

As mentioned before, a transistor is in cutoff when the base-emitter junction is *not* forward-biased. All of the currents are approximately zero, and V_{CE} is approximately equal to V_{CC}:

$$V_{CE(\text{cutoff})} \cong V_{CC} \qquad (5\text{–}14)$$

Conditions in Saturation

As you have learned, when the emitter junction is forward-biased and there is enough base current to produce a maximum collector current, the transistor is saturated. Since V_{CE} is approximately zero at saturation, the collector current is

$$I_{C(\text{sat})} \cong \frac{V_{CC}}{R_C} \qquad (5\text{–}15)$$

The value of base current needed to produce saturation is

$$I_{B(\text{min})} = \frac{I_{C(\text{sat})}}{\beta_{\text{dc}}} \qquad (5\text{–}16)$$

EXAMPLE 5–9

(a) For the transistor switching circuit in Figure 5–21, what is V_{CE} when $V_{IN} = 0$ V?

(b) What minimum value of I_B is required to saturate this transistor if the β_{dc} is 200?

(c) Calculate the maximum value of R_B when $V_{IN} = 5$ V.

FIGURE 5–21

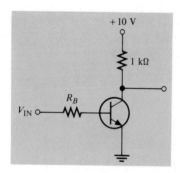

Solution

(a) When $V_{IN} = 0$ V, the transistor is *off* and $V_{CE} = V_{CC} = 10$ V.

(b) When the transistor is saturated, $V_{CE} \cong 0$ V, so

$$I_{C(\text{sat})} = \frac{V_{CC}}{R_C} = \frac{10 \text{ V}}{1 \text{ k}\Omega} = 10 \text{ mA}$$

$$I_B = \frac{I_{C(\text{sat})}}{\beta_{\text{dc}}} = \frac{10 \text{ mA}}{200} = 0.05 \text{ mA}$$

This is the value of I_B necessary to drive the transistor to the point of saturation. *Any further increase in I_B will drive the transistor deeper into saturation but will not increase I_C.*

(c)   When the transistor is saturated, $V_{BE} = 0.7$ V. The voltage across R_B is $V_{IN} - 0.7$ V $= 4.3$ V. The maximum value of R_B needed to allow a minimum I_B of 0.05 mA is calculated by Ohm's law as follows:

$$R_B = \frac{V_{IN} - 0.7 \text{ V}}{I_B} = \frac{4.3 \text{ V}}{0.05 \text{ mA}} = 86 \text{ k}\Omega$$

---

1.   When a transistor is used as a switching device, it is operated in either _____ or _____ .

2.   When does the collector current reach its maximum value?

3.   When is the collector current approximately zero?

4.   Name the two conditions that produce saturation.

5.   When is V_{CE} equal to V_{CC}?

SECTION REVIEW 5–5

TRANSISTOR TESTING

5–6

Ohmmeter Check of Transistor Junctions

The ohmmeter provides a simple test for open or shorted junctions. *The base-emitter and base-collector junctions are each treated as a diode for this test.* The junction should show a *low* resistance when forward-biased and a very *high* resistance when reverse-biased. The internal battery of the ohmmeter must provide the bias voltage.

Figure 5–22(a) shows a *forward-biased* check of an npn transistor. Notice the polarities of the ohmmeter leads. Part (b) of the figure demonstrates a *reverse-biased* check. For a pnp transistor, the meter polarities are reversed from those shown in Figure 5–22.

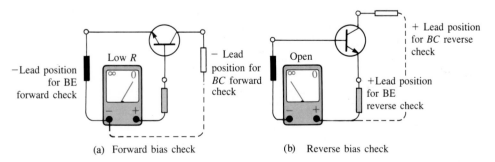

(a) Forward bias check   (b) Reverse bias check

FIGURE 5–22
Ohmmeter check of npn transistor junctions.

Leakage Measurement

Very small *leakage* currents exist in all transistors and in most cases are small enough to neglect. When a transistor is connected as shown in Figure 5–23(a) with the base open ($I_B = 0$), it is in cutoff. Ideally $I_C = 0$; but actually there is a small current from collector to emitter, as mentioned earlier, called I_{CEO} (collector-to-emitter current with base open). This leakage current is usually in the nA range for silicon. A faulty transistor will often have excessive leakage current and can be checked in a transistor tester, which connects an ammeter as shown in part (a). Another leakage current in transistors is the reverse collector-to-base current, I_{CBO}. This is measured with the emitter open, as shown in part (b). If it is excessive, a shorted collector-base junction is likely.

FIGURE 5–23
Leakage current test circuits.

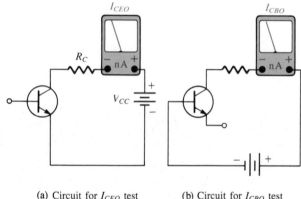

(a) Circuit for I_{CEO} test   (b) Circuit for I_{CBO} test

Gain Measurement

In addition to leakage tests, the typical transistor tester also checks the β_{dc}. A known value of I_B is applied and the resulting I_C is measured. The reading will indicate the value of the I_C/I_B ratio, although in some units only a relative indication is given. Most testers provide for an in-circuit β_{dc} check, so that a suspected device does not have to be removed from the circuit for testing. Figure 5–24 shows two typical transistor testers.

Curve Tracers

The *curve tracer* is an oscilloscope type of instrument that can display transistor characteristics such as a family of collector curves. In addition to the measurement and display of various transistor characteristics, diode curves can also be displayed, as well as the β_{dc}. A typical curve tracer is shown in Figure 5–25.

SECTION REVIEW 5–6

1. The positive lead of an ohmmeter is on a transistor's base, the negative lead on the emitter, and a low resistance reading is obtained. If the transistor is good, is it a pnp or an npn type?

2. Name two types of transistor leakage currents.

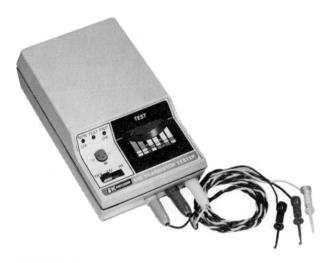

FIGURE 5–24
Transistor testers.

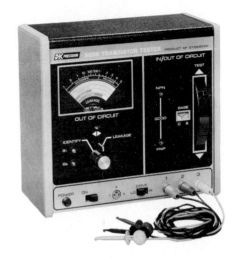

FIGURE 5–25
Curve tracer. (Courtesy Tektronix, Inc.)

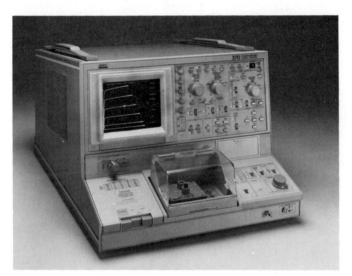

PACKAGES AND TERMINAL IDENTIFICATION

5–7

Transistors are available in a wide range of package types for various applications. Those with mounting studs or heat sinks are power transistors. Low and medium power transistors are usually found in smaller metal or plastic cases. Still another package classification is for high-frequency devices. Figure 5–26 shows some typical transistor packages with some examples of construction views and pin arrangements. Generally, the heat sink mounting or stud on power transistors is the collector terminal. On the metal "top hat" cases, the tab is closest to the emitter lead and the collector is often connected to the case.

(a)

B C E

(b)

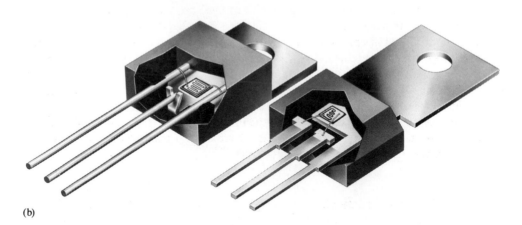

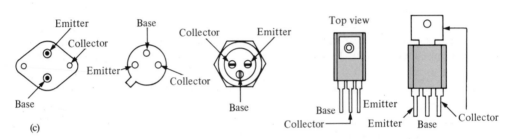

(c)

FIGURE 5–26
Typical transistor packages, construction views, and examples of pin arrangements.

1. Identify the leads on the transistors in Figure 5–27.

FIGURE 5–27

(a)   (b)   (c)

A SYSTEM APPLICATION

5–8

As you have seen, bipolar transistors are important as amplifying devices. To be useful, all types of communications systems, as well as many other kinds of systems, require amplification of signals. The superhet receiver is a good example of how transistors fit into a system application. Figure 5–28 on p. 162 shows a typical AM superheterodyne receiver. The circuitry is representative of what you will find in many receivers, although the details of circuit implementation vary depending on the manufacturer. You can see that this particular system uses eight bipolar transistors.

Notice that this receiver does not have an RF amplifier. The signal from the antenna is fed directly into the mixer (Q_2), which provides amplification. Like this one, many AM receivers do not have an RF amplifier. The absence of an RF amplifier essentially means that the receiver is less sensitive and cannot pick up many weaker signals. Notice also that there are two IF amplifier stages in this system (Q_3 and Q_4). (Receivers generally have two or more IF amplifiers.) The transistors Q_5 and Q_6 are the audio amplifiers, and transistors Q_7 and Q_8 are arranged as a push-pull power amplifier which drives the 8 Ω speaker.

The purpose of presenting this receiver schematic at this point is to indicate how individual transistors fit into an overall system. You do not yet have the background to completely understand the details of this system, but as you progress through the book, you will learn enough about individual devices and circuits to understand a system such as this and other, more complex electronic systems.

SUMMARY

☐ A bipolar transistor consists of three regions: emitter, base, and collector.
☐ Current in a transistor consists of both electron and hole flow, thus the term *bipolar*.
☐ A bipolar transistor has two pn junctions: one between the emitter and base, and another between the base and collector.
☐ The two types of bipolar transistors are the npn and the pnp.
☐ The base current is normally much less than either the emitter current or the collector current.
☐ The dc alpha (α_{dc}) is the ratio of collector current to emitter current and has a value close to 1.
☐ The dc beta (β_{dc}) is the ratio of collector current to base current. This current gain can have values from less than 20 to several hundred.

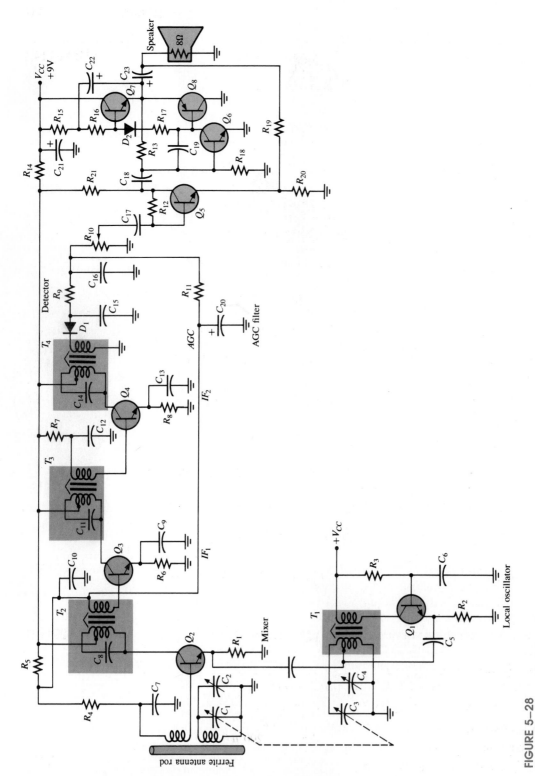

FIGURE 5–28
A typical AM superheterodyne receiver schematic.

			FORMULAS
(5–1)	$I_E = I_C + I_B$	Transistor currents	
(5–2)	$\beta_{dc} = \dfrac{I_C}{I_B}$ $I_C = \beta I_B$	dc current gain	
(5–3)	$\alpha_{dc} = \dfrac{I_C}{I_E}$ $I_B = I_E(1 - \alpha)$	dc alpha	
(5–4)	$\alpha_{dc} = \dfrac{\beta_{dc}}{\beta_{dc} + 1}$	β_{dc}-to-α_{dc} conversion	
(5–5)	$\beta_{dc} = \dfrac{\alpha_{dc}}{1 - \alpha_{dc}}$	α_{dc}-to-β_{dc} conversion	
(5–6)	$V_{BE} \cong 0.7 \text{ V}$	Base-to-emitter voltage (silicon)	
(5–7)	$I_B = \dfrac{V_{BB} - V_{BE}}{R_B}$	Base current	
(5–8)	$V_{CE} = V_{CC} - I_C R_C$	Collector-to-emitter voltage (common-emitter)	
(5–9)	$V_{CB} = V_{CE} - V_{BE}$	Collector-to-base voltage	
(5–10)	$I_C = \dfrac{P_{D(max)}}{V_{CE}}$	Maximum I_C for given V_{CE}	
(5–11)	$V_{CE} = \dfrac{P_{D(max)}}{I_C}$	Maximum V_{CE} for given I_C	
(5–12)	$A_v = \dfrac{V_{out}}{V_{in}}$	ac voltage gain	
(5–13)	$A_v \cong \dfrac{R_C}{r_e}$	Approximate ac voltage gain	
(5–14)	$V_{CE(cutoff)} \cong V_{CC}$	Cutoff condition	
(5–15)	$I_{C(sat)} \cong \dfrac{V_{CC}}{R_C}$	Saturation condition	
(5–16)	$I_{B(min)} = \dfrac{I_{C(sat)}}{\beta_{dc}}$	Minimum base current for saturation	

SELF-TEST

1. In an npn transistor, the n-type semiconductor regions are the _____ and _____ .

2. For normal operation of a pnp transistor, the base must be (+ or −) with respect to the emitter, and (+ or −) with respect to the collector.

3. What is the exact value of I_C for $I_E = 5.34$ mA and $I_B = 475$ μA?

4. What is the α_{dc} when $I_C = 8.23$ mA and $I_E = 8.69$ mA?

5. A certain transistor has an $I_C = 25$ mA and an $I_B = 200$ μA. Determine the β_{dc}.

6. What is the β_{dc} of a transistor if α_{dc} is 0.96?

7.   A certain transistor has a maximum power rating of 0.5 W. If V_{CE} is 8 V, how much collector current can there be without exceeding the power rating?

8.   A 10 mV input signal is applied to a transistor and a 1.5 V output signal is measured. What is the voltage gain?

9.   A base voltage is applied to a transistor with the emitter grounded (0 V). The collector voltage is approximately zero. Is the transistor in cutoff or saturation?

10.   An ohmmeter is connected to a pnp transistor with the positive lead on the base and the negative lead on the emitter. The meter shows a very high resistance. Is this a good indication?

11.   Name two checks that most transistor testers perform on a device under test.

12.   If you have a computer available, enter the program in Example 5–2. Run the program, entering variable values from the example, and observe the results.

PROBLEMS

Section 5–1

5–1   The majority carriers in the base region of an npn transistor are _____.

5–2   Explain the purpose of a thin, lightly doped base.

Section 5–2

5–3   Why is the base current in a transistor so much less than the collector current?

5–4   In a certain transistor circuit, the base current is 2 percent of the 30 mA emitter current. Determine the collector current.

Section 5–3

5–5   A certain transistor exhibits an α_{dc} of 0.96. Determine I_C when $I_E = 9.35$ mA.

5–6   A base current of 50 μA is applied to the transistor in Figure 5–29, and a voltage of 5 V is dropped across R_C. Determine the β_{dc} of the transistor.

FIGURE 5–29

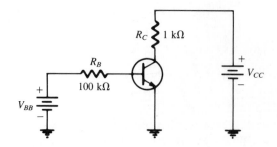

5–7   Calculate α_{dc} for the transistor in problem 5–6.

5–8   Determine each current in Figure 5–30. What is the β_{dc}?

FIGURE 5–30

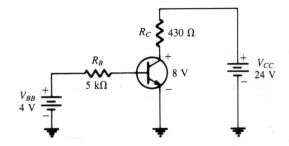

5–9 Find V_{CE}, V_{BE}, and V_{CB} in both circuits of Figure 5–31.

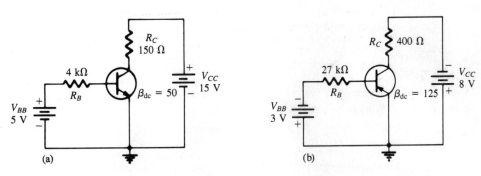

(a)   (b)

FIGURE 5–31

5–10 Determine whether or not the transistors in Figure 5–31 are saturated.

5–11 Find I_B, I_E, and I_C in Figure 5–32. $\alpha_{dc} = 0.98$.

FIGURE 5–32

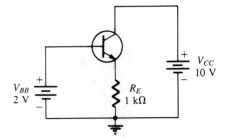

5–12 Determine the terminal voltages of each transistor with respect to ground for each circuit in Figure 5–33 on p. 166. Also determine V_{CE}, V_{BE}, and V_{BC}.

5–13 If the β_{dc} in Figure 5–33(a) changes from 100 to 150 due to a temperature increase, what is the change in collector current?

5–14 A certain transistor is to be operated at a collector current of 50 mA. How high can V_{CE} go without exceeding a $P_{D(max)}$ of 1.2 W?

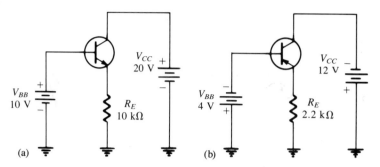

(a)   (b)

FIGURE 5–33

5–15 The power dissipation derating factor for a certain transistor is 1 mW/°C. The $P_{D(max)}$ is 0.5 W at 25°C. What is $P_{D(max)}$ at 100°C?

Section 5–4

5–16 A transistor amplifier has a voltage gain of 50. What is the output voltage when the input voltage is 100 mV?

5–17 To achieve an output of 10 V with an input of 300 mV, what voltage gain is required?

5–18 A 50 mV signal is applied to the base of a properly biased transistor with $r_e = 10 \ \Omega$ and $R_C = 500 \ \Omega$. Determine the signal voltage at the collector.

Section 5–5

5–19 Determine $I_{C(sat)}$ for the transistor in Figure 5–34. What is the value of I_B necessary to produce saturation? What minimum value of V_{IN} is necessary for saturation?

FIGURE 5–34

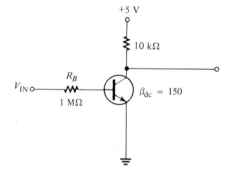

5–20 The transistor in Figure 5–35 has a β_{dc} of 50. Determine the value of R_B required to insure saturation when V_{IN} is 5 V. What must V_{IN} be to cut off the transistor?

FIGURE 5–35

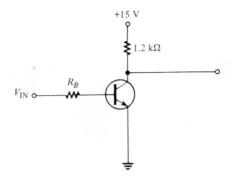

5–21 Which, if any, of the transistors in Figure 5–36 are bad?

FIGURE 5–36

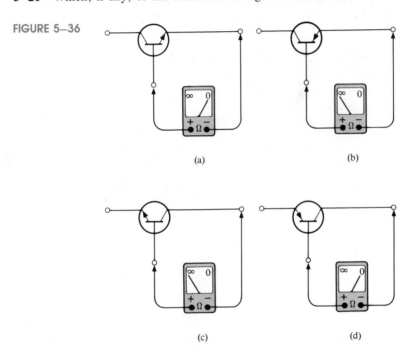

(a)

(b)

(c)

(d)

5–22 Is the transistor in Figure 5–37 an npn or a pnp device? Assume that it is good.

FIGURE 5–37

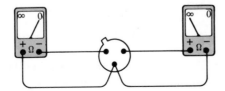

ANSWERS TO SECTION REVIEWS

Section 5–1
1. pnp, npn.
2. Emitter, base, and collector.

Section 5–2
1. Forward, reverse.
2. 1.001 mA.

Section 5–3
1. 80.
2. 35.
3. True.
4. β_{dc} increases with I_C to a certain value, and then decreases.
5. 40 mA.

Section 5–4
1. The process of increasing the amplitude of an electrical signal.
2. 20.
3. 60.

Section 5–5
1. Cutoff, saturation.
2. At the point of saturation.
3. At cutoff.
4. The base-emitter junction is forward-biased, sufficient base current.
5. At cutoff.

Section 5–6
1. npn.
2. I_{CBO}, I_{CEO}.

Section 5–7
1. See Figure 5–38.

FIGURE 5–38

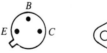 

Example 5–4: Not saturated

Example 5–6: $V_{CC(\max)} = 44.55$ V, $V_{CE(\max)}$

Example 5–7: 4.55 W

Bipolar Transistor Biasing

In this chapter you will learn

☐ The meaning of the term *bias* in transistor applications

☐ How the bias affects the operation of a bipolar transistor

☐ The meaning of *linear operation* in relation to transistor characteristic curves and load lines

☐ How to bias a transistor so that it can operate as a linear amplifier

☐ How the dc bias point (Q-point) affects the linear operation of a transistor

☐ Various types of dc bias circuits and their characteristics

☐ How the type of bias affects the stability of a transistor circuit

☐ How to troubleshoot biased transistor circuits

☐ How biasing is used in a specific system application

Chapter 6

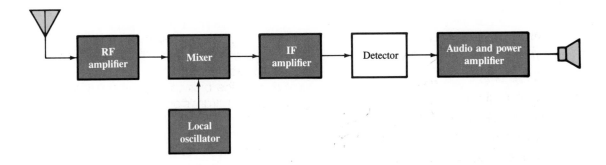

As you saw in the last chapter, a transistor must be properly biased in order to operate as an amplifier. The purpose of the dc bias is to establish a steady level of transistor current and voltage, called the *dc operating point,* or *quiescent point.*

In this chapter we will examine several methods of establishing dc bias. Silicon bipolar junction transistors are used for illustration unless specified otherwise. Without proper dc bias, transistor amplifiers, oscillators, and other circuits cannot function. Bias circuits are an integral part of any system, and the superhet receiver is no exception. Most of the functional blocks of our representative AM system contain resistor bias networks that allow the transistors to do their job, as indicated by the shaded areas.

PURPOSE OF BIASING

6–1

As mentioned, a transistor must be dc-biased in order to operate as an amplifier. A dc *operating point* must be set so that signal variations at the input terminal are amplified and accurately reproduced at the output terminal, as shown in Figure 6–1(a). Improper biasing can cause distortion in the output signal, as illustrated in parts (b) and (c). In part (a) of the figure, the output signal is an amplified *replica* of the input signal. The output signal swings *equally* above and below the dc level of the output. Part (b) illustrates *limiting* of the positive portion of the output voltage as a result of a dc operating point being too close to cutoff. Part (c) shows limiting of the negative portion of the output voltage as a result of a dc operating point being too close to saturation.

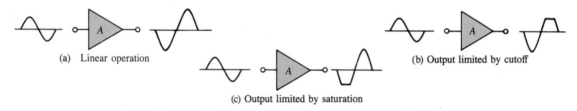

(a) Linear operation   (c) Output limited by saturation   (b) Output limited by cutoff

FIGURE 6–1
Linear and nonlinear operation of an inverting amplifier.

1. Improper bias causes _____ of the output signal.
2. Name two types of distortion.

THE dc OPERATING POINT

6–2

As you learned in Chapter 5, when you bias a transistor, you establish a certain current and voltage condition. This means, for example, that at the dc operating point, I_C and V_{CE} have specified values. The dc operating point is often referred to as the *Q-point* (quiescent point).

Graphical Analysis

The transistor in Figure 6–2(a) is biased with V_{CC} and V_{BB} to obtain certain values of I_B, I_C, I_E, and V_{CE}. The collector characteristic curves for this particular transistor are shown in part (b), and we will use these to graphically illustrate the effects of dc bias. To do this we will set various values of I_B and observe what happens to I_C and V_{CE}.

To start, V_{BB} is adjusted to produce an I_B of 300 μA, as shown in Figure 6–3(a) on p. 174. Since $I_C = \beta_{dc}I_B$, the collector current is 30 mA, as indicated, and $V_{CE} = V_{CC} - I_C R_C = 10 \text{ V} - 6 \text{ V} = 4 \text{ V}$. This Q-point is shown on the graph of Figure 6–3(d) as Q_1.

Now, V_{BB} is increased to produce an I_B of 400 μA, an I_C of 40 mA, and a resulting V_{CE} of 2 V, as shown in part (b). The Q-point for this condition is indicated by Q_2 on the

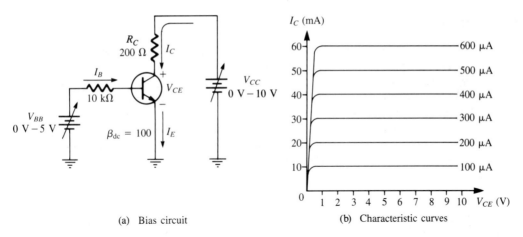

R_C
200 Ω I_C

I_B

10 kΩ V_{CE}

V_{CC}
0 V – 10 V

V_{BB}
0 V – 5 V

$\beta_{dc} = 100$ I_E

I_C (mA)

60 ———————————————— 600 μA
50 ———————————————— 500 μA
40 ———————————————— 400 μA
30 ———————————————— 300 μA
20 ———————————————— 200 μA
10 ———————————————— 100 μA
0   1 2 3 4 5 6 7 8 9 10 V_{CE} (V)

(a) Bias circuit   (b) Characteristic curves

FIGURE 6–2
Biased transistor.

graph. Finally, V_{BB} is reduced to give an I_B of 200 μA, an I_C of 20 mA, and a V_{CE} of 6 V, as in part (c). Q_3 is the corresponding Q-point on the graph.

dc Load Line

Notice that when I_B increases, I_C increases and V_{CE} decreases. When I_B decreases, I_C decreases and V_{CE} increases. So, as V_{BB} is adjusted up or down, the dc operating point of the transistor moves along a sloping *straight line,* called the *dc load line,* connecting each Q-point. At any point along the line, values of I_B, I_C, and V_{CE} can be picked off the graph, as shown in Figure 6–4 on p. 175.

Notice that the load line intersects the V_{CE} axis at 10 V, the point where $V_{CE} = V_{CC}$. This is the transistor *cutoff* point because I_B and I_C are 0 (ideally). Actually, there is a small I_{CBO} at cutoff as indicated, and therefore V_{CE} is slightly less than 10 V.

Next, notice that the load line intersects the I_C axis at 50 mA ideally. This is the transistor *saturation* point because I_C is maximum (ideally 50 mA) at the point where $V_{CE} = 0$ V and $I_C = V_{CC}/R_C$. Actually, there is a small voltage ($V_{CE(sat)}$) across the transistor, and $I_{C(sat)}$ is slightly less than 50 mA, as indicated.

Linear Operation

The region along the load line including all points *between saturation and cutoff* is known as the *linear region* of the transistor's operation. As long as the transistor is operated in this region, the output voltage is a linear reproduction of the input. For example, assume a sine wave voltage is superimposed on V_{BB}, causing the base current to vary 100 μA above and below the Q-point value of 300 μA. This, in turn, causes the collector current to vary 10 mA above and below its Q-point value of 30 mA. As a result of the variation in collector current, the collector-to-emitter voltage varies 2 V above and below its Q-point value of 4 V. This action is shown in Figure 6–5 on p. 175.

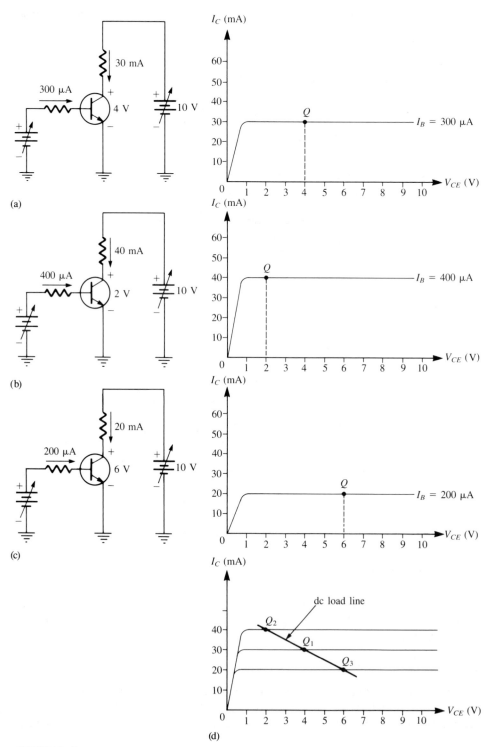

FIGURE 6–3
Q-point adjustments.

FIGURE 6–4
dc load line.

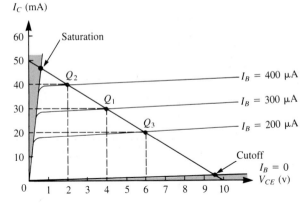

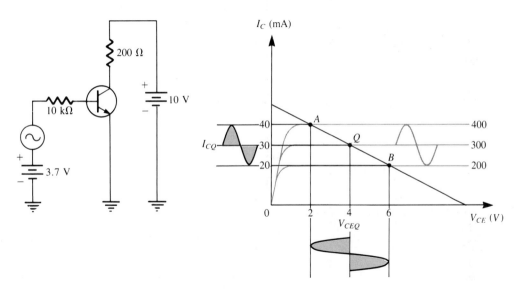

FIGURE 6–5
Variation in collector-to-emitter voltage.

Point *A* on the load line corresponds to the positive peak of the input sine wave. Point *B* corresponds to the negative peak, and the Q-point corresponds to the 0 value of the sine wave, as indicated.

Distortion of the Output

As mentioned in section 6–1, under certain input signal conditions the location of the Q-point on the load line can cause one peak of the output signal to be *limited,* as shown in parts (a) and (b) of Figure 6–6. In each case the input signal is excessive for the Q-point

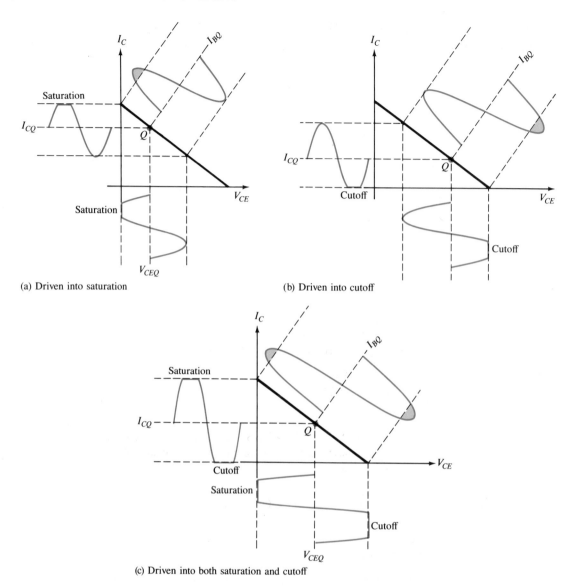

(a) Driven into saturation

(b) Driven into cutoff

(c) Driven into both saturation and cutoff

FIGURE 6–6
Saturation and cutoff.

and is driving the transistor into cutoff or saturation during a portion of the input cycle. When both peaks are limited as in Figure 6–6(c), the transistor is being driven into both saturation and cutoff by an excessively large input signal. When only the positive peak is limited, the transistor is being driven into cutoff but not saturation. When only the negative peak is limited, the transistor is being driven into saturation but not cutoff.

FIGURE 6–11
Effect of I_{CBO}.

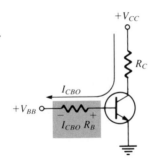

1. Name the two dc quantities that define the Q-point of a biased transistor.
2. What is the main disadvantage of fixed bias?

**SECTION
REVIEW
6–3**

EMITTER BIAS

6–4

This type of bias circuit uses both a positive and a negative supply voltage, as shown in Figure 6–12. In this circuit, the base is at approximately 0 volts, and the $-V_{EE}$ supply forward-biases the base-emitter junction. Expressions for the currents and voltages are as follows:

$$V_B \cong 0 \qquad \text{(6–5)}$$

$$V_E \cong -V_{BE} \qquad \text{(6–6)}$$

$$I_E = \frac{V_E - V_{EE}}{R_E} \qquad \text{(6–7)}$$

$$I_C \cong I_E \qquad \text{(6–8)}$$

$$V_C = V_{CC} - I_C R_C \qquad \text{(6–9)}$$

$$V_{CE} = V_C - V_E \qquad \text{(6–10)}$$

FIGURE 6–12
A transistor with emitter bias.

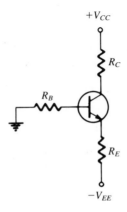

EXAMPLE 6–3

Find I_E, I_C, and V_{CE} in Figure 6–13.

FIGURE 6–13

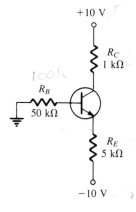

Solution

$$V_E \cong -V_{BE} = -0.7 \text{ V}$$

$$I_E = \frac{V_{BE} - V_{EE}}{R_E}$$

$$= \frac{-0.7 \text{ V} - (-10 \text{ V})}{5 \text{ k}\Omega}$$

$$= \frac{9.3 \text{ V}}{5 \text{ k}\Omega}$$

$$= 1.86 \text{ mA}$$

$$I_C \cong I_E = 1.86 \text{ mA}$$

$$V_C = V_{CC} - I_C R_C$$

$$= 10 \text{ V} - (1.86 \text{ mA})(1 \text{ k}\Omega)$$

$$= 8.14 \text{ V}$$

$$V_{CE} = V_C - V_E$$

$$= 8.14 \text{ V} - (-0.7 \text{ V})$$

$$= 8.84 \text{ V}$$

Practice Exercise 6–3

Find I_E, I_C, and V_{CE} for the circuit in Figure 6–13 with the following component values: $R_B = 100 \text{ k}\Omega$, $R_C = 680 \text{ }\Omega$, $R_E = 3.3 \text{ k}\Omega$, $V_{CC} = +15 \text{ V}$, and $V_{EE} = -15 \text{ V}$.

Stability of Emitter Bias

Applying Kirchhoff's law around the base-emitter circuit in Figure 6–12, we get the following equation:

$$I_B R_B + V_{BE} + I_E R_E = V_{EE}$$

Since

$$I_B \cong \frac{I_E}{\beta_{dc}}$$

then

$$I_E\left(\frac{R_B}{\beta_{dc}} + R_E\right) = V_{EE} - V_{BE}$$

$$I_E = \frac{V_{EE} - V_{BE}}{R_E + R_B/\beta_{dc}} \qquad \text{(6–11)}$$

If $R_E \gg R_B/\beta_{dc}$, the equation becomes

$$I_E \cong \frac{V_{EE} - V_{BE}}{R_E} \qquad \text{(6–12)}$$

A further simplification can be made if $V_{EE} \gg V_{BE}$:

$$I_E \cong \frac{V_{EE}}{R_E} \qquad \text{(6–13)}$$

This shows that the emitter current is essentially independent of β_{dc} and V_{BE}, as long as the conditions mentioned above are satisfied. Of course, if I_E is independent of β_{dc} and V_{BE}, then the Q-point is not affected appreciably by variations in these parameters. Thus, emitter bias provides a reasonably stable bias point.

EXAMPLE 6–4

Determine how much the Q-point in Figure 6–14 (p. 184) will change over a temperature range where β_{dc} increases from 50 to 100 and V_{BE} decreases from 0.7 V to 0.6 V.

Solution

For $\beta_{dc} = 50$ and $V_{BE} = 0.7$ V:

$$I_C \cong I_E = \frac{V_{EE} - V_{BE}}{R_E + R_B/\beta_{dc}}$$

$$= \frac{20 \text{ V} - 0.7 \text{ V}}{10 \text{ k}\Omega + 10 \text{ k}\Omega/50}$$

$$= 1.892 \text{ mA}$$

$$V_C = V_{CC} - I_C R_C$$

$$= 20 \text{ V} - (1.892 \text{ mA})(5 \text{ k}\Omega)$$

$$= 10.54 \text{ V}$$

$$V_E = -0.7 \text{ V}$$

FIGURE 6–14

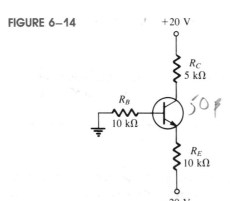

Therefore,

$$V_{CE} = V_C - V_E$$
$$= 10.54 \text{ V} - (-0.7 \text{ V})$$
$$= 11.24 \text{ V}$$

For $\beta_{dc} = 100$ and $V_{BE} = 0.6$ V:

$$I_C \cong I_E = \frac{V_{EE} - V_{BE}}{R_E + R_B/\beta_{dc}}$$
$$= \frac{20 \text{ V} - 0.6 \text{ V}}{10 \text{ k}\Omega + 10 \text{ k}\Omega/100}$$
$$= 1.921 \text{ mA}$$
$$V_C = V_{CC} - I_C R_C$$
$$= 20 \text{ V} - (1.921 \text{ mA})(5 \text{ k}\Omega)$$
$$= 10.395 \text{ V}$$
$$V_E = -0.6 \text{ V}$$

Therefore,

$$V_{CE} = 10.395 \text{ V} - (-0.6 \text{ V})$$
$$= 10.995 \text{ V}$$

The percent change in I_C as β_{dc} changes from 50 to 100 is

$$\frac{1.921 \text{ mA} - 1.892 \text{ mA}}{1.892 \text{ mA}} \times 100\% = 1.53\%$$

The percent change in V_{CE} is

$$\frac{11.24 \text{ V} - 10.995 \text{ V}}{11.24 \text{ V}} \times 100\% = 2.18\%$$

Keep in mind that the small change in the Q-point resulted from a doubling of β_{dc} and a 0.1 V change in V_{BE}.

Practice Exercise 6–4
Determine how much the Q-point in Figure 6–14 changes over a temperature range where β_{dc} increases from 20 to 75 and V_{BE} decreases from 0.75 V to 0.59 V. The supply voltages are ±10 V.

Emitter-Biased pnp

Figure 6–15 shows a pnp transistor with emitter bias. The basic difference is that the polarities of the supply voltages are reversed from those of the npn. The operation and analysis are basically the same.

FIGURE 6–15
Emitter-biased pnp transistor.

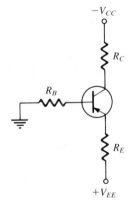

1. In an emitter-biased circuit with an npn silicon transistor, what are the approximate base and emitter voltages with respect to ground?

2. The emitter bias provides much better stability than base bias (T or F).

3. An emitter-biased circuit (npn) has a negative supply voltage of -15 V and an emitter resistor of 10 kΩ. Determine the approximate emitter current.

SECTION REVIEW 6–4

VOLTAGE-DIVIDER BIAS

6–5

The voltage-divider bias is the most widely used arrangement for linear transistor circuits. A bias voltage at the base is developed by a resistive voltage-divider as shown in Figure 6–16. At point A, there are two current paths to ground: one through R_2 and the other through the base-emitter junction of the transistor.

   If the base current is much smaller than the current through R_2, the bias circuit can be viewed as a simplified voltage-divider consisting of R_1 and R_2, as indicated in Figure 6–17(a). If I_B is not small enough to neglect compared to I_2, then the dc *input resistance*, $R_{IN(base)}$, looking in at the base of the transistor must be considered. $R_{IN(base)}$ appears in parallel with R_2, as shown in Figure 6–17(b).

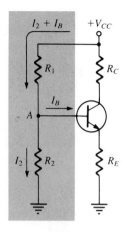

FIGURE 6–16
Voltage-divider bias.

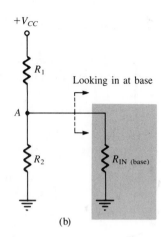

FIGURE 6–17
Simplified voltage-divider.

Input Resistance at the Base

To develop an expression for the dc input resistance at the base of a transistor, we will use the diagram in Figure 6–18. V_{IN} is applied between base and ground, and I_{IN} is the current into the base as shown. By Ohm's law,

$$R_{IN(base)} = \frac{V_{IN}}{I_{IN}} \qquad (6\text{–}14)$$

Kirchhoff's law applied to the base-emitter circuit yields

$$V_{IN} = V_{BE} + I_E R_E$$

With the assumption that $V_{BE} \ll I_E R_E$, the equation reduces to

$$V_{IN} \cong I_E R_E$$

FIGURE 6–18
dc input resistance.

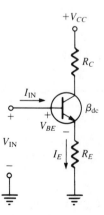

Now, since $I_E \cong \beta_{dc} I_B$,

$$V_{IN} \cong \beta_{dc} I_B R_E$$

Since the input current is the base current ($I_{IN} = I_B$), substitution into equation (6–14) gives

$$R_{IN(base)} \cong \frac{\beta_{dc} I_B R_E}{I_B}$$

So,

$$R_{IN(base)} \cong \beta_{dc} R_E \quad\quad\quad\quad\quad \textbf{(6–15)}$$

EXAMPLE 6–5

Determine the dc input resistance at the base of the transistor in Figure 6–19.
$\beta_{dc} = 125$.

FIGURE 6–19

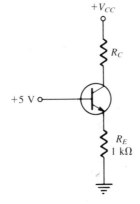

$+V_{CC}$

R_C

$+5$ V

R_E
1 kΩ

Solution

$$R_{IN(base)} \cong \beta_{dc} R_E$$
$$= (125)(1 \text{ k}\Omega)$$
$$= 125 \text{ k}\Omega$$

Analysis of a Voltage-Divider Bias Circuit

A voltage-divider biased npn transistor is shown in Figure 6–20. We begin the analysis by determining the voltage at the base using the voltage-divider method.

$$V_B = \frac{R_2 \| \beta_{dc} R_E}{R_1 + (R_2 \| \beta_{dc} R_E)} V_{CC} \quad\quad\quad \textbf{(6–16)}$$

FIGURE 6–20

A pnp transistor with voltage-divider bias.

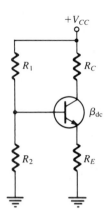

If $\beta_{dc}R_E >> R_2$, then the formula simplifies to

$$V_B \cong \left(\frac{R_2}{R_1 + R_2}\right)V_{CC} \tag{6–17}$$

Once the base voltage is known, the emitter voltage is a V_{BE} drop less (0.7 V for silicon and 0.3 V for germanium).

$$V_E = V_B - V_{BE} \tag{6–18}$$

Now the emitter current is found using Ohm's law.

$$I_E = \frac{V_E}{R_E} \tag{6–19}$$

Once I_E is known, all other circuit values can be found.

$$I_C \cong I_E$$
$$V_C = V_{CC} - I_C R_C$$
$$V_{CE} = V_C - V_E$$

V_{CE} can be expressed in terms of I_E as follows, since $I_C \cong I_E$.

$$V_{CE} \cong V_{CC} - I_E R_C - I_E R_E$$
$$V_{CE} \cong V_{CC} - I_E(R_C + R_E) \tag{6–20}$$

EXAMPLE 6–6

Determine V_{CE} and I_C in Figure 6–21, where $\beta_{dc} = 100$ for the silicon transistor.

Solution

$$R_{IN(base)} = \beta_{dc}R_E = (100)(500 \ \Omega) = 50 \ k\Omega$$

FIGURE 6-21

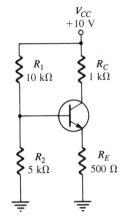

A common rule-of-thumb is that if two resistors are in parallel and one is at least ten times the other, the total resistance is taken to be approximately equal to the smallest value.

In this case, $R_{IN(base)} = 10R_2$, so we can choose to neglect $R_{IN(base)}$, although, of course, some accuracy is lost. You should rework this example taking $R_{IN(base)}$ into account and compare the difference. Proceeding with the analysis, we get

$$V_B \cong \left(\frac{R_2}{R_1 + R_2}\right)V_{CC} = \left(\frac{5\ k\Omega}{15\ k\Omega}\right)10\ V = 3.33\ V$$

So

$$V_E = V_B - V_{BE} = 3.33\ V - 0.7\ V = 2.63\ V$$

and

$$I_E = \frac{V_E}{R_E} = \frac{2.63\ V}{500\ \Omega} = 5.26\ mA$$

Therefore

$$I_C \cong 5.26\ mA$$

and

$$\begin{aligned}V_{CE} &\cong V_{CC} - I_E(R_C + R_E)\\ &= 10\ V - 5.26\ mA(1.5\ k\Omega)\\ &= 2.11\ V\end{aligned}$$

Since $V_{CE} > 0$ V, we know that the transistor is not in saturation.

The following program can be used to compute the values of the voltages and currents in a circuit similar to that in Figure 6-21 for various values of β_{dc}, V_{CC}, and resistances.

```
10  CLS
20  PRINT "THIS PROGRAM COMPUTES VB, VE, VCE, IE, AND IC
30  PRINT "FOR THE CIRCUIT IN FIGURE 6-21 GIVEN R1, R2,
40  PRINT "RC, RE, VCC, AND DC BETA."
50  PRINT:PRINT:PRINT
60  INPUT "TO CONTINUE PRESS 'ENTER'";X
70  CLS
80  INPUT "R1 IN OHMS";R1
90  INPUT "R2 IN OHMS";R2
100 INPUT "RC IN OHMS";RC
110 INPUT "RE IN OHMS";RE
120 INPUT "VCC IN VOLTS";VCC
130 INPUT "DC BETA";B
140 CLS
150 RIN=B*RE
160 IF RIN>=10*R2 THEN R=R2 ELSE R=R2*RIN/(R2+RIN)
170 VB=(R/(R+R1))*VCC
180 VE=VB-.7
190 IE=VE/RE
200 IC=IE
210 VCE=VCC-IC*(RC+RE)
220 IF VE<=0 THEN PRINT "TRANSISTOR IS CUTOFF"
230 IF VE<=0 THEN GOTO 250
240 IF VCE<=0 THEN PRINT "TRANSISTOR IS SATURATED" ELSE 260
250 END
260 PRINT "VB =";VB;"V"
270 PRINT "VE =";VE;"V"
280 PRINT "VCE =";VCE;"V"
290 PRINT "IC =";IC;"A"
300 PRINT "IE =";IE;"A"
```

Stability of Voltage-Divider Bias

First, we will get an *equivalent* base-emitter circuit for Figure 6–20 using *Thevenin's theorem*. Looking out from the base terminal, the bias circuit can be redrawn as shown in Figure 6–22(a). Applying Thevenin's theorem to the circuit left of point A, we get the following results:

$$R_{TH} = \frac{R_1 R_2}{R_1 + R_2}$$

and

$$V_{TH} = \left(\frac{R_2}{R_1 + R_2}\right) V_{CC}$$

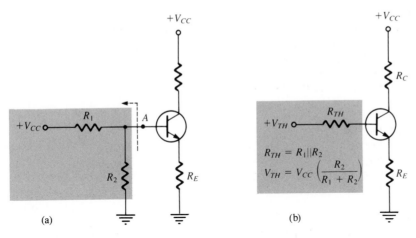

FIGURE 6–22
Thevenizing the bias circuit.

The Thevenin equivalent is shown in Figure 6–22(b). Writing Kirchhoff's equation around the equivalent base-emitter loop gives

$$V_{TH} = I_B R_{TH} + V_{BE} + I_E R_E$$

Substituting I_E/β_{dc} for I_B,

$$V_{TH} = I_E(R_E + R_{TH}/\beta_{dc}) + V_{BE}$$

or

$$I_E = \frac{V_{TH} - V_{BE}}{R_E + R_{TH}/\beta_{dc}}$$

If $R_E \gg R_{TH}/\beta_{dc}$, then

$$I_E \cong \frac{V_{TH} - V_{BE}}{R_E} \qquad\qquad \textbf{(6–21)}$$

This shows that I_E is essentially independent of β_{dc} (notice that β_{dc} does not appear in the equation) under the condition indicated. This is easy to achieve in practice, thus making voltage-divider bias popular because good stability is achieved with a single-polarity supply voltage.

Voltage-Divider Biased pnp

As you know, a pnp transistor requires bias polarities opposite to the npn. This can be accomplished with a *negative* collector supply voltage, as in Figure 6–23(a), or with a *positive* emitter supply voltage, as in Figure 6–23(b). In a schematic diagram, the pnp is

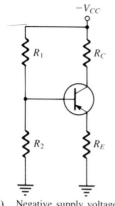

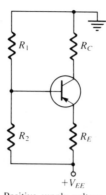

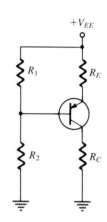

(a)   Negative supply voltage   (b)   Positive supply voltage

FIGURE 6–23
Voltage-divider biased pnp transistor.

FIGURE 6–24

often drawn upside down so that the supply voltage line can be drawn across the top of the schematic and ground at the bottom, as in Figure 6–24. The analysis procedure is basically the same as for an npn circuit, as demonstrated in the following steps with reference to Figure 6–24. Assuming that $\beta_{dc}R_E \gg R_2$, the base voltage is

$$V_B = \left(\frac{R_1}{R_1 + R_2}\right)V_{EE} \qquad\qquad (6\text{–}22)$$

and

$$V_E = V_B + V_{BE} \qquad\qquad (6\text{–}23)$$

So

$$I_E = \frac{V_{EE} - V_E}{R_E} \qquad\qquad (6\text{–}24)$$

and

$$V_C = I_C R_C \qquad\qquad (6\text{–}25)$$
$$V_{EC} = V_E - V_C \qquad\qquad (6\text{–}26)$$

Example 6–7

Find I_C and V_{EC} in Figure 6–25.

Solution

Let's check to see if $R_{IN(base)}$ can be neglected.

$$R_{IN(base)} = \beta_{dc}R_E = (150)(2\ k\Omega) = 300\ k\Omega$$

FIGURE 6–25

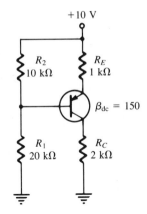

Since 300 kΩ is more than ten times R_2, the condition $\beta_{dc}R_E \gg R_2$ is met. First calculate V_B.

$$V_B = \left(\frac{R_1}{R_1 + R_2}\right)V_{EE} = \left(\frac{20 \text{ k}\Omega}{30 \text{ k}\Omega}\right)10 \text{ V} = 6.67 \text{ V}$$

Then

$$V_E = 6.67 \text{ V} + 0.7 \text{ V} = 7.37 \text{ V}$$

and

$$I_E = \frac{V_{EE} - V_E}{R_E} = \frac{10 \text{ V} - 7.37 \text{ V}}{1 \text{ k}\Omega} = 2.63 \text{ mA}$$

From I_E we can get I_C and V_{CE} as follows:

$$I_C \cong I_E = 2.63 \text{ mA}$$

and

$$V_C = I_C R_C = (2.63 \text{ mA})(2 \text{ k}\Omega) = 5.26 \text{ V}$$

Therefore

$$\begin{aligned} V_{EC} &= V_E - V_C \\ &= 7.37 \text{ V} - 5.26 \text{ V} \\ &= 2.11 \text{ V} \end{aligned}$$

Practice Exercise 6–7
Determine I_C and V_{EC} in Figure 6–25 if $\beta_{dc} = 90$ and $V_{CC} = 15$ V.

**SECTION
REVIEW
6–5**

1. What is the base voltage in Figure 6–26?

FIGURE 6–26

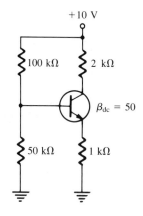

+10 V

100 kΩ   2 kΩ

$\beta_{dc} = 50$

50 kΩ   1 kΩ

2. Determine I_E in Figure 6–26.

3. If the transistor in Figure 6–26 is replaced with one having a β_{dc} of 250, how is the base voltage affected?

COLLECTOR-FEEDBACK BIAS

6–6

Another type of bias circuit, known as *collector feedback* or *voltage feedback,* is shown in Figure 6–27. Notice that the base resistor R_B is connected to the collector rather than to V_{CC}, as in the base bias arrangement discussed earlier. The *collector voltage* provides the bias for the base-emitter junction.

    This *negative feedback* connection provides a very stable Q-point by reducing the effect of variations in β_{dc}. It is also a simple circuit in terms of the components required. The operation of the circuit is as follows. As you know, β_{dc} increases with temperature, which causes I_C to increase. An increase in I_C produces more voltage drop across R_C, lowering the collector voltage. This, in turn, reduces the I_B, which tends to offset the original increase in I_C. The result is that the circuit tends to maintain a stable value of collector current, keeping the Q-point fixed. The reverse action occurs when the temperature decreases. Figure 6–28 illustrates this feedback stabilizing action.

FIGURE 6–27
Collector-feedback bias.

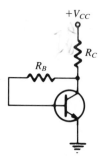

+V_{CC}

R_C

R_B

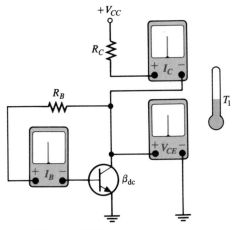

(a) Stabilized at initial temperature

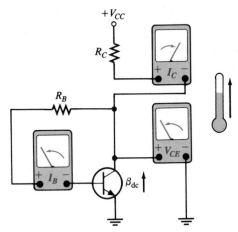

(b) Initial response to temperature rise

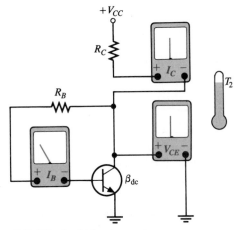

(c) Stabilized at higher temperature

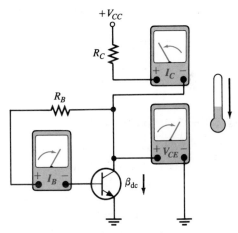

(d) Initial response to temperature drop

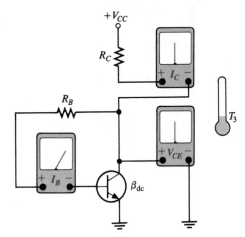

(e) Stabilized at lower temperature

FIGURE 6–28
Collector-feedback stabilization over temperature changes.

Analysis of Collector Feedback

By Ohm's law, the base current can be expressed as

$$I_B = \frac{V_C - V_{BE}}{R_B} \tag{6-27}$$

The collector voltage is

$$V_C \cong V_{CC} - I_C R_C$$

Also,

$$I_B = \frac{I_C}{\beta_{dc}}$$

Substituting in equation (6–27), we get

$$\frac{I_C}{\beta_{dc}} = \frac{V_{CC} - I_C R_C - V_{BE}}{R_B}$$

Rearranging,

$$\frac{I_C R_B}{\beta_{dc}} + I_C R_C = V_{CC} - V_{BE}$$

Solving for I_C,

$$I_C(R_C + R_B/\beta_{dc}) = V_{CC} - V_{BE}$$

$$I_C = \frac{V_{CC} - V_{BE}}{R_C + R_B/\beta_{dc}} \tag{6-28}$$

Example 6–8

Calculate the Q-point values (I_C and V_{CE}) for the circuit in Figure 6–29.

FIGURE 6–29

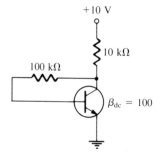

Solution
Using equation (6–28):

$$I_C = \frac{V_{CC} - V_{BE}}{R_C + R_B/\beta_{dc}} = \frac{10 \text{ V} - 0.7 \text{ V}}{10 \text{ k}\Omega + 100 \text{ k}\Omega/100} = 0.845 \text{ mA}$$

Since the emitter is grounded,

$$V_{CE} = V_C \cong V_{CC} - I_C R_C$$
$$= 10 \text{ V} - (0.845 \text{ mA})(10 \text{ k}\Omega)$$
$$= 1.55 \text{ V}$$

SECTION
REVIEW
6–6

1. Explain how an increase in β_{dc} causes a reduction in base current in a collector-feedback circuit.

2. In a certain collector-feedback circuit, $R_B = 50 \text{ k}\Omega$, $R_C = 2 \text{ k}\Omega$, and $V_{CC} = 15 \text{ V}$. If $I_C = 5 \text{ mA}$, what is I_B?

TROUBLESHOOTING BIAS CIRCUITS

6–7

In a biased transistor circuit, either the transistor can fail or one or more resistors in the bias circuit can fail. We will examine several possibilities in this section using the voltage-divider bias arrangement for illustration.

Most circuit failures result from open resistors, internally open transistor leads and junctions, or shorted junctions. In general, these failures will produce an *apparent* cutoff or saturation condition when voltage is measured at the collector. Figure 6–30 indicates certain failures that will produce a collector voltage equal to V_{CC}, thus making it appear that the transistor is in cutoff. (It actually may be.)

Now, let's examine these failures in more detail using specific values. If the resistor R_1 opens, the base is at 0 volts because of R_2 to ground, and the transistor is in cutoff since $I_B = 0$. These voltage levels are shown in Figure 6–31(a).

An open emitter resistor, of course, prevents current in the transistor except for a small I_{CBO}. Therefore, the collector voltage is approximately V_{CC}, the base voltage of 3.33 V is determined by the voltage-divider, and the emitter voltage is 0.7 V below V_B, as indicated in Figure 6–31(b).

An internally open emitter lead or open base-emitter junction will cause the same voltage readings at the collector and base as did the open R_E. However, an internally open

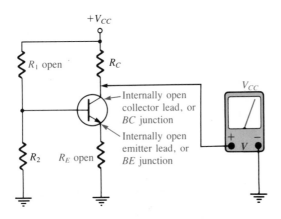

FIGURE 6–30
Typical failures that produce an apparent off condition.

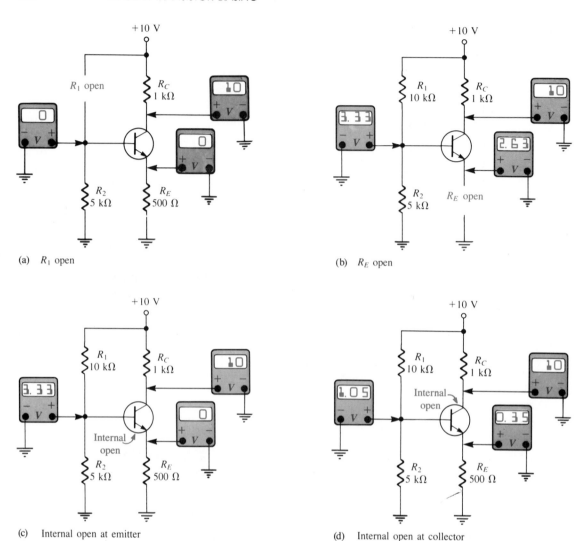

FIGURE 6–31
Open component failures and indications.

emitter causes the external emitter terminal to be at ground potential because there is no current through R_E. This condition is shown in Figure 6–31(c). With an internally open collector lead or open base-collector junction, the collector terminal is at V_{CC}. The base is at 1.05 V because $R_{in(base)}$ is not $\beta_{dc}R_E$ but simply R_E (in series with V_{BE}). The emitter is 0.7 V less than the base, as shown in Figure 6–31(d).

We will now consider the two open-resistor conditions shown in Figure 6–32. An open collector resistor will cause the collector terminal of the transistor to be approximately 0.7 V below the base. You may think that the collector should be at 0 volts, but the positive base voltage of 1.05 V *forward-biases* the base-collector junction so that, when

FIGURE 6–32

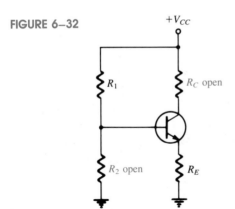

an instrument is connected to the collector, a voltage reading is obtained, as indicated in Figure 6–33(a), which is 0.7 V below the base voltage.

An open R_2 increases the bias voltage on the base because of the voltage-divider action of R_1 and $R_{IN(base)}$. The maximum V_B possible *if saturation does not occur* is developed as follows.

$$R_{IN(base)} = \beta_{dc}R_E = (100)(500 \ \Omega) = 50 \ k\Omega$$

$$V_{B(max)} = \left(\frac{R_{IN(base)}}{R_1 + R_{IN(base)}}\right)V_{CC}$$

$$= \left(\frac{50 \ k\Omega}{60 \ k\Omega}\right)10 \ V$$

$$= 8.33 \ V$$

This excessive base bias *tries* to produce an emitter current of $V_E/R_E = 7.63 \ V/500 \ \Omega = 15.26 \ mA$. Since $I_{C(sat)} = V_{CC}/(R_C + R_E) = 10 \ V/1.5 \ k\Omega = 6.67 \ mA$, the transistor

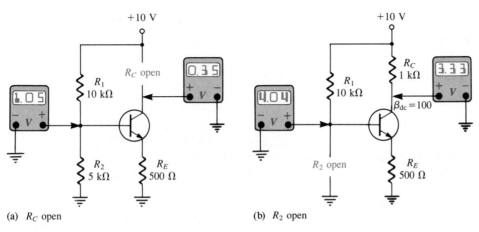

(a) R_C open

(b) R_2 open

FIGURE 6–33
Open component failures.

saturates well before the 15.26 mA value. Thus V_B is limited to 4.04 V, which is 0.7 V greater than the saturation value of V_E, as shown by the following steps.

$$V_{E(sat)} \cong I_{C(sat)}R_E$$
$$= (6.67 \text{ mA})(500 \ \Omega)$$
$$= 3.34 \text{ V}$$
$$V_B = 3.34 \text{ V} + 0.7 \text{ V}$$
$$= 4.04 \text{ V}$$

This condition is illustrated in Figure 6–33(b).

SECTION REVIEW 6–7

1. How do you determine when a transistor is saturated? Cut off?
2. In a voltage-divider biased transistor circuit, you measure V_{CC} at the collector and an emitter voltage 0.7 V less than the base voltage. Is the transistor functioning in cutoff, or is R_E open?
3. What symptoms does an open R_C produce?

A SYSTEM APPLICATION

6–8

A bias circuit is required for each transistor circuit in a receiver system. As mentioned, various types of bias circuits, such as base bias, collector feedback bias, or voltage-divider bias, may be found throughout a given system or in different systems. Voltage-divider bias is one of the more common methods used in bipolar transistor circuits.

Figure 6–34 shows a portion of a typical AM superhet receiver that includes two IF amplifier stages with AGC, a diode detector, and an audio amplifier. Notice that the two IF stages are coupled by tuned transformers. The transformers have variable cores so that the system can be fine tuned to the intermediate frequency of 455 kHz. The audio amplifier is capacitively coupled from the output of the detector by C_7. The potentiometer R_8 is a variable volume control.

Each of the amplifier stages are *voltage-divider* biased. The supply voltage, V_{CC}, is applied to the collector of each of the IF transistors through a tap on the transformer primary and to the collector of the audio transistor through R_{12}. Resistors R_3, R_6, and R_{11} are the emitter resistors. The base bias voltage for each transistor is derived from voltage-dividers formed by R_1 and R_2, R_4 and R_5, and R_9 and R_{10}. The transformer secondaries in the IF stages present shorts to the dc bias voltages and do not affect the bias. Notice that the base bias voltage for the first IF amplifier, Q_1, is derived from the AGC circuit. This permits the gain to be controlled by the level of the detected signal, as you will learn in Chapter 8.

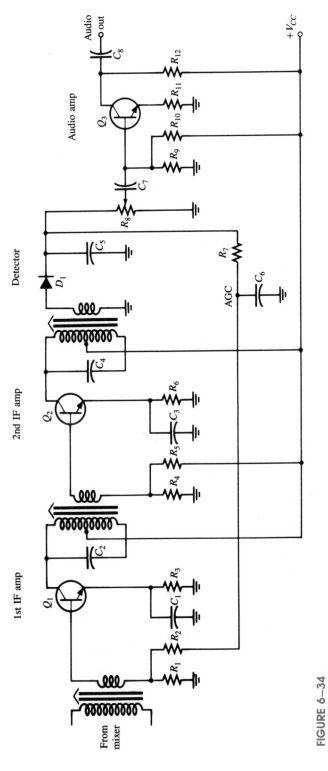

FIGURE 6–34

A portion of a typical superheterodyne receiver.

FORMULAS   **Base bias**

(6–1) $I_B = \dfrac{V_{CC} - V_{BE}}{R_B}$

(6–2) $I_C = \beta_{dc}I_B$

(6–3) $V_{CE} = V_{CC} - I_C R_C$

(6–4) $V_{CE} = V_{CC} - \beta_{dc}I_B R_C$

Emitter bias

(6–5) $V_B \cong 0$

(6–6) $V_E \cong -V_{BE}$

(6–7) $I_E = \dfrac{V_E - V_{EE}}{R_E}$

(6–8) $I_C \cong I_E$

(6–9) $V_C = V_{CC} - I_C R_C$

(6–10) $V_{CE} = V_C - V_E$

(6–11) $I_E = \dfrac{V_{EE} - V_{BE}}{R_E + R_B/\beta_{dc}}$

(6–12) $I_E \cong \dfrac{V_{EE} - V_{BE}}{R_E}, \; R_E \gg R_B/\beta_{dc}$

(6–13) $I_E \cong \dfrac{V_{EE}}{R_E}, \; V_{EE} \gg V_{BE}$

Voltage-divider bias (npn)

(6–14) $R_{IN(base)} = \dfrac{V_{IN}}{I_{IN}}$

(6–15) $R_{IN(base)} \cong \beta_{dc}R_E$

(6–16) $V_B = \dfrac{R_2\|\beta_{dc}R_E}{R_1 + (R_2\|\beta_{dc}R_E)}V_{CC}$

(6–17) $V_B \cong \left(\dfrac{R_2}{R_1 + R_2}\right)V_{CC}, \; \beta_{dc}R_E \gg R_2$

(6–18) $V_E = V_B - V_{BE}$

(6–19) $I_E = \dfrac{V_E}{R_E}$

(6–20) $V_{CE} \cong V_{CC} - I_E(R_C + R_E)$

(6–21) $I_E \cong \dfrac{V_{TH} - V_{BE}}{R_E}, \; R_E \gg R_{TH}/\beta_{dc}$

Voltage-divider bias (pnp)

$$(6\text{--}22) \quad V_B = \left(\frac{R_1}{R_1 + R_2}\right)V_{EE}, \quad \beta_{dc}R_E >> R_2$$

$$(6\text{--}23) \quad V_E = V_B + V_{BE}$$

$$(6\text{--}24) \quad I_E = \frac{V_{EE} - V_E}{R_E}$$

$$(6\text{--}25) \quad V_C = I_C R_C$$

$$(6\text{--}26) \quad V_{EC} = V_E - V_C$$

Collector-feedback bias

$$(6\text{--}27) \quad I_B = \frac{V_C - V_{BE}}{R_B}$$

$$(6\text{--}28) \quad I_C = \frac{V_{CC} - V_{BE}}{R_C + R_B/\beta_{dc}}$$

SUMMARY

- ☐ The purpose of biasing is to establish a proper dc operating point (Q-point).
- ☐ The Q-point of a circuit is defined by specific values for I_C and V_{CE}. These values are sometimes called the *coordinates* of the Q-point.
- ☐ A dc load line passes through the Q-point on a transistor's collector curves and intersects the vertical axis at approximately $I_{C(sat)}$ and the horizontal axis at $V_{CE(off)}$.
- ☐ The linear region of a transistor lies along the load line below saturation and above cutoff.
- ☐ The base bias circuit arrangement has poor stability because its Q-point varies widely with β_{dc}.
- ☐ Emitter bias generally provides good Q-point stability but requires both positive and negative supply voltages.
- ☐ Voltage-divider bias provides good Q-point stability with a single-polarity supply voltage.
- ☐ The dc input resistance looking in at the base of a bipolar transistor is $\beta_{dc}R_E$.
- ☐ Collector-feedback bias provides good stability with negative feedback from collector to base.

SELF-TEST

1. The output (collector voltage) of a biased transistor amplifier is shown in Figure 6–35. Is the transistor biased too close to cutoff or too close to saturation?

FIGURE 6–35

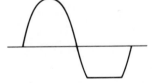

2. What is the Q-point for a base-biased transistor with $I_B = 150 \ \mu A$, $\beta_{dc} = 75$, $V_{CC} = 18$ V, and $R_C = 1$ kΩ?

3. What is the saturation value of collector current in problem 2?

4. What is the cutoff value of V_{CE} in problem 2?

5. Determine I_B, I_C, and V_{CE} for a base-biased silicon transistor circuit with the following values: $\beta_{dc} = 90$, $V_{CC} = 12$ V, $R_B = 22$ kΩ, and $R_C = 100$ Ω.

6. If β_{dc} in problem 5 doubles over temperature, what are the Q-point values?

7. Sketch an emitter-biased npn transistor circuit with the following values: $V_{CC} = 10$ V, $V_{EE} = -10$ V, $R_C = 500$ Ω, $R_E = 1$ kΩ, and $R_B = 47$ kΩ.

8. For the circuit of problem 7, find V_B, V_E, V_C, V_{CE}, I_E, and I_C.

9. What is the resistance looking in at the base of a transistor with $\beta_{dc} = 125$ and $R_E = 300$ Ω?

10. (a) Determine V_B in Figure 6–36.
    (b) If R_E is doubled, what is the value of V_B?

FIGURE 6–36

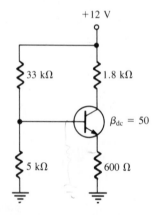

11. (a) Find the Q-point values for Figure 6–36.
    (b) Find the minimum power rating of the transistor in Figure 6–36.

12. If V_{CC} in Figure 6–36 is increased, does I_C increase? Does V_{CE} increase?

13. Sketch a collector-feedback circuit using an npn silicon transistor with $V_{CC} = 12$ V, $R_C = 1.2$ kΩ, and $R_B = 47$ kΩ. Determine the collector voltage and the collector current if $\beta_{dc} = 200$.

14. A certain voltage-divider biased transistor has a V_{CC} of 8 V. If 8 V is measured at the collector terminal and 0 V at the base, what is the most likely failure?

15. The emitter of a voltage-divider biased transistor is internally open. If V_{CC} is 10 V and V_B is 3 V, what voltages will you measure at the collector and the emitter?

Section 6–1

6–1 Limiting on the positive portion of the output waveform of an inverting amplifier indicates
    (a) Transistor in cutoff
    (b) Transistor in saturation
    (c) Distortion of the output
    (d) Linear operation
    (e) (a) and (b)
    (f) (b) and (c)

6–2 The best location of the Q-point on the load line is most probably
    (a) Near saturation
    (b) Near cutoff
    (c) Midway between saturation and cutoff

Section 6–2

6–3 Determine the intercept points of the dc load line on the vertical and horizontal axes of the collector-characteristic curves for the circuit in Figure 6–37.

6–4 Assume that you wish to bias the transistor in Figure 6–37 with $I_B = 20$ μA. To what voltage must you change the V_{BB} supply? What are I_C and V_{CE} at the Q-point, given that $\beta_{dc} = 50$?

6–5 Design a biased transistor circuit using $V_{BB} = V_{CC} = 10$ V, for a Q-point of $I_C = 5$ mA and $V_{CE} = 4$ V. Assume $\beta_{dc} = 100$. The design involves finding R_B, R_C, and the *minimum* power rating of the transistor. (The actual power rating should be greater.) Sketch the circuit.

6–6 Determine whether the transistor in Figure 6–38 is biased in cutoff, saturation, or the linear region. Keep in mind that $I_C = \beta_{dc}I_B$ is valid only in the linear region.

Section 6–3

6–7 You have two base-biased circuits connected for testing. They are identical except that one is biased with a separate V_{BB} source and the other is biased with the base

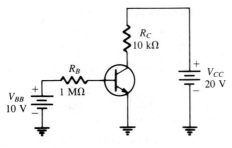

FIGURE 6–37

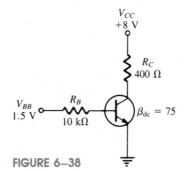

FIGURE 6–38

resistor connected to V_{CC}. Ammeters are connected to measure collector current in each circuit. You vary the V_{CC} supply voltage and observe that the collector current varies in one circuit, but not in the other. In which circuit does the collector current change? Explain your observation.

6–8 The data sheet for a particular transistor specifies a minimum β_{dc} of 50 and a maximum β_{dc} of 125. What range of Q-point values can be expected if an attempt is made to mass-produce the circuit in Figure 6–39? Is this range acceptable if the Q-point must remain in the transistor's linear region?

FIGURE 6–39

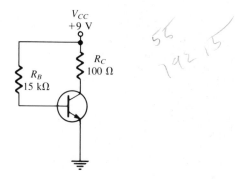

6–9 The base-biased circuit in Figure 6–39 is subjected to a temperature variation from 0°C to 70°C. The β_{dc} decreases by 50 percent at 0°C and increases by 75 percent at 70°C from its nominal value of 110 at 25°C. What are the changes in I_C and V_{CE} over the temperature range of 0°C to 70°C?

Section 6–4

6–10 Analyze the circuit in Figure 6–40 to determine the correct voltage at the transistor terminals with respect to ground.

FIGURE 6–40

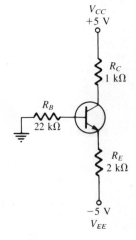

6–11 To what value can R_E in Figure 6–40 be reduced without the transistor going into saturation?

6–12 Taking V_{BE} into account in Figure 6–40, how much will I_E change with a temperature increase from 25°C to 100°C? The V_{BE} is 0.7 V at 25°C and decreases 2.5 mV per degree Celsius. Assume β_{dc} has no effect.

6–13 When can the effect of a change in β_{dc} be neglected in the emitter-biased circuit? Explain why.

6–14 Determine I_C and V_{CE} in the pnp emitter-biased circuit of Figure 6–41.

FIGURE 6–41

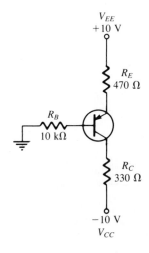

Section 6–5

6–15 What is the minimum value of β_{dc} in Figure 6–42 that makes $R_{IN(base)} \geq 10\,R_2$?

FIGURE 6–42

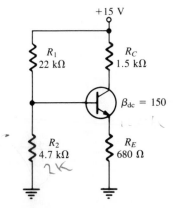

6–16 The bias resistor R_2 in Figure 6–42 is replaced by a 15 kΩ potentiometer. What minimum resistance setting causes saturation?

6–17 If the potentiometer described in problem 6–16 is set at 2 kΩ, what are the values for I_C and V_{CE}?

6–18 Determine all transistor terminal voltages with respect to ground in Figure 6–43. Do not neglect β_{dc} or V_{BE}.

FIGURE 6–43

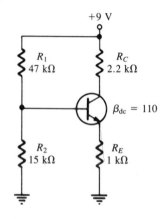

+9 V

R_1
47 kΩ

R_C
2.2 kΩ

$\beta_{dc} = 110$

R_2
15 kΩ

R_E
1 kΩ

6–19 Show the connections required to replace the transistor in Figure 6–43 with a pnp device.

Section 6–6

6–20 Determine V_B, V_C, and I_C in Figure 6–44.

FIGURE 6–44

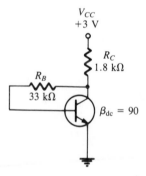

V_{CC}
+3 V

R_C
1.8 kΩ

R_B
33 kΩ

$\beta_{dc} = 90$

6–21 What value of R_C can be used to decrease I_C in problem 6–20 by 25 percent?

6–22 What is the minimum power rating for the transistor in problem 6–21?

Section 6–7

6–23 Determine the most probable failures, if any, in each circuit of Figure 6–45, based on the indicated measurements.

6–24 Assume the emitter becomes shorted to ground in Figure 6–46 by a solder splash or stray wire clipping. What do the meters read?

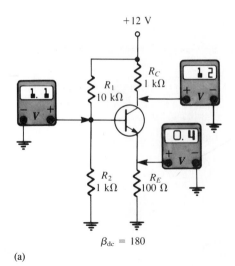

(a)

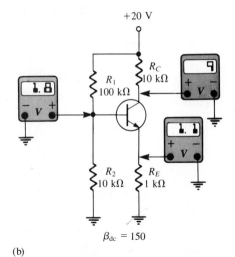

(b)

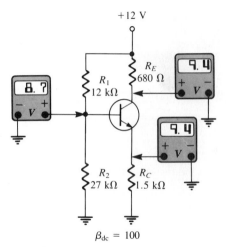

(c)

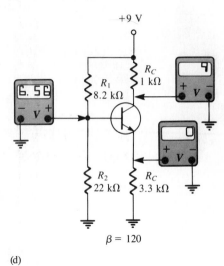

(d)

FIGURE 6–45

FIGURE 6–46

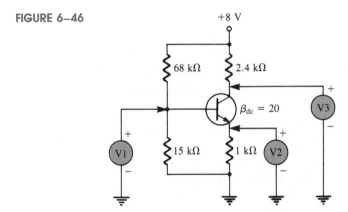

ANSWERS TO SECTION REVIEWS

Section 6–1
1. Limiting (distortion).
2. Saturation, cutoff.

Section 6–2
1. The dc operating point of a transistor is defined by the values of I_C and V_{CE}.
2. $I_C = 0$, cutoff.
3. True.

Section 6–3
1. I_C, V_{CE}.
2. Q-point varies with β_{dc}.

Section 6–4
1. $V_B \cong 0$ V, $V_E \cong -0.7$ V.
2. True.
3. 1.43 mA.

Section 6–5
1. 2 V.
2. 1.3 mA.
3. V_B increases.

Section 6–6
1. I_C increases with β_{dc}, causing a reduction in V_C and therefore less voltage across R_B, thus less I_B.
2. 86 μA.

Section 6–7
1. $V_{CE} \cong 0$, $V_{CE} = V_{CC}$.
2. R_E open.
3. V_C is 0.7 V less than V_B.

Example 6–1: $I_{CQ} = 18.6$ mA, $V_{CEQ} = 5$ V, $I_{b(\text{peak})} = 50$ μA

Example 6–2: $\Delta I_C = 150\%$, $\Delta V_{CE} = 59\%$

Example 6–3: $I_E = 4.33$ mA, $I_C \cong 4.33$ mA, $V_{CE} = 12.7$ V

Example 6–4: $\Delta I_C = 5.68\%$, $\Delta V_{CE} = 6.46\%$

Example 6–7: $I_C \cong 4.66$ mA, $V_{EC} = 1.02$ V

Field-Effect Transistors and Biasing

In this chapter you will learn

☐ The basic construction and operation of junction field-effect transistors (JFETs)

☐ How to use the basic JFET data sheet parameters

☐ Why FETs are voltage-controlled devices

☐ The construction and operation of metal oxide semiconductor field-effect transistors (MOSFETs)

☐ The characteristics of depletion-enhancement and enhancement MOSFETs

☐ The characteristics of VMOS

☐ How to use the basic MOSFET data sheet parameters

☐ How to bias FETs

☐ How FETs are used in a specific system application

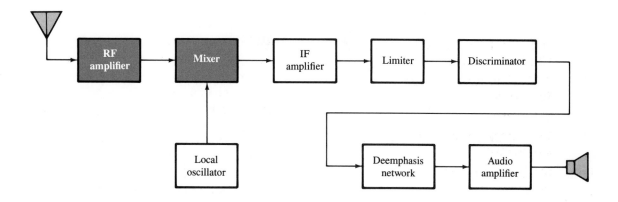

Bipolar transistors were covered in previous chapters. Now it is time to turn our attention to the second major transistor category, field-effect transistors (FETs). FETs are *unipolar* devices because, unlike the bipolar transistor, they operate with only majority carriers. There are two main types of FETs: the *junction field-effect transistor* (JFET) and the *metal oxide semiconductor field-effect transistor* (MOSFET). We will discuss both types.

Recall that the bipolar transistor is a *current-controlled* device; that is, the base current controls the amount of collector current. The FET is different: It is a *voltage-controlled* device, where the voltage at one of the terminals controls the amount of current through the device. FETs are often used rather than bipolar transistors in AM and FM receivers as the gain elements in the RF amplifier and mixer circuits because of certain characteristics we will study in this chapter. As mentioned, AM receivers often do not have an RF amplifier, but quality FM receivers almost universally have at least one RF amplifier stage. The high input impedance, the square-law characteristic, and the availability of dual-gate FETs make them an important device in many communications systems.

THE JUNCTION FIELD-EFFECT TRANSISTOR (JFET)

7–1

Depending on their structure, JFETs fall into either of two categories, *n-channel* or *p-channel*. Figure 7–1(a) shows the basic structure of an n-channel JFET. Wire leads are connected to each end of the n-channel; the *drain* is at the upper end and the *source* is at the lower end. Two p-type regions are diffused in the n-type material to form a channel, and *both* p-type regions are connected to the gate lead. In the remaining structure diagrams, the interconnection of *both* p-type regions is omitted for simplicity, with a connection to only one shown. A p-channel JFET is shown in Figure 7–1(b).

FIGURE 7–1

Structure of the two types of JFET.

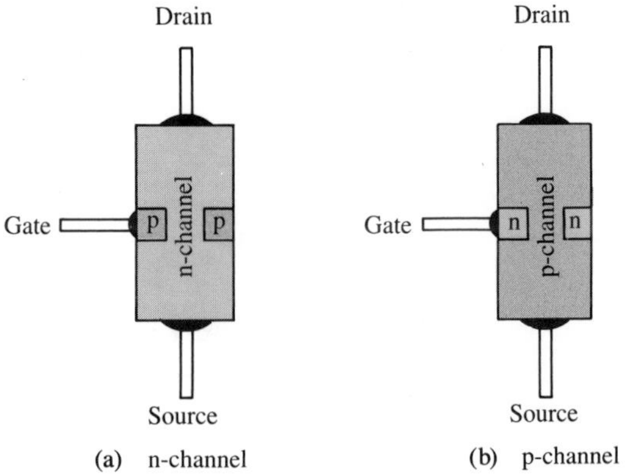

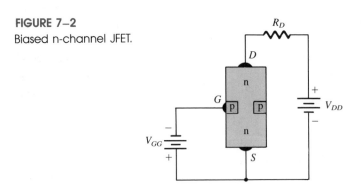

(a)   n-channel   (b)   p-channel

Basic Operation

To illustrate the operation of a JFET, bias voltages are shown applied to an n-channel device in Figure 7–2. V_{DD} provides a drain-to-source voltage and supplies current from drain to source. V_{GG} sets the reverse-bias voltage between the gate and the source, as shown.

FIGURE 7–2

Biased n-channel JFET.

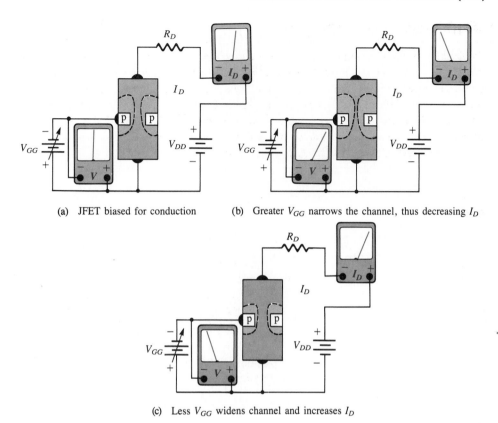

(a) JFET biased for conduction   (b) Greater V_{GG} narrows the channel, thus decreasing I_D

(c) Less V_{GG} widens channel and increases I_D

FIGURE 7–3
Effects of V_{GG} on channel width and drain current.

The JFET is always operated with the gate-to-source pn junction reverse-biased. Reverse-biasing of the gate-source junction with a negative gate voltage produces a *deple-tion* region in the n-channel and thus increases its resistance. The channel width can be controlled by varying the gate voltage, whereby the amount of drain current, I_D, can also be controlled. This concept is illustrated in Figure 7–3. The shaded areas represent the depletion region created by the reverse bias. It is wider toward the drain end of the channel because the reverse-bias voltage between the gate and the drain is greater than that be-tween the gate and the source. We will discuss JFET characteristic curves and some important parameters in the next section.

JFET Symbols

The schematic symbols for both n-channel and p-channel JFETs are shown in Figure 7–4. Notice that the arrow on the gate points "in" for n-channel and "out" for p-channel.

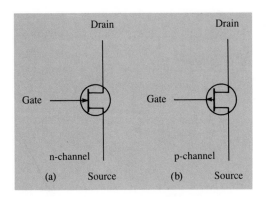

(c) Typical JFET packages (Courtesy of Motorola)

FIGURE 7–4
JFET schematic symbols and packages.

---

1. Name the three terminals of a JFET.
2. An n-channel JFET requires a (positive, negative, or 0) V_{GS}.
3. How is the drain current controlled in a JFET?

JFET CHARACTERISTICS AND PARAMETERS

7–2

Drain Curves and Pinch-Off

First consider the case where the gate-to-source voltage is 0 ($V_{GS} = 0$ V). This is produced by shorting the gate to the source, as in Figure 7–5(a). As V_{DD} (and thus V_{DS}) is increased from 0, I_D will increase proportionally, as shown in the graph of Figure 7–5(b) between points A and B. In this region, the channel resistance is essentially constant because the depletion region is not large enough to have significant effect. This is called the *ohmic region* because V_{DS} and I_D are related by Ohm's law.

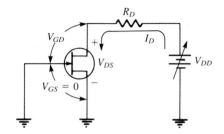

(a)   JFET with $V_{GS} = 0$ V and a variable V_{DS}

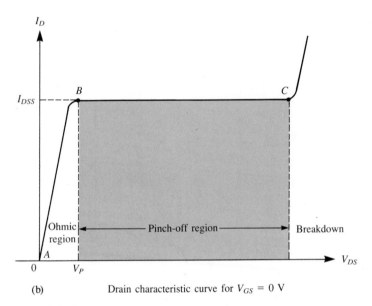

(b)   Drain characteristic curve for $V_{GS} = 0$ V

FIGURE 7–5
Generation of drain characteristic curve.

At point B, the curve levels off and I_D becomes a relatively constant value called I_{DSS}. It is at this point that the reverse-bias voltage across the *gate-to-drain* junction (V_{GD}) produces a depletion region sufficient to narrow the channel so that its resistance begins to *increase* significantly. The value of V_{GD} at this point is called the *pinch-off voltage V_P. In the case where the gate bias voltage is zero, $-V_P = V_{DS}$ at the point of pinch-off, because V_{GS} and V_{GD} are equal.* In general, however,

$$V_P = V_{GS} - V_{DS(P)} \tag{7–1}$$

where $V_{DS(P)}$ is the value of V_{DS} at pinch-off for a given value of V_{GS}.

V_P *is a constant value for a given JFET* and represents a fixed parameter. The value of V_{DS} at pinch-off is a variable that depends on the gate-to-source bias voltage, V_{GS}. As you can see from the curve for $V_{GS} = 0$ V in Figure 7–5(b), a continued increase in V_{DS}

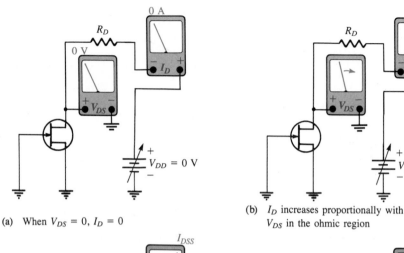

(a) When $V_{DS} = 0$, $I_D = 0$

(b) I_D increases proportionally with V_{DS} in the ohmic region

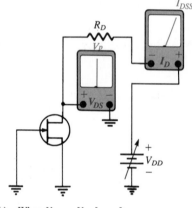

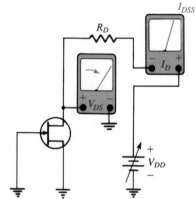

(c) When $V_{DS} = V_P$, $I_D = I_{DSS}$

(d) As V_{DS} increases further, I_D remains at I_{DSS}

FIGURE 7–6
JFET action that produces the characteristic curve for $V_{GS} = 0$ V.

above point B produces an essentially constant I_D equal to a specific value called I_{DSS}. I_{DSS} *is the maximum value of* I_D *when* $V_{GS} = 0$. At point C, breakdown occurs and I_D increases rapidly with irreversible damage to the device very likely. JFETs are always operated below the breakdown point and within the pinch-off region (between points B and C). The JFET action that produces the characteristic curve for $V_{GS} = 0$ V is illustrated in Figure 7–6.

Now consider a negative gate bias voltage, for example $V_{GS} = -1$ V, as shown in Figure 7–7(a). As V_{DD} is increased from 0, pinch-off occurs at a lower value of V_{DS}, as shown in Figure 7–7(b). The reason for this is that the pinch-off voltage V_P is constant, and therefore, for a certain negative gate voltage, the drain voltage must only reach a value sufficient to make the gate-to-drain voltage equal to V_P in order to produce pinch-

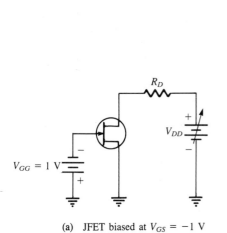
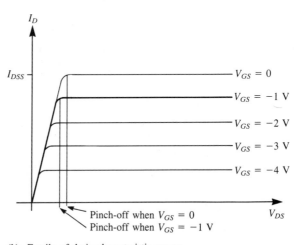

(a) JFET biased at $V_{GS} = -1$ V

(b) Family of drain characteristic curves

FIGURE 7–7

Pinch-off occurs at a lower V_{DS} when V_{GS} goes from 0 V to -1 V, and to successively more negative values.

off. As V_{GS} is set to increasingly negative values, a family of characteristic curves is produced, as shown in Figure 7–7(b).

V_{GS} Controls I_D

Increasingly negative values of V_{GS} cause pinch-off to occur at successively lower values of V_{DS}, resulting in lower values of I_D. So, the amount of drain current is controlled by V_{GS}. I_{DSS}, the maximum drain current, occurs for $V_{GS} = 0$ V and decreases as V_{GS} is made more negative (n-channel), as shown in Figure 7–8. A p-channel device requires positive values of V_{GS}.

Cutoff

As you have seen, for an n-channel JFET, the more negative V_{GS} is, the smaller I_D in the pinch-off region becomes. When V_{GS} is made sufficiently negative, I_D is reduced to 0. This is caused by the widening of the depletion region to a point where it completely closes the channel. The value of V_{GS} at the cutoff point is designated $V_{GS(off)}$.

Equation (7–1)' indicates that for any given n-channel JFET, cutoff occurs when $V_{GS} = V_P$. Since the pinch-off voltage V_P is a constant for a given JFET, when $V_{GS} = V_P$ the drain-to-source voltage $V_{DS(P)}$ must be 0. Since there is *no voltage drop* between the drain and source, I_D *must* be 0. Even though V_{DS} may increase above 0 V, I_D remains essentially constant at near 0 A. Cutoff is illustrated in Figure 7–9.

Do not confuse *cutoff* with *pinch-off*. The *pinch-off voltage V_P* is the value of the gate-to-drain voltage V_{GD} at which the *drain current reaches a constant value for a given value of V_{GS}*. The *cutoff voltage $V_{GS(off)}$* is the value of V_{GS} at which the *drain current is*

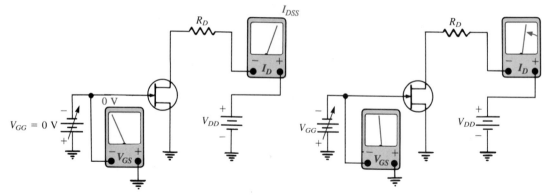

(a) $V_{GS} = 0$ V, $I_D = I_{DSS}$

(b) V_{GS} negative, I_D decreases

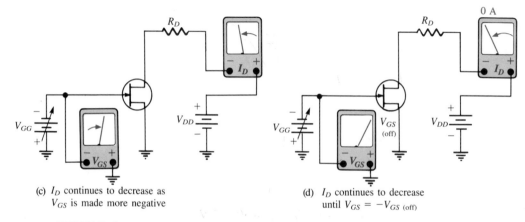

(c) I_D continues to decrease as V_{GS} is made more negative

(d) I_D continues to decrease until $V_{GS} = -V_{GS \text{ (off)}}$

FIGURE 7–8
V_{GS} controls I_D.

FIGURE 7–9
JFET action at cutoff.

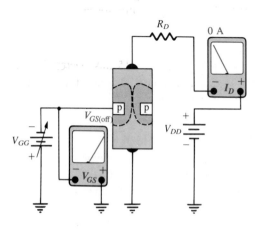

$0.$ I_D is 0 only when the magnitude of V_{GS} is equal to or greater than the magnitude of V_P. I_D is nonzero for less negative values of V_{GS}.

EXAMPLE 7–1

For the JFET in Figure 7–10, $V_P = -8$ V and $I_{DSS} = 12$ mA.

(a) Determine the value of V_{DS} at which pinch-off begins.

(b) If the gate is grounded, what is the value of I_D for $V_{DD} = 12$ V when V_{DS} is above pinch-off?

FIGURE 7–10

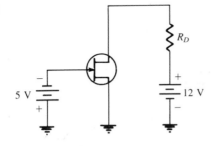

Solution

(a) By rearranging equation (7–1), we get

$$V_{DS(P)} = V_{GS} - V_P$$

Substituting $V_P = -8$ V and $V_{GS} = -5$ V,

$$V_{DS(P)} = -5 \text{ V} - (-8 \text{ V}) = 3 \text{ V}$$

This is the value of V_{DS} at the beginning of pinch-off.

(b) When $V_{GS} = 0$ V, $I_D = I_{DSS} = 12$ mA for any value of V_{DS} between pinch-off (8 V) and breakdown.

Practice Exercise 7–1

In Figure 7–10, $V_P = -6$ V and $I_{DSS} = 15$ mA. Determine V_{DS} at pinch-off. What is I_D when $V_{DD} = 10$ V if the gate is grounded and V_{DS} is above pinch-off?

V_P and $V_{GS(off)}$ Are Equivalent

I_D is 0 when $V_{GS} = V_P$. Because a 0 drain current corresponds to an *off* condition, the magnitude of V_P is equivalent to the magnitude of $V_{GS(off)}$. Most FET data sheets give a value only for $V_{GS(off)}$ and not for V_P. But once we know $V_{GS(off)}$, we also know V_P. For example, if $V_{GS(off)} = -5$ V, then $V_P = -5$ V.

EXAMPLE 7–2

A particular p-channel JFET has a $V_{GS(off)} = 4$ V. What is I_D when $V_{GS} = 6$ V?

Solution

The p-channel JFET requires a positive gate-to-source voltage. The more positive the voltage, the less the drain current. When $V_{GS} = 4$ V, I_D is 0. Any further increase in V_{GS} keeps the JFET cut off, so I_D remains 0.

JFET Transfer Characteristic

You have learned that a range of V_{GS} values from 0 to $V_{GS(off)}$ controls the amount of drain current. For an n-channel JFET, $V_{GS(off)}$ is negative, and for a p-channel JFET, $V_{GS(off)}$ is positive. Because V_{GS} does control I_D, the relationship between these two quantities is very important. Figure 7–11 is a typical transfer characteristic curve which illustrates graphically the relationship between V_{GS} and I_D.

FIGURE 7–11
JFET transfer characteristic curve.

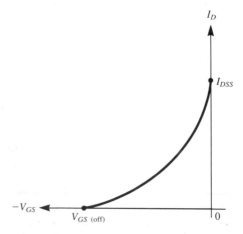

Notice that the bottom end of the curve is at a point on the V_{GS}-axis equal to $V_{GS(off)}$, and the top end of the curve is at a point on the I_D-axis equal to I_{DSS}. This curve, of course, shows that the operating limits of a JFET are

$$I_D = 0 \text{ when } V_{GS} = V_{GS(off)}$$
$$I_D = I_{DSS} \text{ when } V_{GS} = 0$$

The transfer characteristic curve can be developed from the drain characteristic curves by plotting values of I_D for the values of V_{GS} taken from the family of drain curves in the pinch-off region, as illustrated in Figure 7–12 for a specific set of curves. Notice that each point on the transfer characteristic curve corresponds to specific values of V_{GS} and I_D on the drain curves. For example, when $V_{GS} = -2$ V, $I_D = 7.2$ mA. Also, for this specific JFET, $V_{GS(off)} = -5$ V and $I_{DSS} = 12$ mA.

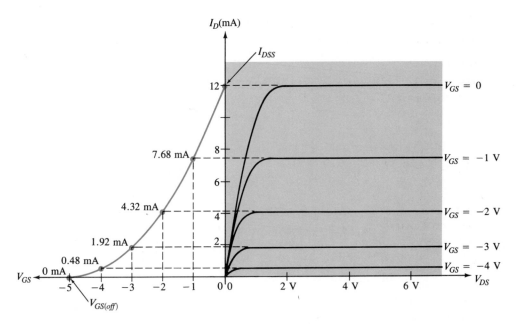

FIGURE 7–12

Example of the development of a JFET transfer characteristic curve (left) from the JFET drain characteristic curves (right).

A JFET transfer characteristic curve is actually parabolic in shape and can therefore be expressed mathematically as

$$I_D = I_{DSS}\left[1 - \frac{V_{GS}}{V_{GS(off)}}\right]^2 \qquad (7\text{–}2)$$

With equation (7–2), I_D can be determined for any V_{GS} if $V_{GS(off)}$ and I_{DSS} are known. These quantities are usually available from the data sheet for a given JFET. Notice the squared term in the equation. Because of its form, a parabolic relationship is known as a *square law*, and therefore, JFETs and MOSFETs are often referred to as *square-law devices*.

EXAMPLE 7–3

The data sheet for a certain JFET indicates that $I_{DSS} = 15$ mA and $V_{GS(off)} = -5$ V. Determine the drain current for $V_{GS} = 0$ V, -1 V, and -4 V.

Solution

For $V_{GS} = 0$ V, $I_D = I_{DSS} = 15$ mA. For $V_{GS} = -1$ V, we use equation (7–2):

$$I_D = I_{DSS}\left[1 - \frac{V_{GS}}{V_{GS(off)}}\right]^2 = (15 \text{ mA})\left(1 - \frac{-1 \text{ V}}{-5 \text{ V}}\right)^2$$

$$= (15 \text{ mA})(1 - 0.2)^2 = (15 \text{ mA})(0.64)$$

$$= 9.6 \text{ mA}$$

For $V_{GS} = -4$ V:

$$I_D = (15 \text{ mA})\left(1 - \frac{-4 \text{ V}}{-5 \text{ V}}\right)^2$$

$$= (15 \text{ mA})(1 - 0.8)^2$$

$$= (15 \text{ mA})(0.04)$$

$$= 0.6 \text{ mA}$$

The following program computes I_D for JFETs and DE MOSFETS, given the data sheet parameters I_{DSS} and $V_{GS(off)}$ and a value of V_{GS}. You can input any number of V_{GS} values; the computer will tabulate the corresponding I_D values.

```
10   DIM VBIAS(100)
20   CLS
30   PRINT "COMPUTATION OF DRAIN CURRENT FOR JFETS AND MOSFETS"
40   PRINT
50   PRINT "THE FOLLOWING INPUTS ARE REQUIRED:"
60   PRINT "(1) IDSS FROM THE DATA SHEET"
70   PRINT "(2) VGS(OFF) FROM THE DATA SHEET"
80   PRINT "(3) VGS AS REQUIRED"
90   PRINT:PRINT:PRINT
100  INPUT "TO CONTINUE PRESS 'ENTER'";X
110  CLS
120  INPUT "VALUE OF IDSS IN MILLIAMPS";IDSS
130  INPUT "VALUE OF VGS(OFF) IN VOLTS";VGSOFF
140  INPUT "FOR HOW MANY VALUES OF VGS DO YOU WANT ID COMPUTED";N
150  CLS
160  FOR A=1 TO N
170  INPUT "VALUE OF VGS";VBIAS(A)
180  NEXT
190  CLS
200  PRINT "VGS (VOLTS)", "ID (MA)":PRINT
210  FOR A=1 TO N
220  ID=IDSS*(1-VBIAS(A)/VGSOFF)[2
230  PRINT VBIAS(A),ID
240  NEXT
```

JFET Forward Transconductance

The forward transconductance, g_m, is the change in drain current for a given change in gate-to-source voltage with the drain-to-source voltage constant. It is expressed as a ratio.

$$g_m = \frac{\Delta I_D}{\Delta V_{GS}} \tag{7-3}$$

Other common designations for this parameter are g_{fs} and y_{fs} (forward transfer admittance).

Because the transfer characteristic (transconductance) curve for a JFET is nonlinear, g_m varies in value depending on the location on the curve as set by V_{GS}. g_m has a greater

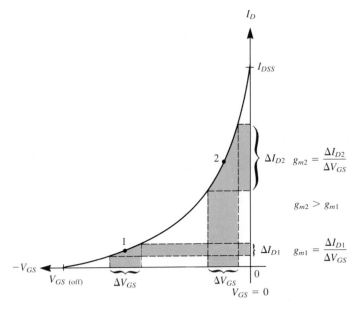

FIGURE 7–13
g_m varies depending on the bias point (V_{GS}).

value near the top of the curve (near $V_{GS} = 0$) than it does near the bottom (near $V_{GS(off)}$), as illustrated in Figure 7–13. A data sheet normally gives the value of g_m at $V_{GS} = 0$ V (g_{m0}). For example, the data sheet for the 2N3823 JFET specifies a minimum g_{m0} of 3500 μS with $V_{DS} = 15$ V.

Given g_{m0}, you can calculate an approximate value for g_m at any point on the transfer characteristic curve using the following formula.

$$g_m = g_{m0}\left[1 - \frac{V_{GS}}{V_{GS(off)}}\right] \tag{7–4}$$

EXAMPLE 7–4

The following information is included on the data sheet for a certain JFET: $I_{DSS} = 20$ mA, $V_{GS(off)} = -8$ V, and $g_{m0} = 4000$ μS. Determine the forward transconductance for $V_{GS} = -4$ V, and find I_D at this point.

Solution
First, g_m is found using equation (7–4).

$$g_m = g_{m0}\left[1 - \frac{V_{GS}}{V_{GS(off)}}\right]$$

$$= (4000 \ \mu S)\left(1 - \frac{-4 \text{ V}}{-8 \text{ V}}\right)$$

$$= 2000 \ \mu S$$

Next, using equation (7–2), I_D at $V_{GS} = -4$ V is calculated.

$$I_D = I_{DSS}\left[1 - \frac{V_{GS}}{V_{GS(off)}}\right]^2$$

$$= (20 \text{ mA})\left(1 - \frac{-4 \text{ V}}{-8 \text{ V}}\right)^2$$

$$= 5 \text{ mA}$$

Practice Exercise 7–4
A given JFET has the following characteristics: $I_{DSS} = 12$ mA, $V_{GS(off)} = -5$ V, and $g_{m0} = 3000$ μS. Find g_m and I_D when $V_{GS} = -2$ V.

Input Resistance and Capacitance

A JFET operates with its gate-source junction reverse-biased. Therefore, the input resistance at the gate is very high. This high input resistance is one advantage of the JFET over the bipolar transistor. (Recall that a bipolar transistor operates with a forward-biased base-emitter junction.) JFET data sheets often specify the input resistance by giving a value for the *gate reverse current* I_{GSS} at a certain gate-to-source voltage. The input resistance can then be determined using the following equation.

$$R_{\text{IN}} = \left|\frac{V_{GS}}{I_{GSS}}\right| \tag{7–5}$$

For example, the 2N3970 data sheet lists a maximum I_{GSS} of 250 pA for $V_{GS} = 20$ V at 25°C. I_{GSS} increases with temperature, so the input resistance decreases.

The input capacitance C_{iss} of a JFET is considerably greater than that of a bipolar transistor because the JFET operates with a reverse-biased pn junction. Recall that a reverse-biased pn junction acts as a capacitor whose capacitance depends on the amount of reverse voltage.

EXAMPLE 7–5

A certain JFET has an I_{GSS} of 1 nA for $V_{GS} = -20$ V. Determine the input resistance.

Solution

$$R_{\text{IN}} = \left|\frac{V_{GS}}{I_{GSS}}\right| = \frac{20 \text{ V}}{1 \text{ nA}} = 20,000 \text{ M}\Omega$$

Drain-to-Source Resistance

You learned from the drain characteristic curve that, above pinch-off, the drain current is relatively constant over a range of drain-to-source voltages. Therefore, a large change in

V_{DS} produces only a very small change in I_D. The ratio of these changes is the drain-to-source resistance of the device, r_{ds}.

$$r_{ds} = \frac{\Delta V_{DS}}{\Delta I_D} \qquad (7\text{-}6)$$

Data sheets often specify this parameter as output conductance, g_{os}, or output admittance, y_{os}. Typical values for r_{ds} are on the order of several thousand ohms.

SECTION REVIEW 7-2

1. The drain-to-source voltage at the pinch-off point of a particular JFET is 7 V. If the gate-to-source voltage is 0, what is V_P?

2. The V_{GS} of a certain n-channel JFET is increased negatively. Does the drain current increase or decrease?

3. What value must V_{GS} have to produce cutoff in a p-channel JFET with a $V_P = 3$ V?

JFET BIASING

7-3

Using some of the FET parameters discussed in the previous sections, we will now see how to dc-bias JFETs. The purpose of biasing is to select a proper dc gate-to-source voltage to establish a desired value of drain current.

Self-Biasing a JFET

Recall that a JFET must be operated such that the gate-source junction is always reverse-biased. This condition requires a negative V_{GS} for an n-channel JFET and a positive V_{GS} for a p-channel JFET. This can be achieved using the *self-bias* arrangements shown in Figure 7-14. Notice that the gate is biased at approximately 0 V by resistor R_G connected to ground. The reverse leakage current I_{GSS} does produce a very small voltage across R_G,

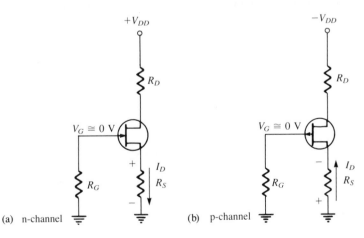

FIGURE 7-14
Self-biased JFETS ($I_S = I_D$ in all FETs).

but this can be neglected in most cases, and it can be assumed that R_G has no voltage drop across it.

For the n-channel JFET in Figure 7–14(a), I_D produces a voltage drop across R_S and makes the source positive with respect to ground. Since $V_G = 0$ and $V_S = I_D R_S$, the gate-to-source voltage is

$$V_{GS} = V_G - V_S = 0 - I_D R_S$$

so

$$V_{GS} = -I_D R_S \qquad (7\text{–}7)$$

For the p-channel JFET shown in Figure 7–14(b), the current through R_S produces a negative voltage at the source and therefore

$$V_{GS} = +I_D R_S \qquad (7\text{–}8)$$

In the following analysis, the n-channel JFET is used for illustration. Keep in mind that analysis of the p-channel JFET is the same except for opposite-polarity voltages. The drain voltage with respect to ground is determined as follows.

$$V_D = V_{DD} - I_D R_D \qquad (7\text{–}9)$$

Since $V_S = I_D R_S$, the drain-to-source voltage is

$$V_{DS} = V_D - V_S$$
$$V_{DS} = V_{DD} - I_D(R_D + R_S) \qquad (7\text{–}10)$$

EXAMPLE 7–6

Find V_{DS} and V_{GS} in Figure 7–15, given that $I_D = 5$ mA.

FIGURE 7–15

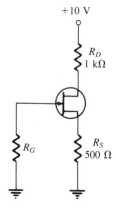

+10 V

R_D
1 kΩ

R_G

R_S
500 Ω

Solution

$$V_S = I_D R_S = (5 \text{ mA})(500 \ \Omega) = 2.5 \text{ V}$$
$$V_D = V_{DD} - I_D R_D = 10 \text{ V} - (5 \text{ mA})(1 \text{ k}\Omega)$$
$$= 10 \text{ V} - 5 \text{ V} = 5 \text{ V}$$

Therefore

$$V_{DS} = V_D - V_S = 5\ \text{V} - 2.5\ \text{V} = 2.5\ \text{V}$$

Since $V_G = 0\ \text{V}$,

$$V_{GS} = V_G - V_S = -2.5\ \text{V}$$

Practice Exercise 7–6
Determine V_{DS} and V_{GS} in Figure 7–15 when $I_D = 8$ mA. Assume that $R_D = 860\ \Omega$, $R_S = 390\ \Omega$, and $V_{DD} = 12\ \text{V}$.

Setting a Bias Point

The basic approach to establishing a JFET bias point is to determine I_D for a desired value of V_{GS}, then calculate the required value of R_S using the relationship derived from equation (7–7) and stated in equation (7–11). The vertical lines indicate an absolute value (no sign).

$$R_S = \left| \frac{V_{GS}}{I_D} \right| \qquad (7\text{–}11)$$

For a desired value of V_{GS}, I_D can be determined in either of two ways: from the transfer characteristic curve for the particular JFET, or from equation (7–2) using I_{DSS} and $V_{GS(off)}$ from the JFET data sheet. The next two examples illustrate these procedures.

EXAMPLE 7–7
Determine the value of R_S required to self-bias a JFET having the transfer characteristic curve shown in Figure 7–16 at a $V_{GS} = -5\ \text{V}$.

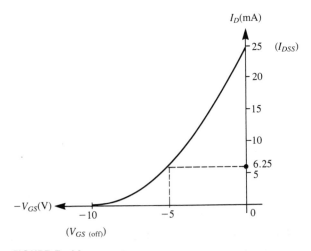

FIGURE 7–16

Solution

From the graph, $I_D = 6.25$ mA when $V_{GS} = -5$ V. Calculating R_S, we get

$$R_S = \left| \frac{V_{GS}}{I_D} \right| = \frac{5 \text{ V}}{6.25 \text{ mA}} = 800 \text{ }\Omega$$

EXAMPLE 7–8

Determine the value of R_S required to self-bias an n-channel JFET with $I_{DSS} = 25$ mA and $V_{GS(off)} = -10$ V. V_{GS} is to be -5 V.

Solution

I_D is calculated using equation (7–2).

$$I_D = I_{DSS} \left[1 - \frac{V_{GS}}{V_{GS(off)}} \right]^2$$

$$= (25 \text{ mA}) \left(1 - \frac{-5 \text{ V}}{-10 \text{ V}} \right)^2$$

$$= (25 \text{ mA})(1 - 0.5)^2$$

$$= 6.25 \text{ mA}$$

Now, R_S can be determined.

$$R_S = \left| \frac{V_{GS}}{I_D} \right| = \frac{5 \text{ V}}{6.25 \text{ mA}} = 800 \text{ }\Omega$$

The result is the same as in Example 7–7 because the same JFET was used.

Practice Exercise 7–8

Find the value of R_S required to self-bias an n-channel JFET with $I_{DSS} = 18$ mA and $V_{GS(off)} = -8$ V. $V_{GS} = -4$ V.

Midpoint Bias

It is often desirable to bias a JFET near the midpoint of its transfer characteristic curve where $I_D = I_{DSS}/2$. Under signal conditions, midpoint bias allows a maximum amount of drain current swing between I_{DSS} and 0. Using equation (7–2), we see that I_D is approximately one-half of I_{DSS} when $V_{GS} = V_{GS(off)}/4$.

$$I_D = I_{DSS}(1 - 0.25)^2 = I_{DSS}(0.75)^2 = 0.5625 \, I_{DSS}$$

So, by selecting $V_{GS} = V_{GS(off)}/4$, we get close to a midpoint bias in terms of I_D. As you can see in the above calculation, I_D is slightly more than one-half of I_{DSS} but close enough to approximate a midpoint bias.

To set the drain voltage at midpoint ($V_D = V_{DD}/2$), a value of R_D is selected to produce the desired voltage drop. R_G is chosen arbitrarily large to prevent loading on the driving stage in a cascaded amplifier arrangement. Example 7–9 illustrates these concepts.

EXAMPLE 7-9

Select resistor values in Figure 7–17 to set up an approximate midpoint bias. The JFET parameters are $I_{DSS} = 15$ mA and $V_{GS(off)} = -8$ V. V_D should be 6 V (one-half of V_{DD}).

FIGURE 7-17

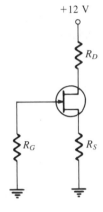

+12 V

R_D

R_G R_S

Solution

For midpoint bias, $I_D \cong I_{DSS}/2 = 7.5$ mA and $V_{GS} \cong V_{GS(off)}/4 = -8$ V$/4 = -2$ V.

$$R_S = \left| \frac{V_{GS}}{I_D} \right| = \frac{2 \text{ V}}{7.5 \text{ mA}}$$

$$= 267 \ \Omega$$

$$V_D = V_{DD} - I_D R_D$$

$$I_D R_D = V_{DD} - V_D$$

$$R_D = \frac{V_{DD} - V_D}{I_D} = \frac{12 \text{ V} - 6 \text{ V}}{7.5 \text{ mA}}$$

$$= 800 \ \Omega$$

R_G can be arbitrarily large if I_{GSS} is assumed to be negligible. R_G essentially establishes the input resistance to the JFET stage because it appears in parallel with the very high resistance of the reverse-biased pn junction.

Practice Exercise 7-9

Select resistor values in Figure 7–17 to set up an approximate midpoint bias. The JFET parameters are $I_{DSS} = 10$ mA and $V_{GS(off)} = -10$ V. $V_{DD} = 15$ V.

---

1. A p-channel JFET must have a (positive, negative) V_{GS}.

2. In a certain self-biased n-channel JFET circuit, $I_D = 8$ mA and $R_S = 1$ kΩ. Determine V_{GS}.

SECTION REVIEW 7-3

THE METAL OXIDE SEMICONDUCTOR FET (MOSFET)

7–4

The MOSFET is the second category of field-effect transistor. It differs from the JFET in that it has no pn junction structure; instead, the gate of the MOSFET is *insulated* from the channel by a silicon dioxide (SiO_2) layer. There are two basic types of MOSFETs: *depletion-enhancement* (DE) and *enhancement-only* (E).

Depletion-Enhancement MOSFET

Figure 7–18 illustrates the basic structure of DE MOSFETs. The drain and source are diffused into the substrate material and then connected by a narrow channel adjacent to the insulated gate. Both n-channel and p-channel devices are shown in the figure. We will use the n-channel device to describe the basic operation. The p-channel operation is the same, except the voltage polarities are opposite those of the n-channel.

The DE MOSFET can be operated in either of two modes: the *depletion mode* or the *enhancement mode*. Since the gate is insulated from the channel, either a positive or a negative gate voltage can be applied. The MOSFET operates in the depletion mode when a *negative* gate-to-source voltage is applied and in the enhancement mode when a *positive* gate-to-source voltage is applied.

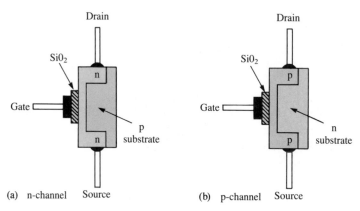

(a) n-channel   Source   (b) p-channel   Source

FIGURE 7–18
Basic structure of DE MOSFETs.

Depletion Mode.   Visualize the gate as one plate of a parallel-plate capacitor and the channel as the other plate. The silicon dioxide insulating layer is the dielectric. With a negative gate voltage, the negative charges on the gate repel conduction electrons from the channel, leaving positive ions in their place. Thereby, the n-channel is *depleted* of some of its electrons, thus decreasing the channel conductivity. The greater the negative voltage on the gate, the greater the depletion of n-channel electrons. At a sufficiently negative gate-to-source voltage, $V_{GS(off)}$, the channel is totally depleted and the drain current is 0. This depletion mode is illustrated in Figure 7–19(a). Like the n-channel JFET, the n-channel DE MOSFET conducts drain current for gate-to-source voltages between $V_{GS(off)}$ and 0. In addition, the DE MOSFET conducts for values of V_{GS} above 0.

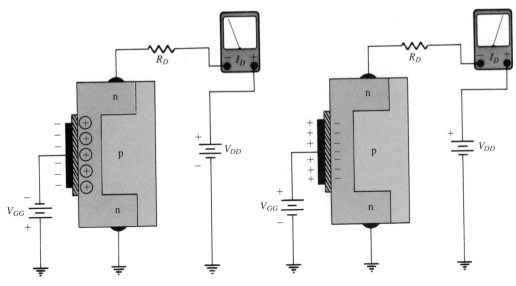

(a) Depletion mode (V_{GS} negative and less than $V_{GS(\text{off})}$)   (b) Enhancement mode (V_{GS} positive)

FIGURE 7–19
Operation of n-channel DE MOSFET.

Enhancement Mode.   With a positive gate voltage, more conduction electrons are attracted into the channel, thus increasing (enhancing) the channel conductivity, as illustrated in Figure 7–19(b).

DE MOSFET Symbols.   The schematic symbols for both the n-channel and the p-channel depletion-enhancement MOSFETs are shown in Figure 7–20. The substrate, indicated by the arrow, is normally (but not always) connected internally to the source. An inward substrate arrow is for n-channel, and an outward arrow is for p-channel.

FIGURE 7–20
DE MOSFET schematic symbols.

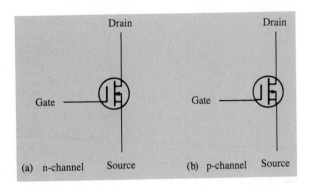

(a)   n-channel   (b) p-channel

Enhancement MOSFET

This type of MOSFET operates *only* in the enhancement mode and has no depletion mode. It differs in construction from the DE MOSFET in that it has *no physical channel*. Notice

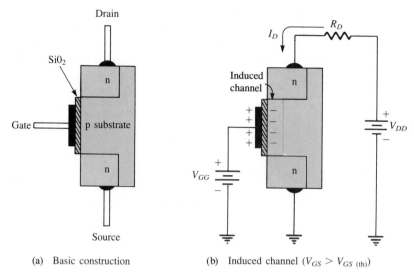

(a) Basic construction

(b) Induced channel ($V_{GS} > V_{GS \text{ (th)}}$)

FIGURE 7–21

E MOSFET construction and operation.

in Figure 7–21(a) that the substrate extends completely to the SiO_2 layer. For an n-channel device, a positive gate voltage above a threshold value *induces* a channel by creating a thin layer of negative charges in the substrate region adjacent to the SiO_2 layer, as shown in Figure 7–21(b). The conductivity of the channel is enhanced by increasing the gate-to-source voltage and thus pulling more electrons into the channel. For any gate voltage below the threshold value, there is no channel.

The schematic symbols for the n-channel and p-channel E MOSFETs are shown in Figure 7–22. The broken lines symbolize the absence of a physical channel.

FIGURE 7–22

E MOSFET schematic symbols.

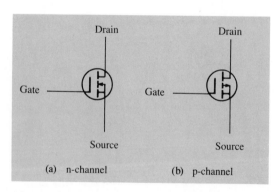

(a) n-channel

(b) p-channel

V MOSFET

The V MOSFET is an enhancement mode device with a physical structure that permits higher current operation and greater drain-to-source voltage than the conventional E MOSFET. Thus, the V MOSFET is particularly suited for high power use and competes

Handling Precautions

Because the gate of a MOSFET is insulated from the channel, the input resistance is extremely high (ideally infinite). The gate leakage current I_{GSS} for a typical MOSFET is in the pA range, whereas the gate reverse current for a typical JFET is in the nA range. The input capacitance, of course, results from the insulated gate structure. Excess static charge can be accumulated because the input capacitance combines with the very high input resistance and can result in damage to the device. To avoid excess charge build-up, certain precautions should be taken.

1. MOS devices should be shipped and stored in conductive foam.
2. All instruments and metal benches used in assembly or test should be connected to earth ground (round prong of wall outlets).
3. The assembler's or handler's wrist should be connected to earth ground with a length of wire and a high-value series resistor.
4. Never remove an MOS device (or any other device, for that matter) from the circuit while the power is on.
5. Do not apply signals while the dc power supply is off.

---

1. Describe the major difference in construction of the DE MOSFET and the E MOS-FET.
2. An FET is also known as a _____-_____ device.

**SECTION
REVIEW
7–5**

MOSFET BIASING

DE MOSFET Bias

7–6

Recall that depletion-enhancement MOSFETs can be operated with either positive or negative values of V_{GS}. A simple bias method is to set $V_{GS} = 0$ so that an ac signal at the gate varies the gate-to-source voltage above and below this bias point. A MOSFET with *0 bias* is shown in Figure 7–26. Since $V_{GS} = 0$, $I_D = I_{DSS}$ as indicated. The drain-to-source voltage is expressed as follows.

$$V_{DS} = V_{DD} - I_{DSS}R_D \qquad (7-13)$$

FIGURE 7–26
A zero-biased DE MOSFET.

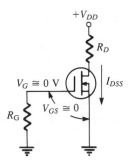

EXAMPLE 7–12

Determine the drain-to-source voltage in the circuit of Figure 7–27. The MOSFET data sheet gives $V_{GS(off)} = -8$ V and $I_{DSS} = 12$ mA.

FIGURE 7–27

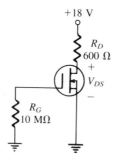

Solution
Since $I_D = I_{DSS} = 12$ mA, the drain-to-source voltage is calculated as follows.

$$V_{DS} = V_{DD} - I_{DSS}R_D$$
$$= 18 \text{ V} - (12 \text{ mA})(600 \text{ } \Omega)$$
$$= 10.8 \text{ V}$$

Practice Exercise 7–12
Find V_{DS} in Figure 7–27 when $V_{GS(off)} = -10$ V and $I_{DSS} = 20$ mA.

E MOSFET Bias

Recall that enhancement-only MOSFETs must have a V_{GS} greater than the threshold value, $V_{GS(th)}$. Figure 7–28 shows two ways to bias an E MOSFET. An n-channel device

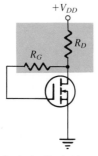

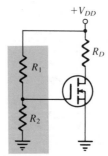

(a) Drain-feedback bias   (b) Voltage-divider bias

FIGURE 7–28
E MOSFET biasing arrangements.

☐ A depletion-enhancement MOSFET can operate with a 0, positive, or negative gate-to-source voltage.
☐ The DE MOSFET has a physical channel between drain and source.
☐ For an n-channel DE MOSFET, negative values of V_{GS} produce the depletion mode and positive values produce the enhancement mode.
☐ The enhancement-only MOSFET (E MOSFET) has no physical channel.
☐ A channel is induced in an E MOSFET by the application of a V_{GS} greater than the threshold value, $V_{GS(th)}$.
☐ Midpoint bias for a DE MOSFET is $I_D = I_{DSS}$, obtained by setting $V_{GS} = 0$.
☐ An E MOSFET has no I_{DSS} parameter.
☐ An n-channel E MOSFET has a positive $V_{GS(th)}$. A p-channel E MOSFET has a negative $V_{GS(th)}$.
☐ A V MOSFET can handle higher power and voltage than a conventional E MOSFET.

---

1. Sketch the schematic diagrams for a p-channel and an n-channel JFET. Label the terminals.

2. Show how to connect bias batteries between the gate and source of the JFETs in Figure 7–32.

SELF-TEST

FIGURE 7–32

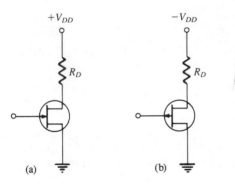

(a)   (b)

3. A certain n-channel JFET has a pinch-off voltage of 5 V. If $V_{GS} = -2$ V, what is V_{DS} at pinch-off?

4. $I_{DSS} = 20$ mA and $V_{GS(off)} = -6$ V for a particular JFET.
    (a) What is I_D when $V_{GS} = 0$ V?
    (b) What is I_D when $V_{GS} = V_{GS(off)}$?
    (c) If V_{GS} is increased from -4 V to -1 V, does I_D increase or decrease?

5. Calculate I_D for $V_{GS} = -2.5$ V, given that $I_{DSS} = 12$ mA and $V_{GS(off)} = -8$ V.

6. Calculate the forward transconductance at $V_{GS} = 2$ V for a JFET with $g_{m0} = 2000$ μS and $V_{GS(off)} = 6$ V. Is this an n-channel or a p-channel device? Why?

7. For a certain JFET, $I_{GSS} = 10$ nA at $V_{GS} = 10$ V. Determine the input resistance.

8. How does a MOSFET differ from a JFET?

9. Sketch the schematic symbols for n-channel and p-channel DE MOSFETs and E MOSFETs. Label the terminals.

10. A certain DE MOSFET is biased at $V_{GS} = 0$. Its data sheet specifies $I_{DSS} = 20$ mA and $V_{GS(off)} = -5$ V. What is the value of the drain current?

11. An n-channel DE MOSFET with a positive V_{GS} is operating in the _____ mode.

12. Describe the basic difference between a DE MOSFET and an E MOSFET.

13. A certain E MOSFET is specified to have an $I_{D(on)} = 4$ mA at $V_{GS} = 8$ V. The threshold voltage is 2 V. Determine the drain current when $V_{GS} = 6$ V.

14. Explain why both types of MOSFETs have an extremely high input resistance at the gate.

15. An n-channel self-biased JFET has a drain current of 12 mA and a 100 Ω source resistor. What is the value of V_{GS}?

16. Determine the value of R_S required to produce a V_{GS} of -4 V when $I_D = 5$ mA.

17. The DE MOSFET in Figure 7–33 has an $I_{DSS} = 15$ mA. What are I_D and V_{DS}?

FIGURE 7–33

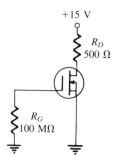

+15 V

R_D
500 Ω

R_G
100 MΩ

18. A certain E MOSFET has a $V_{GS(th)} = 3$ V. What is the minimum V_{GS} for the device to turn on?

PROBLEMS

Section 7–1

7–1 The V_{GS} of a p-channel JFET is increased from 1 V to 3 V.
(a) Does the depletion region narrow or widen?
(b) Does the resistance of the channel increase or decrease?

7–2 Why must the gate-to-source voltage of an n-channel JFET always be either 0 or negative?

Section 7–2

7–3 A JFET has a specified pinch-off voltage of 5 V. When $V_{GS} = 0$, what is V_{DS}?

7–4 A certain JFET is biased such that $V_{GS} = -2$ V. What is the value of V_{DS} if V_P is specified to be 6 V?

7–5 A certain JFET data sheet gives $V_{GS(off)} = -8$ V and $I_{DSS} = 10$ mA.
   (a) Determine the value of V_{DS} at which pinch-off begins when $V_{GS} = -4$ V.
   (b) When $V_{GS} = 0$, what is I_D for values of V_{DS} above pinch-off? $V_{DD} = 15$ V.

7–6 A certain p-channel JFET has a $V_{GS(off)} = 6$ V. What is I_D when $V_{GS} = 8$ V?

7–7 The JFET in Figure 7–34 has a $V_{GS(off)} = -4$ V. Assume that you increase the supply voltage, V_{DD}, beginning at 0 until the ammeter reaches a steady value. What does the voltmeter read at this point?

FIGURE 7–34

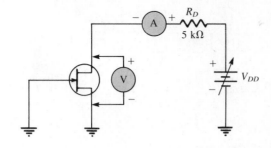

7–8 The following parameters are obtained from a certain JFET data sheet: $V_{GS(off)} = -8$ V and $I_{DSS} = 5$ mA. Determine the values of I_D for each value of V_{GS} ranging from 0 V to -8 V in 1 V steps. Plot the transfer characteristic curve from these data.

7–9 For the JFET in problem 7–8, what value of V_{GS} is required to set up a drain current of 2.25 mA?

7–10 For a particular JFET, $g_{m0} = 3200$ μS. What is g_m when $V_{GS} = -4$ V, given that $V_{GS(off)} = -8$ V?

7–11 Determine the forward transconductance of a JFET biased at $V_{GS} = -2$ V. From the data sheet, $V_{GS(off)} = -7$ V and $g_m = 2000$ μS at $V_{GS} = 0$ V. Also determine the forward transfer admittance, y_{fs}.

7–12 A p-channel JFET data sheet shows that $I_{GSS} = 5$ nA at $V_{GS} = 10$ V. Determine the input resistance.

Section 7–3

7–13 Determine in which mode (depletion or enhancement) each DE MOSFET in Figure 7–35 is biased.

7–14 Each E MOSFET in Figure 7–36 has a $V_{GS(th)}$ of $+5$ V or -5 V, depending on whether it is an n-channel or a p-channel device. Determine whether each MOSFET is *on* or *off*.

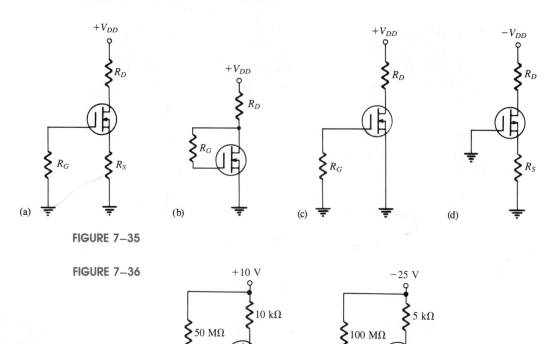

FIGURE 7–35

FIGURE 7–36

Section 7–4

7–15 The data sheet for a certain DE MOSFET gives $V_{GS(off)} = -5$ V and $I_{DSS} = 8$ mA.

    (a) Is this device p-channel or n-channel?

    (b) Determine I_D for values of V_{GS} ranging from -5 V to $+5$ V in increments of 1 V.

    (c) Plot the transfer characteristic curve using the data from part (b).

7–16 Determine I_{DSS}, given $I_D = 3$ mA, $V_{GS} = -2$ V, and $V_{GS(off)} = -10$ V.

7–17 The data sheet for an E MOSFET reveals that $I_{D(on)} = 10$ mA at $V_{GS} = -12$ V and $V_{GS(th)} = -3$ V. Find I_D when $V_{GS} = -6$ V.

Section 7–5

7–18 For each circuit in Figure 7–37, determine V_{DS} and V_{GS}.

7–19 Using the curve in Figure 7–38, determine the value of R_S required for a 9.5 mA drain current.

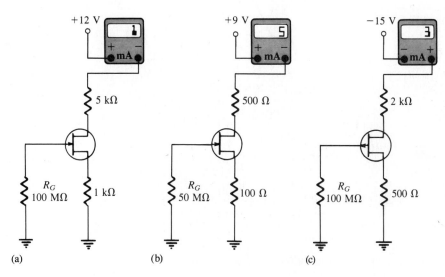

FIGURE 7–37

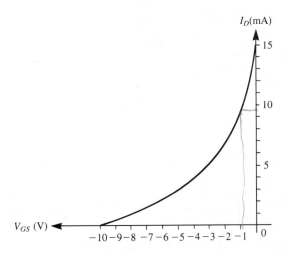

FIGURE 7–38

7–20 Set up a midpoint bias for a JFET with $I_{DSS} = 14$ mA and $V_{GS(off)} = -10$ V. Use a 24 V dc source as the supply voltage. Show the circuit and resistor values. Indicate the values of I_D, V_{GS}, and V_{DS}.

7–21 Determine the total input resistance in Figure 7–39. $I_{GSS} = 20$ nA at $V_{GS} = -10$ V.

FIGURE 7–39

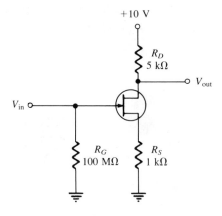

Section 7–6

7–22 Determine V_{DS} for each circuit in Figure 7–40. $I_{DSS} = 8$ mA.

FIGURE 7–40

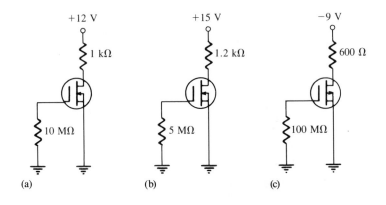

(a)   (b)   (c)

7–23 Find V_{GS} and V_{DS} for the E MOSFETs in Figure 7–41. Data sheet information is listed with each circuit.

FIGURE 7–41

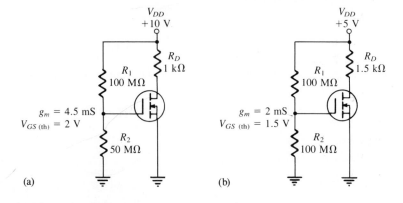

(a)   (b)

7–24 Based on the indicated V_{GS} measurements, determine the drain current and drain-to-source voltage for each circuit in Figure 7–42.

FIGURE 7–42

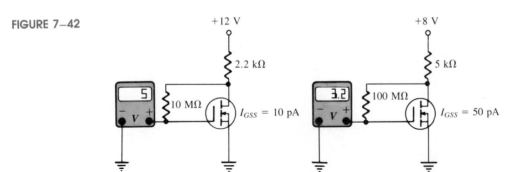

(a)   (b)

7–25 Determine the actual gate-to-source voltage in Figure 7–43 by taking into account the gate leakage current I_{GSS}. Assume that I_{GSS} is 50 pA and I_D is 1 mA under the existing bias conditions.

FIGURE 7–43

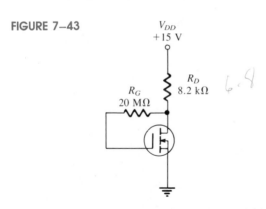

Section 7–1
1. Drain, source, and gate.
2. Negative.
3. By V_{GS}.

Section 7–2
1. −7 V.
2. Decreases.
3. +3 V.

Section 7–3
1. Positive.
2. −8 V.

Section 7–4

1. Depletion-enhancement MOSFET; enhancement-only MOSFET; V MOSFET.
2. Yes.
3. No.

Section 7–5

1. The DE MOSFET has a physical channel, but the E MOSFET does not.
2. Square-law.

Section 7–6

1. I_{DSS}.
2. 2 V.

ANSWERS TO PRACTICE EXERCISES

Example 7–1: $V_{DS} = 1$ V, $I_D = 15$ mA

Example 7–4: $g_m = 1800$ μS, $I_D = 4.32$ mA

Example 7–6: $V_{DS} = 2$ V, $V_{GS} = -3.12$ V

Example 7–8: 889 Ω

Example 7–9: $R_S = 500$ Ω, $R_D = 1.5$ kΩ

Example 7–10: (a) p-channel; (b) 6.48 mA; (c) 35.28 mA

Example 7–11: 1.24 mA

Example 7–12: 6 V

Example 7–13: $V_{GS} = 14.4$ V, $V_{DS} = 9.86$ V

Small-Signal Bipolar Amplifiers

In this chapter you will learn

☐ How a circuit operates as a small-signal amplifier

☐ How a bipolar transistor can be represented by an equivalent circuit

☐ What the h parameters of a transistor are and their meaning

☐ What the r parameters of a transistor are and their meaning

☐ The characteristics of common-emitter, common-collector, and common-base amplifiers

☐ How to analyze an amplifier by reducing it to its equivalent circuit

☐ How the voltage gain, input impedance, and output impedance of an amplifier are determined

☐ How an emitter bypass capacitor affects voltage gain

☐ How loading affects the voltage gain

☐ How the voltage gain can be increased by cascading amplifiers

☐ How to express voltage gain and power gain in decibels

☐ How direct-coupled amplifiers differ from those that are capacitively coupled or transformer coupled

☐ How to troubleshoot amplifiers by signal tracing using an oscilloscope

☐ How bipolar transistor amplifiers are used in a specific system application

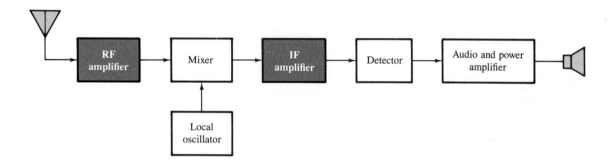

The biasing of a transistor is purely a *dc operation*. The purpose, however, is to establish a Q-point about which variations in current and voltage can occur in response to an *ac signal*. In applications where very small signal voltages must be amplified, such as from an antenna, variations about the Q-point are relatively small. Amplifiers designed to handle these small ac signals are called *small-signal amplifiers*.

In this chapter you will study small-signal amplifiers and their operation using bipolar transistors. In addition, we will introduce transistor equivalent circuits, multistage amplifiers, troubleshooting, and amplifier applications.

Small-signal amplifiers are widely used in all types of communications systems. In the superheterodyne receiver system that we are using as a representative system example, the RF and the IF amplifiers can be classified as small-signal amplifiers. Typically, the RF amplifiers in receiver systems must accept signals as low as around 30 μV in the case of AM and often less than 1 μV in the case of FM. By the time the signal gets to the IF amplifiers it is, of course, much greater than its input level, but still quite small. The IF amplifier section provides the bulk of the system amplification. In some receivers, the audio section consists of a preamplifier, which can also be considered a small-signal device, followed by a power amplifier that handles large signals. You will learn the distinction between small and large signal operation as you study this and later chapters.

SMALL-SIGNAL AMPLIFIER OPERATION

8–1

A biased transistor with an ac source capacitively coupled to the base is shown in Figure 8–1. The coupling capacitor blocks dc and thus prevents the source resistance from changing the bias voltage at the base. The signal voltage causes the base voltage to vary above and below its dc bias level. The resulting variation in base current produces a larger variation in collector current because of the current gain of the transistor.

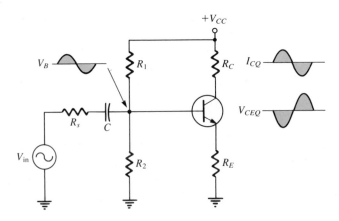

FIGURE 8–1
Voltage-divider biased amplifier driven by an ac source.

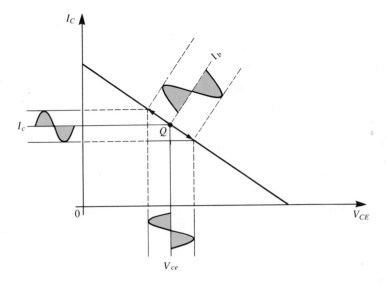

FIGURE 8–2
Graphical operation of the amplifier showing the base current, collector current, and collector-to-emitter voltage swings.

As the collector current increases, the voltage drop across R_C also increases, causing the collector voltage to decrease. The collector current varies above and below its Q-point value *in-phase* with the base current, and the collector-to-emitter voltage varies above and below its Q-point value 180° *out-of-phase* with the base voltage, as illustrated in Figure 8–1.

A Graphical Picture

The operation just described can be illustrated graphically on the collector-characteristic curves, as shown in Figure 8–2. The signal at the base drives the base current equally above and below the Q-point on the load line, as shown by the arrows. Lines projected from the peaks of the base current, across to the I_C-axis, and down to the V_{CE}-axis, indicate the peak-to-peak variations of the collector current and collector-to-emitter voltage, as shown.

EXAMPLE 8–1

The load line operation of a certain amplifier extends 10 μA above and below the Q-point base current value of 50 μA, as shown in Figure 8–3. Determine the resulting peak-to-peak values of collector current and collector-to-emitter voltage from the graph.

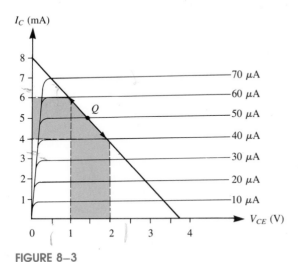

FIGURE 8–3

Solution
Projections on the graph of Figure 8–3 show a peak-to-peak collector current of 2 mA and a peak-to-peak collector-to-emitter voltage of 1 V.

Labels

The labels used for dc quantities in the previous chapters were identified by uppercase (capital) subscripts such as I_C, I_E, V_C, V_{CE}, and so forth. We will use *lowercase subscripts* to indicate *fixed-value* ac quantities of rms, peak, and peak-to-peak currents and voltages, such as I_c, I_e, I_b, V_c, V_{ce}, and so forth. These symbols will represent rms values unless otherwise stated. Also, instantaneous quantities are represented by lowercase letters and subscripts, such as i_c, i_e, i_b, v_e, v_{ce}, and so forth.

In addition to currents and voltages, resistances often have different values when a circuit is analyzed from an ac viewpoint as opposed to a dc viewpoint. Lowercase subscripts are used to identify ac resistance values. For example, R_c is the ac collector resistance, and R_C is the dc collector resistance. You will see the need for this distinction later. Resistance values *internal* to the transistor use a *lowercase r*. An example is the internal ac emitter resistance, r_e.

SECTION REVIEW 8–1

1. When I_b is at its positive peak, I_c is at its _____ peak, and V_{ce} is at its _____ peak.
2. What is the difference between V_{CE} and V_{ce}?

TRANSISTOR ac EQUIVALENT CIRCUITS

8–2

In order to better visualize the operation of a transistor in an amplifier circuit, it is often useful to represent the device by an *equivalent circuit*. An equivalent circuit uses various internal transistor *parameters* to represent the transistor's operation. Two types of equivalent circuit representations are described in this section. One is based on *h parameters* and the other on *r parameters*.

h Parameters

Because they are typically specified on a manufacturer's data sheet, h (hybrid) parameters are important. These parameters are usually specified because they are relatively easy to measure. The four *basic* ac h parameters and their descriptions are given in Table 8–1. Each of the four h parameters carries a second subscript letter to designate the common-emitter (*e*), common-base (*b*), or common-collector (*c*) configuration. These are listed in Table 8–2.

We can describe the three amplifier circuit configurations this way:

☐ When the emitter is connected to ac ground, the input signal is applied to the base, and the output signal is on the collector, the circuit is a *common-emitter* type. (The term *common* refers to the ground point.)
☐ When the collector is connected to ac ground, the input signal is applied to the base, and the output signal is on the emitter, the circuit is a *common-collector* type.

TABLE 8–1
Basic ac h parameters

h parameter	Description	Condition
h_i	Input resistance	Output shorted
h_r	Voltage feedback ratio	Input open
h_f	Forward current gain	Output shorted
h_o	Output conductance	Input open

TABLE 8–2
Subscripts of h parameters

Configuration	h parameters
Common-emitter	$h_{ie}, h_{re}, h_{fe}, h_{oe}$
Common-base	$h_{ib}, h_{rb}, h_{fb}, h_{ob}$
Common-collector	$h_{ic}, h_{rc}, h_{fc}, h_{oc}$

☐ When the base is connected to ac ground, the input signal is applied to the emitter, and the output signal is on the collector, the circuit is a *common-base* type.

We will examine the characteristics of each of these bipolar amplifier configurations later in this chapter.

h Parameter Measurement and Meaning

Each of the h parameters is derived from ac measurements taken from operating transistor characteristic curves. h_i is the impedance (ac resistance) looking in at the input terminal of the transistor *with the output shorted,* as indicated in Figure 8–4(a), for a common-emitter connection. It is the ratio of input voltage to input current, expressed as follows:

$$h_{ie} = \frac{V_b}{I_b} \qquad (8–1)$$

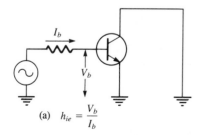

(a) $h_{ie} = \dfrac{V_b}{I_b}$

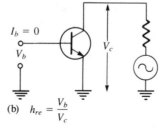

(b) $h_{re} = \dfrac{V_b}{V_c}$

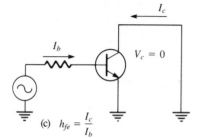

(c) $h_{fe} = \dfrac{I_c}{I_b}$

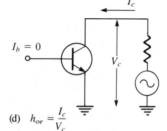

(d) $h_{oe} = \dfrac{I_c}{V_c}$

FIGURE 8–4
ac equivalent circuits for defining h parameters.

h_r is a measure of the amount of output voltage that is reflected (fed back) to the input with the input open. The common-emitter measurement circuit is shown in Figure 8–4(b). It is the ratio of input voltage to output voltage.

$$h_{re} = \frac{V_b}{V_c} \qquad \text{(8–2)}$$

h_f is the forward current gain measured with the output shorted, as shown in Figure 8–4(c). For the common-emitter configuration, h_{fe} is expressed as

$$h_{fe} = \frac{I_c}{I_b} \qquad \text{(8–3)}$$

Finally, h_o is the admittance (conductance) looking in at the output terminal with the input open, as shown in Figure 8–4(d), and it is expressed as

$$h_{oe} = \frac{I_c}{V_c} \qquad \text{(8–4)}$$

Its units are siemens (S). Table 8–3 summarizes the ratios for each amplifier configuration.

TABLE 8–3
h parameter ratios for the amplifier configurations

Common-emitter	Common-base	Common-collector
$h_{ie} = V_b/I_b$	$h_{ib} = V_e/I_b$	$h_{ic} = V_b/I_b$
$h_{re} = V_b/V_c$	$h_{rb} = V_e/V_c$	$h_{rc} = V_b/V_e$
$h_{fe} = I_c/I_b$	$h_{fb} = I_c/I_e$	$h_{fc} = I_e/I_b$
$h_{oe} = I_c/V_c$	$h_{ob} = I_c/V_c$	$h_{oc} = I_e/V_e$

Hybrid Equivalent Circuits

The general form of the h parameter equivalent circuit is shown in Figure 8–5. The input impedance, h_i, appears in series at the input. The reverse voltage ratio, h_r, is multiplied by the output voltage ($h_r v_{out}$) to produce an equivalent voltage source in series with the input. The forward current gain, h_f, is multiplied by the input current ($h_f i_{in}$) and appears as an equivalent current source in the output. The output admittance, h_o, appears across the output terminals. Specifically, there are three hybrid equivalent circuits: one for the

FIGURE 8–5
Generalized h parameter equivalent circuit for a bipolar junction transistor.

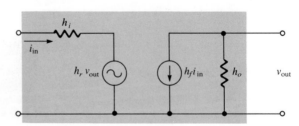

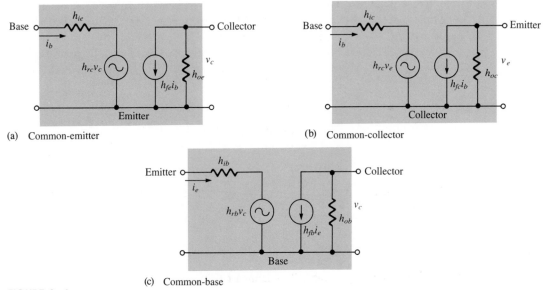

FIGURE 8–6
h parameter equivalent circuits.

common-emitter, one for the common-base, and one for the common-collector configuration, as shown in Figure 8–6.

r Parameters

The h parameters are important because they are given on data sheets, so you need to know what they mean. There is, however, another widely used set of parameters that are perhaps easier to work with than the h parameters. These are the *r parameters*. The five r parameters are given in Table 8–4.

TABLE 8–4
r parameters

r parameter	Description
α	ac Alpha (I_c/I_e)
β	ac Beta (I_c/I_b)
r_e	ac Emitter resistance
r_b	ac Base resistance
r_c	ac Collector resistance

Conversion from h Parameters to r Parameters

The ac current gain parameters, α and β, convert directly from h parameters as follows:

$$\alpha = h_{fb} \qquad \textbf{(8–5)}$$

$$\beta = h_{fe} \qquad \textbf{(8–6)}$$

Recall that we used α_{dc} and β_{dc} in previous chapters. These are dc parameters and sometimes have values different from those of the ac parameters. The difference between β_{dc} and β will be discussed later.

Because data sheets most often provide only *common-emitter* h parameters, the following formulas show how to convert them to the remaining r parameters.

$$r_e = \frac{h_{re}}{h_{oe}} \tag{8-7}$$

$$r_c = \frac{h_{re} + 1}{h_{oe}} \tag{8-8}$$

$$r_b = h_{ie} - \frac{h_{re}}{h_{oe}}(1 + h_{fe}) \tag{8-9}$$

We will use r parameters throughout the text.

r Parameter Equivalent Circuits

An r parameter equivalent circuit is shown in Figure 8–7. For most general analysis work, Figure 8–7 can be simplified as follows: The effect of the ac base resistance r_b is usually small enough to neglect, so that it can be replaced by a short. The ac collector resistance is usually several megohms and can be replaced by an open. The resulting simplified r parameter equivalent circuit is shown in Figure 8–8.

The interpretation of this equivalent circuit in terms of a transistor's ac operation is as follows: A resistance r_e appears between the emitter and base terminals. This is the resistance "seen" by an ac signal when applied to the base or emitter terminals of a forward-biased transistor. The collector effectively acts as a current source of αI_e or, equivalently, βI_b. These factors are shown with a standard transistor symbol in Figure 8–9.

FIGURE 8–7
Generalized r parameter equivalent circuit for a bipolar junction transistor.

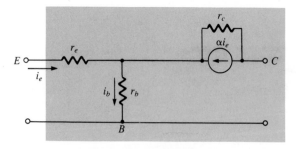

FIGURE 8–8
Simplified r parameter equivalent circuit for a transistor.

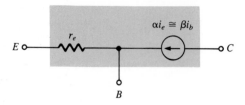

FIGURE 8–9
Relation of transistor symbol to r parameter equivalent.

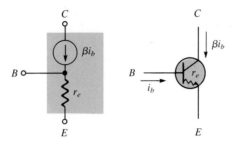

Determining r_e by Formula

Instead of using h parameters to get r_e, we can also use the simple formula of equation
(8–10) to calculate an approximate value at 25°C.

$$r_e \cong \frac{25 \text{ mV}}{I_E} \qquad (8\text{–}10)$$

I_E is the dc emitter current in milliamperes. Although the formula is simple, its derivation
is not and is therefore reserved for Appendix B.

EXAMPLE 8–2

Determine the r_e of a transistor that is operating with a dc emitter current of 2 mA.

Solution

$$I_E = 2 \text{ mA}$$

$$r_e \cong \frac{25 \text{ mV}}{I_E} = \frac{25 \text{ mV}}{2 \text{ mA}} = 12.5 \text{ }\Omega$$

Comparison of the ac Beta (β) to the dc Beta (β_{dc})

For a typical transistor, a graph of I_C versus I_B is nonlinear, as shown in Figure 8–10. If
we pick a Q-point on the curve and cause the base current to vary an amount ΔI_B, then the

FIGURE 8–10
I_C-versus-I_B curve illustrates the difference
between $\beta_{dc} = I_C/I_B$ and $\beta = \Delta I_C/\Delta I_B$.

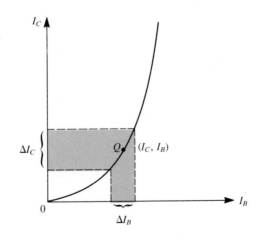

collector current will vary an amount ΔI_C as shown. At different points on the curve, the ratio $\Delta I_C/\Delta I_B$ will be different, and it may also differ from the I_C/I_B ratio at the Q-point. Since $\beta_{dc} = I_C/I_B$ and $\beta = \Delta I_C/\Delta I_B$, the values of these two quantities can differ. Remember that $\beta_{dc} = h_{FE}$ and $\beta = h_{fe}$.

**SECTION
REVIEW
8–2**

1. List the four basic h parameters.
2. List the r parameters.
3. Calculate r_e if $I_E = 10$ mA.

COMMON-EMITTER AMPLIFIERS

8–3

Now that you have an idea of how a transistor can be represented in an ac circuit, a complete amplifier circuit will be examined. Figure 8–11 shows a common-emitter amplifier with voltage-divider bias and *coupling* capacitors on the input and output and a *bypass* capacitor from emitter to ground. The circuit has a combination of dc and ac operation, both of which must be considered.

FIGURE 8–11
A common-emitter amplifier.

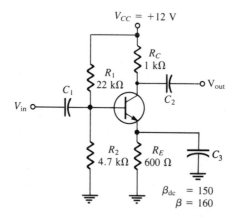

dc Analysis

To analyze the amplifier in Figure 8–11, the dc bias values must first be determined. To do this, a *dc equivalent circuit* is developed by simply *replacing the coupling and bypass capacitors with opens*, as shown in Figure 8–12. The analysis for this dc circuit was covered in Chapter 6, but we will go through it again here. The dc input resistance at the base is

$$R_{IN(base)} = \beta_{dc} R_E = 90 \text{ k}\Omega$$

FIGURE 8–12
dc equivalent circuit for the amplifier in
Figure 8–11.

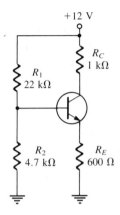

Since this is more than ten times R_2, it can be neglected when calculating the base voltage.

$$V_B = \left(\frac{R_2}{R_1 + R_2}\right)V_{CC} = \left(\frac{4.7\ k\Omega}{26.7\ k\Omega}\right)12\ V = 2.11\ V$$

and

$$V_E = 2.11\ V - 0.7\ V = 1.41\ V$$

Therefore

$$I_E = \frac{V_E}{R_E} = \frac{1.41\ V}{600\ \Omega} = 2.4\ mA$$

Since $I_C \cong I_E$, then

$$V_C = V_{CC} - I_C R_C = 12\ V - 2.4\ V = 9.6\ V$$

Finally,

$$V_{CE} = V_C - V_E = 9.6\ V - 1.41\ V = 8.19\ V$$

ac Equivalent Circuit

To analyze the signal operation of the amplifier in Figure 8–11, an *ac equivalent* circuit is developed as follows: The coupling capacitors C_1 and C_2 are replaced by effective shorts. This is based on the simplifying assumption that $X_C \cong 0$ at the signal frequency. (The bypass capacitor C_3 is omitted for this initial analysis, but we will consider it later.)

The dc source is replaced by a *ground*. This is based on the assumption that the voltage source has an internal resistance of approximately 0, so that no ac voltage is developed across the source terminals. Therefore, the V_{CC} terminal is at a 0 volt ac potential and is called *ac ground*.

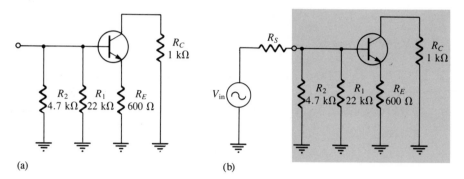

FIGURE 8–13

ac equivalent circuit for the amplifier in Figure 8–11.

The ac equivalent circuit is shown in Figure 8–13(a). Notice in the figure that both R_C and R_1 have one end connected to ac ground because, in the actual circuit, they are connected to V_{CC} (ac ground). In ac analysis, the ac ground and the actual ground are treated as the same point.

Signal (ac) Voltage at the Base

An ac voltage source is shown connected to the input in Figure 8–13(b). If the internal resistance of the ac source is 0, then *all* of the input signal voltage appears at the base terminal. If, however, the ac source has a *nonzero* internal resistance, then three factors must be taken into account in determining the actual signal voltage at the base. These are the *source resistance,* the *bias resistance,* and the *input impedance* at the base. This is illustrated in Figure 8–14(a) and is simplified by combining R_1, R_2, and $R_{in(base)}$ in parallel to get R_{in}, as shown in part (b). As you can see, the input voltage V_{in} is divided down by R_s (source resistance) and R_{in}, so that the signal voltage at the base of the transistor is

$$V_b = \left(\frac{R_{in}}{R_s + R_{in}}\right)V_{in}$$ (8–11)

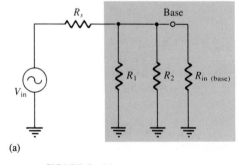

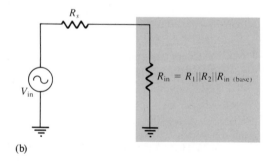

FIGURE 8–14

Equivalent base circuit.

Of course, if $R_s \ll R_{in}$, then $V_b \cong V_{in}$.

Input Impedance

To develop an expression for the input impedance as seen by an ac source looking in at the base, we will use the simplified r parameter model of the transistor. Figure 8–15 shows it connected with external resistors R_E and R_C. The input impedance will be treated as a resistance and designated R_{in}. The input impedance looking in at the base is

$$R_{in(base)} = \frac{V_b}{I_b} \qquad (8-12)$$

$$V_b = I_e(r_e + R_E)$$

and

$$I_b \cong \frac{I_e}{\beta}$$

Substituting and cancelling I_e, we get

$$R_{in(base)} = \beta(r_e + R_E) \qquad (8-13)$$

The total input impedance seen by the source is the parallel combination of R_1, R_2, and $R_{in(base)}$.

$$R_{in} = R_1\|R_2\|R_{in(base)} \qquad (8-14)$$

FIGURE 8–15
r parameter transistor model (inside shaded block) connected to external circuit.

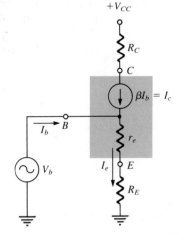

Output Impedance

The output impedance looking in at the collector is approximately equal to the collector resistor.

$$R_{out} \cong R_C \qquad (8-15)$$

Actually, $R_{out} = R_C \| r_c$, but since the ac collector resistance, r_c, is typically much larger than R_C, the approximation is usually valid.

EXAMPLE 8–3

Determine the signal voltage at the base in Figure 8–16. This is the ac equivalent of the amplifier in Figure 8–11, with a 10 mV rms, 300 Ω signal source. I_E was previously found to be 2.4 mA.

FIGURE 8–16

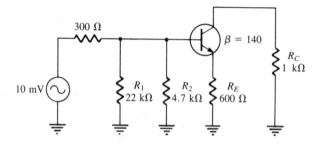

Solution

First,

$$r_e = \frac{25 \text{ mV}}{I_E} = \frac{25 \text{ mV}}{2.4 \text{ mA}} \cong 10.4 \ \Omega$$

Then,

$$R_{in(base)} = \beta(r_e + R_E) = 140(610.4 \ \Omega) = 85.5 \text{ k}\Omega$$

Next, determine the *total* input impedance viewed from the source:

$$\frac{1}{R_{in}} = \frac{1}{R_1} + \frac{1}{R_2} + \frac{1}{R_{in(base)}}$$

$$= \frac{1}{22 \text{ k}\Omega} + \frac{1}{4.7 \text{ k}\Omega} + \frac{1}{85.5 \text{ k}\Omega}$$

and

$$R_{in} = 3.7 \text{ k}\Omega$$

The input signal voltage is divided down by R_s and R_{in}, so the signal voltage at the base is the voltage developed across R_{in}.

$$V_b = \left(\frac{R_{in}}{R_s + R_{in}}\right) V_{in}$$

$$= \left(\frac{3.7 \text{ k}\Omega}{4 \text{ k}\Omega}\right) 10 \text{ mV}$$

$$= 9.25 \text{ mV}$$

As you can see, there is some attenuation (reduction) of the input signal due to the source resistance and amplifier input impedance.

Practice Exercise 8–3

Determine the signal voltage at the base in Figure 8–16 if the source resistance is 75 Ω and another transistor with a beta of 200 is inserted.

Voltage Gain

The ac voltage gain expression is developed using the equivalent circuit in Figure 8–17. The gain is the ratio of output voltage (V_c) to input voltage (V_b).

$$A_v = \frac{V_c}{V_b} \qquad (8\text{–}16)$$

Notice in the figure that $V_c = \alpha I_e R_C \cong I_e R_C$ and $V_b = I_e(r_e + R_E)$. Therefore,

$$A_v = \frac{I_e R_C}{I_e(r_e + R_E)}$$

Cancelling the I_e terms, we get

$$A_v = \frac{R_C}{r_e + R_E} \qquad (8\text{–}17)$$

FIGURE 8–17
ac voltage gain.

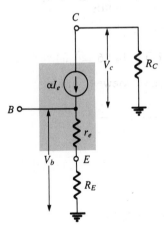

Equation (8–17) is the voltage gain from base to collector. To get the *overall* gain of the amplifier from *signal input* to collector, the attenuation of the input circuit must be included. The *attenuation* from input to base multiplied by the *gain* from base to collector is the overall amplifier gain. You can see this better with an example. Suppose a 10 mV signal is applied to the input, and the input network is such that the base voltage is 5 mV. The attenuation is therefore 5 mV/10 mV = 0.5. Now assume the amplifier has a voltage gain from base to collector of 20. The output voltage is 5 mV × 20 = 100 mV. Therefore, the overall gain is 100 mV/10 mV = 10 and is equal to the attenuation times the

FIGURE 8–18
Base circuit attenuation and overall gain.

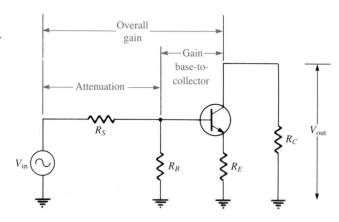

gain $(0.5 \times 20 = 10)$. (This is illustrated in Figure 8–18.) The expression for the attenuation in the base circuit is

$$\frac{V_b}{V_{in}} = \frac{R_{in}}{R_s + R_{in}} \qquad (8\text{–}18)$$

The overall gain A'_v is the product of A_v and V_b/V_{in}.

$$A'_v = A_v \left(\frac{V_b}{V_{in}} \right) \qquad (8\text{–}19)$$

Emitter Bypass Capacitor Increases Gain

When the bypass capacitor is connected across R_E, as shown in Figure 8–19(a), the emitter is effectively at ac ground, as shown in part (b), because the capacitor is selected large enough so that X_C is much less than R_E and effectively appears as a short at the signal frequency. It is important to recognize that the bypass capacitor does not alter the dc bias of the transistor because it is open to dc. Since the emitter resistor R_E is assumed to

FIGURE 8–19
Bypassing R_E with C_3.

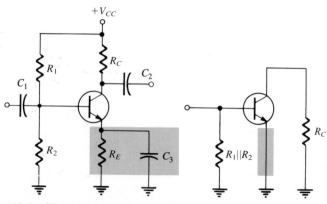

(a) Amplifier with emitter bypass capacitor   (b) ac equivalent

be shorted by the capacitor at the signal frequency, the ac voltage gain expression becomes

$$A_v = \frac{R_C}{r_e} \qquad (8\text{--}20)$$

EXAMPLE 8–4

Calculate the base-to-collector voltage gain of the amplifier in Figure 8–11 with and without an emitter bypass capacitor.

Solution
Recall that $r_e \cong 10.4 \ \Omega$ for this particular amplifier. Without C_3, the gain is

$$A_v \cong \frac{R_C}{r_e + R_E} = \frac{1 \ k\Omega}{610.4 \ \Omega} = 1.64$$

With C_3, the gain is

$$A_v \cong \frac{R_C}{r_e} = \frac{1 \ k\Omega}{10.4 \ \Omega} = 96.15$$

What a difference a capacitor can make!

Practice Exercise 8–4
Determine the base-to-collector voltage gain in Figure 8–11 with R_E bypassed, for the following circuit values: $R_C = 1.8 \ k\Omega$, $R_E = 1 \ k\Omega$, $R_1 = 33 \ k\Omega$, and $R_2 = 6.8 \ k\Omega$.

Effect of an ac Load on Voltage Gain

When a load R_L is connected to the output through the coupling capacitor C_2, as shown in Figure 8–20(a), the collector resistance at the signal frequency is effectively R_C in parallel

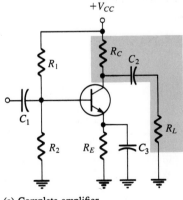

(a) Complete amplifier

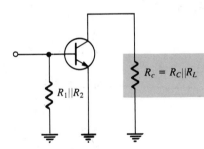

(b) ac equivalent

FIGURE 8–20
A common-emitter amplifier with an ac (capacitively) coupled load.

with R_L. Remember, the upper end of R_C is effectively at *ac ground*. The ac equivalent circuit is shown in Figure 8–20(b). The total ac collector resistance is

$$R_c = \frac{R_C R_L}{R_C + R_L} \qquad (8\text{–}21)$$

Replacing R_C with R_c in the voltage gain expression gives

$$A_v = \frac{R_c}{r_e} \qquad (8\text{–}22)$$

Since $R_c < R_C$, the voltage gain is reduced. Of course, if $R_L \gg R_C$, then $R_c \cong R_C$. So the load has very little effect on the gain.

EXAMPLE 8–5

Calculate the base-to-collector voltage gain of the amplifier in Figure 8–11 when a 5 kΩ load resistance is connected. Assume the emitter is effectively bypassed.

Solution
The ac collector resistance is

$$R_c = \frac{R_C R_L}{R_C + R_L} = \frac{(1 \text{ k}\Omega)(5 \text{ k}\Omega)}{6 \text{ k}\Omega} = 833 \ \Omega$$

Therefore,

$$A_v \cong \frac{R_c}{r_e} = \frac{833 \ \Omega}{10.4 \ \Omega} = 80.1$$

The unloaded gain was 96.15 in Example 8–4.

Practice Exercise 8–5
Determine the base-to-collector voltage gain in Figure 8–11 when a 10 kΩ load resistance is connected from collector to ground. Change the resistance values as follows: $R_C = 1.8$ kΩ, $R_E = 1$ kΩ, $R_1 = 33$ kΩ, and $R_2 = 6.8$ kΩ. The emitter resistor is bypassed.

Gain Stability

The internal emitter resistance of a transistor, r_e, varies considerably with temperature, and since $A_v = R_c/r_e$, the voltage gain also varies. A technique of partially bypassing R_E is often used to minimize the dependency on r_e and thus on temperature. Figure 8–21 shows that only a portion of the total emitter resistor is bypassed by C_3. The effect of this method is to reduce the gain but make it essentially independent of variations in r_e. The voltage gain for the circuit in Figure 8–21 is

$$A_v = \frac{R_c}{r_e + R_{E_1}} \qquad (8\text{–}23)$$

FIGURE 8–21

Partially bypassing the emitter resistance for gain stability.

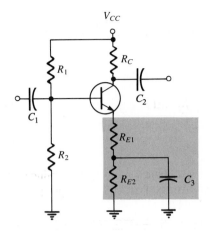

If $R_{E_1} \gg r_e$, then $A_v \cong R_c/R_{E_1}$.

   The total emitter resistance $R_{E_1} + R_{E_2}$ is used in dc biasing, but only R_{E_1} is used in the ac operation to "swamp out" r_e, thus making its effect on circuit operation and particularly the voltage gain negligible.

Phase Inversion

As mentioned earlier, the output voltage at the collector of a common-emitter amplifier is 180° out-of-phase with the input voltage at the base. The phase *inversion* is sometimes indicated by a negative voltage gain, $-A_v$. The next example pulls together the concepts covered so far as they relate to the common-emitter amplifier.

EXAMPLE 8–6

For the amplifier in Figure 8–22, determine the total output voltage (dc and ac).

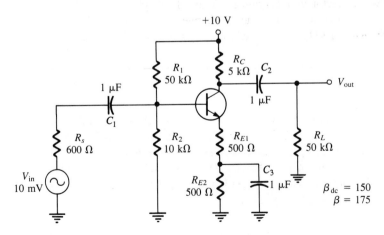

FIGURE 8–22

Solution

We will first determine the dc bias values. Refer to the dc equivalent circuit in Figure 8–23.

$$R_{\text{IN(base)}} = \beta_{\text{dc}}(R_{E_1} + R_{E_2}) = 150(1 \text{ k}\Omega) = 150 \text{ k}\Omega$$

FIGURE 8–23

dc equivalent for the circuit in Figure 8–22.

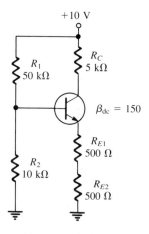

Since $R_{\text{IN(base)}}$ is fifteen times larger than R_2, we will neglect it in the dc base voltage calculation.

$$V_B = \left(\frac{10 \text{ k}\Omega}{50 \text{ k}\Omega + 10 \text{ k}\Omega}\right) 10 \text{ V} = 1.67 \text{ V}$$

$$V_E = V_B - 0.7 \text{ V} = 0.97 \text{ V}$$

$$I_E = \frac{V_E}{R_{E_1} + R_{E_2}} = \frac{0.97 \text{ V}}{1 \text{ k}\Omega} = 0.97 \text{ mA}$$

$$V_C = V_{CC} - I_C R_C \cong 10 \text{ V} - (0.97 \text{ mA})(5 \text{ k}\Omega) = 5.15 \text{ V} \quad \leftarrow$$

Next is the ac analysis, based on the ac equivalent circuit in Figure 8–24. The first thing to do in the ac analysis is calculate r_e.

$$r_e \cong \frac{25 \text{ mV}}{I_E} = \frac{25 \text{ mV}}{0.97 \text{ mA}} = 25.77 \text{ }\Omega$$

Next, we determine the attenuation in the base circuit. Looking from the 600 Ω source, the total R_{in} is

$$R_{\text{in}} = R_1 \| R_2 \| R_{\text{in(base)}}$$

$$R_{\text{in(base)}} = \beta(r_e + R_E) = 175(525.77 \text{ }\Omega) \cong 92 \text{ k}\Omega$$

Therefore,

$$R_{\text{in}} = 10 \text{ k}\Omega \| 50 \text{ k}\Omega \| 92 \text{ k}\Omega = 7.64 \text{ k}\Omega$$

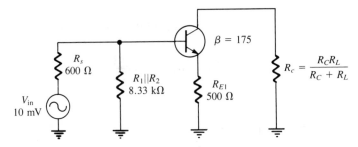

FIGURE 8–24
ac equivalent for the circuit in Figure 8–22.

The attenuation is

$$\frac{V_b}{V_{in}} = \frac{R_{in}}{R_s + R_{in}} = \frac{7.64 \text{ k}\Omega}{600 \text{ }\Omega + 7.64 \text{ k}\Omega} \doteq 0.93$$

Before A_v can be determined, we must know R_c.

$$R_c = \frac{R_C R_L}{R_C + R_L} = \frac{(5 \text{ k}\Omega)(50 \text{ k}\Omega)}{5 \text{ k}\Omega + 50 \text{ k}\Omega} = 4.55 \text{ k}\Omega$$

The voltage gain from base to collector is

$$A_v = \frac{R_c}{r_e + R_{E_1}} = \frac{4.55 \text{ k}\Omega}{525.77 \text{ k}\Omega} = 8.65$$

The overall gain is

$$A'_v = (0.93)(8.65) = 8.04$$

The source produces 10 mV, so the ac output voltage at the collector is

$$V_{out} = A'_v V_{in} = (8.04)(10 \text{ mV}) = 80.4 \text{ mV}$$

The total output voltage is the signal of 80.4 mV rms riding on a dc level of 5.15 V, as shown in Figure 8–25, where the peak values are shown.

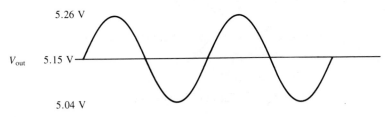

FIGURE 8–25
Output voltage for Figure 8–22.

Current Gain

The current gain from base to collector is I_c/I_b or β. However, the overall current gain of the amplifier is

$$A_i = \frac{I_c}{I_{in}} \tag{8-24}$$

I_{in} is the total current from the source, part of which is base current and part of which is in the bias network, as shown in Figure 8–26. The total input current is

$$I_{in} = \frac{V_{in}}{R_{in}} \tag{8-25}$$

FIGURE 8–26
Total ac input current (directions shown are for the positive half-cycle of V_{in}).

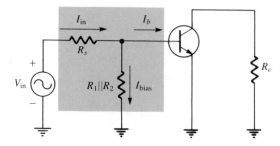

Power Gain

The power gain is the product of the voltage gain and the current gain.

$$A_p = A'_v A_i \tag{8-26}$$

SECTION REVIEW 8–3

1. If $r_e = 15\ \Omega$ and $R_c = 1500\ \Omega$, what is the ac voltage gain of a common-emitter amplifier with R_E bypassed?

2. In question 1, if $R_E = 100\ \Omega$ and there is no bypass capacitor, what is the voltage gain?

3. A certain amplifier has an $R_C = 1\ k\Omega$, and the collector is coupled to a 10 $k\Omega$ load. What is the effective ac collector resistance?

COMMON-COLLECTOR AMPLIFIERS

8–4

The common-collector amplifier is usually referred to as an *emitter-follower*. Figure 8–27 shows a voltage-divider biased emitter-follower. Notice that the input is applied to the base through a coupling capacitor, and the output is at the emitter. There is no collector resistor.

FIGURE 8-27
Emitter-follower with voltage-divider bias.

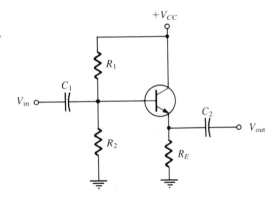

Voltage Gain

As in all amplifiers, the voltage gain is $A_v = V_{out}/V_{in}$. For the emitter-follower, V_{out} is $I_e R_E$ and V_{in} is $I_e(r_e + R_E)$, as shown in Figure 8–28. Therefore, the voltage gain is $I_e R_E / I_e(r_e + R_E)$. The currents cancel, and the base-to-emitter voltage gain expression simplifies to

$$A_v = \frac{R_E}{r_e + R_E} \qquad (8\text{–}27)$$

Notice here that the gain is always *slightly less than 1*. If $R_E \gg r_e$, then a good approximation is $A_v \cong 1$. Since the output voltage is at the emitter, it is *in-phase* with the base or input voltage, so there is no inversion. Because of this and because the voltage gain is approximately 1, the output voltage closely *follows* the input voltage; thus the term *emitter-follower*.

FIGURE 8-28
Emitter-follower model for voltage gain derivation.

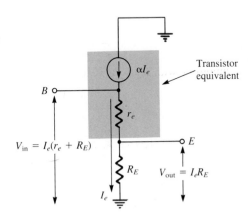

Input Impedance

The emitter-follower is characterized by a *high input impedance;* this is what makes it a very useful circuit. Because of the high input impedance, it can be used as a *buffer* to minimize loading effects when one circuit is driving another. The derivation of the input impedance looking in at the base is similar to that for the common-emitter amplifier. In this case, however, the emitter resistor is *never* bypassed.

$$R_{in(base)} = \frac{V_b}{I_b} = \frac{I_e(r_e + R_E)}{I_b}$$

$$\cong \frac{\beta I_b(r_e + R_E)}{I_b}$$

$$R_{in(base)} \cong \beta(r_e + R_E) \tag{8–28}$$

If $R_E \gg r_e$, then the input impedance at the base is

$$R_{in(base)} \cong \beta R_E \tag{8–29}$$

The bias resistors in Figure 8–27 appear in parallel with $R_{in(base)}$ looking from the input source, just as in the common-emitter circuit.

$$R_{in} = R_1 \| R_2 \| R_{in(base)} \tag{8–30}$$

Output Impedance

The output impedance for the emitter-follower is approximated as follows:

$$R_{out} \cong \left(\frac{R_s}{\beta}\right) \| R_E \tag{8–31}$$

The derivation of this expression is relatively involved and several simplifying assumptions have been made, as shown in Appendix B. The output impedance is very low, making the emitter-follower useful for driving low impedance loads.

Current Gain

The overall gain for the emitter-follower is I_e / I_{in}. I_{in} can be calculated as V_{in} / R_{in}. If the parallel combination of the bias resistors R_1 and R_2 is much greater than $R_{in(base)}$, then most of the input current goes into the base; thus the current gain of the amplifier approaches the current gain of the transistor, β. This is because very little signal current is in the bias resistors. Stated concisely, if

$$R_1 \| R_2 \gg \beta R_E$$

then

$$A_i \cong \beta \tag{8–32}$$

Otherwise,

$$A_i = \frac{I_e}{I_{in}} \tag{8–33}$$

β is the maximum achievable current gain in both common-collector and common-emitter amplifiers.

Power Gain

The common-collector power gain is the product of the voltage gain and the current gain. For the emitter-follower, the power gain is approximately equal to the current gain because the voltage gain is approximately 1.

$$A_p = A_v A_i$$

Since $A_v \cong 1$,

$$A_p \cong A_i \qquad\qquad \textbf{(8–34)}$$

EXAMPLE 8–7

Determine the total input impedance of the emitter-follower in Figure 8–29. Also find the voltage gain, current gain, and power gain. Assume $\beta = 175$.

FIGURE 8–29

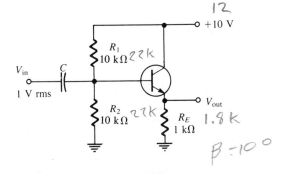

12

R_1 10 kΩ 22k

V_{in} C 1 V rms

R_2 10 kΩ 22k

R_E 1 kΩ 1.8 k

V_{out}

+10 V

β = 100

Solution
The approximate impedance looking in at the base is

$$R_{in(base)} \cong \beta R_E = (175)(1 \text{ k}\Omega) = 175 \text{ k}\Omega$$

The total input impedance is

$$R_{in} = R_1 \| R_2 \| R_{in(base)}$$
$$= 10 \text{ k}\Omega \| 10 \text{ k}\Omega \| 175 \text{ k}\Omega = 4.86 \text{ k}\Omega$$

The voltage gain is

$$A_v \cong 1$$

By calculating r_e, a more precise value of A_v can be determined if necessary.

$$I_E = \frac{V_E}{R_E} \cong \frac{4.3 \text{ V}}{1 \text{ k}\Omega} = 4.3 \text{ mA}$$

and

$$r_e = \frac{25 \text{ mV}}{I_E} = \frac{25 \text{ mV}}{4.3 \text{ mA}} = 5.8 \ \Omega$$

So

$$A_v = \frac{R_E}{r_e + R_E} = \frac{1 \text{ k}\Omega}{1.0058 \text{ k}\Omega} = 0.994$$

The difference is hardly worth the trouble in most cases. The current gain is

$$A_i = \frac{I_e}{I_{in}}$$

$$I_e = \frac{V_e}{R_E} = \frac{A_v V_b}{R_E} \cong \frac{1 \text{ V}}{1 \text{ k}\Omega} = 1 \text{ mA}$$

$$I_{in} = \frac{V_{in}}{R_{in}} = \frac{1 \text{ V}}{4.86 \text{ k}\Omega} = 0.21 \text{ mA}$$

$$A_i = \frac{1 \text{ mA}}{0.21 \text{ mA}} = 4.76$$

The power gain is

$$A_p \cong A_i = 4.76$$

Practice Exercise 8–7

Find the input impedance, voltage gain, current gain, and power gain for the circuit in Figure 8–29 with the following values: $V_{CC} = 12$ V, $R_1 = R_2 = 22$ kΩ, $R_E = 1.8$ kΩ, and $\beta = 100$.

The Darlington Pair

No Loading

As you have seen, β is a major factor in determining the input impedance. The β of the transistor limits the maximum achievable input impedance you can get from a given emitter-follower circuit.

One way to boost input impedance is to use a *Darlington pair,* as shown in Figure 8–30. The collectors of two transistors are connected, and the emitter of the first drives the base of the second. This configuration achieves β multiplication as shown in the following steps. The emitter current of the first transistor is

$$I_{e_1} \cong \beta_1 I_b$$

This emitter current becomes the base current for the second transistor, producing a second emitter current of

$$I_{e_2} \cong \beta_2 I_{e_1}$$
$$I_{e_2} \cong \beta_1 \beta_2 I_b$$

Therefore, the effective current gain of the Darlington pair is

$$\beta = \beta_1 \beta_2 \qquad\qquad (8\text{–}35)$$

The input impedance is $\beta_1 \beta_2 R_E$.

EXAMPLE 8–9

A given cascaded amplifier arrangement has the following voltage gains: $A_{v1} = 10$, $A_{v2} = 15$, and $A_{v3} = 20$. What is the overall gain? Also express each gain in dB and determine the total dB voltage gain.

Solution

$$A'_v = A_{v1}A_{v2}A_{v3} = (10)(15)(20) = 3000$$
$$A_{v1} \text{ (dB)} = 20 \log 10 = 20 \text{ dB}$$
$$A_{v2} \text{ (dB)} = 20 \log 15 = 23.52 \text{ dB}$$
$$A_{v3} \text{ (dB)} = 20 \log 20 = 26.02 \text{ dB}$$
$$A'_v \text{ (dB)} = 20 \text{ dB} + 23.52 \text{ dB} + 26.02 \text{ dB} = 69.54 \text{ dB}$$

Practice Exercise 8–9

In a certain multistage amplifier, the individual stages have the following voltage gains: $A_{v1} = 25$, $A_{v2} = 5$, and $A_{v3} = 12$. What is the overall gain? Express each gain in dB and determine the total dB voltage gain.

Multistage Analysis

For purposes of illustration, the two-stage capacitively-coupled amplifier in Figure 8–34 is used. Notice that both stages are identical common-emitter amplifiers with the output of the first stage *capacitively coupled* to the input of the second stage. Capacitive coupling prevents the dc bias of one stage from affecting that of the other. Notice, also, that the transistors are designated Q_1 and Q_2.

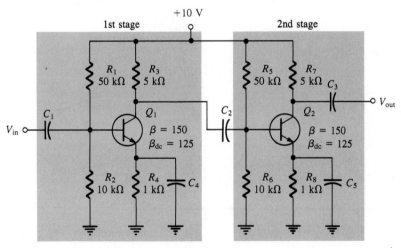

FIGURE 8–34
Two-stage amplifier.

$r_{in2} = Load 1$

Loading Effects

In determining the gain of the first stage, we must consider the loading effect of the second stage. Because the coupling capacitor C_2 appears as a short to the signal frequency, the total input impedance of the second stage presents an ac load to the first stage. Looking from the collector of Q_1, the two biasing resistors, R_5 and R_6, appear in parallel with the input impedance at the base of Q_2. In other words, the signal at the collector of Q_1 "sees" R_3 and R_5, R_6, and $R_{in(base)}$ of the second stage all in parallel to ac ground. Thus the *effective ac collector resistance* of Q_1 is the total of all these in parallel, as Figure 8–35 illustrates. The voltage gain of the first stage is reduced by the loading of the second stage, because the effective ac collector resistance of the first stage is less than the actual value of its collector resistor, R_3. Remember that $A_v = R_C/r_e$.

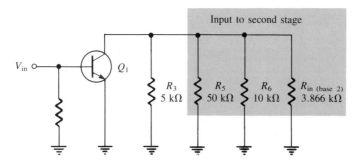

FIGURE 8–35
ac equivalent of first stage in Figure 8–34, showing loading from second stage.

Voltage Gain of the First Stage

The ac collector resistance of the first stage is

$$R_{c1} = R_3\|R_5\|R_6\|R_{in(base2)}$$

(Keep in mind that lowercase subscripts denote ac quantities such as for R_c.)

You can verify that $I_E = 0.97$ mA, $r_e = 25.77$ Ω, and $R_{in(base2)} = 3.866$ kΩ. The effective ac collector resistance of the first stage is as follows:

$$R_{c1} = 5\ k\Omega\|50\ k\Omega\|10\ k\Omega\|3.866\ k\Omega$$
$$= 1.73\ k\Omega$$

Therefore, the base-to-collector voltage gain of the first stage is

$$A_{v1} = \frac{R_{c1}}{r_e} = \frac{1.73\ k\Omega}{25.77\ \Omega} = 67$$

Voltage Gain of the Second Stage

The second stage has no load resistor, so the ac collector resistance is R_7, and the gain is

$$A_{v2} = \frac{R_7}{r_e} = \frac{5 \text{ k}\Omega}{25.77 \text{ }\Omega} = 194$$

Compare this to the gain of the first stage, and notice how much the loading effect of the second stage reduced the gain.

Overall Voltage Gain

The overall amplifier gain is

$$A'_v = A_{v1}A_{v2} = (67)(194) \cong 13,000$$

If an input signal of, say, 100 μV is applied to the first stage and if the attenuation of the input base circuit is neglected, an output from the second stage of (100 μV)(13,000) = 1.3 V will result. The overall gain can be expressed in dB as follows:

$$A'_v \text{ (dB)} = 20 \log(13,000) = 82.28 \text{ dB}$$

dc Voltage Levels in the Capacitively-Coupled Multistage Amplifier

Since both stages in Figure 8–34 are identical, the dc voltages for Q_1 and Q_2 are the same. The dc base voltage for Q_1 and Q_2 is

$$V_B \cong \left(\frac{R_2}{R_1 + R_2}\right)10 \text{ V} = \left(\frac{10 \text{ k}\Omega}{60 \text{ k}\Omega}\right)10 \text{ V} = 1.67 \text{ V}$$

The dc emitter and collector voltages are as follows:

$$V_E = V_B - 0.7 \text{ V} = 0.97 \text{ V}$$

$$I_E = \frac{V_E}{R_4} = \frac{0.97 \text{ V}}{1 \text{ k}\Omega} = 0.97 \text{ mA}$$

$$I_C \cong I_E = 0.97 \text{ mA}$$

$$V_C = V_{CC} - I_C R_3 = 10 \text{ V} - (0.97 \text{ mA})(5 \text{ k}\Omega) = 5.15 \text{ V}$$

Direct-Coupled Multistage Amplifiers

A basic two-stage, direct-coupled amplifier is shown in Figure 8–36. Notice that there are no coupling or bypass capacitors in this circuit. Because of the direct coupling, this type of amplifier has a better low-frequency response than the capacitively-coupled type because the reactance of coupling and bypass capacitors at very low frequencies becomes excessive for practical capacitor values. The increased reactance produces signal loss and gain reduction.

FIGURE 8–36
A basic two-stage direct-coupled amplifier.

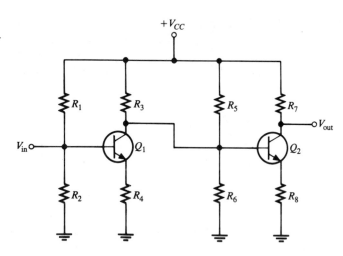

Direct-coupled amplifiers, on the other hand, can be used to amplify low frequencies all the way down to dc (0 Hz) without loss of gain because there are no capacitive
reactances in the circuit. The disadvantage of direct-coupled amplifiers is that small
changes in the dc bias voltages from temperature effects or power-supply variation are
amplified by the succeeding stages, which can result in a significant drift in the dc levels
throughout the circuit.

Transformer-Coupled Multistage Amplifiers

A basic transformer-coupled two-stage amplifier is shown in Figure 8–37. Transformer
coupling is often used in high-frequency amplifiers such as those in the RF (radio fre-

FIGURE 8–37
A basic two-stage transformer-coupled
amplifier.

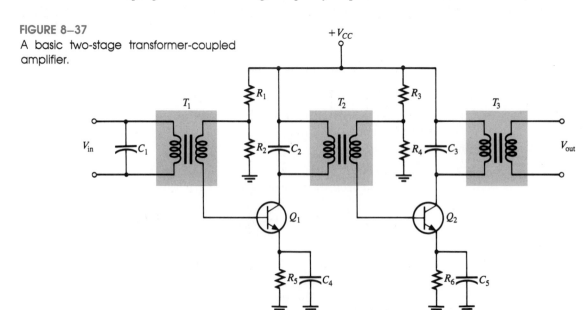

quency) and IF (intermediate frequency) sections of radio and TV receivers. At lower frequency ranges such as audio, the size of transformers is usually prohibitive. Capacitors are usually connected across one or both windings of the transformers to obtain resonance and increased selectivity for the band of frequencies to be amplified.

1. What does the term *stage* mean?

2. How is the overall gain of a multistage amplifier determined?

3. Express a voltage gain of 500 in dB.

SECTION
REVIEW
8–6

TROUBLESHOOTING

8–7

In working with any circuit, you must first know how it is supposed to work before you can troubleshoot it for a failure. The multistage capacitively-coupled amplifier discussed in the last section is used to illustrate a typical troubleshooting procedure.

The proper signal levels and dc voltage levels for the capacitively-coupled two-stage amplifier (determined in the previous section) are shown in Figure 8–38. The amplifier circuit is built on a PC board and connected to a test box, as shown in Figure 8–39. The input, output, supply voltage, and ground are routed through a connector in the test box. A step-by-step check of the circuit is shown in the figure using the *signal tracing* method of troubleshooting. A failure is assumed to exist in the circuit; as we go through the troubleshooting procedure, we will isolate the fault.

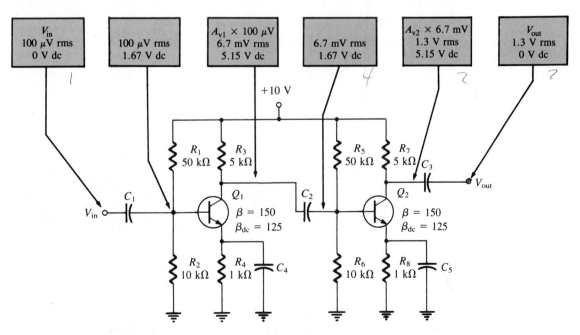

FIGURE 8–38
Two-stage amplifier with proper voltage levels indicated.

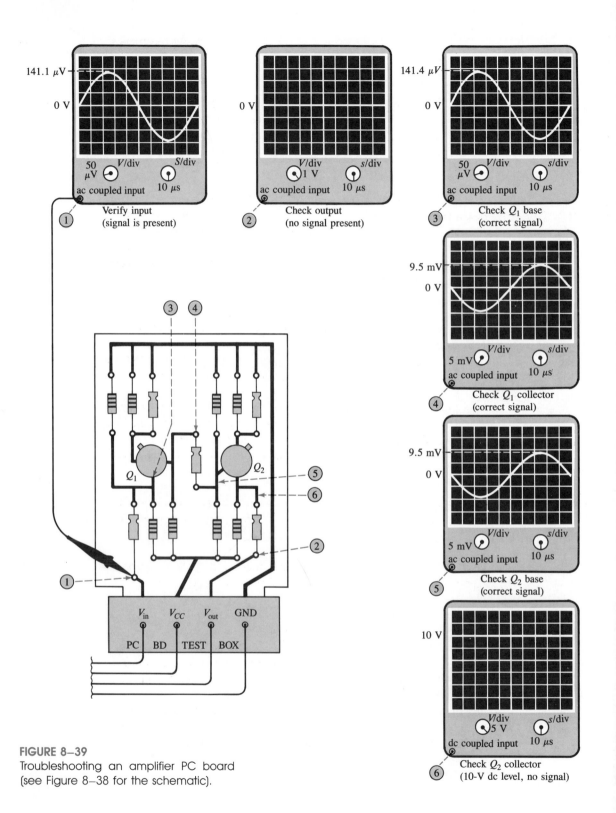

141.1 μV

0 V

50 μV — V/div

S/div

ac coupled input   10 μs

① Verify input
(signal is present)

0 V

V/div — 1 V

s/div

ac coupled input   10 μs

② Check output
(no signal present)

141.4 μV

0 V

50 μV — V/div

s/div

ac coupled input   10 μs

③ Check Q_1 base
(correct signal)

9.5 mV

0 V

5 mV — V/div

s/div

ac coupled input   10 μs

④ Check Q_1 collector
(correct signal)

9.5 mV

0 V

5 mV — V/div

s/div

ac coupled input   10 μs

⑤ Check Q_2 base
(correct signal)

10 V

V/div — 5 V

s/div

dc coupled input   10 μs

⑥ Check Q_2 collector
(10-V dc level, no signal)

Q_1 Q_2

V_{in} V_{CC} V_{out} GND

PC | BD | TEST | BOX

FIGURE 8–39
Troubleshooting an amplifier PC board
(see Figure 8–38 for the schematic).

Tracing the Signal Through the Circuit

In this particular example, an oscilloscope is used for tracing the signal and for checking dc levels. The Volts/division (V/div) and the Seconds/division (s/div) settings are used to illustrate the procedure; they may or may not correspond to settings available on your laboratory scope. The purpose of this procedure is to verify that the amplifier is operating properly and, if it is not, to determine the failure and isolate the faulty component.

Step 1. Connect the scope probe to the input point on the PC board as indicated in Figure 8–39 to verify that a proper input signal is present. Since the input signal has been set to an rms value of 100 μV (141.5 V peak) and a frequency of 10 kHz ($T = 100$ μs), the settings on the scope are selected to display at least one cycle of the sine wave (10 μs/div $\times$ 10 div = 100 μs) and to allow the full peak-to-peak waveform to appear. As you see in Figure 8–39, the proper waveform is displayed. More cycles can be displayed if desired.

Step 2. Connect the scope probe to the output point on the PC board as shown to verify that a proper output signal is present. The V/div switch is set at 1 V/div because an rms output voltage of 1.3 V (1.84 V peak) with a 0 V dc value is expected. The scope shows that there is no output signal. Therefore, the amplifier is not operating properly.

Step 3. We now go back to the input and begin to trace the signal through the amplifier (we have verified that there is a proper input signal). Connect the scope probe to the base of transistor Q_1 as indicated. The capacitor C_1 couples the input signal to the base of Q_1, and thus the same signal voltage is displayed. Notice that the oscilloscope input is selected for ac coupling, so the dc bias voltage at the base is not indicated.

Step 4. Connect the probe to the collector of Q_1 and set the V/div switch to 5 mV. The scope displays a signal with a peak value of 9.5 mV (6.7 mV rms) as expected. This verifies that the first stage of the amplifier is operating properly.

Step 5. Connect the probe to the base of Q_2 to verify that the Q_1 collector signal is passing through the coupling capacitor C_2. The scope display shows that it is.

Step 6. Connect the probe to the collector of Q_2. The scope shows that the signal is absent at this point. This means either that Q_2 is internally open so that the base signal is not getting through to the collector or that the collector is shorted to ac ground. As a further check, the scope input is switched to dc coupling so that the dc voltage at the Q_2 collector can be measured and the V/div switch is set to 5 V. As shown, the horizontal trace is deflected two divisions from the center axis (0 V), indicating the presence of 10 V dc on the Q_2 collector. This suggests that the base-collector junction is open or the collector lead is internally open and, therefore, the collector is pulled up to V_{CC} through R_7. The next step is to replace Q_2 and see if the amplifier works properly.

SECTION REVIEW 8-7

1. Name an important troubleshooting technique.

2. Assume that the base-emitter junction of Q_2 in Figure 8–36 shorts.
   (a) Will the ac signal at the base of Q_2 change? In what way?
   (b) Will the dc level at the base of Q_2 change? In what way?

A SYSTEM APPLICATION

8–8

You have learned that the voltage gain of a transistor amplifier is dependent on the collector impedance and the internal ac emitter resistance, r_e. A good illustration of how the dependence of the gain on r_e is used in a practical application is an amplifier with AGC (automatic gain control). Figure 8–40(a) shows a typical IF amplifier with the AGC derived from the detector circuit. The AGC controls the gain of the amplifier to maintain a relatively constant volume of sound for both weak and strong signals without having to readjust the volume control each time you change the station or when the station's signal strength changes because of weather, ionospheric conditions, or, in the case of a car radio, distance from the transmitter.

The AGC voltage level is obtained from the detector output. Recall that the detector removes the IF and leaves the relatively low-frequency audio signal. Further filtering by R_{AGC} and C_{AGC} smoothes out the audio and produces a slowly varying dc level that changes with the average variations in the strength of the received signal, as illustrated in Figure 8–40(b). The AGC filter has a long enough time constant that only average signal strength changes are used to control the amplifier gain. In this particular circuit, a negative AGC level is produced (the direction of the detector diode determines this). A larger negative AGC voltage lowers the forward bias on the base of the transistor and reduces the dc emitter current. A smaller negative AGC voltage increases the bias, causing an increase in the dc emitter current.

Recall that the ac emitter resistance is dependent on the dc emitter current by the relationship $r_e = 25 \text{ mV}/I_E$. When the strength of the received signal increases, the AGC voltage becomes more negative and lowers the transistor bias voltage. This causes r_e to increase because of a decrease in I_E. The increase in r_e lowers the voltage gain, which compensates for the increased signal strength. When the strength of the received signal decreases, the gain is raised. This action keeps the audio output constant.

FORMULAS

h parameters

(8–1) $\quad h_{ie} = \dfrac{V_b}{I_b}$ $\qquad\qquad$ Input impedance, common-emitter

(8–2) $\quad h_{re} = \dfrac{V_b}{V_c}$ $\qquad\qquad$ Reverse voltage ratio, common-emitter

(8–3) $\quad h_{fe} = \dfrac{I_c}{I_b}$ $\qquad\qquad$ Current gain, common-emitter

(8–4) $\quad h_{oe} = \dfrac{I_c}{V_c}$ $\qquad\qquad$ Output admittance, common-emitter

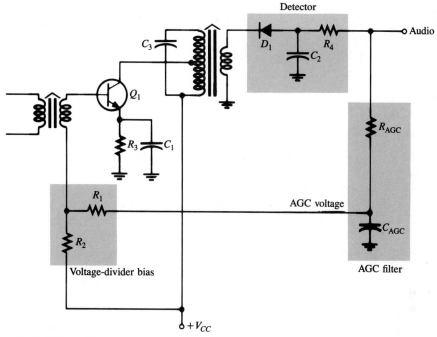

(a) Typical IF amplifier and AGC circuit

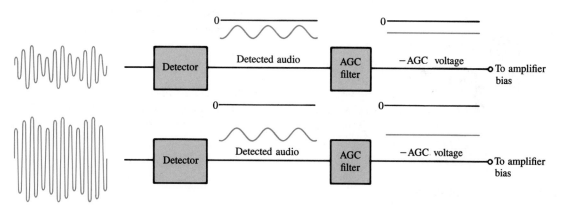

(b) When signal strength increases, the AGC voltage
becomes more negative, and vice versa

FIGURE 8–40
Automatic gain control in a superheterodyne receiver.

Conversions from h parameters to r parameters

(8–5) $\alpha = h_{fb}$ ac alpha

(8–6) $\beta = h_{fe}$ ac beta

(8–7) $r_e = \dfrac{h_{re}}{h_{oe}}$ ac emitter resistance

(8–8) $r_c = \dfrac{h_{re} + 1}{h_{oe}}$ ac collector resistance

(8–9) $r_b = h_{ie} - \dfrac{h_{re}}{h_{oe}}(1 + h_{fe})$ ac base resistance

(8–10) $r_e \cong \dfrac{25 \text{ mV}}{I_E}$ ac emitter resistance

Common-emitter

(8–11) $V_b = \left(\dfrac{R_{\text{in}}}{R_s + R_{\text{in}}}\right) V_{\text{in}}$ Base signal voltage

(8–12) $R_{\text{in(base)}} = \dfrac{V_b}{I_b}$ Input impedance at base

(8–13) $R_{\text{in(base)}} = \beta(r_e + R_E)$ Input impedance at base

(8–14) $R_{\text{in}} = R_1 \| R_2 \| R_{\text{in(base)}}$ Total input impedance, voltage-divider bias

(8–15) $R_{\text{out}} \cong R_C$ Output impedance

(8–16) $A_v = \dfrac{V_c}{V_b}$ Voltage gain, base-to-collector

(8–17) $A_v = \dfrac{R_C}{r_e + R_E}$ Voltage gain, base-to-collector, unloaded, unbypassed R_E

(8–18) $\dfrac{V_b}{V_{\text{in}}} = \dfrac{R_{\text{in}}}{R_s + R_{\text{in}}}$ Attenuation, base circuit

(8–19) $A'_v = A_v\left(\dfrac{V_b}{V_{\text{in}}}\right)$ Voltage gain, input-to-collector

(8–20) $A_v = \dfrac{R_C}{r_e}$ Voltage gain, base-to-collector, unloaded, bypassed R_E

(8–21) $R_c = \dfrac{R_C R_L}{R_C + R_L}$ ac collector resistance

(8–22) $A_v = \dfrac{R_c}{r_e}$ Voltage gain, base-to-collector, loaded, bypassed R_E

(8–23) $A_v = \dfrac{R_c}{r_e + R_{E_1}}$ Voltage gain, base-to-collector, partially bypassed R_E (swamping)

(8–24) $A_i = \dfrac{I_c}{I_{in}}$ Current gain

(8–25) $I_{in} = \dfrac{V_{in}}{R_{in}}$ Input signal current

(8–26) $A_p = A'_v A_i$ Power gain

Common-collector

(8–27) $A_v = \dfrac{R_E}{r_e + R_E}$ Voltage gain, base-to-emitter

(8–28) $R_{in(base)} \cong \beta(r_e + R_E)$ Input impedance at base

(8–29) $R_{in(base)} \cong \beta R_E$ Input impedance at base when $r_e \ll R_E$

(8–30) $R_{in} = R_1 \| R_2 \| R_{in(base)}$ Total input impedance, voltage-divider bias

(8–31) $R_{out} \cong \left(\dfrac{R_s}{\beta}\right) \| R_E$ Output impedance

(8–32) $A_i \cong \beta$ Current gain when $R_1 \| R_2 \gg \beta R_E$

(8–33) $A_i = \dfrac{I_e}{I_{in}}$ Current gain

(8–34) $A_p \cong A_i$ Power gain

(8–35) $\beta = \beta_1 \beta_2$ Darlington pair

Common-base

(8–36) $A_v \cong \dfrac{R_C}{r_e}$ Voltage gain, emitter-to-collector

(8–37) $R_{in(emitter)} = r_e$ Input impedance at emitter

(8–38) $R_{out} \cong R_C$ Output impedance

(8–39) $A_i \cong 1$ Current gain

(8–40) $A_p \cong A_v$ Power gain

Multistage amplifier

(8–41) $A'_v = A_{v1} A_{v2} A_{v3} \cdots A_{vn}$ Overall gain

(8–42) $A_v \,(dB) = 20 \log A_v$ dB voltage gain

(8–43) $A'_v \,(dB) = A_{v1} \,(dB) + A_{v2} \,(dB)$ Overall dB voltage gain
$\qquad\qquad + \cdots + A_{vn} \,(dB)$

SUMMARY

☐ A small-signal amplifier uses only a small portion of its load line under signal conditions.

☐ h parameters are important because they are usually specified on manufacturers' data sheets.

☐ r parameters are easily identifiable with a transistor's circuit operation.

☐ A common-emitter amplifier has the advantages of good voltage, current, and power gains, but the disadvantage of a relatively low input impedance.

☐ A common-collector amplifier has the advantages of a high input impedance and good current gain, but its voltage gain is approximately 1.

☐ The common-base amplifier has a good voltage gain, but it has a very low input impedance and its current gain is approximately 1.

☐ A *Darlington pair* provides beta multiplication for increased input impedance.

☐ The total gain of a multistage amplifier is the product of the individual gains (sum of dB gains).

☐ *Signal tracing* is a troubleshooting technique whereby a signal is checked point by point, beginning at the input and progressing toward the output.

SELF-TEST

1. Refer to Figure 8–3 and determine the peak-to-peak values of I_c and V_{ce} if the base current varies 15 μA above and 15 μA below the Q-point.

2. The following ac rms values are measured with the output of a common-emitter connection shorted: $I_b = 5\ \mu$A, $I_c = 800\ \mu$A, and $V_b = 100\ \mu$V. Determine h_{ie} and h_{fe}.

3. A certain manufacturer's data sheet provides the following h parameter data: $h_{ie} = 200\ \Omega$, $h_{fe} = 100$, $h_{re} = 0.001$, and $h_{oe} = 50\ \mu$S. From these data, determine the following r parameters: β, r_e, and r_c.

4. The dc emitter current of a certain biased transistor is 3 mA. What is the approximate r_e?

5. Draw the dc equivalent circuit and the ac equivalent circuit for the amplifier in Figure 8–41.

FIGURE 8–41

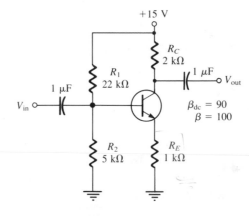

6. Find the dc emitter current and collector voltage for the circuit in Figure 8–41.

7. Determine the following values for the amplifier in Figure 8–41: **(a)** $R_{in(base)}$; **(b)** R_{in}; **(c)** A_v.

8. Connect a bypass capacitor across R_E in Figure 8–41, and repeat problem 7.

9. Connect a 10 kΩ load resistor to the output in Figure 8–41, and repeat problem 8.

10. What is the total input impedance of the emitter-follower in Figure 8–42? Neglect r_e.

FIGURE 8–42

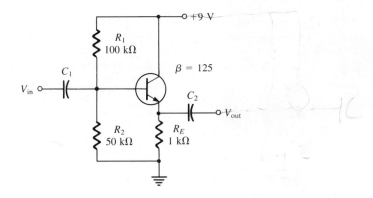

11. If an identical transistor is connected in Figure 8–42 to form a Darlington pair, what is the total input impedance?

12. Use the formula 25 mV/I_E to determine r_e for both transistors in the Darlington configuration of problem 11. $\beta_{dc} = \beta$.

13. Determine the input impedance, voltage gain, current gain, and power gain in Figure 8–43.

FIGURE 8–43

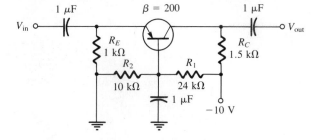

14. What is the overall gain of a four-stage cascaded amplifier arrangement if each stage has a voltage gain of 15? Express the overall gain in dB.

15. Connect two amplifiers identical to the one in Figure 8–41 in cascade. Bypass the emitter resistors and determine the overall gain.

PROBLEMS

Section 8–1

8–1 What is the lowest value of dc collector current to which a transistor having the characteristic curves in Figure 8–3 can be biased and still retain linear operation with a peak-to-peak base current swing of 20 μA?

8–2 What is the highest value of I_C under the conditions described in problem 8–1?

Section 8–2

8–3 What are the hybrid parameters that can be measured with each of the test circuits in Figure 8–44, and what is the value of each?

FIGURE 8–44

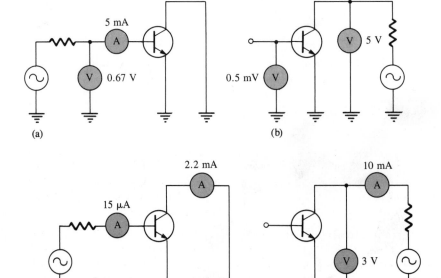

8–4 Use the h parameter values obtained in problem 8–3 to determine the r parameters β, α, r_e, r_b, and r_c.

8–5 A certain transistor has a dc beta (h_{FE}) of 130. If the dc base current is 10 μA, determine r_e. Assume $\alpha_{dc} = 0.99$.

8–6 At the dc bias point of a certain transistor circuit, $I_B = 15$ μA and $I_C = 2$ mA. Also, a variation in I_B of 3 μA about the Q-point produces a variation in I_C of 0.35 mA about the Q-point. Determine β_{dc} and β.

Section 8–3

8–7 Determine the following dc values for the amplifier in Figure 8–45.

    (a) V_B **(d)** I_C

    (b) V_E **(e)** V_C

    (c) I_E **(f)** V_{CE}

FIGURE 8–45

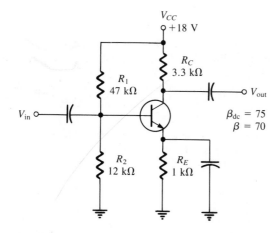

8–8 Determine the following ac values for the amplifier in Figure 8–45.
 (a) $R_{in(base)}$
 (b) R_{in}
 (c) A_v
 (d) A_i
 (e) A_p

8–9 Assume that a 600 Ω, 12 μV rms voltage source is driving the amplifier in Figure 8–45. Determine the overall gain by taking into account the attenuation in the base circuit, and find the *total* output voltage (ac and dc). What is the phase relationship of the collector signal voltage to the base signal voltage?

8–10 The amplifier in Figure 8–46 has a variable gain control, using a 100 Ω potentiometer for R_E with the wiper ac-grounded. As the potentiometer is adjusted, more or less of R_E is bypassed to ground, thus varying the gain. The total R_E remains constant to dc, keeping the bias fixed. Determine the maximum and minimum gains for this amplifier.

FIGURE 8–46

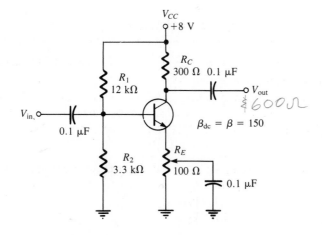

8–11 If a load resistance of 600 Ω is placed on the output of the amplifier in Figure 8–46, what are the maximum and minimum gains?

8–12 Find the overall maximum voltage gain, current gain, and power gain for the amplifier in Figure 8–46 with a 1 kΩ load, if it is being driven by a 300 Ω source.

8–13 Modify the schematic to show how you would "swamp out" the temperature effects of r_e in Figure 8–45 by making R_e at least ten times larger than r_e. Keep the same total R_E. How does this affect the gain?

Section 8–4

8–14 Determine the *exact* voltage gain for the emitter-follower in Figure 8–47.

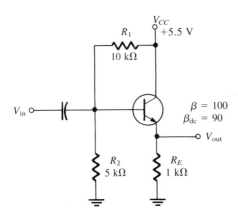

FIGURE 8–47

8–15 What is the total input impedance in Figure 8–47? What is the dc output voltage?

8–16 A load resistance is capacitively coupled to the emitter in Figure 8–47. In terms of signal operation, the load appears in parallel with R_E and reduces the effective emitter resistance. How does this affect the voltage gain?

8–17 In problem 8–16, what value of R_L will cause the voltage gain to drop to 0.9?

8–18 For the circuit in Figure 8–48, determine the following:
(a) Q_1 and Q_2 dc terminal voltages
(b) Overall β
(c) r_e for each transistor
(d) Total input impedance

8–19 Find the overall current gain A_i in Figure 8–48.

Section 8–5

8–20 What is the main disadvantage of the common-base amplifier compared to the common-emitter and the emitter-follower?

8–21 Find $R_{in(emitter)}$, A_v, A_i, and A_p for the amplifier in Figure 8–49.

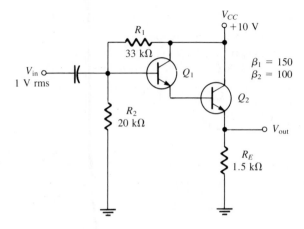

FIGURE 8–48

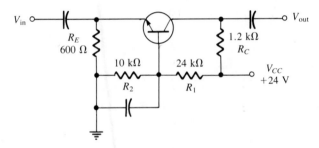

FIGURE 8–49

8–22 Match the following generalized characteristics with the appropriate amplifier configuration.
   (a) Unity current gain, good voltage gain, very low input impedance.
   (b) Good current gain, good voltage gain, low input impedance.
   (c) Good current gain, unity voltage gain, high input impedance.

Section 8–6

8–23 Each of two cascaded amplifier stages has an $A_v = 20$. What is the overall gain?

8–24 Each of three cascaded amplifier stages has a dB voltage gain of 10. What is the overall dB voltage gain? What is the actual overall voltage gain?

8–25 For the two-stage, capacitively-coupled amplifier in Figure 8–50, find the following values:
   (a) Voltage gain of each stage
   (b) Overall voltage gain
   (c) Express the gains found above in dB

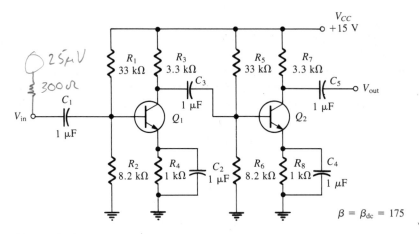

FIGURE 8–50

8–26 If the multistage amplifier in Figure 8–50 is driven by a 75 Ω, 50 μV source and loaded with an $R_L = 19$ kΩ, determine
   (a) Voltage gain of each stage
   (b) Overall voltage gain
   (c) Express the gains found above in dB.

8–27 Figure 8–51 shows a *direct-coupled* (that is, with no coupling capacitors between stages) two-stage amplifier. The dc bias of the first stage sets the dc bias of the second. Determine all dc voltages for both stages and the overall ac voltage gain.

FIGURE 8–51

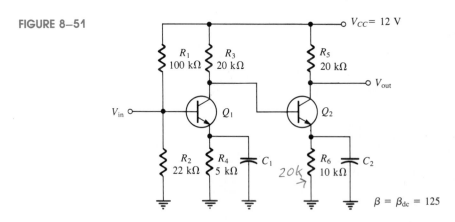

8–28 Refer to Figure 8–50 and determine the general effect of each of the following failures:
   (a) C_2 opens
   (b) C_3 opens
   (c) C_4 opens

(d) C_2 shorts

(e) Base-collector junction of Q_1 opens

(f) Base-emitter of Q_2 opens

8–29 Assume that you must troubleshoot the amplifier in Figure 8–50. Set up a table of *test point* values, input, output, and all transistor terminals that include both dc and rms values that you expect to observe when a 300 Ω test signal source with a 25 μV rms output is used.

8–30 Express the following voltage gains in dB:

(a) 12

(b) 50

(c) 100

(d) 2500

8–31 Express the following dB voltage gains as actual voltage ratios:

(a) 3 dB

(b) 6 dB

(c) 10 dB

(d) 20 dB

(e) 40 dB

Section 8–7

8–32 Assume that the coupling capacitor C_2 is shorted in Figure 8–34. What dc voltage will appear at the collector of Q_1?

8–33 Assume that R_5 opens in Figure 8–34. Will Q_2 be in cutoff or in conduction? What dc voltage will you observe at the Q_2 collector?

Section 8–1

1. Positive, negative.

2. V_{CE} is a dc quantity; V_{ce} is an ac quantity.

Section 8–2

1. h_i, h_r, h_f, h_o.

2. α, β, r_e, r_c, r_b.

3. 2.5 Ω.

Section 8–3

1. 100.

2. 13.

3. 909 Ω.

Section 8–4

1. Emitter-follower.

2. 1.

3. High input impedance.

Section 8–5
1. True.
2. True.

Section 8–6
1. One amplifier in a cascaded arrangement.
2. Product of individual gains.
3. 53.98 dB.

Section 8–7
1. Signal tracing.
2. (a) It will disappear because it is shorted to ground through the base-emitter junction and C_5.
   (b) Yes. It will decrease.

ANSWERS TO PRACTICE EXERCISES

Example 8–3: 9.8 mV

Example 8–4: 93.75

Example 8–5: 79.2

Example 8–7: $R_{in} = 10.37$ kΩ, $A_v = 0.995$, $A_i = 5.8$, $A_p = 5.8$

Example 8–9: 1500, 27.96 dB, 13.98 dB, 21.58 dB, 63.52 dB

Small-Signal FET Amplifiers

In this chapter you will learn
- [ ] How FETs are used as amplifiers
- [ ] How an FET can be represented by an equivalent circuit
- [ ] The factors that affect the voltage gain of an FET amplifier
- [ ] The characteristics of common-source, common-drain, and common-gate amplifiers
- [ ] How to troubleshoot an FET amplifier
- [ ] How FET amplifiers are used in a specific system application

Chapter 9

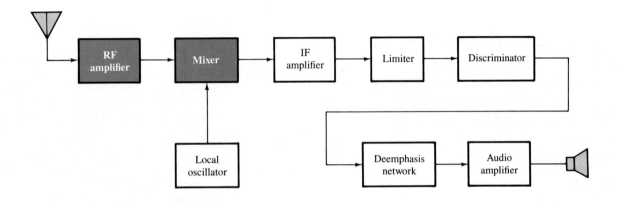

In Chapter 8 you studied amplifiers using bipolar transistors. In this chapter, we will examine FET amplifiers. Many concepts that relate to amplifiers using bipolar transistors apply as well to FET amplifiers. The three FET amplifier configurations are *common-source, common-drain,* and *common-gate*. These are analogous to the common-emitter, common-collector, and common-base configurations of bipolar amplifiers.

The format of this chapter is roughly parallel to that of Chapter 8, so that you can relate FET and bipolar amplifiers and see the major differences in operation and analysis. As indicated in Chapter 7, FETs have characteristics that make them preferable to BJTs in certain applications. For example, current state-of-the-art FM receiver systems use FETs in the RF amplifier and mixer sections, as indicated by the shaded blocks. Although the mixer is not classified as an amplifier, it usually provides some voltage gain in most systems.

SMALL-SIGNAL AMPLIFIER OPERATION

9–1

JFET

A self-biased n-channel JFET with an ac source capacitively coupled to the gate is shown in Figure 9–1(a). The resistor, R_G, serves two purposes: It keeps the gate at approximately 0 V dc (because I_{GSS} is extremely small), and its large value (usually several megohms) prevents loading of the ac signal source. The bias voltage is created by the drop across R_S. The bypass capacitor, C_2, keeps the source of the FET effectively at ac ground.

The signal voltage causes the gate-to-source voltage to swing above and below its Q-point value, causing a swing in drain current. As the drain current increases, the voltage drop across R_D also increases, causing the drain voltage to decrease. The drain current swings above and below its Q-point value *in-phase* with the gate-to-source voltage. The drain-to-source voltage swings above and below its Q-point value 180° *out-of-phase* with the gate-to-source voltage, as illustrated in Figure 9–1(b).

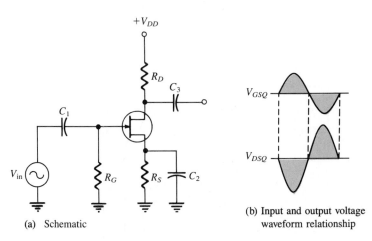

(a) Schematic

(b) Input and output voltage
waveform relationship

FIGURE 9–1
JFET common-source amplifier.

A Graphical Picture

The operation just described for an n-channel JFET can be illustrated graphically on both the transfer characteristic curve and the drain characteristic curve in Figure 9–2. Figure 9–2(a) shows how a sinusoidal variation, V_{gs}, produces a corresponding variation in I_d. As V_{gs} swings from the Q-point to a more negative value, I_d decreases from its Q-point value. As V_{gs} swings to a less negative value, I_d increases. Figure 9–2(b) shows a view of the same operation using the drain curves. The signal at the gate drives the drain current equally above and below the Q-point on the load line, as indicated by the arrows. Lines projected from the peaks of the gate voltage across to the I_D-axis and down to the V_{DS}-axis indicate the peak-to-peak variations of the drain current and drain-to-source voltage, as shown.

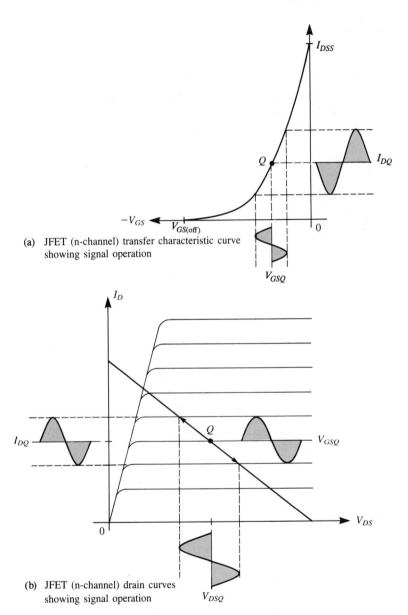

(a) JFET (n-channel) transfer characteristic curve
    showing signal operation

(b) JFET (n-channel) drain curves
    showing signal operation

FIGURE 9–2
JFET characteristic curves.

DE MOSFET

A zero-biased n-channel DE MOSFET with an ac source capacitively coupled to the gate
is shown in Figure 9–3. The gate is at approximately 0 V dc and the source terminal is
at ground, thus making $V_{GS} = 0$ V.

FIGURE 9–3

Zero-biased DE MOSFET common-source amplifier.

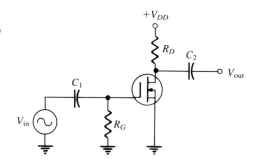

The signal voltage causes V_{gs} to swing above and below its 0 value, producing a swing in I_d as shown in Figure 9–4. The negative swing in V_{gs} produces the depletion mode, and I_d decreases. The positive swing in V_{gs} produces the enhancement mode, and I_d increases. Note that the enhancement mode is to the right of the vertical axis ($V_{GS} = 0$), and the depletion mode is to the left.

FIGURE 9–4

Depletion-enhancement operation of DE MOSFET shown on transfer characteristic curve.

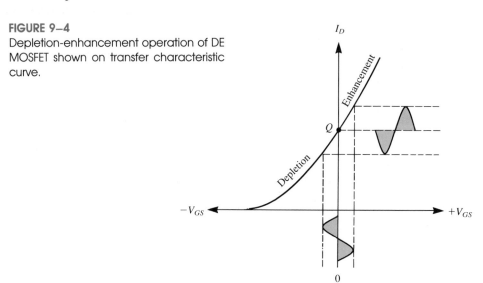

E MOSFET

A voltage-divider biased n-channel E MOSFET with an ac signal source capacitively coupled to the gate is shown in Figure 9–5. The gate is biased with a positive voltage such that $V_{GS} > V_{GS(th)}$.

As with the JFET and DE MOSFET, the signal voltage produces a swing in V_{gs} above and below its Q-point value. This, in turn, causes a swing in I_d, as illustrated in Figure 9–6. Operation is entirely in the enhancement mode.

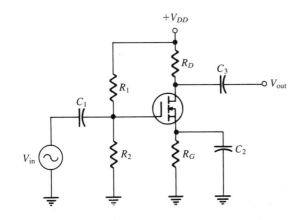

FIGURE 9–5
Common-source E MOSFET amplifier with voltage-divider bias.

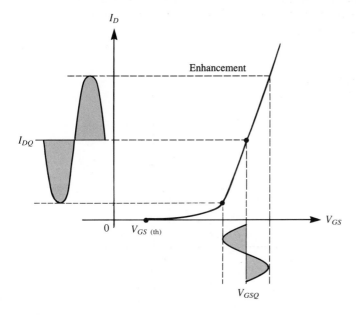

FIGURE 9–6
E MOSFET (n-channel) operation shown on transfer characteristic curve.

EXAMPLE 9–1

Transfer characteristic curves for an n-channel JFET, DE MOSFET, and E MOSFET are shown in Figure 9–7. Determine the peak-to-peak variation in I_d when V_{gs} is varied ± 1 V about its Q-point value for each curve.

Solution
(a) The JFET Q-point is at $V_{GS} = -2$ V and $I_D = 2.3$ mA. From the graph in Figure 9–7(a), $I_D = 3.3$ mA when $V_{GS} = -1$ V, and $I_D = 1.5$ mA when $V_{GS} = -3$ V. The peak-to-peak drain current is therefore 1.8 mA.

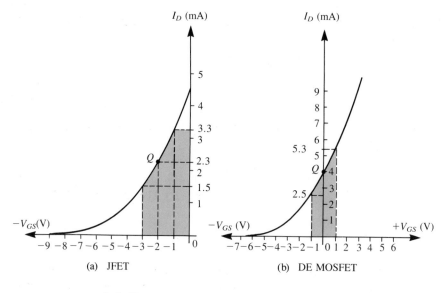

(a) JFET   (b) DE MOSFET

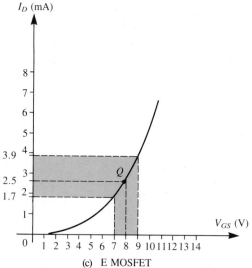

(c) E MOSFET

FIGURE 9–7

(b) The DE MOSFET Q-point is at $V_{GS} = 0$ V and $I_D = I_{DSS} = 4$ mA. From the graph in Figure 9–7(b), $I_D = 2.5$ mA when $V_{GS} = -1$ V, and $I_D = 5.3$ mA when $V_{GS} = +1$ V. The peak-to-peak drain current is therefore 2.8 mA.

(c) The E MOSFET Q-point is at $V_{GS} = +8$ V and $I_D = 2.5$ mA. From the graph in Figure 9–7(c), $I_D = 3.9$ mA when $V_{GS} = +9$ V, and $I_D = 1.7$ mA when $V_{GS} = +7$ V. The peak-to-peak drain current is therefore 2.2 mA.

1. When V_{gs} is at its positive peak, I_d is at its _____ peak and V_d is at its _____ peak.

2. What is the difference between V_{gs} and V_{GS}?

FET AMPLIFICATION

9–2

The transconductance was defined as $g_m = \Delta I_D / \Delta V_{GS}$. In terms of ac quantities, $g_m = I_d / V_{gs}$. By rearranging, we get

$$I_d = g_m V_{gs} \tag{9–1}$$

This equation says that the *output* current I_d equals the *input* voltage V_{gs} multiplied by the transconductance g_m.

Equivalent Circuit

An FET equivalent circuit representing the relationship in equation (9–1) is shown in Figure 9–8. In part (a), the resistance, r_{gs}, appears between gate and source, and a current source equal to $g_m V_{gs}$ appears between drain and source. Also, the internal drain-to-source resistance, r_{ds}, is included. In part (b), a simplified ideal model is shown. The resistance, r_{gs}, is assumed to be infinitely large so that there is an open circuit between gate and source. Also, r_{ds} is assumed large enough to neglect ($r_{ds} \cong 10\,R_d$).

FIGURE 9–8
FET equivalent circuits.

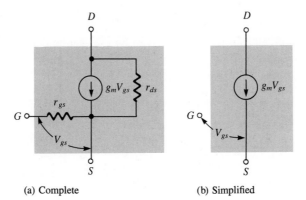

(a) Complete   (b) Simplified

Voltage Gain

An FET ideal equivalent circuit with an external ac drain resistance is shown in Figure 9–9. The ac voltage gain of this circuit is V_{out}/V_{in}, where $V_{in} = V_{gs}$. The voltage gain expression is therefore

$$A_v = \frac{V_{ds}}{V_{gs}} \tag{9–2}$$

FIGURE 9–9
Simplified FET equivalent circuit with external drain resistor.

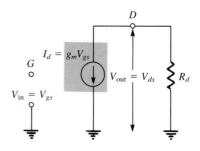

From the equivalent circuit, we see that

$$V_{ds} = I_d R_d$$

From the definition of transconductance in Chapter 7,

$$V_{gs} = \frac{I_d}{g_m}$$

Substituting these two expressions into equation (9–2),

$$A_v = \frac{I_d R_d}{I_d/g_m} = \frac{g_m I_d R_d}{I_d}$$

$$A_v = g_m R_d \tag{9-3}$$

EXAMPLE 9–2

A certain JFET has a $g_m = 4$ mS. With an external ac drain resistance of 1.5 kΩ, what is the ideal voltage gain?

Solution

$$A_v = g_m R_d = (4 \text{ mS})(1.5 \text{ k}\Omega) = 6$$

Effect of r_{ds} on Gain

If the internal drain-to-source resistance of the FET is taken into account, it appears in parallel with R_d, as indicated in Figure 9–10. The resulting gain expression is stated in

FIGURE 9–10
FET equivalent circuit including the internal drain-to-source resistance, r_{ds}.

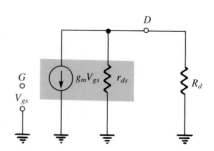

equation (9–4). As you can see, if r_{ds} is not sufficiently greater than R_d ($r_{ds} \geq 10\, R_d$), the gain is reduced from the ideal case of equation (9–3) to the following.

$$A_v = g_m \left(\frac{R_d r_{ds}}{R_d + r_{ds}} \right) \qquad (9\text{–}4)$$

EXAMPLE 9–3

The JFET in Example 9–2 has an $r_{ds} = 10\ \text{k}\Omega$. Determine the voltage gain when r_{ds} is taken into account.

Solution

The r_{ds} is effectively in parallel with the external drain resistance R_d. Therefore, equation (9–4) is used.

$$A_v = g_m \left(\frac{R_d r_{ds}}{R_d + r_{ds}} \right)$$

$$= (4\ \text{mS}) \left[\frac{(1.5\ \text{k}\Omega)(10\ \text{k}\Omega)}{1.5\ \text{k}\Omega + 10\ \text{k}\Omega} \right]$$

$$= (4\ \text{mS})(1.3\ \text{k}\Omega) = 5.2$$

The gain is reduced from a value of 6 (Example 9–2) because r_{ds} is in parallel with R_d.

Practice Exercise 9–3

A JFET has a $g_m = 6\ \text{mS}$, an $r_{ds} = 5\ \text{k}\Omega$, and an external ac drain resistance of $1\ \text{k}\Omega$. What is the voltage gain?

Effect of External Source Resistance on Gain

Including an external resistance from the FET's source terminal to ground results in the equivalent circuit of Figure 9–11. Examination of this circuit shows that the total input voltage between the gate and ground is

$$V_{in} = V_{gs} + I_d R_s$$

The output voltage taken across R_d is

$$V_{out} = I_d R_d$$

FIGURE 9–11

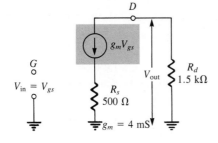

Therefore, the voltage gain is developed as follows.

$$A_v = \frac{V_{out}}{V_{in}} = \frac{I_d R_d}{V_{gs} + I_d R_s}$$

$$= \frac{g_m V_{gs} R_d}{V_{gs} + g_m V_{gs} R_s} = \frac{g_m V_{gs} R_d}{V_{gs}(1 + g_m R_s)}$$

$$A_v = \frac{g_m R_d}{1 + g_m R_s} \qquad\qquad (9\text{--}5)$$

EXAMPLE 9–4

An FET equivalent circuit is shown in Figure 9–11. Determine the voltage gain when the output is taken across R_d. Neglect r_{ds}.

Solution
There is an external source resistor, so the voltage gain is

$$A_v = \frac{g_m R_d}{1 + g_m R_s}$$

$$= \frac{(4 \text{ mS})(1.5 \text{ k}\Omega)}{1 + (4 \text{ mS})(500 \text{ }\Omega)}$$

$$= \frac{6}{1 + 2} = \frac{6}{3}$$

$$= 2$$

This is the same circuit as in Example 9–2 except for R_s. As you can see, R_s reduces the voltage gain. The voltage gain was ideally 6 in Example 9–2.

Practice Exercise 9–4
For the circuit in Figure 9–11, $g_m = 3.5$ mS, $R_s = 330 \text{ }\Omega$, and $R_d = 1.8$ kΩ. Find the voltage gain when the output is taken across R_d. Neglect r_{ds}.

SECTION REVIEW 9–2

1. An FET circuit has a $g_m = 2500 \text{ }\mu\text{S}$ and an $R_d = 10$ kΩ. Ideally, what voltage gain can it produce?

2. Two FETs have the same g_m. One has an $r_{ds} = 50$ kΩ and the other has an $r_{ds} = 100$ kΩ under the same conditions. Which FET can produce the higher voltage gain when used in a circuit with $R_d = 10$ kΩ?

COMMON-SOURCE AMPLIFIER

9–3

Now that you have an idea of how an FET functions as an amplifying device, we will look at a complete amplifier circuit.

Common-Source JFET Amplifiers

Figure 9–12 shows a common-source amplifier with a self-biased n-channel JFET. There are coupling capacitors on the input and output in addition to the source bypass capacitor. The circuit has a combination of dc and ac operations.

FIGURE 9–12
JFET common-source amplifier.

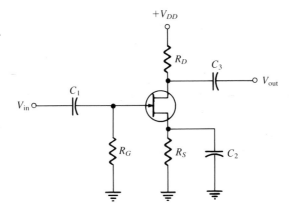

dc Analysis

To analyze the amplifier in Figure 9–12, the dc bias values must first be determined. To do this, a dc equivalent circuit is developed by replacing all capacitors with opens, as shown in Figure 9–13. Analysis of dc bias was covered in Chapter 7, but we will go through it again here. First, I_D must be determined before any analysis can be done. If the circuit is biased at the midpoint of load line, I_D can be calculated using I_{DSS} from the FET data sheet.

$$I_D = \frac{I_{DSS}}{2}$$

(9–6)

FIGURE 9–13
dc equivalent circuit for the amplifier in Figure 9–12.

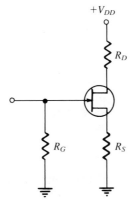

Otherwise, I_D must be known before any other dc calculations can be made. Determination of I_D from circuit parameter values is tedious because equation (9–7) must be solved for I_D. (This equation is derived by substitution of $V_{GS} = I_D R_S$ into equation (9–2).) Solution of the equation for I_D involves expanding it into a *quadratic* form and then finding the root of the quadratic. This is developed in Appendix B.

$$I_D = I_{DSS}\left[1 - \frac{I_D R_S}{V_{GS(\text{off})}}\right]^2 \qquad\qquad \textbf{(9–7)}$$

To make the solution of equation (9–7) easier, a BASIC computer program is provided below. The required inputs are the data sheet values of I_{DSS} and $V_{GS(\text{off})}$ and the value of R_S taken from the circuit. Once I_D is determined, the dc analysis can proceed using the following relationships:

$$V_S = V_{GS} = I_D R_S$$
$$V_D = V_{DD} - I_D R_D$$
$$V_{DS} = V_D - V_S$$

The following program computes I_D from equation (9–7) for self-biased JFETs:

```
10  CLS
20  PRINT "THIS PROGRAM COMPUTES JFET DRAIN CURRENT"
30  PRINT
40  PRINT "THE REQUIRED INPUTS ARE AS FOLLOWS"
50  PRINT "(1) IDSS FROM DATA SHEET"
60  PRINT "(2) VGS(OFF) FROM DATA SHEET"
70  PRINT "(3) RS FROM CIRCUIT DIAGRAM"
80  PRINT:PRINT:PRINT
90  INPUT "TO CONTINUE PRESS 'ENTER'";X
100 CLS
110 INPUT "VALUE OF IDSS IN AMPS";IDSS
120 INPUT "VALUE OF VGS(OFF) IN VOLTS";VGSOFF
130 INPUT "VALUE OF RS IN OHMS";RS
140 CLS
150 A=RS[2*IDSS/VGSOFF[2
160 B=-(1+2*RS*IDSS/ABS(VGSOFF))
170 C=IDSS
180 DI=(-B-SQR(B[2-4*A*C))/(2*A)
190 PRINT "ID =";ABS(DI);"A"
```

ac Equivalent Circuit

To analyze the signal operation of the amplifier in Figure 9–12, an *ac equivalent circuit* is developed as follows. The capacitors are replaced by effective shorts, based on the simplifying assumption that $X_C \cong 0$ at the signal frequency. The dc source is replaced by a ground, based on the assumption that the voltage source has a 0 internal resistance. The V_{DD} terminal is at a 0 volt ac potential and therefore acts as an *ac ground*.

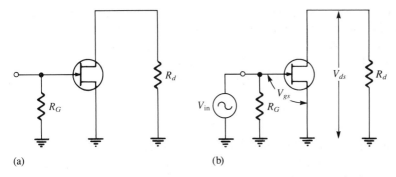

FIGURE 9–14
ac equivalent for the amplifier in Figure 9–12.

The ac equivalent circuit is shown in Figure 9–14(a). Notice that the $+V_{DD}$ end of R_d and the source terminal are both effectively at ac ground. Recall that in ac analysis, the ac ground and the actual circuit ground are treated as the same point.

Signal Voltage at the Gate

An ac voltage source is shown connected to the input in Figure 9–14(b). Since the input impedance to the FET is extremely high, practically all of the input voltage from the signal source appears at the gate with very little voltage dropped across the internal source resistance.

$$V_{gs} = V_{in}$$

The Output Voltage

Recall that the voltage gain expression $A_v = g_m R_d$ was developed in the last section. The output signal voltage V_{ds} at the drain is

$$V_{out} = V_{ds} = A_v V_{gs}$$

or

$$V_{out} = g_m R_d V_{in} \qquad\qquad (9\text{–}8)$$

where $R_d = R_D$ with no load resistor and $R_d = R_D \| R_L$ with a load resistor.

EXAMPLE 9–5

What is the total output voltage of the amplifier in Figure 9–15? The g_m is 4500 μS, I_{DSS} is 8 mA, and $V_{GS(off)}$ is -10 V.

Solution

First, let's find the dc output voltage using the computer program. When executed with the parameter values given, the computer determines that $I_D = 2$ mA. From this,

FIGURE 9–15

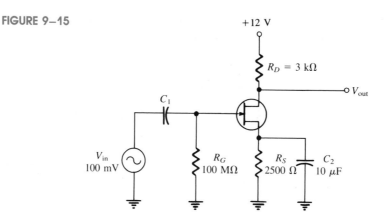

V_D is calculated.

$$V_D = V_{DD} - I_D R_D$$
$$= 12 \text{ V} - (2 \text{ mA})(3 \text{ k}\Omega)$$
$$= 6 \text{ V}$$

Next, the ac output voltage is found.

$$V_{out} = g_m R_D V_{in}$$
$$= (4500 \ \mu\text{S})(3 \text{ k}\Omega)(100 \text{ mV})$$
$$= 1.35 \text{ V rms}$$

The total output voltage is an ac signal with a peak-to-peak value of $1.35 \text{ V} \times 2.828 = 3.82 \text{ V}$, riding on a dc level of 6 V.

Effect of an ac Load on Gain

When a load is connected to the amplifier's output through a coupling capacitor, as shown in Figure 9–16(a), the drain resistance at the signal frequency is effectively R_D in parallel

FIGURE 9–16
JFET amplifier and its ac equivalent.

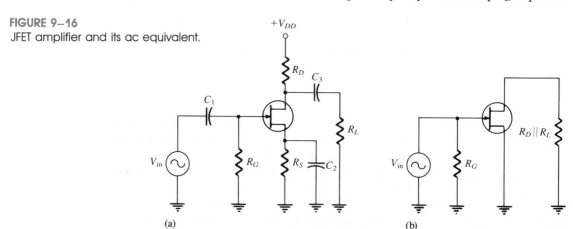

(a)   (b)

with R_L. Remember that the upper end of R_D is at ac ground. The ac equivalent circuit is shown in Figure 9–16(b). The total ac drain resistance is

$$R_d = \frac{R_D R_L}{R_D + R_L} \tag{9–9}$$

The effect of R_L is to reduce the voltage gain, as the next example illustrates.

EXAMPLE 9–6

If a 5 kΩ load resistor is ac coupled to the output of the amplifier in Example 9–5, what is the resulting rms output voltage?

Solution
The ac drain resistance is

$$R_d = \frac{R_D R_L}{R_D + R_L} = \frac{(3 \text{ k}\Omega)(5 \text{ k}\Omega)}{8 \text{ k}\Omega} = 1.88 \text{ k}\Omega$$

Calculation of V_{out} yields

$$V_{\text{out}} = g_m R_d V_{\text{in}} = (4500 \text{ } \mu\text{S})(1.88 \text{ k}\Omega)(100 \text{ mV}) = 846 \text{ mV rms}$$

The *unloaded* ac output voltage was 1.35 V rms in Example 9–5.

Practice Exercise 9–6
If a 10 kΩ load resistor is ac coupled to the output of the amplifier in Example 9–5 and the JFET is replaced with one having a $g_m = 3000 \text{ } \mu\text{S}$, what is the resulting rms output voltage?

Phase Inversion

The output voltage (at the drain) is 180° out-of-phase with the input voltage (at the gate). The phase inversion is sometimes denoted by a negative voltage gain, $-A_v$. Recall that the common-emitter bipolar amplifier also exhibited a phase inversion.

Input Impedance

Because the input to a common-source amplifier is at the gate, the input impedance is extremely high. Ideally, it approaches infinity and can be neglected. As you know, the high input impedance is produced by the reverse-biased pn junction in a JFET and by the insulated gate structure in a MOSFET. The actual input impedance seen by the signal source is the gate-to-ground resistor R_G in parallel with the FET's input inpedance, V_{GS}/I_{GSS}. The reverse leakage current I_{GSS} is typically given on the data sheet for a specific value of V_{GS} so that the input impedance of the device can be calculated.

EXAMPLE 9–7

What input impedance is seen by the signal source in Figure 9–17? $I_{GSS} = 30$ nA at $V_{GS} = 10$ V.

FIGURE 9–17

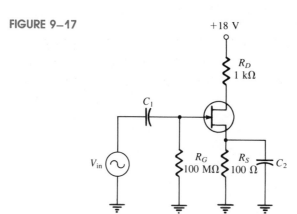

Solution

The input resistance at the gate of the JFET is

$$R_{IN(gate)} = \frac{V_{GS}}{I_{GSS}} = \frac{10 \text{ V}}{30 \text{ nA}} = 333 \text{ M}\Omega$$

The input impedance seen by the signal source is

$$R_{in} = R_G \| R_{IN(gate)} = 100 \text{ M}\Omega \| 333 \text{ M}\Omega = 76.9 \text{ M}\Omega$$

Common-Source MOSFET Amplifiers

In the previous discussion, the JFET was used to illustrate common-source amplifier analysis. The considerations for MOSFET amplifiers are similar, with the exception of the biasing arrangements required. A DE MOSFET is usually biased at $V_{GS} = 0$ and an E MOSFET at a V_{GS} greater than threshold.

Figure 9–18 shows a common-source amplifier using a DE MOSFET. The dc analysis of this amplifier is somewhat easier than for a JFET because $I_D = I_{DSS}$ at $V_{GS} = 0$.

FIGURE 9–18

DE MOSFET common-source amplifier.

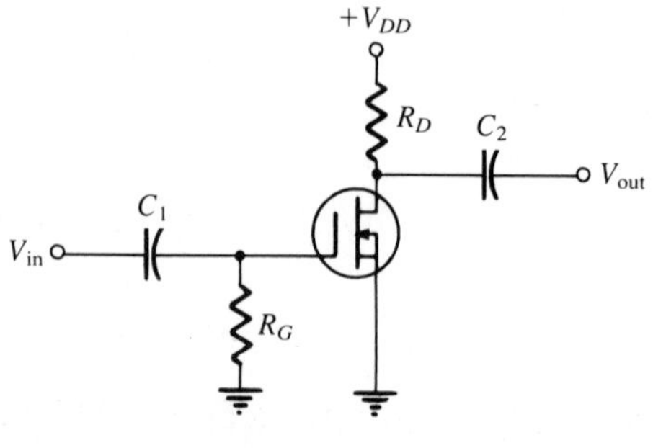

Once I_D is known, the analysis involves calculating only V_D.

$$V_D = V_{DD} - I_D R_D$$

The ac analysis is the same as for the JFET amplifier.

EXAMPLE 9–8

The DE MOSFET used in the amplifier of Figure 9–19 has an I_{DSS} of 12 mA and a g_m of 3.2 mS. Determine both the dc and ac output voltages. $V_{in} = 500$ mV.

FIGURE 9–19

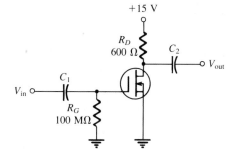

Solution

Since the amplifier is zero-biased,

$$I_D = I_{DSS} = 12 \text{ mA}$$

and, therefore,

$$V_D = V_{DD} - I_D R_D = 15 \text{ V} - (12 \text{ mA})(600 \ \Omega) = 7.8 \text{ V}$$

The signal voltage output is

$$V_{out} = g_m R_D V_{in} = (3.2 \text{ mS})(600 \ \Omega)(500 \text{ mV}) = 960 \text{ mV}$$

Practice Exercise 9–8

If a DE MOSFET with $g_m = 5$ mS and $I_{DSS} = 10$ mA replaces the one in Figure 9–19, what are the ac and dc output voltages when $V_{in} = 500$ mV?

Figure 9–20 shows a common-source amplifier using an E MOSFET. This circuit uses voltage-divider bias to achieve a V_{GS} above threshold. The general dc analysis proceeds as follows.

$$V_{GS} = \left(\frac{R_2}{R_1 + R_2} \right) V_{DD}$$

$$I_D = K[V_{GS} - V_{GS(th)}]^2$$

$$V_{DS} = V_{DD} - I_D R_D$$

The ac analysis is the same as for the JFET and the DE MOSFET circuits.

FIGURE 9–20
E MOSFET common-source amplifier.

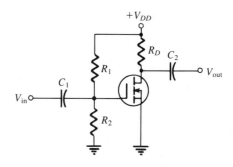

EXAMPLE 9–9

A common-source amplifier using an E MOSFET is shown in Figure 9–21. Find V_{GS}, I_D, V_{DS}, and the ac output voltage. $I_{D(on)} = 5$ mA at $V_{GS} = 10$ V, $V_{GS(th)} = 4$ V, and $g_m = 5.5$ mS. $V_{in} = 50$ mV.

FIGURE 9–21

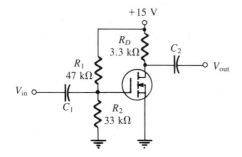

Solution

$$V_{GS} = \left(\frac{R_2}{R_1 + R_2}\right)V_{DD} = \left(\frac{33 \text{ k}\Omega}{80 \text{ k}\Omega}\right)15 \text{ V} = 6.19 \text{ V}$$

Using $V_{GS} = 10$ V,

$$K = \frac{I_{D(on)}}{[V_{GS} - V_{GS(th)}]^2} = \frac{5 \text{ mA}}{(10 \text{ V} - 4 \text{ V})^2} = 0.139 \text{ mA/V}^2$$

Therefore,

$$I_D = K[V_{GS} - V_{GS(th)}]^2 = 0.139 \text{ mA/V}^2(6.19 \text{ V} - 4 \text{ V})^2 = 0.667 \text{ mA}$$
$$V_{DS} = V_{DD} - I_D R_D = 15 \text{ V} - (0.667 \text{ mA})(3.3 \text{ k}\Omega) = 12.8 \text{ V}$$

The ac output voltage is

$$V_{out} = g_m R_D V_{in}$$
$$= (5.5 \text{ mS})(3.3 \text{ k}\Omega)(50 \text{ mV})$$
$$= 0.908 \text{ V rms}$$

Practice Exercise 9–9

In Figure 9–21, $I_{D(on)}$ = 8 mA at V_{GS} = 12 V, $V_{GS(th)}$ = 3 V, and g_m = 3.5 mS. Find V_{GS}, I_D, V_{DS}, and the ac output voltage. V_{in} = 50 mV.

---

1. What factors determine the voltage gain of a common-source FET amplifier?

2. A certain amplifier has an R_D = 1 kΩ. When a load resistance of 1 kΩ is capacitively coupled to the drain, how much does the gain change?

COMMON-DRAIN AMPLIFIER

A common-drain JFET amplifier is shown in Figure 9–22. Self-biasing is used in this circuit. The input signal is applied to the gate through a coupling capacitor, and the output is at the source terminal. There is no drain resistor. This circuit is, of course, analogous to the bipolar emitter-follower and is sometimes called a *source-follower*.

9–4

FIGURE 9–22
JFET common-drain amplifier (source-follower).

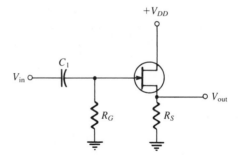

Voltage Gain

As in all amplifiers, the voltage gain is $A_v = V_{out}/V_{in}$. For the source-follower, V_{out} is I_dR_S and V_{in} is $V_{gs} + I_dR_S$, as shown in Figure 9–23. Therefore, the gate-to-source voltage

FIGURE 9–23
Voltages in a common-drain amplifier.

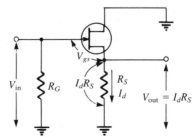

gain is $I_dR_S/(V_{gs} + I_dR_S)$. Substituting $I_d = g_mV_{gs}$ into the expression gives the following result.

$$A_v = \frac{g_mV_{gs}R_S}{V_{gs} + g_mV_{gs}R_S}$$

Cancelling V_{gs}, we get

$$A_v = \frac{g_mR_S}{1 + g_mR_S} \tag{9–10}$$

Notice here that the gain is always slightly *less than 1*. If $g_mR_S \gg 1$, then a good approximation is $A_v \cong 1$. Since the output voltage is at the source, it is in-phase with the gate (input) voltage.

Input Impedance

Because the input signal is applied to the gate, the input impedance seen by the input signal source is extremely high, just as in the common-source amplifier configuration. The gate resistor R_G, in parallel with the input impedance looking in at the gate, is the total input impedance.

EXAMPLE 9–10

Determine the voltage gain of the amplifier in Figure 9–24(a) using the data sheet information in Figure 9–24(b). Also determine the input impedance. Assume maximum data sheet values where available.

Solution
From the data sheet, $g_m = y_{fs} = 8000\ \mu S$. The gain is

$$A_v = \frac{g_mR_S}{1 + g_mR_S} = \frac{(8000\ \mu S)(10\ k\Omega)}{1 + (8000\ \mu S)(10\ k\Omega)} = 0.988$$

From the data sheet, $I_{GSS} = 10\ nA$ at $V_{GS} = 15\ V$. Therefore,

$$R_{IN(gate)} = \frac{15\ V}{10\ nA} = 1500\ M\Omega$$

$$R_{IN} = R_G\|R_{IN(gate)} = 100\ M\Omega\|1500\ M\Omega = 93.75\ M\Omega$$

SECTION REVIEW 9–4

1. What is the maximum voltage gain of a common-drain amplifier?
2. What factors influence the voltage gain?

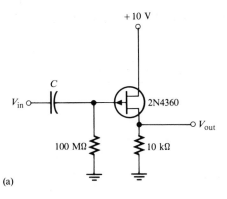

(a)

*ELECTRICAL CHARACTERISTICS (T_A = 25°C unless otherwise noted)

Characteristic	Symbol	Min	Max	Unit
OFF CHARACTERISTICS				
Gate-Source Breakdown Voltage (I_G = 10 μAdc, V_{DS} = 0)	$V_{(BR)GSS}$	20	–	Vdc
Gate-Source Cutoff Voltage (V_{DS} = -10 Vdc, I_D = 1.0 μAdc)	$V_{GS(off)}$	0.7	10	Vdc
Gate Reverse Current (V_{GS} = 15 Vdc, V_{DS} = 0) (V_{GS} = 15 Vdc, V_{DS} = 0, T_A = 65°C)	I_{GSS}	– –	10 0.5	nAdc μAdc
ON CHARACTERISTICS				
Zero-Gate Voltage Drain Current (Note 1) (V_{DS} = -10 Vdc, V_{GS} = 0)	I_{DSS}	3.0	30	mAdc
Gate-Source Voltage (V_{DS} = -10 Vdc, I_D = 0.3 mAdc)	V_{GS}	0.4	9.0	Vdc
SMALL SIGNAL CHARACTERISTICS				
Drain-Source "ON" Resistance (V_{GS} = 0, I_D = 0, f = 1.0 kHz)	$r_{ds(on)}$	–	700	Ohms
Forward Transadmittance (Note 1) (V_{DS} = -10 Vdc, V_{GS} = 0, f = 1.0 kHz)	$|y_{fs}|$	2000	8000	μmhos
Forward Transconductance (V_{DS} = -10 Vdc, V_{GS} = 0, f = 1.0 MHz)	$Re(y_{fs})$	1500	–	μmhos
Output Admittance (V_{DS} = -10 Vdc, V_{GS} = 0, f = 1.0 kHz)	$|y_{os}|$	–	100	μmhos
Input Capacitance (V_{DS} = -10 Vdc, V_{GS} = 0, f = 1.0 MHz)	C_{iss}	–	20	pF
Reverse Transfer Capacitance (V_{DS} = -10 Vdc, V_{GS} = 0, f = 1.0 MHz)	C_{rss}	–	5.0	pF
Common-Source Noise Figure (V_{DS} = -10 Vdc, I_D = 1.0 mAdc, R_G = 1.0 Megohm, f = 100 Hz)	NF	–	5.0	dB
Equivalent Short-Circuit Input Noise Voltage (V_{DS} = -10 Vdc, I_D = 1.0 mAdc, f = 100 Hz, BW = 15 Hz)	E_n	–	0.19	$μV/\sqrt{Hz}$

(b)

FIGURE 9–24

COMMON-GATE AMPLIFIER

9–5

A typical common-gate amplifier is shown in Figure 9–25. The gate is effectively at ac ground because of capacitor C_2. The input signal is applied at the source terminal through C_1. The output is coupled through C_3 from the drain terminal.

FIGURE 9–25
JFET common-gate amplifier.

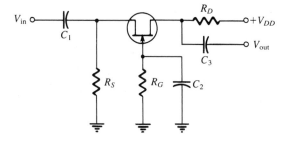

Voltage Gain

The voltage gain from source to drain is developed as follows.

$$A_v = \frac{V_{out}}{V_{in}} = \frac{V_d}{V_{gs}} = \frac{I_d R_d}{V_{gs}} = \frac{g_m V_{gs} R_d}{V_{gs}}$$

$$A_v = g_m R_d \qquad\qquad (9\text{–}11)$$

where $R_d = R_D$ with no load resistor, and $R_d = R_D \| R_L$ with a load connected. Notice that the gain expression is the same as for the common-source JFET amplifier.

Input Impedance

As you have seen, both the common-source and common-drain configurations have extremely high input impedances because the gate is the input terminal. In contrast, the common-gate configuration has a low input impedance, as shown in the following steps. First, the input current (source current) is equal to the drain current.

$$I_{in} = I_s = I_d = g_m V_{gs}$$

The input voltage equals V_{gs}.

$$V_{in} = V_{gs}$$

The input impedance at the source terminal is therefore

$$R_{in(source)} = \frac{V_{in}}{I_{in}} = \frac{V_{gs}}{g_m V_{gs}}$$

$$R_{in(source)} = \frac{1}{g_m} \qquad\qquad (9\text{–}12)$$

If, for example, g_m has a value of 4000 μS, then

$$R_{in(source)} = \frac{1}{4000}\ \mu S = 250\ \Omega$$

EXAMPLE 9–11

Determine the voltage gain and input impedance of the amplifier in Figure 9–26.

FIGURE 9–26

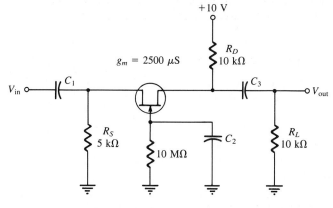

Solution

This common-gate amplifier has a load resistor, so the effective drain impedance is $R_D \| R_L$ and the gain is

$$A_v = g_m(R_D \| R_L) = (2500 \ \mu S)(10 \text{ k}\Omega \| 10 \text{ k}\Omega) = 12.5$$

The input impedance at the source terminal is

$$R_{in(source)} = \frac{1}{g_m} = \frac{1}{2500 \ \mu S} = 400 \ \Omega$$

The signal source actually sees R_S in parallel with $R_{in(source)}$, so the total input impedance is

$$R_{in} = 400 \ \Omega \| 5 \text{ k}\Omega = 370 \ \Omega$$

Summary of Gain and Input Impedance Characteristics

A summary of the gain and input impedance characteristics for the three FET amplifier configurations is given in Table 9–1.

TABLE 9–1
FET amplifier gain and input impedance

	Common-source	Common-drain	Common-gate
Voltage gain A_v	$g_m R_d$	$\dfrac{g_m R_S}{1 + g_m R_S}$	$g_m R_d$
Input impedance	$\left(\dfrac{V_{GS}}{I_{GSS}}\right) \| R_G$	$\left(\dfrac{V_{GS}}{I_{GSS}}\right) \| R_G$	$\left(\dfrac{1}{g_m}\right) \| R_S$

1.   What is a major difference between a common-gate amplifier and the other two
     configurations?

2.   What common factor determines the voltage gain and the input impedance of a
     common-gate amplifier?

TROUBLESHOOTING

9–6

The technician who understands the basics of circuit operation and who can, if necessary,
perform basic analysis on a given circuit is much more valuable than the individual who is
limited to carrying out routine test procedures. In this example of troubleshooting, we will
discuss how to test a circuit board that has only a schematic with no specified test procedure or voltage levels. Basic knowledge of how the circuit operates and the ability to do
a quick circuit analysis are thus useful in this case.

   Assume that you are given a circuit board pulled from the audio amplifier section of
a sound system and told simply that it is not working properly. The first step is to obtain
the system schematic and locate this particular circuit on it. Notice that the circuit is a
two-stage FET amplifier, as shown in Figure 9–27. The problem is approached in the
following sequence.

Step 1.   Determine what the voltage levels in the circuit should be so that you know what
to look for. First, pull a data sheet on the particular transistor (both Q_1 and Q_2 are found to
be the same type of transistor from the label on the case) and determine the g_m so that you
can calculate the voltage gain. Assume that for this particular device, a typical g_m of
5000 μS is specified. Now, calculate the expected voltage gain of each stage (notice they
are identical).

$$A_v = g_m R_2 = (5000\ \mu\text{S})(1.5\ \text{k}\Omega) = 7.5$$

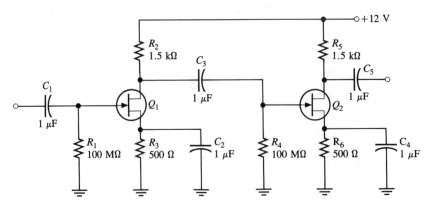

FIGURE 9–27
Two-stage FET amplifier circuit.

Chapter 10

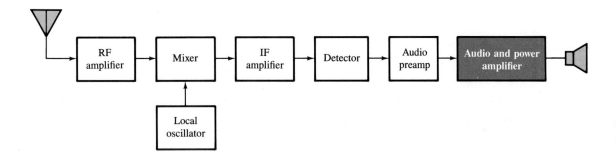

Power amplifiers are *large-signal* amplifiers, which generally means that a much larger portion of the load line is used during signal operation than in a small-signal amplifier. In this chapter we will cover the general classes of large-signal amplifiers: class A, class B, and class C. These amplifier classifications are based on the percentage of an input cycle for which the amplifier operates in its linear region.

The emphasis in large-signal amplifiers is *power amplification*. Power amplifiers are normally used as the final stage of a receiving or transmitting system to provide signal power to speakers or to a transmitting antenna, for example. The final-stage audio amplifier in AM and FM receivers is typically a push-pull power amplifier.

CLASS A AMPLIFIERS

10–1   When an amplifier, whether common-emitter, common-collector, or common-base, is biased such that it operates in the linear region for 360° of the input cycle, it is a *class A* amplifier. This is illustrated in Figure 10–1, where the output waveform is an amplified replica of the input and may be either in-phase or 180° out-of-phase with the input. All the small-signal amplifiers discussed in previous chapters are class A.

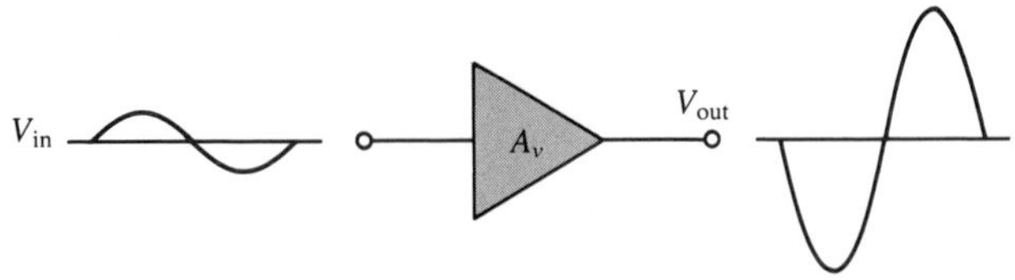

FIGURE 10–1
Class A amplifier (inverting).

Q-Point Is Centered for Maximum Output Signal

When the Q-point is at the center of the ac load line (midway between saturation and cutoff), a maximum class A signal can be obtained. This is graphically portrayed in the load line in Figure 10–2(a). Ideally, the collector current can vary from its Q-point value I_{CQ}, up to its saturation value $I_{c(sat)}$, and down to its cutoff value of 0. This operation is indicated in Figure 10–2(b).

The *peak* value of the collector current is I_{CQ}, and the *peak* value of the collector-to-emitter voltage is V_{CEQ}. This is the largest signal possible from a class A amplifier. If the input signal is too large, the amplifier is driven further than this and will go into cutoff and saturation, as illustrated in Figure 10–3 on p. 348.

Noncentered Q-Point Limits Output Swing

If the Q-point is not centered on the load line, the output signal is limited to less than the possible maximum. Figure 10–4(a) on p. 348 shows a load line with the Q-point moved away from center toward cutoff. The output swing is limited by cutoff in this case. The collector current can swing only down to near 0 and an *equal amount* above I_{CQ}. The collector-to-emitter voltage can swing only up to its cutoff value and an *equal amount* below V_{CEQ}. If the amplifier is driven any further than this, it will *go into cutoff*, as shown in Figure 10–4(b), and the waveforms will appear to be clipped off on one peak.

Figure 10–5(a) on p. 349 shows a load line with the Q-point moved away from center toward saturation. In this case, the output swing is limited by saturation. The collector current can swing only up to near saturation and an *equal amount* below I_{CQ}. The collector-to-emitter voltage can swing only down to near its saturation value and an *equal amount* above V_{CEQ}. If the amplifier is driven any further than this, it will *go into saturation*, as shown in Figure 10–5(b).

Large-Signal Load Line Operation

Recall that an amplifier such as that shown in Figure 10–6 on p. 349 can be represented in terms of either its dc or its ac equivalent.

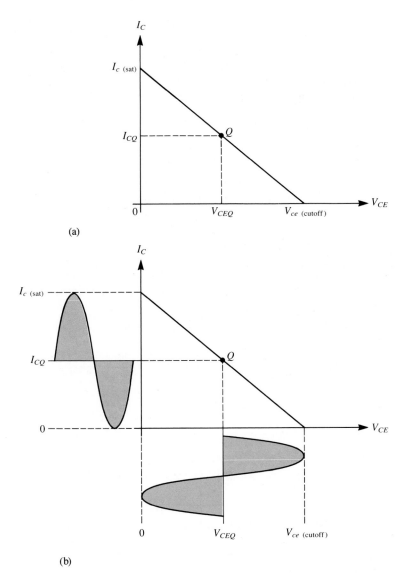

(a)

(b)

FIGURE 10–2
Maximum class A output (centered Q-point).

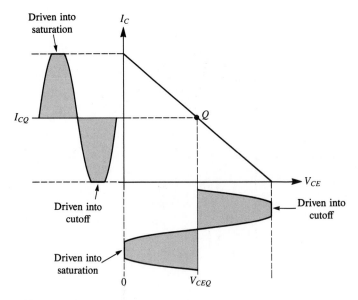

FIGURE 10–3
Waveforms are clipped off at cutoff and saturation because amplifier is overdriven (too large an input signal) about a centered Q-point.

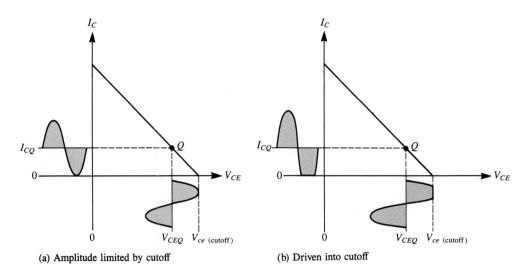

(a) Amplitude limited by cutoff   (b) Driven into cutoff

FIGURE 10–4
Q-point closer to cutoff.

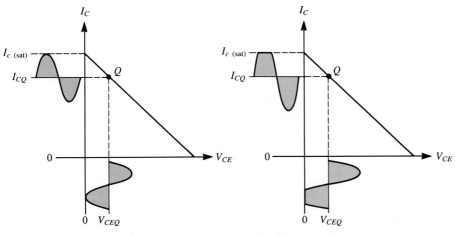

(a) Amplitude limited by saturation   (b) Driven into saturation

FIGURE 10–5
Q-point closer to saturation.

FIGURE 10–6
Common-emitter amplifier.

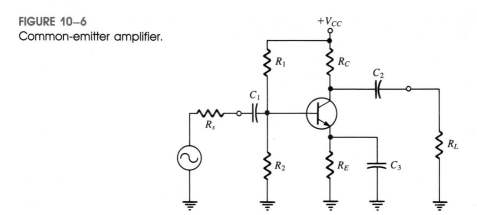

Using the dc equivalent in Figure 10–7(a), we can determine the dc load line as follows: $I_{C(\text{sat})}$ occurs when $V_{CE} \cong 0$, so

$$I_{C(\text{sat})} \cong \frac{V_{CC}}{R_C + R_E}$$

$V_{CE(\text{cutoff})}$ occurs when $I_C \cong 0$, so

$$V_{CE(\text{cutoff})} \cong V_{CC}$$

The dc load line is shown in Figure 10–7(b).

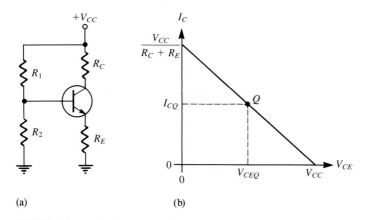

(a)   (b)

FIGURE 10–7
dc equivalent circuit and dc load line for the amplifier in Figure 10–6.

From the ac viewpoint, the circuit in Figure 10–6 looks different than from the dc viewpoint. The collector load is different and the emitter resistance is 0; therefore, the *ac load* line is different from the *dc load* line. How much collector current can there be under ac conditions before saturation occurs? To answer this question, we will refer to the ac equivalent circuit and ac load line in Figure 10–8. I_{CQ} and V_{CEQ} are the dc Q-point coordinates. Going from the Q-point to the saturation point, the collector-to-emitter voltage changes from V_{CEQ} to 0; that is, $\Delta V_{CE} = V_{CEQ}$. The swing in collector current going from the Q-point to saturation is therefore

$$\Delta I_C = \frac{\Delta V_{CE}}{R_C \| R_L} = \frac{V_{CEQ}}{R_c}$$

The maximum (saturation) ac collector current is

$$I_{c(\text{sat})} = I_{CQ} + \Delta I_C$$

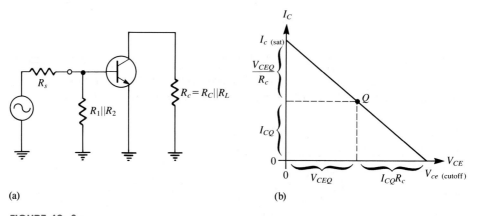

(a)   (b)

FIGURE 10–8
ac equivalent circuit and ac load line for the amplifier in Figure 10–6.

Thus,

$$I_{c(\text{sat})} = I_{CQ} + \frac{V_{CEQ}}{R_c} \qquad (10\text{--}1)$$

Going from the Q-point to the cutoff point, the *collector current* swings from I_{CQ} to near 0; that is, $\Delta I_C = I_{CQ}$. The swing in collector-to-emitter voltage going from the Q-point to cutoff is therefore

$$\Delta V_{CE} = (\Delta I_C)R_c = I_{CQ}R_c$$

The cutoff value of ac collector-to-emitter voltage is

$$V_{ce(\text{cutoff})} = V_{CEQ} + I_{CQ}R_c \qquad (10\text{--}2)$$

These results are shown on the ac load line of Figure 10–9. The corresponding dc load line is shown for comparison.

FIGURE 10–9
dc and ac load lines.

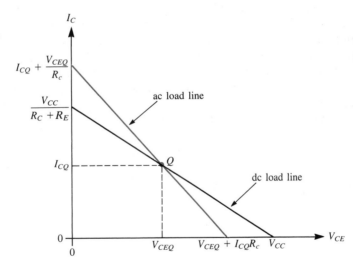

EXAMPLE 10–1
Determine the collector current and the collector-to-emitter voltage at the points of saturation and cutoff in Figure 10–10 under signal (ac) operation. Assume $X_{C_1} = X_{C_2} = X_{C_3} \cong 0$.

Solution
The Q-point values for this amplifier are determined as follows:

$$V_{BQ} = \left(\frac{5\ \text{k}\Omega}{15\ \text{k}\Omega}\right) 10\ \text{V} = 3.33\ \text{V}, \text{ neglecting } R_{\text{IN(base)}}$$

$$I_{EQ} = \frac{3.33\ \text{V} - 0.7\ \text{V}}{500\ \Omega} = 5.26\ \text{mA}$$

FIGURE 10–10

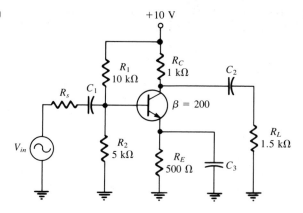

Therefore,

$$I_{CQ} \cong 5.26 \text{ mA}$$

and

$$V_{CQ} \cong 10 \text{ V} - (5.26 \text{ mA})(1 \text{ k}\Omega) = 4.74 \text{ V}$$

Therefore,

$$V_{CEQ} = V_{CQ} - I_{EQ}R_E = 4.74 \text{ V} - 2.63 \text{ V} = 2.11 \text{ V}$$

The point of *saturation* under *ac* conditions is determined as follows:

$$V_{ce(\text{sat})} \cong 0 \text{ V}$$

$$I_{c(\text{sat})} = I_{CQ} + \frac{V_{CEQ}}{R_c}$$

The ac collector resistance is

$$R_c = \frac{R_C R_L}{R_C + R_L} = \frac{(1 \text{ k}\Omega)(1.5 \text{ k}\Omega)}{2.5 \text{ k}\Omega} = 600 \text{ }\Omega$$

Thus

$$I_{c(\text{sat})} = 5.26 \text{ mA} + \frac{2.11 \text{ V}}{600 \text{ }\Omega}$$

$$= 5.26 \text{ mA} + 3.52 \text{ mA}$$

$$= 8.78 \text{ mA}$$

The point of *cutoff* under *ac* conditions is determined as follows:

$$I_{c(\text{cutoff})} = 0 \text{ A}$$

$$V_{ce(\text{cutoff})} = V_{CEQ} + I_{CQ}R_c$$

$$= 2.11 \text{ V} + (5.26 \text{ mA})(600 \text{ }\Omega)$$

$$= 5.27 \text{ V}$$

These results show that the collector current can swing up to almost 8.78 mA or the collector-to-emitter voltage can swing up to almost 5.27 V without the peaks being clipped.

Practice Exercise 10–1

Determine I_c and V_{ce} at the points of saturation and cutoff in Figure 10–10 for the following circuit values: $V_{CC} = 15$ V and $\beta = 150$.

As described earlier, a *centered* Q-point allows a maximum unclipped output swing. A closer look at Example 10–1 shows that the Q-point is *not centered,* and therefore, the unclipped output swing is somewhat less than it could be if the Q-point were centered on the ac load line. This is examined further in Example 10–2.

EXAMPLE 10–2

Figure 10–11 shows the ac load line for the circuit in Example 10–1. Notice that the Q-point is not centered. The maximum output swings are:

$$\Delta I_C = I_{c(\text{sat})} - I_{CQ}$$
$$= 8.78 \text{ mA} - 5.26 \text{ mA}$$
$$= 3.52 \text{ mA}$$

and

$$\Delta V_{CE} = V_{CEQ} = 2.11 \text{ V}$$

Determine the maximum output swings for collector current and collector-to-emitter voltage when the Q-point is centered, assuming the same load line is maintained.

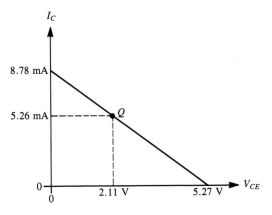

FIGURE 10–11

Solution

The maximum output swings for a centered Q-point are

$$\Delta I_C = I_{CQ} = \frac{I_{c(sat)}}{2} = \frac{8.78 \text{ mA}}{2} = 4.39 \text{ mA}$$

$$\Delta V_{CE} = V_{CEQ} = \frac{V_{ce(cutoff)}}{2} = \frac{5.27 \text{ V}}{2} = 2.635 \text{ V}$$

Large-Signal Voltage Gain

The voltage gain of a class A large-signal amplifier is determined in the same way as for a small-signal amplifier with the exception that the formula $r_e \cong 25 \text{ mV}/I_E$ is not valid for the large-signal amplifier. This is because the signal swings over a large portion of the transconductance curve. Since $r_e = \Delta V_{BE}/\Delta I_C$, the value is different for large-signal operation than it is for small-signal conditions.

The *large-signal ac emitter resistance, r'_e,* can be determined graphically from the transconductance curve, as shown in Figure 10–12, using the relationship in equation (10–3). Note that the prime mark (') distinguishes the large-signal parameter from the small-signal r_e.

$$r'_e = \frac{\Delta V_{BE}}{\Delta I_C} \tag{10–3}$$

The voltage-gain formula for a common-emitter, large-signal amplifier is therefore

$$A_v = \frac{R_c}{r'_e} \tag{10–4}$$

FIGURE 10–12

Determination of r'_e from the transconductance curve.

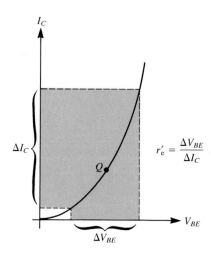

EXAMPLE 10-3

Find the large-signal voltage gain of the amplifier in Figure 10–13. Assume that r'_e has been found to be 8 Ω from graphical data.

FIGURE 10–13

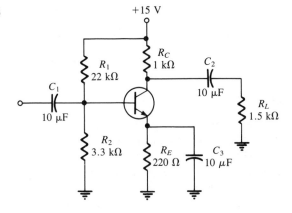

Solution

$$R_c = \frac{(1 \text{ k}\Omega)(1.5 \text{ k}\Omega)}{2.5 \text{ k}\Omega} = 600 \text{ }\Omega$$

$$A_v = \frac{R_c}{r'_e} = \frac{600 \text{ }\Omega}{8 \text{ }\Omega} = 75$$

Practice Exercise 10-3

Find the large-signal voltage gain of the amplifier in Figure 10–13 if the supply voltage is changed to 9 V, R_L is changed to 1 kΩ, and r'_e is 10 Ω.

Nonlinear Distortion

When the collector current swings over a large portion of the transconductance curve, distortion can occur on the negative half-cycle. This is caused by the greater nonlinearity on the lower end of the curve, as shown in Figure 10–14. This distortion can be sufficiently reduced by keeping the collector current on the more linear part of the curve (at higher values of I_{CQ} and V_{BEQ}).

Power Gain

The main purpose of a large-signal amplifier is to achieve power gain. If we assume that the large-signal current gain A_i is approximately equal to β_{dc}, then the power gain for a common-emitter amplifier is

$$A_p = A_i A_v = \beta_{dc} A_v$$

$$A_p = \beta_{dc} \frac{R_c}{r'_e} \tag{10-5}$$

FIGURE 10–14
Nonlinear distortion.

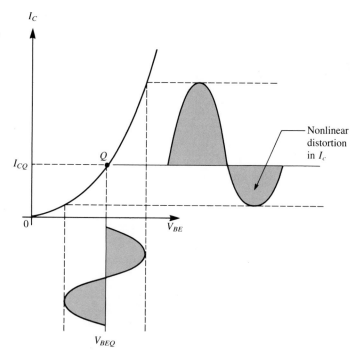

Quiescent Power

The power dissipation of a transistor with no signal input is the product of its Q-point current and voltage.

$$P_{DQ} = I_{CQ}V_{CEQ} \qquad \text{(10–6)}$$

As you will see, the quiescent power is the *maximum* power that the class A transistor must handle; therefore, its *power rating* should exceed this value.

Output Power

In general, for any Q-point location on the ac load line, the output power of a common-emitter amplifier is the product of the rms collector current and the rms collector-to-emitter voltage.

$$P_{\text{out}} = V_{ce}I_c \qquad \text{(10–7)}$$

Let's now consider the output power for three cases of Q-point location.

Q-Point Closer to Saturation. When the Q-point is closer to saturation, the maximum collector-to-emitter voltage swing is V_{CEQ}, and the maximum collector current swing is V_{CEQ}/R_c, as shown in Figure 10–15(a). The output power is therefore

$$P_{\text{out}} = (0.707 \, V_{CEQ}/R_c)(0.707 \, V_{CEQ})$$

$$P_{\text{out}} = \frac{0.5 \, V_{CEQ}^2}{R_c} \qquad \text{(10–8)}$$

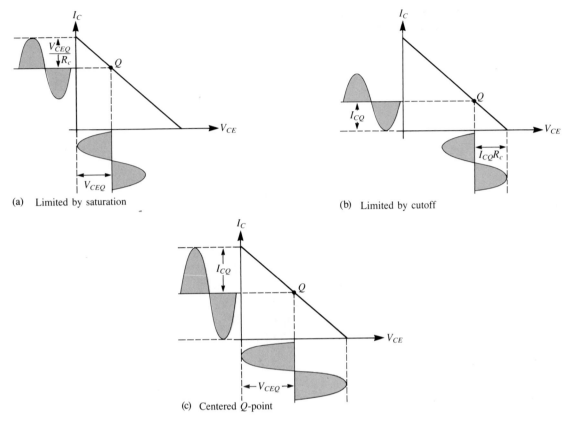

(a)   Limited by saturation

(b)   Limited by cutoff

(c)   Centered Q-point

FIGURE 10–15
ac load line operation showing limitations of output voltage swings.

Q-Point Closer to Cutoff.   When the Q-point is closer to cutoff, the maximum collector current swing is I_{CQ}, and the collector-to-emitter voltage swing is $I_{CQ}R_c$, as shown in Figure 10–15(b). The output power is therefore

$$P_{out} = (0.707\ I_{CQ})(0.707\ I_{CQ}R_c)$$
$$P_{out} = 0.5\ I_{CQ}^2R_c \qquad\qquad\qquad \textbf{(10–9)}$$

Q-Point Centered.   When the Q-point is centered, the maximum collector current swing is I_{CQ}, and the maximum collector-to-emitter voltage swing is V_{CEQ}, as shown in Figure 10–15(c). The output power is therefore

$$P_{out} = (0.707\ V_{CEQ})(0.707\ I_{CQ})$$
$$P_{out} = 0.5\ V_{CEQ}I_{CQ} \qquad\qquad\qquad \textbf{(10–10)}$$

This is the *maximum* ac output power from a class A amplifier under signal conditions. Notice that it is one-half the quiescent power dissipation.

Efficiency

Efficiency of an amplifier is the ratio of ac output power to dc input power. The dc input power is the dc supply voltage times the current drawn from the supply.

$$P_{dc} = V_{CC}I_{CC} \qquad (10\text{--}11)$$

The average supply current I_{CC} equals I_{CQ}, and the supply voltage V_{CC} is twice V_{CEQ} when the Q-point is centered. The maximum efficiency is therefore

$$\eta_{max} = \frac{P_{out}}{P_{dc}} = \frac{0.5\,V_{CEQ}I_{CQ}}{V_{CC}I_{CC}}$$

$$= \frac{0.5\,V_{CEQ}I_{CQ}}{2\,V_{CEQ}I_{CQ}} = \frac{0.5}{2}$$

$$\eta_{max} = 0.25 \qquad (10\text{--}12)$$

This 25 percent is the highest possible efficiency available from a class A amplifier and is approached only when the Q-point is at the center of the ac load line.

EXAMPLE 10–4

Determine the following values for the amplifier in Figure 10–16 when operated with the maximum possible output signal.
(a) Minimum transistor power rating.
(b) ac output power.
(c) Efficiency.

FIGURE 10–16

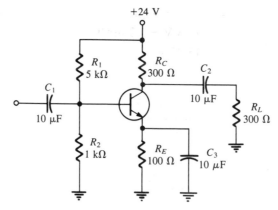

Solution
The dc values are first determined.

$$V_B = \left(\frac{R_2}{R_1 + R_2}\right)V_{CC} = \left(\frac{1\text{ k}\Omega}{6\text{ k}\Omega}\right)24\text{ V} = 4\text{ V, neglecting } R_{in(base)}$$

$$I_E = \frac{V_E}{R_E} = \frac{3.3\text{ V}}{100\ \Omega} = 33\text{ mA}$$

Therefore,

$$I_{CQ} \cong 33 \text{ mA}$$

and

$$V_C = V_{CC} - I_{CQ}R_C = 24 \text{ V} - (33 \text{ mA})(300 \text{ }\Omega)$$
$$= 14.1 \text{ V}$$
$$V_{CEQ} = V_C - V_E = 14.1 \text{ V} - 3.3 \text{ V}$$
$$= 10.8 \text{ V}$$

(a) The transistor power rating must be greater than $P_D = V_{CEQ}I_{CQ} = (10.8 \text{ V})(33 \text{ mA}) = 0.356 \text{ W}$.

(b) To make a calculation of ac output power under a *maximum* signal condition, we must know the location of the Q-point relative to center. This will tell us whether I_{CQ} or V_{CEQ} is the limiting factor if the Q-point is not centered. The ac load line values are as follows:

$$I_{c(\text{sat})} = I_{CQ} + \frac{V_{CEQ}}{R_c}$$

$$= 33 \text{ mA} + \frac{10.8 \text{ V}}{150 \text{ }\Omega}$$

$$= 105 \text{ mA}$$

and

$$V_{ce(\text{cutoff})} = V_{CEQ} + I_{CQ}R_c$$
$$= 10.8 \text{ V} + (33 \text{ mA})(150 \text{ }\Omega)$$
$$= 15.75 \text{ V}$$

A *centered* Q-point is at

$$I_{CQ} = \frac{105 \text{ mA}}{2} = 52.5 \text{ mA}$$

and

$$V_{CEQ} = \frac{15.75 \text{ V}}{2} \cong 7.88 \text{ V}$$

These are shown on the ac load line in Figure 10–17. The *actual* Q-point for this amplifier is closer to *cutoff*, as shown in the figure. Therefore the maximum collector current swing is I_{CQ}, and the ac output power is

$$P_{\text{out}} = 0.5 \, I_{CQ}^2 R_c$$
$$= 0.5(33 \text{ mA})^2(150 \text{ }\Omega)$$
$$= 81.68 \text{ mW}$$

FIGURE 10–17

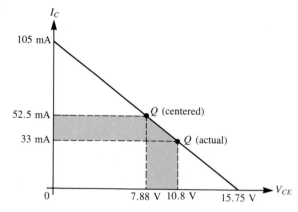

(c) The efficiency is

$$\eta = \frac{P_{out}}{P_{dc}} = \frac{P_{out}}{V_{CC}I_{CC}} = \frac{P_{out}}{V_{CC}I_{CQ}}$$

$$= \frac{81.68 \text{ mW}}{(24 \text{ V})(33 \text{ mA})} = 0.103$$

This efficiency of 10.3 percent is considerably less than the maximum possible efficiency (25 percent) because the actual Q-point is not centered.

The following program computes efficiency.

```
10   DIM RL(100):CLS
20   PRINT "THIS PROGRAM COMPUTES CLASS A EFFICIENCY FOR VARIOUS
30   PRINT "LOADS IN AN AMPLIFIER OF THE TYPE IN FIGURE 10-16."
40   PRINT:PRINT
50   PRINT "ENTER RESISTOR VALUE IN OHMS."
60   INPUT "R1";R1
70   INPUT "R2";R2
80   INPUT "RC";RC
90   INPUT "RE";RE
100  CLS
110  INPUT "DC SUPPLY VOLTAGE, VCC";VCC
120  INPUT "FOR HOW MANY VALUES OF RL DO YOU WANT THE EFFICIENCY
     COMPUTED";N
130  FOR A=1 TO N
140  INPUT "RL";RL(A)
150  NEXT
160  CLS
170  VB=R2/(R1+R2)*VCC
180  VE=VB-.7
190  ICQ=VE/RE
200  PRINT "RL", "EFFICIENCY":PRINT
210  FOR A=1 TO N
220  TRC=RC*RL(A)/(RC+RL(A))
230  P=.5*ICQ[2*TRC
```

```
240 EFF=P/(VCC*ICQ)
250 IF EFF>.25 THEN PRINT "AMPLIFIER IS NOT OPERATING CLASS A"
    ELSE PRINT RL(A), EFF
260 NEXT
```

**SECTION
REVIEW
10–1**

1. Basically, why does the ac load line differ from the dc load line?

2. What is the optimum Q-point location for class A amplifiers?

3. What is the maximum efficiency of a class A amplifier?

4. A certain amplifier has a centered Q-point of $I_{CQ} = 10$ mA and $V_{CEQ} = 7$ V. What is the *maximum* ac output power?

CLASS B PUSH-PULL AMPLIFIERS

10–2

When an amplifier is biased such that it operates in the linear region for 180° of the input cycle and is in cutoff for 180°, it is a class B amplifier. This is illustrated in Figure 10–18, where the output waveform is shown relative to the input.

FIGURE 10–18
Class B amplifier (noninverting).

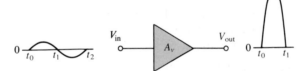

Q-Point Is at Cutoff

The class B amplifier is biased at cutoff so that $I_{CQ} = 0$ and $V_{CEQ} = V_{CE(cutoff)}$. It is brought out of cutoff and operates in its linear region when the input signal drives it into conduction. This is illustrated in Figure 10–19 with an emitter-follower circuit. Obviously, *the output is not a replica of the input.* Therefore, a two-transistor configuration, known as a *push-pull amplifier,* is necessary to get a sufficiently good reproduction of the input waveform.

FIGURE 10–19
Common-collector class B amplifier.

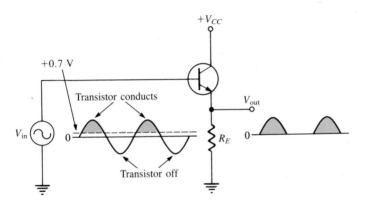

Push-Pull Operation

Figure 10–20 shows one type of push-pull class B amplifier using two emitter-followers. This is a *complementary* amplifier, because one emitter-follower uses an npn transistor and the other a pnp, which conduct on *opposite* alternations of the input cycle. Notice that

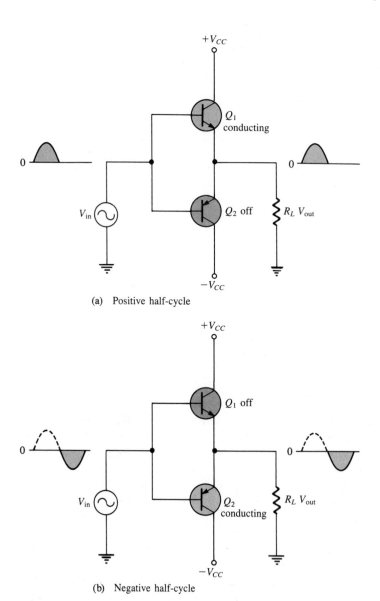

(a)   Positive half-cycle

(b)   Negative half-cycle

FIGURE 10–20
Class B push-pull operation.

there is no dc base bias voltage ($V_B = 0$); thus only the signal voltage drives the transistors into conduction. Q_1 conducts during the positive half of the input cycle, and Q_2 conducts during the negative half.

Crossover Distortion

When the dc base voltage is 0, the input signal voltage must exceed V_{BE} before a transistor conducts. Because of this, there is a time interval between the positive and negative alternations of the input when neither transistor is conducting, as shown in Figure 10–21. The resulting distortion in the output waveform is quite common and is called *crossover distortion*.

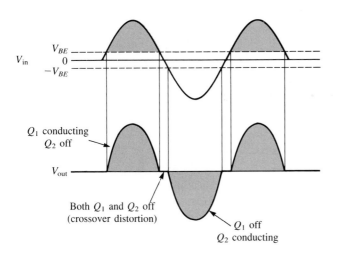

FIGURE 10–21
Illustration of crossover distortion in class B push-pull amplifier. The transistors conduct only during portions of the input indicated by the shaded areas.

Biasing the Push-Pull Amplifier

To eliminate crossover distortion, both transistors in the push-pull arrangement must be slightly above cutoff when there is no signal. This can be done with a voltage-divider arrangement, as shown in Figure 10–22(a). It is, however, difficult to maintain a stable bias point with this circuit due to changes in V_{BE} over temperature changes. (The requirement for dual-polarity power supplies is eliminated when R_L is capacitively coupled.) A more suitable arrangement is shown in Figure 10–22(b). When the diode characteristics of D_1 and D_2 are closely matched to the transconductance characteristics of the transistors, a stable bias is maintained.

The dc equivalent circuit of the push-pull amplifier is shown in Figure 10–23. R_1 and R_2 are of equal value; therefore the voltage at point A between the two diodes is $V_{CC}/2$. Assuming that both diodes and both transistors are identical, the drop across D_1 equals the V_{BE} of Q_1, and the drop across D_2 equals the V_{BE} of Q_2. As a result, the voltage at the emitters is also $V_{CC}/2$, and therefore, $V_{CEQ_1} = V_{CEQ_2} = V_{CC}/2$, as indicated. Because both transistors are biased near cutoff, $I_{CQ} \cong 0$.

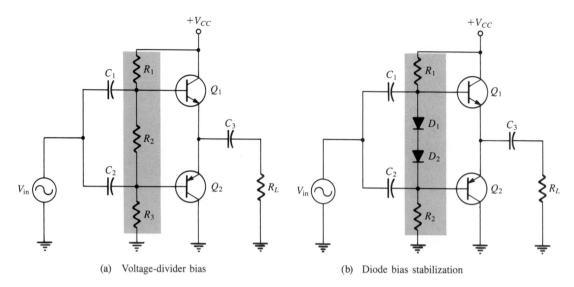

(a)   Voltage-divider bias   (b)   Diode bias stabilization

FIGURE 10–22
Biasing the push-pull amplifier to eliminate crossover distortion.

FIGURE 10–23
dc equivalent of push-pull amplifier.

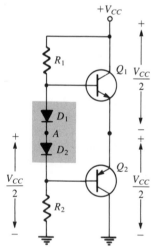

EXAMPLE 10–5

Determine the dc voltages at the bases and emitters of Q_1 and Q_2 in Figure 10–24. Also determine V_{CEQ} for each transistor.

Solution
The equivalent circuit for the bias network is shown in Figure 10–25(a).

$$I_T = \frac{V_{CC} - 2 V_D}{R_1 + R_2} = \frac{20 \text{ V} - 1.4 \text{ V}}{2 \text{ k}\Omega} = 9.3 \text{ mA}$$

$$V_{B_1} = V_{CC} - I_T R_1 = 20 \text{ V} - (9.3 \text{ mA})(1 \text{ k}\Omega) = 10.7 \text{ V}$$

FIGURE 10–24

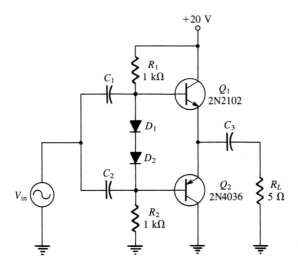

FIGURE 10–25

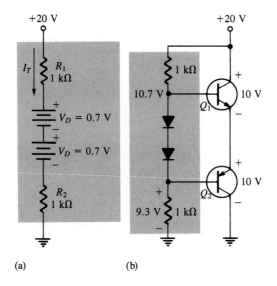

(a)   (b)

and

$$V_{B_2} = V_{B_1} - 2\,V_D = 10.7\text{ V} - 1.4\text{ V} = 9.3\text{ V}$$
$$V_{E_1} = V_{E_2} = 10.7\text{ V} - 0.7\text{ V} = 10\text{ V}$$

Therefore,

$$V_{CEQ_1} = V_{CEQ_2} = \frac{V_{CC}}{2} = \frac{20\text{ V}}{2}$$
$$= 10\text{ V}$$

These values are shown in Figure 10–25(b).

FIGURE 10–26
ac push-pull operation.

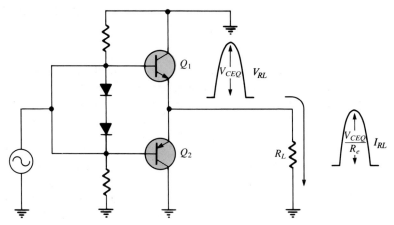

(a) Q_1 conducting with maximum signal output

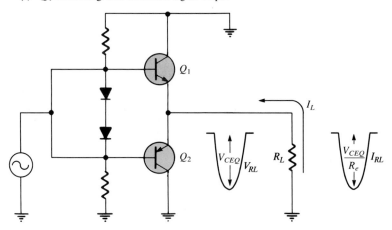

(b) Q_2 conducting with maximum signal output

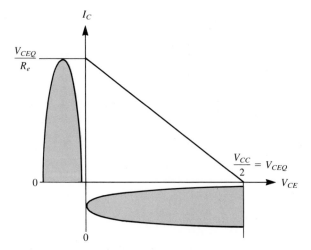

(c) ac load line for each transistor.

ac Operation

Under maximum conditions, transistors Q_1 and Q_2 are alternately driven from near cutoff to near saturation. During the positive alternation of the input signal, the Q_1 emitter is driven from its Q-point value of $V_{CC}/2$ to near V_{CC}, producing a positive peak voltage approximately equal to V_{CEQ}. At the same time, the Q_1 current swings from its Q-point value near 0 to near-saturation value, as shown in Figure 10–26(a).

During the negative alternation of the input signal, the Q_2 emitter is driven from its Q-point value of $V_{CC}/2$ to near 0, producing a negative peak voltage approximately equal to V_{CEQ}. Also, the Q_2 current swings from near 0 to near-saturation value, as shown in Figure 10–26(b). In terms of the ac load line operation, the V_{ce} of both transistors swings from $V_{CC}/2$ to 0, and the current swings from 0 to $I_{c(sat)}$, as shown in Figure 10–26(c). Because the peak voltage across each transistor is V_{CEQ}, the ac saturation current is

$$I_{c(sat)} = \frac{V_{CEQ}}{R_L} \qquad (10\text{--}13)$$

Since $I_e \cong I_c$ and the output current is the emitter current, the peak output current is also V_{CEQ}/R_L.

EXAMPLE 10–6

Determine the maximum peak values for the output voltage and current in Figure 10–27.

FIGURE 10–27

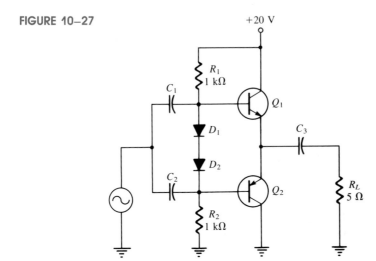

Solution
The maximum peak output voltage is

$$V_{\text{out(peak)}} \cong V_{CEQ} = \frac{V_{CC}}{2} = \frac{20 \text{ V}}{2} = 10 \text{ V}$$

The maximum peak output current is

$$I_{out(peak)} \cong I_{c(sat)} = \frac{V_{CEQ}}{R_L} = \frac{10 \text{ V}}{5 \text{ }\Omega} = 2 \text{ A}$$

Practice Exercise 10–6
Find the maximum peak values for the output voltage and current in Figure 10–27 if V_{CC} is lowered to 15 V and the load resistance is changed to 10 Ω.

Maximum Output Power

It has been shown that the maximum peak output current is approximately $I_{c(sat)}$, and the maximum peak output voltage is approximately V_{CEQ}. The maximum average output power is therefore

$$P_{out} = V_{out(rms)}I_{out(rms)}$$

Since

$$V_{out(rms)} = 0.707 \ V_{out(peak)} = 0.707 \ V_{CEQ}$$

and

$$I_{out(rms)} = 0.707 \ I_{out(peak)} = 0.707 \ I_{c(sat)}$$

then

$$P_{out} = 0.5 \ V_{CEQ}I_{c(sat)} \tag{10–14}$$

Substituting $V_{CC}/2$ for V_{CEQ}, we get

$$P_{out} = 0.25 \ V_{CC}I_{c(sat)} \tag{10–15}$$

dc Input Power

The input power comes from the V_{CC} supply and is

$$P_{dc} = V_{CC}I_{CC} \tag{10–16}$$

Since each transistor draws current for a half-cycle, the current is a *half-wave* signal with an average value of

$$I_{CC} = \frac{I_{c(sat)}}{\pi} \tag{10–17}$$

So

$$P_{dc} = \frac{V_{CC}I_{c(sat)}}{\pi} \tag{10–18}$$

Efficiency

The great advantage of push-pull class B amplifiers over class A is a much higher efficiency. This advantage usually overrides the difficulty of biasing the class B push-pull amplifier to eliminate crossover distortion. The efficiency is again defined as the ratio of ac output power to dc input power.

$$\text{Efficiency} = \frac{P_{\text{out}}}{P_{\text{dc}}}$$

The maximum efficiency for a class B amplifier is designated η_{max} and is developed as follows, starting with equation (10–15).

$$P_{\text{out}} = 0.25 \, V_{CC}I_{c(\text{sat})}$$

$$\eta_{\text{max}} = \frac{P_{\text{out}}}{P_{\text{dc}}} = \frac{0.25 \, V_{CC}I_{c(\text{sat})}}{V_{CC}I_{c(\text{sat})}/\pi} = 0.25\pi$$

$$\eta_{\text{max}} = 0.785 \qquad\qquad\qquad \textbf{(10–19)}$$

or

$$\eta_{\text{max}} = 78.5\%$$

Recall that the maximum efficiency for class A is 0.25 (25 percent).

EXAMPLE 10–7

Find the maximum ac output power and the dc input power of the amplifier in Figure 10–28.

FIGURE 10–28

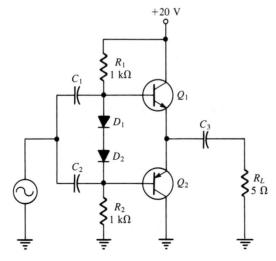

Solution

$I_{c(\text{sat})}$ for this same circuit was found to be 2 A in Example 10–6.

$$P_{\text{out}} = 0.25 \ V_{CC}I_{c(\text{sat})}$$
$$= 0.25(20 \ \text{V})(2 \ \text{A})$$
$$= 10 \ \text{W}$$

$$P_{\text{dc}} = \frac{V_{CC}I_{c(\text{sat})}}{\pi}$$
$$= \frac{(20 \ \text{V})(2 \ \text{A})}{\pi}$$
$$= 12.73 \ \text{W}$$

Practice Exercise 10–7

Determine the maximum ac output power and the dc input power in Figure 10–28 for $V_{CC} = 15$ V and $R_L = 10 \ \Omega$.

| **SECTION REVIEW 10–2** | 1. Where is the Q-point for a class B amplifier? |

SECTION REVIEW 10–2

1. Where is the Q-point for a class B amplifier?
2. What causes crossover distortion?
3. What is the maximum efficiency of a push-pull class B amplifier?
4. Explain the purpose of the push-pull configuration for class B.

CLASS C AMPLIFIERS

10–3

Class C amplifiers are biased so that conduction occurs for much less than 180°, as Figure 10–29 illustrates. Class C amplifiers are more efficient than either class A or push-pull class B. This means that more output power can be obtained from class C operation. Because the output waveform is severely distorted, class C amplifiers are normally limited to applications as tuned amplifiers at radio frequencies (RF).

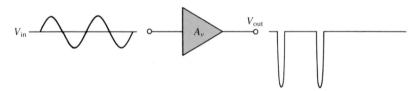

FIGURE 10–29
Class C amplifier.

Basic Operation

A basic common-emitter class C amplifier with a resistive load is shown in Figure 10–30(a). It is biased *below cutoff* with the $-V_{BB}$ supply. The ac source voltage has a

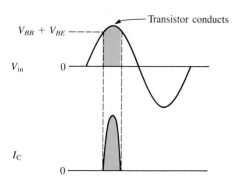

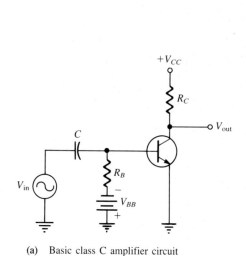

(a) Basic class C amplifier circuit

(b) Input voltage and output current waveforms

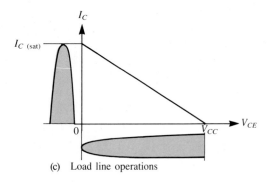

(c) Load line operations

FIGURE 10–30
Class C operation.

peak value that is slightly greater than $V_{BB} + V_{BE}$ so that the base voltage exceeds the barrier potential of the base-emitter junction for a short time near the positive peak of each cycle, as illustrated in Figure 10–30(b). During this short interval, the transistor is turned on. When the entire ac load line is used, as shown in Figure 10–30(c), the maximum collector current is approximately $I_{C(sat)}$, and the minimum collector voltage is approximately $V_{CE(sat)}$.

Power Dissipation

The power dissipation of the transistor in a class C amplifier is low because it is on for only a small percentage of the input cycle. Figure 10–31(a) shows the collector current pulses. The time between the pulses is the *period* (T) of the ac input voltage. To avoid complex mathematics, we will use ideal pulse approximations for the collector current and the collector voltage during the *on*-time of the transistor, as shown in Figure 10–31(b). Using this simplification, the current amplitude is $I_{C(sat)}$ and the voltage is $V_{CE(sat)}$ during the time the transistor is on, if the output swings over the entire load line. The power dissipation during the on-time is therefore

$$P_{D(on)} = V_{CE(sat)}I_{C(sat)}$$

FIGURE 10–31
Class C waveforms.

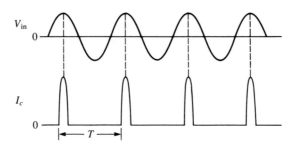

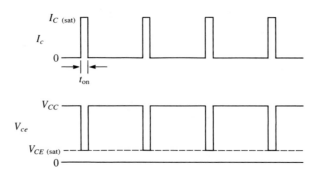

(a)   Collector current pulses

(b)   Ideal class C waveforms

The transistor is on for a short time, t_{on}, and *off* for the rest of the input cycle. Therefore, the power dissipation averaged over the entire cycle is

$$P_{D(avg)} = \left(\frac{t_{on}}{T}\right)P_{D(on)}$$

$$P_{D(avg)} = \left(\frac{t_{on}}{T}\right)V_{CE(sat)}I_{C(sat)} \tag{10–20}$$

EXAMPLE 10–8

A class C amplifier is driven by a 200 kHz signal. The transistor is *on* for 1 μs, and the amplifier is operating over 100 percent of its load line. If $I_{C(sat)} = 100$ mA and $V_{CE(sat)} = 0.2$ V, what is the average power dissipation?

Solution

$$T = \frac{1}{200 \text{ kHz}} = 5 \ \mu s$$

Therefore,

$$P_{D(\text{avg})} = \left(\frac{t_{\text{on}}}{T}\right)V_{CE(\text{sat})}I_{C(\text{sat})}$$
$$= (0.2)(0.2 \text{ V})(100 \text{ mA})$$
$$= 4 \text{ mW}$$

Tuned Operation

Because the collector voltage (output) is not a replica of the input, the resistively loaded class C amplifier is of no value in linear applications. It is therefore necessary to use a class C amplifier with a parallel *resonant circuit* (tank), as shown in Figure 10–32(a). The resonant frequency of the tank circuit is determined by the formula $f_r = 1/(2\pi\sqrt{LC})$. The short pulse of collector current on each cycle of the input initiates and sustains the oscillation of the tank circuit so that an output sine wave voltage is produced, as illustrated in Figure 10–32(b).

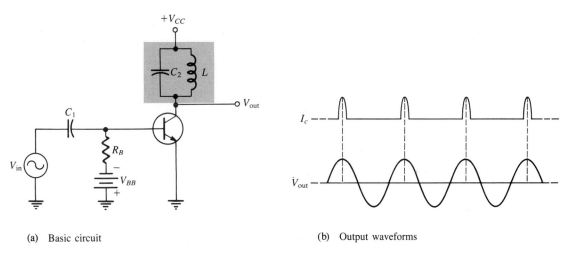

(a) Basic circuit

(b) Output waveforms

FIGURE 10–32
Tuned class C amplifier.

The current pulse charges the capacitor to approximately $+V_{CC}$, as shown in Figure 10–33(a). After the pulse, the capacitor quickly discharges, thus charging the inductor. Then, after the capacitor completely discharges, the inductor's magnetic field collapses and then quickly recharges C to near V_{CC} in a direction opposite to the previous charge. This completes one half-cycle of the oscillation, as shown in parts (b) and (c) of Figure 10–33. Next, the capacitor discharges again, quickly increasing the inductor's magnetic field. The inductor then quickly recharges the capacitor back to a positive peak less than

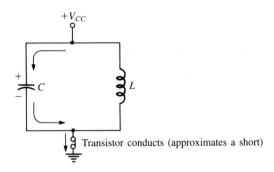

(a)   C charges to $+V_{CC}$ at the input peak when
      transistor is conducting

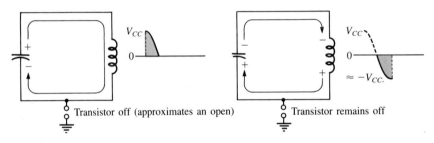

(b)   C discharges to 0 volts

(c)   L recharges C in opposite direction

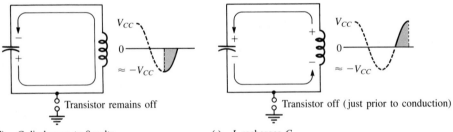

(d)   C discharges to 0 volts

(e)   L recharges C

FIGURE 10–33
Resonant circuit action.

the previous one, due to energy loss in the winding resistance. This completes the second half-cycle, as shown in parts (d) and (e). The peak-to-peak output voltage is therefore approximately equal to 2 V_{CC}.

The amplitude of each successive cycle of the oscillation will be less than that of the previous cycle because of energy loss in the resistance of the tank circuit, as shown in Figure 10–34(a), and the oscillation will eventually die out. However, the regular recur-

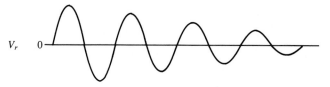

(a) Oscillation dies out due to energy loss

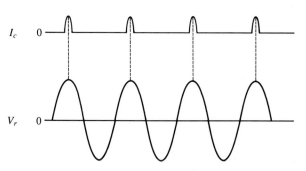

(b) Oscillation is sustained by short pulses of collector current

rences of the collector current pulse re-energizes the resonant circuit and sustains the oscillations at a constant amplitude. When the tank circuit is tuned to the frequency of the input signal, re-energizing occurs on each cycle of the tank voltage V_r, as shown in Figure 10–34(b).

Maximum Output Power

Since the voltage developed across the tank circuit has a peak-to-peak value of approximately 2 V_{CC}, the maximum output power can be expressed as

$$P_{out} = \frac{V_{rms}^2}{R_c} = \frac{(0.707 \ V_{CC})^2}{R_c}$$

$$P_{out} = \frac{0.5 \ V_{CC}^2}{R_c} \tag{10-21}$$

R_c is the equivalent parallel resistance of the collector tank circuit and represents the parallel combination of the coil resistance and the load resistance. It usually has a low value. The total power that must be supplied to the amplifier is

$$P_T = P_{out} + P_{D(avg)}$$

Therefore, the efficiency is

$$\eta = \frac{P_{out}}{P_{out} + P_{D(avg)}} \tag{10-22}$$

When $P_{out} \gg P_{D(avg)}$, the class C efficiency closely approaches 100 percent.

EXAMPLE 10–9

Suppose the class C amplifier described in Example 10–8 has a V_{CC} equal to 24 V, and the R_c is 100 Ω. Determine the efficiency.

Solution

$$P_{D(avg)} = 4 \text{ mW (from Example 10–8)}$$

$$P_{out} = \frac{0.5 \, V_{CC}^2}{R_c} = \frac{0.5(24 \text{ V})^2}{100 \, \Omega} = 2.88 \text{ W}$$

Therefore,

$$\eta = \frac{P_{out}}{P_{out} + P_{D(avg)}} = \frac{2.88 \text{ W}}{2.88 \text{ W} + 4 \text{ mW}} = 0.9986$$

or

$$\eta = 99.86\%$$

Clamper Bias for a Class C Amplifier

Figure 10–35 shows a class C amplifier with a base bias clamping circuit. The base-emitter junction functions as a diode. When the input signal goes positive, capacitor C_1 is charged to the peak value with the polarity shown in Figure 10–36(a). This action produces an average voltage at the base of approximately $-V_p$. This places the transistor in cutoff except at the positive peaks, when the transistor conducts for a short interval. For good clamping action, the R_1C_1 time constant of the clamping circuit must be much greater than the period of the input signal. Parts (b) through (f) illustrate the bias clamping action in more detail. During the time up to the positive peak of the input ($t_0 - t_1$), the capacitor charges to $V_p - 0.7$ V through the base-emitter diode, as shown in part (b). During the time from t_1 to t_2, as shown in part (c), the capacitor discharges very little because of the large RC time constant. The capacitor, therefore, maintains an average charge slightly less than $V_p - 0.7$ V.

FIGURE 10–35
Tuned class C amplifier with clamper bias.

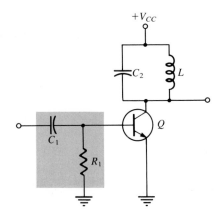

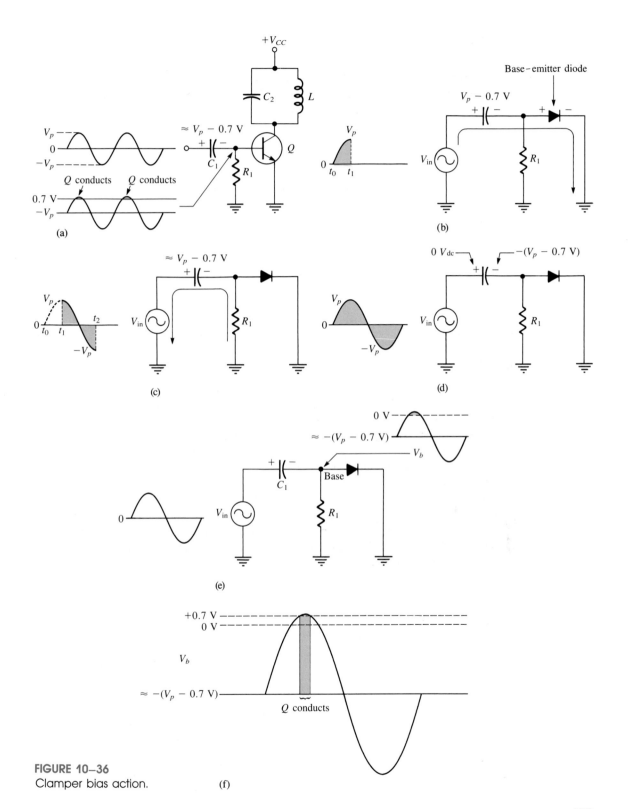

FIGURE 10–36
Clamper bias action.

(a)

(b)

(c)

(d)

(e)

(f)

Since the dc value of the input signal is 0 (positive side of C_1), the dc voltage at the base (negative side of C_1) is slightly more positive than $-(V_p - 0.7$ V), as indicated in part (d). As shown in part (e), the capacitor couples the ac input signal through to the base so that the voltage at the transistor's base is the ac signal riding on a dc level slightly more positive than $-(V_p - 0.7$ V). Near the positive peaks of the input voltage, the base voltage goes slightly above 0.7 V and causes the transistor to conduct for a short time, as shown in part (f).

SECTION REVIEW 10–3

1. A class C amplifier is normally biased in _____ .
2. What is the purpose of the tuned circuit in a class C amplifier?
3. A certain class C amplifier has a power dissipation of 100 mW and an output power of 1 W. What is its efficiency?

TROUBLESHOOTING

10–4

As an example of isolating a component failure in a circuit, we will use a class A amplifier with the output monitored by an oscilloscope, as illustrated in Figure 10–37. As shown, the amplifier should have a normal sine wave output when a sinusoidal input signal is applied.

Now we will consider several incorrect output waveforms and the most likely causes in each case. In Figure 10–38(a), the scope displays a dc level equal to the dc supply voltage, indicating that the transistor is in cutoff. The two most likely causes of this condition are: (1) the transistor has an open pn junction, or (2) R_4 is open, preventing collector and emitter current.

In Figure 10–38(b), the scope displays a dc level at the collector approximately equal to the emitter voltage. The two probable causes of this indication are: (1) the transistor is shorted from collector to emitter, or (2) R_2 is open, causing the transistor to be biased in saturation. In the second case, a sufficiently large input signal can bring the transistor out of saturation on its negative peaks, resulting in short pulses on the output. In Figure 10–38(c), the scope displays an output waveform that indicates the transistor is cut

FIGURE 10–37
Class A amplifier with proper output display.

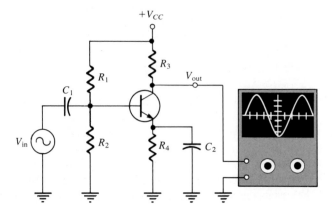

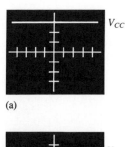

(a)

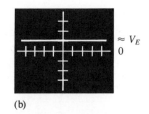

(b)

FIGURE 10–38

Oscilloscope displays of output voltage for the amplifier in Figure 10–37, illustrating several types of failures.

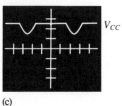

(c)

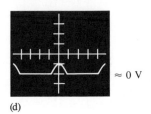

(d)

off except during a small portion of the input cycle. Possible causes of this indication are: (1) the Q-point has shifted down due to a drastic out-of-tolerance change in a resistor value, or (2) R_1 is open, biasing the transistor in cutoff. The display shows that the input signal is sufficient to bring it out of cutoff for a small portion of the cycle. In Figure 10–38(d), the scope displays an output waveform that indicates the transistor is saturated except during a small portion of the input cycle. Again, it is possible that a resistance change has caused a drastic shift in the Q-point up toward saturation, or R_2 is open, causing the transistor to be biased in saturation, and the input signal is bringing it out of saturation for a small portion of the cycle.

1. What would you check for if you noticed clipping at both peaks of the output waveform?

2. A significant loss of gain in the amplifier of Figure 10–37 would most likely be caused by what type of failure?

SECTION REVIEW 10–4

A SYSTEM APPLICATION

10–5

In a typical AM or FM receiver system, the audio section consists of an audio voltage amplifier stage (sometimes called an audio preamp) followed by a power amplifier, which is the final output stage. The audio preamp boosts the audio signal from the detector to a sufficient level to drive the output power amplifier. The power amplifier then drives the speaker. Some receivers use a class A power amplifier for the audio output, but most systems, particularly those that are battery operated, use a class B push-pull amplifier. The high efficiency of a class B push-pull amplifier compared to a class A makes the class B a logical choice when the current drain on a battery is of concern. Recall that the class B amplifier is biased near cutoff and draws very little quiescent current.

A typical audio circuit is shown in Figure 10–39. The audio signal from the detector is fed through the volume control potentiometer and capacitively coupled by C_1 to the

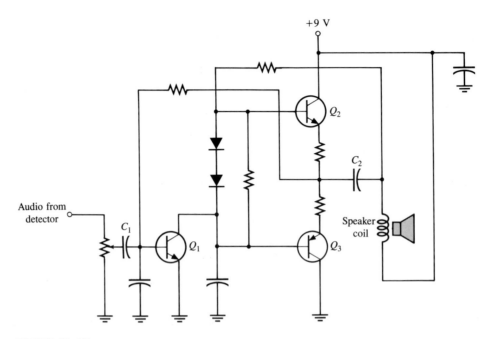

FIGURE 10–39
A typical audio amplifier section.

base of the audio preamplifier Q_1. The output from the collector of Q_1 provides the input to the push-pull amplifier formed by Q_2 and Q_3. The output of the push-pull amplifier is capacitively coupled through C_2 to the speaker.

FORMULAS

Large-signal class A amplifiers

(10–1)	$I_{c(sat)} = I_{CQ} + \dfrac{V_{CEQ}}{R_c}$	Saturation current
(10–2)	$V_{ce(cutoff)} = V_{CEQ} + I_{CQ}R_c$	Cutoff voltage
(10–3)	$r'_e = \dfrac{\Delta V_{BE}}{\Delta I_C}$	ac emitter resistance
(10–4)	$A_v = \dfrac{R_c}{r'_e}$	Voltage gain
(10–5)	$A_p = \beta_{dc}\dfrac{R_c}{r'_e}$	Power gain
(10–6)	$P_{DQ} = I_{CQ}V_{CEQ}$	Quiescent power
(10–7)	$P_{out} = V_{ce}I_c$	Output power

(10–8) $P_{out} = \dfrac{0.5 \, V_{CEQ}^2}{R_c}$ Output power (Q-point closer to saturation)

(10–9) $P_{out} = 0.5 \, I_{CQ}^2 R_c$ Output power (Q-point closer to cutoff)

(10–10) $P_{out} = 0.5 \, V_{CEQ} I_{CQ}$ Output power (Q-point centered)

(10–11) $P_{dc} = V_{CC} I_{CC}$ dc input power

(10–12) $\eta_{max} = 0.25$ Maximum efficiency

Large-signal class B push-pull amplifiers

(10–13) $I_{c(sat)} = \dfrac{V_{CEQ}}{R_L}$ Saturation current

(10–14) $P_{out} = 0.5 \, V_{CEQ} I_{c(sat)}$ Output power

(10–15) $P_{out} = 0.25 \, V_{CC} I_{c(sat)}$ Output power

(10–16) $P_{dc} = V_{CC} I_{CC}$ dc input power

(10–17) $I_{CC} = \dfrac{I_{c(sat)}}{\pi}$ dc supply current

(10–18) $P_{dc} = \dfrac{V_{CC} I_{c(sat)}}{\pi}$ dc input power

(10–19) $\eta_{max} = 0.785$ Maximum efficiency

Large-signal class C amplifiers

(10–20) $P_{D(avg)} = \left(\dfrac{t_{on}}{T}\right) V_{CE(sat)} I_{C(sat)}$ Average power dissipation

(10–21) $P_{out} = \dfrac{0.5 \, V_{CC}^2}{R_c}$ Output power

(10–22) $\eta = \dfrac{P_{out}}{P_{out} + P_{D(avg)}}$ Efficiency

---

SUMMARY

☐ A class A amplifier operates entirely in the linear region of the transistor's characteristic curves. The transistor conducts during the entire input cycle.

☐ The Q-point must be centered on the load line for maximum class A output signal swing.

☐ The maximum efficiency of a class A amplifier is 25 percent.

☐ A class B amplifier operates in the linear region for half of the input cycle (180°), and is in cutoff for the other half.

☐ The Q-point is at cutoff for class B operation.

☐ Class B amplifiers are normally operated in a push-pull configuration in order to produce an output that is a replica of the input.

☐ The maximum efficiency of a class B amplifier is 78.5 percent.

☐ A class C amplifier operates in the linear region for only a small part of the input cycle.

☐ The class C amplifier is biased below cutoff.

☐ Class C amplifiers are normally operated as tuned amplifiers to produce a sinusoidal output.

☐ The maximum efficiency of a class C amplifier is higher than that of either class A or class B amplifiers. Under conditions of low power dissipation and high output power, the efficiency can approach 100 percent.

SELF-TEST

1. When the Q-point of a class A amplifier (inverting) is nearer saturation than cutoff, on which peak of the output voltage will clipping first appear as the input signal voltage is increased?

2. Determine the saturation value of the ac collector current for an amplifier with Q-point values of $I_{CQ} = 2$ mA and $V_{CEQ} = 3$ V. The ac collector resistance is 3 kΩ.

3. Determine the cutoff value of the ac collector-to-emitter voltage for the amplifier in problem 2.

4. What is the maximum peak-to-peak collector voltage for the amplifier in problem 2?

5. The large-signal ac emitter resistance of the transistor in a certain class A amplifier is 18 Ω. Determine the large-signal voltage gain for $R_c = 500$ Ω.

6. A certain class A amplifier has a current gain of 75 and a voltage gain of 50. What is the power gain?

7. A class A amplifier is biased with a centered Q-point at $V_{CEQ} = 5$ V and $I_{CQ} = 10$ mA. Determine the maximum output power. What is the efficiency under this condition?

8. The class B amplifier in Figure 10–40 with a dc supply voltage (V_{CC}) of 20 V is biased at cutoff. What are the approximate values of I_C, V_C, and V_B?

FIGURE 10–40

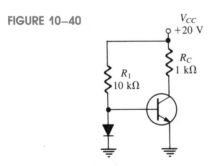

9. The emitters of a certain push-pull class B emitter-follower amplifier have a Q-point value of 10 V. If $R_e = 50$ Ω, what is the value of $I_{c(sat)}$? What is the value of V_{CC}?

10. What is the output power of the amplifier in problem 9 under maximum signal conditions? The dc input power? The efficiency?

11. Calculate the average current drawn from the dc supply when $I_{c(sat)}$ = 20 mA.

12. Why is the power dissipation of a class C amplifier normally very low?

13. A 1 MHz input signal is applied to a class C amplifier. What is the average transistor power dissipation if $V_{CE(sat)}$ = 0.3 V and $I_{C(sat)}$ = 200 mA during an on-time of 0.5 μs?

14. The amplifier in problem 13 has a V_{CC} of 20 V, and the equivalent parallel resistance of the tank circuit is 100 Ω. Calculate the maximum output power.

15. What is the efficiency of the amplifier in problems 13 and 14?

Section 10–1

10–1 Determine the approximate values $I_{C(sat)}$ and $V_{CE(cutoff)}$ in Figure 10–41. Assume $V_{CE(sat)} \cong 0$ V.

FIGURE 10–41

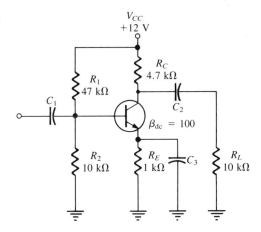

10–2 Sketch the ac equivalent circuit for the amplifier in Figure 10–41. Calculate the saturation value of the collector current under signal conditions.

10–3 Find the cutoff value of the ac collector-to-emitter voltage in Figure 10–41.

10–4 What is the maximum peak value of collector current that can be realized in each circuit of Figure 10–43? What is the maximum peak value of output voltage?

10–5 In Figure 10–42, assume β_{dc} for each transistor increases by 50 percent with a certain temperature rise. How is the maximum output voltage affected?

10–6 Find the large-signal voltage gain for each circuit in Figure 10–43.

10–7 Determine the maximum rms value of input voltage that can be applied to each amplifier in Figure 10–42 without producing a clipped output voltage.

10–8 Determine the minimum power rating for each of the transistors in Figure 10–44.

FIGURE 10-42

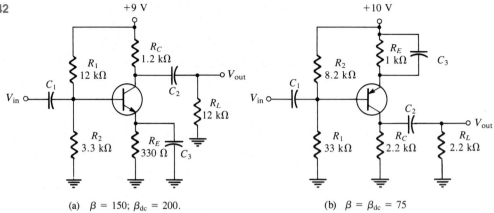

(a) $\beta = 150$; $\beta_{dc} = 200$.

(b) $\beta = \beta_{dc} = 75$

FIGURE 10-43

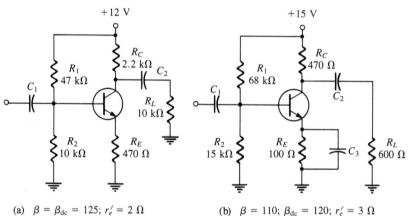

(a) $\beta = \beta_{dc} = 125$; $r'_e = 2\ \Omega$

(b) $\beta = 110$; $\beta_{dc} = 120$; $r'_e = 3\ \Omega$

FIGURE 10-44

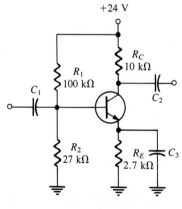

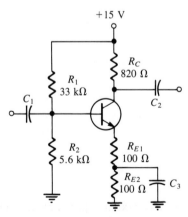

(a) $\beta = \beta_{dc} = 90$; $r'_e = 10\ \Omega$

(b) $\beta = \beta_{dc} = 175$; $r'_e = 1.2\ \Omega$

10–9 Find the maximum output signal power and efficiency for each amplifier in Figure 10–44.

Section 10–2

10–10 What dc voltage would you expect to measure with each dc meter in Figure 10–45?

FIGURE 10–45

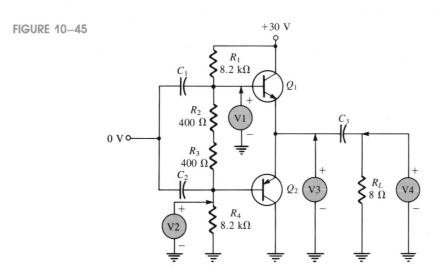

10–11 Determine the dc voltages at the bases and emitters of Q_1 and Q_2 in Figure 10–46. Also determine V_{CEQ} for each transistor.

FIGURE 10–46

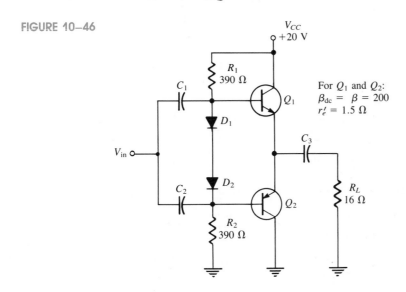

For Q_1 and Q_2:
$\beta_{dc} = \beta = 200$
$r'_e = 1.5 \ \Omega$

10–12 Determine the maximum peak output voltage and peak load current for the circuit in Figure 10–46.

10–13 Find the maximum signal power achievable with the class B push-pull amplifier in Figure 10–46. Find the dc input power.

10–14 The efficiency of a certain class B push-pull amplifier is 0.71, and the dc input power is 16.25 W. What is the ac output power?

10–15 A certain class B push-pull amplifier has an R_e of 8 Ω and a V_{CEQ} of 12 V. Determine I_{CC}, P_{dc}, and P_{out}. Assuming this amplifier is operating under maximum output conditions, what is V_{CC}?

10–16 In Figure 10–47, what rms input voltage will produce a maximum output voltage swing? Use an ideal voltage gain of 1 for the emitter-followers.

FIGURE 10–47

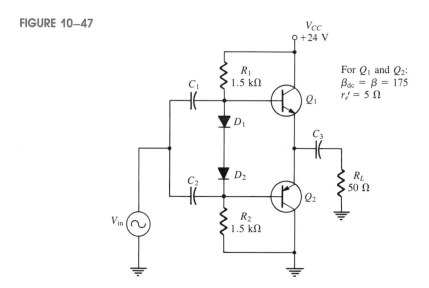

10–17 Sketch the waveforms you would expect to see with a scope across R_L in Figure 10–47 if Q_1 were open from collector to emitter. If Q_2 were open from collector to emitter.

Section 10–3

10–18 A certain class C amplifier transistor is on for 10 percent of the input cycle. If $V_{CE(sat)} = 0.18$ V and $I_{C(sat)} = 25$ mA, what is the average power dissipation for maximum output?

10–19 What is the resonant frequency of a tank circuit with $L = 10$ mH and $C = 0.001$ μF?

10–20 What is the peak-to-peak output voltage of a tuned class C amplifier with $V_{CC} = 12$ V?

10–21 Determine the efficiency of the class C amplifier described in problem 10–18 if $V_{CC} = 15$ V and the equivalent parallel resistance in the collector tank circuit is 50 Ω.

Section 10–4

10–22 Refer to Figure 10–22(b). What would you expect to observe across R_L if C_1 opened?

10–23 Your oscilloscope displays a half-wave output when connected across R_L in Figure 10–22. What is the probable cause?

Section 10–1

1. The ac load resistance has a value different from that of the dc load resistance.

2. Centered on the load line.

3. 25 percent.

4. 35 mW.

Section 10–2

1. Cutoff.

2. The barrier potential of the base-emitter junction.

3. 78.5 percent.

4. To reproduce both positive and negative alternations of the input signal with greater efficiency.

Section 10–3

1. Cutoff.

2. To produce a sine wave output.

3. 90.9 percent.

Section 10–4

1. Excess input signal voltage.

2. Open bypass capacitor, C_2.

Example 10–1: $V_{ce(sat)} \cong 0$ V, $I_{c(sat)} = 12.1$ mA, $V_{ce(cutoff)} = 7.26$ V, $I_{c(cutoff)} = 0$ A

Example 10–3: 50

Example 10–6: 7.5 V, 0.75 A

Example 10–7: $P_{out} = 2.8$ W, $P_{dc} = 3.58$ W

Amplifier Frequency Response

In this chapter you will learn

☐ The definition and application of Miller's theorem

☐ How the input and output capacitances of an amplifier are determined

☐ The characteristics of an amplifier's frequency response

☐ The meaning of critical frequency in amplifier response

☐ The effects of coupling and bypass capacitors at low frequencies

☐ How the internal transistor capacitances limit the response of the amplifier at high frequencies

☐ How amplifier gain is expressed in decibels

☐ What is meant by midrange gain

☐ What is meant by gain roll-off and what factors determine it

☐ What a Bode plot is

☐ How the low-frequency response of a direct-coupled amplifier differs from that of an amplifier that is capacitively coupled

☐ How to measure an amplifier's frequency response

☐ Why an amplifier's frequency response is important in a specific system application

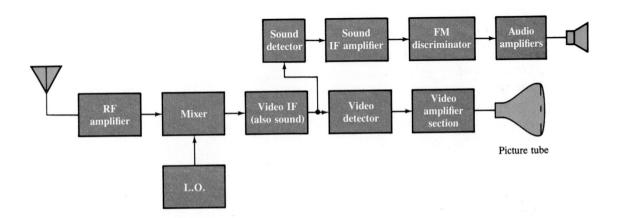

Chapter 11

Picture tube

The previous chapters on amplifiers neglected the effects of signal frequency on amplifier operation. We have always considered the coupling and bypass capacitors in the circuits to be ideal shorts at the signal frequencies and the internal device capacitances to be ideal opens. As you know, capacitive reactance decreases with increasing frequency. Although it is valid to assume $X_C \cong 0$ above a certain frequency for coupling and bypass capacitors and $X_C = \infty$ below a certain frequency for internal device capacitances, a complete picture of an amplifier's response must take into account the full range of frequencies over which the amplifier is expected to operate.

This chapter examines the frequency effects on amplifier gain and phase shift and introduces other, related topics. The coverage applies to both bipolar and FET amplifiers, and we include a mixture of both to illustrate the concepts.

The concept of frequency response is important in all communications systems. In AM and FM receivers it is imperative that the circuits be capable of correct frequency response. For example, the RF amplifier in an FM radio receiver must be able to handle frequencies from 88 MHz to 108 MHz, and the IF amplifiers must work at 10.7 MHz for FM or at 455 kHz for AM. Also, the transistors in the local oscillators of AM and FM receivers must operate at frequencies from 995 kHz to 2095 kHz in the case of AM and frequencies from 98.7 MHz to 118.7 MHz in the case of FM. The audio amplifiers, generally, must respond to frequencies from about 20 Hz up to about 20 kHz.

The basic superheterodyne concept is also used in other than radio receivers. Television is another example of application of the AM and FM superheterodyne principles and the importance of frequency response. The block diagram shows the basic video and audio sections of a TV receiver. TV video signals are amplitude modulated and the accompanying sound is frequency modulated. The frequencies are different from broadcast radio, but the basic principles of reception are the same. In a TV receiver, processing and separation of the video and sound signals depend on the frequency response of the receiver circuits.

GENERAL CONCEPTS

11–1

In our previous studies of amplifiers, the capacitive reactance of the coupling and bypass capacitors was assumed to be 0 at the signal frequency and therefore had no effect on the amplifier's gain or phase shift. Also, the internal transistor capacitances were assumed to be small enough to neglect at the signal frequency. All of these simplifying assumptions are valid and necessary for studying amplifier theory. However, they do give a limited picture of an amplifier's total operation, so in this chapter we consider the frequency effects.

Effect of Coupling Capacitors

Recall from basic circuit theory that $X_C = 1/2\pi fC$. This formula shows that the capacitive reactance varies inversely with frequency. At lower frequencies the reactance is greater, and it decreases as the frequency increases. At lower frequencies, for example audio frequencies below 10 Hz, capacitively coupled amplifiers such as those in Figure 11–1 have less voltage gain than at higher frequencies. The reason is that at lower frequencies,

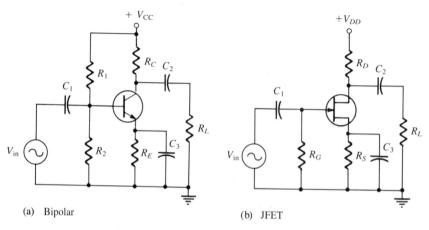

(a) Bipolar   (b) JFET

FIGURE 11–1
Typical capacitively coupled amplifiers.

more signal voltage is dropped across C_1, C_2, and C_3 because their reactances are higher. This is the result of the limitation imposed by availability and physical size of large values of capacitance required in these applications. Also, a phase shift is introduced by the coupling capacitors because C_1 forms a lead network with the R_{in} of the amplifier, and C_2 forms a lead network with R_L and R_C or R_D.

Effect of Bypass Capacitors

At lower frequencies, the bypass capacitor C_3 is not a short. So, the emitter (or FET source terminal) is not at ac ground. X_{C3} in parallel with R_E (or R_S) creates an impedance which reduces the gain. This is illustrated in Figure 11–2.

FIGURE 11–2

Nonzero reactance of the bypass capacitor in parallel with R_E creates an emitter impedance which reduces the voltage gain.

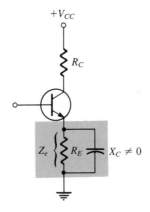

Effect of Internal Transistor Capacitances

At high frequencies, the coupling and bypass capacitors become effective shorts and do not affect the amplifier's response. Undesired internal device capacitances, however, do come into play, reducing the amplifier's gain and introducing phase shift as the signal frequency increases.

Figure 11–3 shows the internal capacitances for both a bipolar transistor and a JFET. In the case of the bipolar transistor, C_{be} is the base-emitter junction capacitance and C_{bc} is the base-collector junction capacitance. In the case of the JFET, C_{gs} is the internal capacitance between gate and source and C_{gd} is the internal capacitance between gate and drain.

FIGURE 11–3

Internal transistor capacitances.

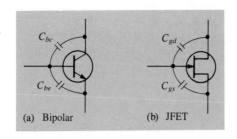

(a) Bipolar   (b) JFET

Data sheets often refer to the bipolar transistor capacitance C_{bc} as the *output capacitance,* often designated C_{ob}. The capacitance C_{be} is often designated as the *input capacitance,* C_{ib}. Data sheets for FETs normally specify *input capacitance* C_{iss} and *reverse transfer capacitance* C_{rss}. From these, C_{gs} and C_{gd} can be calculated, as you will see later.

At lower frequencies, the internal capacitances have a very high reactance because of their low capacitance value (usually only a few picofarads). Therefore, they look like opens and have no effect on the transistor's performance. As the frequency goes up, the capacitive reactances go down, and at some point they begin to have a significant effect on the transistor's gain. When the reactance of C_{be} (or C_{gs}) becomes small enough, a significant amount of the signal voltage is lost due to a voltage-divider effect of the source resistance and the capacitive reactance of C_{be}, as illustrated in Figure 11–4(a). When the reactance of C_{bc} (or C_{gd}) becomes small enough, a significant amount of output signal voltage is fed back out-of-phase with the input (negative feedback), thus effectively reducing the voltage gain. This is illustrated in Figure 11–4(b).

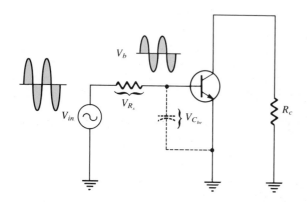

(a) Effect of C_{be}, where V_{in} is reduced by the voltage divider action of R_s and $X_{C_{be}}$

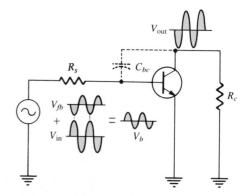

(b) Effect of C_{bc}, where part of V_{out} (V_{fb}) goes back through C_{bc} to the base and reduces the input signal because it is approximately 180° out-of-phase with V_{in}

FIGURE 11–4

ac equivalent circuit for a bipolar amplifier showing effects of the internal capacitances.

SECTION REVIEW 11–1

1. In an ac amplifier, which capacitors affect the low-frequency gain?
2. The high-frequency gain of an amplifier is limited by _____ _____ in addition to stray effects.

MILLER'S THEOREM AND DECIBELS

11–2

Before we get into the analysis of amplifier frequency response, two important topics must be considered: Miller's theorem and the concept of decibels.

Miller's Theorem

Miller's theorem can be used to simplify the analysis of inverting amplifiers at high frequencies where the internal transistor capacitances are important. The capacitance C_{bc} in bipolar transistors (C_{gd} in FETs) between the input (base or gate) and the output (collector or drain) is shown in Figure 11–5 in a generalized form. A_v is the voltage gain of the amplifier at midrange frequencies, and C represents either C_{bc} or C_{gd}.

Miller's theorem states that C effectively appears as a capacitance from input to ground that can be expressed as follows.

$$C_{in(Miller)} = C(A_v + 1) \qquad \text{(11–1)}$$

This shows that C_{bc} (or C_{gd}) has a much greater impact on input capacitance than its actual value. For example, if $C_{bc} = 6$ pF and the amplifier gain is 50, then $C_{in(Miller)} = 306$ pF. Figure 11–6 shows how this effective input capacitance appears in the actual ac equivalent circuit in parallel with C_{be} (or C_{gs}).

Miller's theorem also states that C effectively appears as a capacitance from output to ground that can be expressed as follows.

$$C_{out(Miller)} = C\left(\frac{A_v + 1}{A_v}\right) \qquad \text{(11–2)}$$

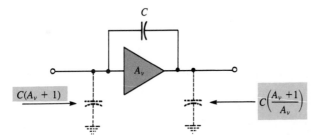

FIGURE 11–5
General case of Miller input and output capacitances.

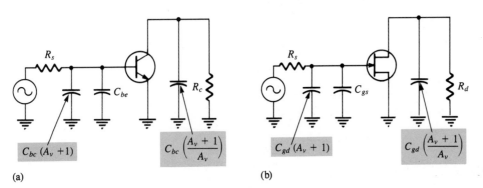

(a)   (b)

FIGURE 11–6
ac equivalent circuits showing effective Miller capacitances.

This formula shows that if the voltage gain is much greater than 1, $C_{out(Miller)}$ is approximately equal to C. Figure 11–6 also shows how this effective output capacitance appears in the ac equivalent circuit for both bipolar transistors and FETs. Equations (11–1) and (11–2) are derived in Appendix B.

EXAMPLE 11–1

Apply Miller's theorem to the amplifier in Figure 11–7 to determine a high-frequency equivalent circuit, given that $C_{bc} = 3$ pF and $C_{be} = 1$ pF.

FIGURE 11–7

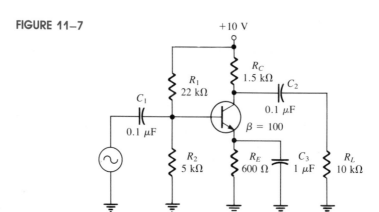

Solution
The voltage gain must first be found before Miller's theorem can be applied. Recall that the voltage gain is expressed as

$$A_v = \frac{R_c}{r_e}$$

R_c is R_C in parallel with R_L.

$$R_c = 1.5 \text{ k}\Omega \| 10 \text{ k}\Omega = 1.3 \text{ k}\Omega$$

To find r_e, I_E must be known.

$$V_B \cong \left(\frac{R_2}{R_1 + R_2}\right)V_{CC} = \left(\frac{5 \text{ k}\Omega}{27 \text{ k}\Omega}\right)10 \text{ V} = 1.85 \text{ V}$$

$$I_E = \frac{V_E}{R_E} = \frac{1.85 \text{ V} - 0.7 \text{ V}}{600 \text{ }\Omega} = 1.92 \text{ mA}$$

$$r_e = \frac{25 \text{ mV}}{I_E} = \frac{25 \text{ mV}}{1.92 \text{ mA}} \cong 13 \text{ }\Omega$$

Therefore,

$$A_v = \frac{1.3 \text{ k}\Omega}{13 \text{ }\Omega} = 100$$

Applying Miller's theorem, we get

$$C_{\text{in(Miller)}} = C_{bc}(A_v + 1) = (3 \text{ pF})(101) = 303 \text{ pF}$$

and

$$C_{\text{out(Miller)}} = C_{bc}\left(\frac{A_v + 1}{A_v}\right) = 3.03 \text{ pF}$$

The high-frequency equivalent circuit for the amplifier in Figure 11–7 is shown in Figure 11–8.

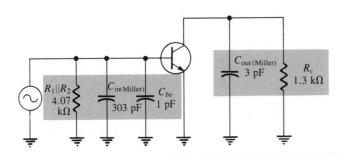

FIGURE 11–8
High-frequency equivalent for the amplifier in Figure 11–7.

Decibels

Chapter 8 introduced the use of decibels in expressing gain. Because of the importance of the decibel unit in amplifier measurements, additional coverage of this topic is necessary before going any further.

The basis for the decibel unit stems from the logarithmic response of the human ear to the intensity of sound. The decibel is a measurement of the ratio of one power to another or one voltage to another. Power gain is expressed in decibels (dB) by the following formula.

$$A_p \text{ (dB)} = 10 \log A_p \qquad\qquad \textbf{(11–3)}$$

where A_p is the actual power gain, $P_{\text{out}}/P_{\text{in}}$. Voltage gain is expressed in decibels by the following formula.

$$A_v \text{ (dB)} = 20 \log A_v \qquad\qquad \textbf{(11–4)}$$

If A_v is greater than 1, the dB gain is *positive*. If A_v is less than 1, the dB gain is *negative* and is usually called *attenuation*.

EXAMPLE 11–2

Express each of the following ratios in dB.

(a) $\dfrac{P_{out}}{P_{in}} = 250$ (b) $\dfrac{P_{out}}{P_{in}} = 100$ (c) $A_v = 10$

(d) $A_p = 0.5$ (e) $\dfrac{V_{out}}{V_{in}} = 0.707$

Solution
(a) A_p (dB) = 10 log(250) = 24 dB
(b) A_p (dB) = 10 log(100) = 20 dB
(c) A_v (dB) = 20 log(10) = 20 dB
(d) A_p (dB) = 10 log(0.5) = −3 dB
(e) A_v (dB) = 20 log(0.707) = −3 dB

Practice Exercise 11–2
Express each of the following ratios in dB: (a) $A_v = 1200$; (b) $A_p = 50$;
(c) $A_v = 125,000$.

0 dB Reference

It is often convenient in amplifier analysis to assign a certain value of gain as the *0 dB reference*. This does not mean that the actual voltage gain is 1 (which is 0 dB); it means that the reference gain, no matter what its actual value, is used as a reference with which to compare other values of gain and is therefore assigned a 0 dB value.

Many amplifiers exhibit a maximum gain over a certain range of frequencies and a reduced gain at frequencies below and above this range. The maximum gain is called the *midrange gain* in this case and is assigned a 0 dB value. Any value of gain below or above midrange can be referenced to 0 dB and expressed as a negative dB value. For example, if the midrange voltage gain of a certain amplifier is 100 and the gain at a certain frequency below midrange is 50, then this reduced gain can be expressed as 20 log(50/100) = 20 log(0.5) = −6 dB. This indicates that it is 6 dB *below* the 0 dB reference. So a halving of the output voltage for a steady input voltage is a 6 dB reduction in the gain. Correspondingly, a doubling of the output voltage is a 6 dB increase in the gain.

Figure 11–9 illustrates the gain-versus-frequency curve of the example amplifier showing several dB points. Table 11–1 shows how doubling or halving voltage ratios translates into dB values. Notice that each step in the doubling or halving of the voltage ratio increases or decreases the dB value by 6 dB. Notice in Table 11–1 that every time the voltage ratio is doubled, the dB value *increases* by 6 dB, and every time the ratio is halved, the dB value *decreases* by 6 dB.

The Critical Frequency

Of particular importance in amplifier analysis is the frequency at which the output power drops to one-half of its midrange value. This corresponds to a 3 dB reduction in the power gain, as expressed by the following formula.

$$A_p \text{ (dB)} = 10 \log(0.5) = -3 \text{ dB}$$

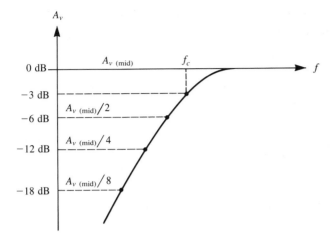

FIGURE 11–9
Normalized gain-versus-frequency curve.

TABLE 11–1
dB values corresponding to doubling and halving
of the voltage ratio

Voltage gain (A_v)	dB (with respect to zero reference)
32	$20 \log(32) = 30$ dB
16	$20 \log(16) = 24$ dB
8	$20 \log(8) = 18$ dB
4	$20 \log(4) = 12$ dB
2	$20 \log(2) = 6$ dB
1	$20 \log(1) = 0$ dB
0.707	$20 \log(0.707) = -3$ dB
0.5	$20 \log(0.5) = -6$ dB
0.25	$20 \log(0.25) = -12$ dB
0.125	$20 \log(0.125) = -18$ dB
0.0625	$20 \log(0.0625) = -24$ dB
0.03125	$20 \log(0.03125) = -30$ dB

Also, when the output voltage is 70.7 percent of its midrange value, it is expressed in dB
as

$$A_v \text{ (dB)} = 20 \log(0.707) = -3 \text{ dB}$$

This tells us that at the critical frequency, the voltage gain is *down 3 dB* from its midrange
value. At this same frequency, the power is one-half of its midrange value.

dBm Measurement

A common unit used in power measurements is the dBm. This unit references power
measurements to a standard reference value of $1\,\text{mW} = 0$ dBm. For example, a 2 mW
power measurement translates into $+3$ dBm, and a 0.5 mW power translates into
-3 dBm. Many power meters have a scale calibrated in dBm.

EXAMPLE 11–3

A certain amplifier has a midrange rms output voltage of 10 V. What is the rms output voltage for each of the following dB gain reductions with a constant rms input voltage?

(a)   −3 dB
(b)   −6 dB
(c)   −12 dB
(d)   −24 dB

Solution
(a)   At −3 dB, $V_{out} = 0.707(10 \text{ V}) = 7.07$ V
(b)   At −6 dB, $V_{out} = 0.5(10 \text{ V}) = 5$ V
(c)   At −12 dB, $V_{out} = 0.25(10 \text{ V}) = 2.5$ V
(d)   At −24 dB, $V_{out} = 0.0625(10 \text{ V}) = 0.625$ V

Practice Exercise 11–3
Determine the output voltage at the following dB levels for a midrange value of 50 V: **(a)**   0 dB; **(b)**   −18 dB; **(c)**   −30 dB.

SECTION REVIEW 11–2

1.   How much increase in actual voltage gain corresponds to +12 dB?
2.   Convert a power gain of 25 to decibels.
3.   What power corresponds to 0 dBm?

LOW-FREQUENCY AMPLIFIER RESPONSE

11–3

In this section, we will examine how the voltage gain and phase shift of a capacitively coupled amplifier are affected by frequencies below which the capacitive reactance becomes significant. At the end of the section a comparison is made to direct-coupled amplifiers.

A typical capacitively coupled common-emitter amplifier is shown in Figure 11–10 (the approach for FETs is similar). Assuming that the coupling and bypass capacitors are ideal *shorts* at the midrange signal frequency, the voltage gain can be calculated as you have done before. Once this value of gain is determined, you have the *midrange gain* of the amplifier as restated in Equation (11–5), where $R_c = R_C \| R_L$.

$$A_{v(mid)} = \frac{R_c}{r_e} \qquad (11\text{–}5)$$

The amplifier in Figure 11–10 has three high-pass RC networks that affect its gain as the frequency is reduced below midrange. These are shown in the *low-frequency* equivalent circuit in Figure 11–11. Unlike the ac equivalent circuit used in previous chapters which represented midrange response ($X_C \cong 0$), the low-frequency equivalent retains the coupling and bypass capacitors because X_C is not small enough to neglect.

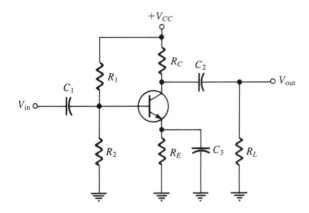

FIGURE 11–10
Typical capacitively coupled amplifier.

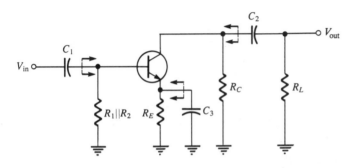

FIGURE 11–11
Low-frequency equivalent of the amplifier in Figure 11–10.

One RC network is formed by the input coupling capacitor C_1 and the *input imped-ance* of the amplifier. The second RC network is formed by the output coupling capacitor C_2, the resistance looking in at the collector, and the load resistance. The third RC network that affects the low-frequency response is formed by the emitter bypass capacitor C_3 and the resistance looking in at the emitter.

The Input RC Network

The RC network for the amplifier in Figure 11–10 that is formed by C_1 and the amplifier's input impedance is shown in Figure 11–12. (Input impedance was discussed in Chapter 8.) As the signal frequency decreases, X_{C1} increases. This causes less voltage to

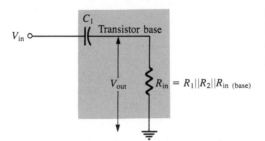

FIGURE 11–12
RC network formed by the input coupling capacitor and the amplifier's input im-pedance.

be applied across the input impedance of the amplifier because more is dropped across X_{C1}. As you can see, this reduces the overall voltage gain of the amplifier. Recall from basic ac circuit theory that the input/output relationship for the RC network in Figure 11–12 (neglecting the internal resistance of the signal source) can be stated as

$$V_{out} = \left(\frac{R_{in}}{\sqrt{R_{in}^2 + X_{C1}^2}} \right) V_{in} \qquad \textbf{(11–6)}$$

As previously mentioned, a critical point in the amplifier's response is generally accepted to occur when the output voltage is 70.7 percent of the input ($V_{out} = 0.707\ V_{in}$). This condition occurs in the input RC network when $X_{C1} = R_{in}$, as shown in the following steps using equation (11–6).

$$V_{out} = \left(\frac{R_{in}}{\sqrt{R_{in}^2 + R_{in}^2}} \right) V_{in}$$

$$= \left(\frac{R_{in}}{\sqrt{2\ R_{in}^2}} \right) V_{in}$$

$$= \left(\frac{R_{in}}{\sqrt{2}\ R_{in}} \right) V_{in}$$

$$= \frac{1}{\sqrt{2}} V_{in}$$

$$= 0.707\ V_{in}$$

In terms of dB measurement, the amplifier voltage gain is

$$A_v\ (dB) = 20\ \log\left(\frac{V_{out}}{V_{in}} \right) = 20\ \log(0.707) = -3\ dB$$

This particular condition is often called the *−3 dB point* of the amplifier response because the overall gain is 3 dB less than at midrange frequencies because of the attenuation of the input RC network. The frequency f_c at which this condition occurs is called the *lower critical frequency* (also known as the *corner* or *break frequency*) and can be calculated as follows.

$$X_{C1} = R_{in}$$

$$\frac{1}{2\pi f_c C_1} = R_{in}$$

$$f_c = \frac{1}{2\pi R_{in} C_1} \qquad \textbf{(11–7)}$$

EXAMPLE 11-4

For an input RC network in a certain amplifier, $R_{in} = 1$ kΩ and $C_1 = 1$ μF.
(a) Determine the lower critical frequency.
(b) What is the attenuation of the RC network at the lower critical frequency?
(c) If the midrange voltage gain of the amplifier is 100, what is the gain at the lower critical frequency?

Solution

(a) $f_c = \dfrac{1}{2\pi R_{in} C_1} = \dfrac{1}{2\pi(1 \text{ k}\Omega)(1 \text{ }\mu\text{F})} = 159$ Hz

(b) At f_c, $X_{C1} = R_{in}$. Therefore, $V_{out}/V_{in} = 0.707$ from equation (11–6).

(c) $A_v = 0.707 A_{v(mid)} = 0.707(100) = 70.7$

Practice Exercise 11-4

For an input RC network in a certain amplifier, $R_{in} = 10$ kΩ and $C_1 = 2.2$ μF.
(a) What is f_c?
(b) What is the attenuation at f_c?
(c) If $A_{v(mid)} = 500$, what is A_v at f_c?

Gain Roll-Off at Low Frequencies

As you have seen, the input RC network reduces the overall gain of an amplifier by 3 dB when the frequency is reduced to the critical value f_c. As the frequency continues to decrease, the overall gain also continues to decrease. In fact, *for each time the frequency is reduced by ten below f_c, there is a 20 dB reduction in gain,* as shown in the following steps.

Let's take a frequency that is one-tenth of the critical frequency ($f = 0.1 f_c$). Since $X_{C1} = R_{in}$ at f_c, then $X_{C1} = 10 R_{in}$ at $0.1 f_c$ because of the inverse relationship of X_{C1} and f. The attenuation of the RC network is therefore

$$\frac{V_{out}}{V_{in}} = \frac{R_{in}}{\sqrt{R_{in}^2 + X_{C1}^2}} = \frac{R_{in}}{\sqrt{R_{in}^2 + (10 R_{in})^2}}$$

$$= \frac{R_{in}}{\sqrt{R_{in}^2 + 100 R_{in}^2}} = \frac{R_{in}}{\sqrt{R_{in}^2(1 + 100)}}$$

$$= \frac{R_{in}}{R_{in}\sqrt{101}} = \frac{1}{\sqrt{101}} \cong \frac{1}{10} = 0.1$$

The dB attenuation is

$$20 \log(V_{out}/V_{in}) = 20 \log(0.1) = -20 \text{ dB}$$

A ten-fold change in frequency is called a *decade*. So, for the input RC network, the attenuation is reduced by 20 dB for each decade decrease in frequency. This causes the overall gain to *drop* 20 dB per decade. For example, if the frequency is reduced to one-hundredth of f_c (a two-decade decrease), the amplifier gain drops 20 dB for *each* decade, giving a total decrease in gain of -20 dB $+ (-20$ dB$) = -40$ dB. This is illustrated in Figure 11–13, which is a graph of dB gain versus frequency. It is the low-frequency response curve for the amplifier showing the effect of the input RC network on the voltage gain.

FIGURE 11–13

dB gain versus frequency for an RC lead network.

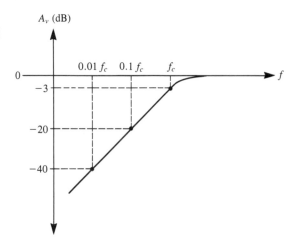

EXAMPLE 11–5

The midrange voltage gain of a certain amplifier is 100. The input RC network has a lower critical frequency of 1 kHz. Determine the actual voltage gain at $f = 1$ kHz, $f = 100$ Hz, and $f = 10$ Hz.

Solution

When $f = 1$ kHz, the gain is 3 dB less than at midrange. At -3 dB, the gain is reduced by a factor of 0.707.

$$A_v = (0.707)(100) = 70.7$$

When $f = 100$ Hz, $f = 0.1\ f_c$, so the gain is 17 dB less than at -3 dB. The gain at -20 dB is one-tenth of that at the midrange frequencies.

$$A_v = (0.1)(100) = 10$$

When $f = 10$ Hz, $f = 0.01\ f_c$, so the gain is 20 dB less than at -20 dB. The gain at -40 dB is one-tenth of that at -20 dB or one-hundredth that at the midrange frequencies.

$$A_v = (0.1)(10) = 1$$

Practice Exercise 11–5

The midrange gain of an amplifier is 300. The lower critical frequency of the input RC network is 400 Hz. Determine the actual voltage gain at 400 Hz, 40 Hz, and 4 Hz.

Phase Shift

In addition to reducing the voltage gain, the input RC network also causes an increasing phase shift through the amplifier as the frequency decreases. At midrange frequencies, the phase shift through the RC network is approximately 0 because $X_{C1} \cong 0$. At lower frequencies, higher values of X_{C1} cause a phase shift to be introduced, and the output voltage of the RC network *leads* the input voltage. As discussed in ac circuit theory, the phase angle is expressed as

$$\theta = \arctan\left(\frac{X_{C1}}{R_{\text{in}}}\right)$$ **(11–8)**

For midrange frequencies, $X_{C1} \cong 0$, so

$$\theta = \arctan\left(\frac{0}{R_{\text{in}}}\right) = \arctan(0) = 0°$$

At the critical frequency, $X_{C1} = R_{\text{in}}$, and

$$\theta = \arctan\left(\frac{R_{\text{in}}}{R_{\text{in}}}\right) = \arctan(1) = 45°$$

A decade below the critical frequency, $X_{C1} = 10\,R_{\text{in}}$, and

$$\theta = \arctan\left(\frac{10\,R_{\text{in}}}{R_{\text{in}}}\right) = \arctan(10) = 84.29°$$

This analysis tells us that the phase shift through the input RC network approaches 90° as the frequency is reduced a few decades below midrange. A plot of phase angle versus frequency is shown in Figure 11–14. The net result is that the voltage at the base of the transistor *leads* the input signal voltage below midrange, as shown in Figure 11–15.

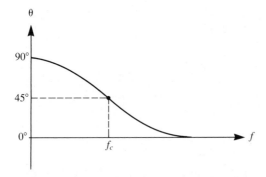

FIGURE 11–14

Phase angle versus frequency for an RC lead network.

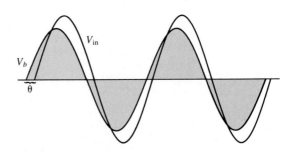

FIGURE 11–15

Input RC network causes base voltage to lead input voltage (below midrange).

The Output RC Network

The second high-pass RC network in the amplifier of Figure 11–10 is formed by the coupling capacitor C_2 and the resistance looking in at the collector, as shown in Figure 11–16(a). In determining the output impedance, the transistor is treated as an ideal current source (with infinite internal resistance), and the upper end of R_C is effectively ac ground, as shown in Figure 11–16(b). Therefore, Thevenizing the circuit to the left of the capacitor produces an equivalent voltage source and series resistance, as shown in part (c). The critical frequency for this RC network is

$$f_c = \frac{1}{2\pi(R_C + R_L)C_2} \tag{11–9}$$

The effect of the output RC network on the amplifier gain is similar to that of the input RC network. As the signal frequency decreases, X_{C2} increases. This causes less voltage across the load resistance because more is dropped across X_{C2}. The signal voltage is reduced by a factor of 0.707 when the frequency is reduced to the lower critical value, f_c. This corresponds to a 3 dB reduction in voltage gain.

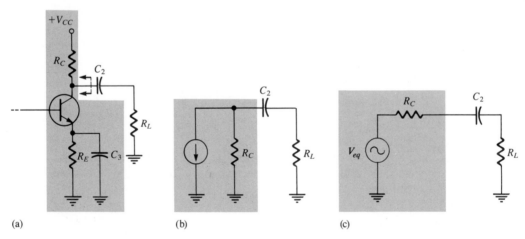

(a)　　　　(b)　　　　(c)

FIGURE 11–16
Development of low-frequency output RC network.

EXAMPLE 11–6

For an output RC network in a certain amplifier, $R_C = 10$ kΩ, $C_2 = 0.1$ μF, and $R_L = 10$ kΩ.
(a) Determine the critical frequency.
(b) What is the attenuation of the RC network at the critical frequency?
(c) If the midrange voltage gain of the amplifier is 50, what is the gain at the critical frequency?

Solution

(a) $f_c = \dfrac{1}{2\pi(R_C + R_L)C_2} = \dfrac{1}{2\pi(20 \text{ k}\Omega)(0.1 \ \mu\text{F})} = 79.58 \text{ Hz}$

(b) For the midrange frequencies, $X_{C2} = 0$, thus the attenuation of the network is

$$\frac{V_{out}}{V_{in}} = \frac{R_L}{R_C + R_L} = \frac{10 \text{ k}\Omega}{20 \text{ k}\Omega} = 0.5$$

or, in dB, $20 \log(0.5) = -6$ dB. This shows that, in this case, the midrange gain is reduced by 6 dB because of the load resistor. At the critical frequency, $X_{C2} = R_C + R_L$.

$$\frac{V_{out}}{V_{in}} = \frac{R_L}{\sqrt{(R_C + R_L)^2 + X_{C2}^2}} = \frac{10 \text{ k}\Omega}{\sqrt{(20 \text{ k}\Omega)^2 + (20 \text{ k}\Omega)^2}} = 0.354$$

or, in dB, $20 \log(0.354) = -9$ dB. As you can see, the gain at f_c is 3 dB less than the gain at midrange.

(c) $A_v = 0.707 A_{v(\text{mid})} = 0.707(50) = 35.35$

Practice Exercise 11–6
The output RC network in a certain amplifier has the following values: $R_C = 3.9 \text{ k}\Omega$, $C_2 = 1 \ \mu\text{F}$, and $R_L = 8.2 \text{ k}\Omega$.
(a) Find the critical frequency.
(b) What is the attenuation at f_c?
(c) If $A_{v(\text{mid})}$ is 100, what is the gain at f_c?

Phase Shift

The phase shift in the output RC network is

$$\theta = \arctan\left(\frac{X_{C2}}{R_C + R_L}\right) \tag{11-10}$$

As previously discussed, $\theta \cong 0°$ in the midrange of frequencies and approaches 90° as the frequency approaches 0 (X_{C2} approaches infinity). At f_c, the phase shift is 45°.

The Bypass Network

The third RC network affecting the low-frequency gain of the amplifier in Figure 11–10 includes the bypass capacitor C_3. For midrange frequencies, it is assumed that $X_{C3} \cong 0$ and effectively shorts the emitter to ground so that the amplifier gain is R_c/r_e, as you already know. As the frequency is reduced, X_{C3} increases and no longer provides a sufficiently low impedance to effectively place the emitter at ac ground. This is illustrated in Figure 11–17. Because the impedance from emitter to ground increases, the gain decreases: recall the formula $A_v = R_c/(r_e + R_e)$. In this case, R_e becomes Z_e, an impedance formed by R_E in parallel with X_{C3}.

FIGURE 11–17

At low frequencies, X_{C_3} in parallel with R_E creates an impedance that reduces the voltage gain.

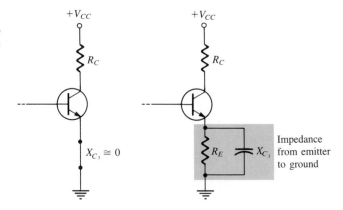

The bypass RC network is formed by C_3 and the resistance looking in at the emitter R_{out}, as shown in Figure 11–18(a). The impedance looking in at the emitter is derived as follows. First, Thevenin's theorem is applied looking from the base of the transistor toward the input V_{in}, as shown in Figure 11–18(b). This results in an equivalent resistance R_{TH} in series with the base, as shown in part (c). The output impedance looking in at the emitter is measured with the input grounded, as shown in part (d), and is expressed as follows.

$$R_{out} = \frac{V_e}{I_e} + r_e \cong \frac{V_b}{\beta I_b} + r_e = \frac{I_b R_{TH}}{\beta I_b} + r_e$$

$$R_{out} = \frac{R_{TH}}{\beta} + r_e \tag{11–11}$$

Looking from the capacitor, C_3, $R_{TH}/\beta + r_e$ is in parallel with R_E, as shown in part (e). Thevenizing again, we get the equivalent RC network shown in part (f). The critical frequency for the bypass network is

$$f_c = \frac{1}{2\pi[(r_e + R_{TH}/\beta)\|R_E]C_3} \tag{11–12}$$

EXAMPLE 11–7

Determine the critical frequency of the bypass network in Figure 11–19 ($r_e = 17\ \Omega$).

Solution

Thevenize the base circuit (looking from the base toward the input source):

$$R_{TH} = R_1\|R_2\|R_s = 60\ \text{k}\Omega\|20\ \text{k}\Omega\|1\ \text{k}\Omega \cong 1\ \text{k}\Omega$$

The resistance looking in at the emitter is

$$R_{out} = r_e + \frac{R_{TH}}{\beta} = 17\ \Omega + 10\ \Omega = 27\ \Omega$$

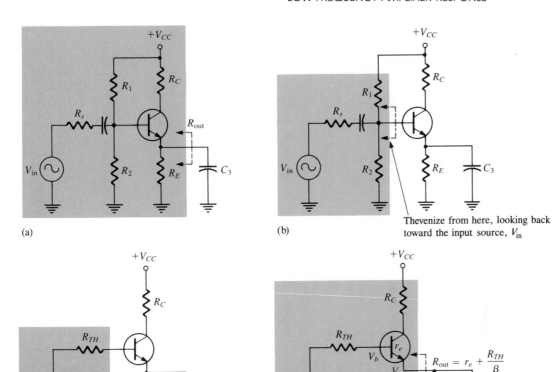

(a)

(b)

Thevenize from here, looking back
toward the input source, V_{in}

(c)

(d)

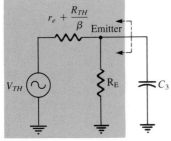

(e)

(f)

FIGURE 11–18
Development of the bypass RC network.

FIGURE 11–19

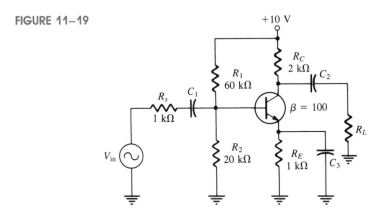

The resistance of the equivalent RC network is $R_{out} \| R_E$.

$$R_{out} \| R_E = 27 \ \Omega \| 1000 \ \Omega = 26.3 \ \Omega$$

The critical frequency is

$$f_c = \frac{1}{2\pi(R_{out}\|R_E)C_3} = \frac{1}{2\pi(26.3 \ \Omega)(100 \ \mu F)} = 60.5 \ Hz$$

Practice Exercise 11–7
In Figure 11–19, the source resistance, R_s, is 50 Ω, and the transistor beta is 150. Determine the critical frequency of the bypass network.

Bode Plot

A plot of dB gain versus frequency on semilog graph paper is called a Bode plot. A generalized Bode plot for an RC network like that shown in Figure 11–20(a) appears in part (b) of the figure. The *ideal* response curve is drawn with a solid line. Notice that it is flat (0 dB) down to the critical frequency, at which point the gain drops at −20 dB/decade as shown. Above f_c are the midrange frequencies. The actual response curve is shown with the dashed line. Notice that it decreases gradually in midrange and is down to −3 dB at the critical frequency. Often, the ideal response is used to simplify amplifier analysis. As previously mentioned, the critical frequency at which the curve "breaks" into a −20 dB/decade drop is often called the *lower break frequency*.

Total Low-Frequency Response

Now that we have individually examined the three high-pass RC networks that affect the amplifier's gain at low frequencies, we will look at the *combined* effect of all three networks. Each network has a critical frequency determined by the RC values. The critical frequencies are not necessarily all equal. If one of the RC networks has a critical (break) frequency *higher* than the other two, then it is the *dominant* network. The dominant network determines the frequency at which the overall gain of the amplifier begins to drop

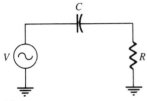

(a) RC lead network

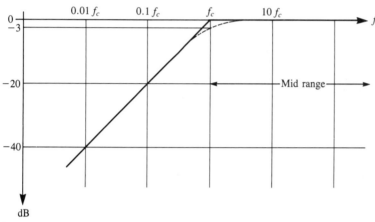

(b) Bode plot

FIGURE 11–20
RC network and its low-frequency response.

at -20 dB/decade. The other networks cause an *additional* -20 dB/decade roll-off below their respective critical (break) frequencies.

To get a better picture of what happens at low frequencies, refer to the Bode plot in Figure 11–21, which shows the superimposed *ideal* responses for the three RC networks (dashed lines). In this example, each RC network has a *different* critical frequency. The *input* RC network is dominant (highest f_c), and the bypass network has the lowest f_c. The overall response is the solid line.

Here is what happens. As the frequency is reduced from midrange, the first "break point" occurs at $f_{c(input)}$ and the gain begins to drop at -20 dB/decade. This constant roll-off rate continues until $f_{c(output)}$ is reached. At this break point, the output RC network adds another -20 dB/decade to make a total roll-off of -40 dB/decade. This constant roll-off continues until $f_{c(bypass)}$ is reached. At this break point, the bypass RC network adds still another -20 dB/decade, making the gain roll-off at -60 dB/decade.

If all three RC networks have the *same* critical frequency, the response curve has one break point at that value of f_c, and the gain rolls off at -60 dB/decade, as shown in Figure 11–22. Keep in mind that the ideal response curves have been used. Actually, the midrange gain does not extend down to the dominant critical frequency, but is really at -9 dB below the midrange gain at that point (-3 dB for each RC network).

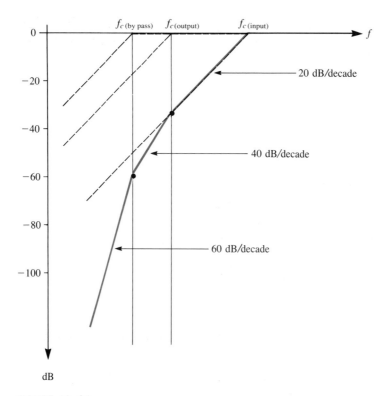

FIGURE 11–21
Composite Bode plot for three low-frequency RC networks with different critical frequencies.

FIGURE 11–22
Composite Bode plot where all three networks have same f_c.

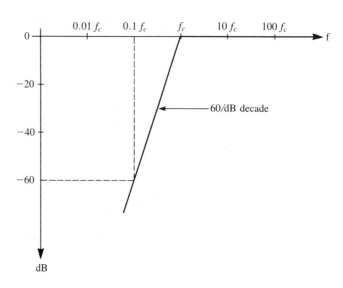

EXAMPLE 11-8

Determine the total low-frequency response of the amplifier in Figure 11–23. $\beta = 100$ and $r_e = 13.9 \ \Omega$.

FIGURE 11-23

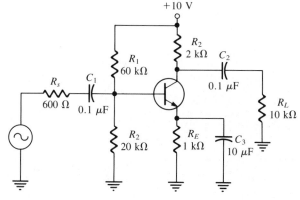

FIGURE 11-23

Solution

Each RC network is analyzed to determine its critical frequency. For the input RC network:

$$R_{in} = R_1 \| R_2 \| \beta r_e = 60 \text{ k}\Omega \| 20 \text{ k}\Omega \| 1.39 \text{ k}\Omega = 1.272 \text{ k}\Omega$$

$$f_{c(input)} = \frac{1}{2\pi(R_s + R_{in})C_1}$$

$$= \frac{1}{2\pi(600 \ \Omega + 1.272 \text{ k}\Omega)(0.1 \ \mu F)}$$

$$= 850 \text{ Hz}$$

For the bypass RC network:

$$R_{TH} = R_1 \| R_2 \| R_s = 60 \text{ k}\Omega \| 20 \text{ k}\Omega \| 600 \ \Omega \cong 600 \ \Omega$$

$$R_{out} = \frac{R_{TH}}{\beta} + r_e = \frac{600 \ \Omega}{100} + 13.9 \ \Omega = 19.9 \ \Omega$$

$$f_{c(bypass)} = \frac{1}{2\pi(R_{out} \| R_E)C_3}$$

$$= \frac{1}{2\pi(19.9 \ \Omega \| 1 \text{ k}\Omega)(10 \ \mu F)}$$

$$= \frac{1}{2\pi(19.5 \ \Omega)(10 \ \mu F)} = 816.18 \text{ Hz}$$

For the output RC network:

$$f_{c(output)} = \frac{1}{2\pi(R_C + R_L)C_2}$$

$$= \frac{1}{2\pi(2 \text{ k}\Omega + 10 \text{ k}\Omega)(0.1 \ \mu F)}$$

$$= 132.63 \text{ Hz}$$

The above analysis shows that the input network produces the dominant lower critical frequency. The midrange gain of the amplifier is

$$A_{v(mid)} = \frac{R_c}{r_e} = \frac{2 \text{ k}\Omega \| 10 \text{ k}\Omega}{13.9 \ \Omega} = 119.9$$

$$A_{v(mid)} = 20 \log(119.9) = 41.58 \text{ dB}$$

The Bode plot of the low-frequency response of this amplifier is shown in Figure 11–24.

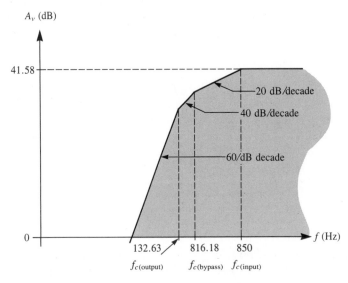

FIGURE 11–24
Bode plot for the amplifier in Figure 11–23.

Low-Frequency Response of Direct-Coupled Amplifiers

Recall from Chapter 8 that direct-coupled amplifiers have no coupling or bypass capacitors, which allows the frequency response to extend down to dc (0 Hz). Direct-coupled amplifiers are popular and commonly used in linear ICs because of their simplicity and because they can effectively amplify signals with frequencies less than 10 Hz (the frequency to which capacitively coupled amplifiers are generally limited because of the size of the capacitors required).

FIGURE 11–25
Basic direct-coupled amplifier stage.

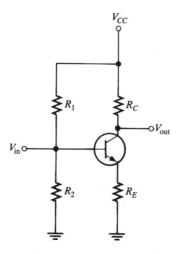

A direct-coupled amplifier has the same gain for a dc input voltage as it does for a signal with a midrange frequency. The gain of a direct-coupled amplifier such as the one in Figure 11–25 is determined by the ratio R_C/R_E and remains constant because there are no frequency-dependent components.

---

1. A certain amplifier exhibits three critical frequencies in its low-frequency response: $f_{c1} = 130$ Hz, $f_{c2} = 167$ Hz, and $f_{c3} = 75$ Hz. Which is the dominant critical frequency?

2. If the midrange gain of the amplifier in question 1 is 50 dB, what is the gain at the dominant f_c?

3. A certain RC network has an $f_c = 235$ Hz, above which the attenuation is 0 dB. What is the dB attenuation at 23.5 Hz?

SECTION REVIEW 11–3

HIGH-FREQUENCY AMPLIFIER RESPONSE

11–4

You have seen how the coupling and bypass capacitors affect the voltage gain of an amplifier, such as the one in Figure 11–26(a), at lower frequencies where the capacitive reactances are significant. Above the lower critical frequency, the effects of the capacitors are minimal and can usually be neglected. If the frequency is increased sufficiently, a point is reached where the transistor's internal capacitances begin to have a significant effect on the gain. (These capacitances were discussed in the first section of this chapter.) We use a bipolar transistor amplifier to illustrate the principles, but the approach for FET amplifiers is similar. The basic differences are the specifications of the internal capacitances and the input impedance. This coverage applies to both capacitively coupled and direct-coupled amplifiers.

A high-frequency ac equivalent circuit for the amplifiers in Figure 11–26(a) and 11–26(b) is shown in part (c). Notice that the coupling and bypass capacitors are treated effectively as shorts, and the two internal capacitances, C_{be} and C_{bc}, which are significant

FIGURE 11–26
Amplifier and its high-frequency equivalent circuit.

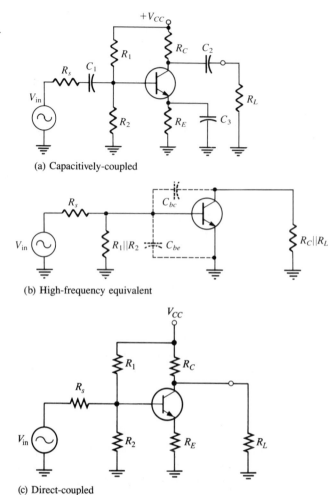

(a) Capacitively-coupled

(b) High-frequency equivalent

(c) Direct-coupled

only at high frequencies, appear in the diagram. As previously mentioned, C_{be} is sometimes called the *input capacitance* and C_{bc} the *output capacitance*. C_{be} (C_{ib}) is specified
on data sheets at a certain value of V_{BE}. For example, a 2N2222A has a C_{be} of 25 pF at
$V_{BE} = 0.5$ V dc, $I_C = 0$, and $f = 100$ kHz. Also, C_{bc} (C_{ob}) is specified at a certain value
of V_{CB}. The 2N2222A has a maximum C_{bc} of 8 pF at $V_{CB} = 10$ V dc.

Using Miller's Theorem for High-Frequency Analysis

By applying Miller's theorem to the circuit in Figure 11–26(c) and using the midrange
gain, we can get a circuit that can be analyzed for high-frequency response. Looking in
from the signal source, the capacitance C_{bc} effectively appears in the Miller input capacitance from base to ground.

$$C_{in(Miller)} = C_{bc}(A_v + 1) \qquad (11\text{–}13)$$

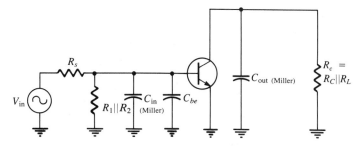

FIGURE 11–27
High-frequency equivalent circuit after applying Miller's theorem.

C_{be} simply appears as a capacitance to ac ground, as shown in Figure 11–27, in parallel with $C_{in(Miller)}$. Looking in at the collector, C_{bc} effectively appears in the Miller output capacitance from collector to ground. As shown in Figure 11–27, it appears in parallel with R_c.

$$C_{out(Miller)} = C_{bc}\left(\frac{A_v + 1}{A_v}\right) \qquad (11\text{–}14)$$

These two Miller capacitances create a high-frequency input RC network and a high-frequency output RC network. These two networks differ from the low-frequency input and output lead networks in that the capacitances go to ground and therefore act as *lag* networks (low-pass filters).

The Input RC Network

At high frequencies, the input network appears as in Figure 11–28(a), where βr_e is the input impedance at the base of the transistor because the bypass capacitor effectively shorts the emitter to ground. By combining C_{be} and $C_{in(Miller)}$ in parallel and repositioning, the simplified network in part (b) results. Next, by Thevenizing the circuit to the left of the capacitor, as indicated, the input RC network is reduced to the equivalent form in part (c).

As the frequency increases, the capacitive reactance becomes smaller. This causes the signal voltage at the base to decrease; thus the amplifier's gain decreases. The reason for this is that the capacitance and resistance act as a voltage-divider and, as the frequency increases, more voltage is dropped across the resistance and less across the capacitance. At the critical frequency, the gain is 3 dB less than its midrange value. Just as with the low-frequency response, the critical frequency f_c is the frequency at which the capacitive reactance is equal to the resistance.

$$X_{C_T} = R_s \| R_1 \| R_2 \| \beta r_e \qquad (11\text{–}15)$$

Therefore,

$$\frac{1}{2\pi f_c C_T} = R_s \| R_1 \| R_2 \| \beta r_e$$

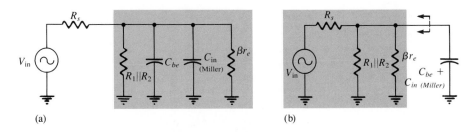

(a)   (b)

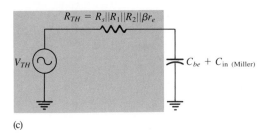

(c)

FIGURE 11–28
Development of high-frequency input RC network.

and

$$f_c = \frac{1}{2\pi(R_s\|R_1\|R_2\|\beta r_e)C_T}$$ **(11–16)**

where R_s is the resistance of the signal source and $C_T = C_{be} + C_{in(Miller)}$. As the frequency goes above f_c, the input RC network rolls off the gain at a rate of -20 dB/decade just as in the low-frequency response.

EXAMPLE 11–9

Derive the high-frequency input RC network for the amplifier in Figure 11–29. Also determine the critical frequency. The transistor's data sheet provides the following: $\beta = 125$, $C_{be} = 20$ pF, and $C_{bc} = 3$ pF.

FIGURE 11–29

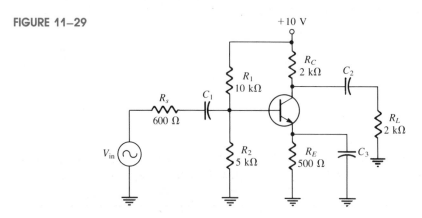

Solution

The first thing to do is find r_e.

$$V_B = \left(\frac{5 \text{ k}\Omega}{15 \text{ k}\Omega}\right) 10 \text{ V} = 3.33 \text{ V}$$

$$V_E = V_B - 0.7 \text{ V} = 2.63 \text{ V}$$

$$I_E = \frac{V_E}{R_E} = \frac{2.63 \text{ V}}{500 \text{ }\Omega} = 5.27 \text{ mA}$$

$$r_e = \frac{25 \text{ mV}}{I_E} = 4.75 \text{ }\Omega$$

The resistance of the input network is

$$R_s\|R_1\|R_2\|\beta r_e = 600 \text{ }\Omega\|10 \text{ k}\Omega\|5 \text{ k}\Omega\|125(4.75 \text{ }\Omega) = 273.9 \text{ }\Omega$$

Next, in order to determine the capacitance, the midrange gain of the amplifier must be determined so that Miller's theorem can be applied.

$$A_{v(\text{mid})} = \frac{R_c}{r_e} = \frac{1 \text{ k}\Omega}{4.75 \text{ }\Omega} = 210.53$$

Applying Miller's theorem we get

$$C_{\text{in(Miller)}} = C_{bc}(A_{v(\text{mid})} + 1) = (3 \text{ pF})(211.53) = 634.59 \text{ pF}$$

The total input capacitance is $C_{\text{in(Miller)}}$ in parallel with C_{be}.

$$C_T = C_{\text{in(Miller)}} + C_{be} = 634.59 \text{ pF} + 20 \text{ pF} = 654.59 \text{ pF}$$

The resulting high-frequency input RC network is shown in Figure 11–30. The critical frequency is

$$f_c = \frac{1}{2\pi(273.9 \text{ }\Omega)(654.59 \text{ pF})} = 887.7 \text{ kHz}$$

FIGURE 11–30
High-frequency input RC network for the amplifier in Figure 11–29.

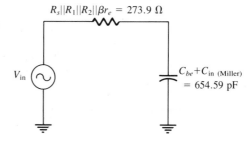

Practice Exercise 11–9

Determine the high-frequency input RC network for Figure 11–29 and find its critical frequency if a transistor with the following specifications is used: $\beta = 75$, $C_{be} = 15 \text{ pF}$, $C_{bc} = 2 \text{ pF}$.

Phase Shift

Because the output voltage of a high-frequency input RC network is across the capacitor, it acts as a lag network. That is, the output of the network *lags* the input. The phase angle is expressed as

$$\theta = \arctan\left(\frac{X_{C_T}}{R_s \| R_1 \| R_2 \| \beta r_e}\right) \tag{11-17}$$

At the critical frequency, the phase angle is 45° with the signal voltage at the base of the transistor lagging the input signal. As the frequency increases above f_c, the phase angle increases above 45° and approaches 90° when the frequency is sufficiently high.

The Output RC Network

The high-frequency output RC network is formed by the Miller output capacitance and the resistance looking in at the collector, as shown in Figure 11–31(a). In determining the output resistance, the transistor is treated as a current source (open) and one end of R_C is effectively ac ground, as shown in part (b). By rearranging the position of the capacitance in the diagram and Thevenizing the circuit to the left, as shown in part (c), we get the equivalent circuit in part (d). The equivalent output RC network consists of a resistance equal to R_C and R_L in parallel and a capacitance as follows.

$$C_{\text{out(Miller)}} = C_{bc}\left(\frac{A_v + 1}{A_v}\right)$$

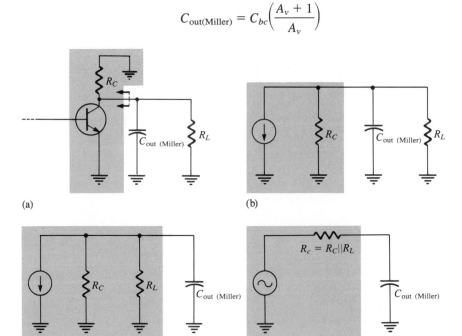

(a)   (b)

(c)   (d)

FIGURE 11–31
Development of high-frequency output RC network.

If the voltage gain is at least 10, the above formula can be approximated as

$$C_{out(Miller)} \cong C_{bc} \tag{11–18}$$

The critical frequency is determined with the following equation, where $R_c = R_C \| R_L$.

$$f_c = \frac{1}{2\pi R_c C_{out(Miller)}} \tag{11–19}$$

Just as in the input RC network, the output network reduces the gain by 3 dB at the critical frequency. When the frequency goes above the critical value, the gain drops at a −20 dB/decade rate. The phase shift introduced by the output RC network is

$$\theta = \arctan\left(\frac{R_c}{X_{C_{out(Miller)}}}\right) \tag{11–20}$$

EXAMPLE 11–10

Determine the critical frequency of the amplifier in Example 11–9 (Figure 11–29) due to the output RC network.

Solution
The Miller output capacitance is as follows.

$$C_{out(Miller)} = C_{bc}\left(\frac{A_v + 1}{A_v}\right) = (3 \text{ pF})\left(\frac{210.53 + 1}{210.53}\right) \cong 3 \text{ pF}$$

The equivalent resistance is

$$R_c = R_C \| R_L = 2 \text{ k}\Omega \| 2 \text{ k}\Omega = 1 \text{ k}\Omega$$

The equivalent network is shown in Figure 11–32, and the critical frequency is as follows ($C_{out(Miller)} \cong C_{bc}$).

$$f_c = \frac{1}{2\pi R_c C_{bc}} = \frac{1}{2\pi (1 \text{ k}\Omega)(3 \text{ pF})} = 53.1 \text{ MHz}$$

FIGURE 11–32

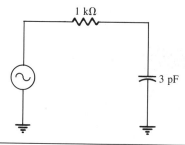

1 kΩ

3 pF

Total High-Frequency Response

As you have seen, two RC networks created by the internal transistor capacitances influence the high-frequency response of an amplifier. As the frequency increases and reaches

FIGURE 11–33

High-frequency Bode plots.

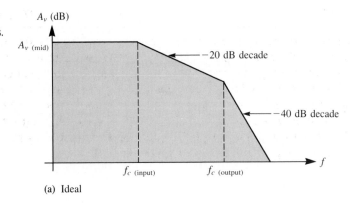

(a) Ideal

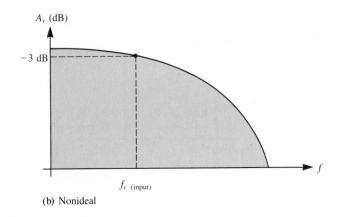

(b) Nonideal

the high end of its midrange values, one of the RC networks will cause the amplifier's gain to begin dropping off. The frequency at which this occurs is the dominant critical frequency; it is the *lower* of the two critical frequencies. An ideal high-frequency Bode plot is shown in Figure 11–33(a). It shows the first break point at $f_{c(input)}$ where the voltage gain begins to roll off at -20 dB/decade. At $f_{c(output)}$, the gain begins dropping at -40 dB/decade because each RC network is providing a -20 dB/decade roll-off. Figure 11–33(b) shows a nonideal Bode plot where the gain is actually -3 dB below midrange at $f_{c(input)}$. Other possibilities are that the output RC network is dominant or that both networks have the same critical frequency.

SECTION REVIEW 11–4

1. What determines the high-frequency response of an amplifier?

2. If an amplifier has a voltage gain of 80 and the transistor's C_{bc} is 4 pF, what is the Miller input capacitance?

3. A certain amplifier has $f_{c(input)} = 3.5$ MHz and $f_{c(output)} = 8.2$ MHz. Which network dominates the high-frequency response?

TOTAL AMPLIFIER RESPONSE

In the previous sections you learned how each RC network in an amplifier affects the frequency response. In this section, we will bring these concepts together and examine the total response of typical amplifiers and the specifications relating to their performance.

Figure 11–34(b) shows an ideal total generalized response curve (Bode plot) for an amplifier of the type shown in Figure 11–34(a). As previously discussed, the three break points at the lower critical frequencies—f_{c1}, f_{c2}, and f_{c3}—are a result of the three low-frequency RC lead networks. The break points at the upper critical frequencies, f_{c4} and f_{c5}, are due to the two high-frequency RC lag networks.

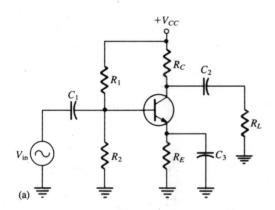

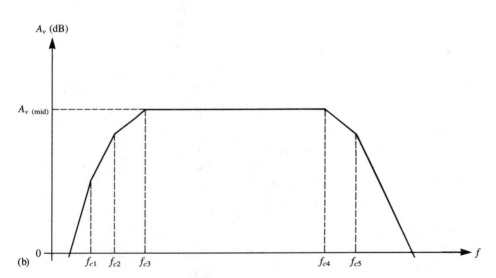

FIGURE 11–34
Amplifier and its generalized ideal response curve (Bode plot).

Of particular interest are the two dominant critical frequencies f_{c3} and f_{c4} in Figure 11–34(b). These two frequencies are where the gain of the amplifier is 3 dB below its midrange value. From now on, these frequencies are referred to as the *lower cutoff frequency* (f_{cl}) and the *upper cutoff frequency* (f_{ch}).

Bandwidth

An amplifier normally operates with frequencies between f_{cl} and f_{ch}. As you know, when the input signal frequency is at f_{cl} or f_{ch}, the output signal voltage level is 70.7 percent of its midrange value (-3 dB). If the signal frequency drops below f_{cl}, the gain and thus the output signal level drops at 20 dB/decade until the next critical frequency is reached. The same is true when the signal frequency goes above f_{ch}.

The range (band) of frequencies lying between f_{cl} and f_{ch} constitute the *bandwidth* of the amplifier, as illustrated in Figure 11–35. Only the dominant critical frequencies appear in the response curve because they determine the bandwidth. Also, sometimes the other critical frequencies are far enough away from the dominant frequencies that they play no significant role in the total amplifier response and can be neglected for simplicity. The bandwidth is expressed as

$$BW = f_{ch} - f_{cl} \qquad (11–21)$$

Ideally, all signal frequencies lying in an amplifier's bandwidth are amplified equally. For example, if a 10 mV rms signal is applied to an amplifier with a gain of 20, it is amplified to 200 mV rms. Likewise, a 50 mV rms signal is amplified to 1 V rms.

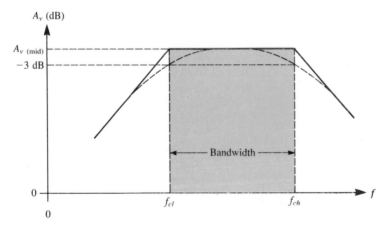

FIGURE 11–35
Response curve illustrating bandwidth.

EXAMPLE 11–11

What is the bandwidth of an amplifier having an f_{cl} of 200 Hz and an f_{ch} of 2 kHz?

Solution

$$BW = f_{ch} - f_{cl} = 2000 \text{ Hz} - 200 \text{ Hz} = 1800 \text{ Hz}$$

Notice that bandwidth has the unit of hertz.

Gain-Bandwidth Product

Let's assume that the lower cutoff frequency of a particular amplifier is much less than the upper cutoff frequency.

$$f_{cl} << f_{ch}$$

The bandwidth can then be approximated as

$$BW = f_{ch} - f_{cl} \cong f_{ch}$$

The simplified Bode plot for this condition is shown in Figure 11–36. Notice that f_{cl} is neglected, and the bandwidth equals f_{ch}. Beginning at f_{ch}, the gain rolls off until *unity gain* (0 dB) is reached. The frequency at which the amplifier's gain is 1 is called the *unity-gain frequency*, f_T. The significance of f_T is that it always equals the product of the voltage gain times the bandwidth and is a constant for a given transistor.

$$f_T = A_{v(\text{mid})}BW \qquad\qquad (11\text{–}22)$$

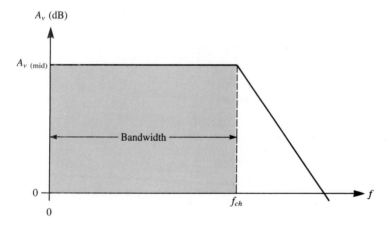

FIGURE 11–36
Simplified response curve where f_{cl} is negligible compared to f_{ch}.

For the case shown in Figure 11–36, $f_T = A_{v(mid)}f_{ch}$. For example, if a transistor data sheet specifies $f_T = 100$ MHz, this means that the transistor is capable of producing a voltage gain of 1 up to 100 MHz, or a gain of 100 up to 1 MHz, or any combination of gain and bandwidth that produces a product of 100 MHz.

EXAMPLE 11–12

A certain transistor has an f_T of 175 MHz. When this transistor is used as an amplifier with a voltage gain of 50, what bandwidth can be achieved?

Solution

$$f_T = A_{v(mid)}BW$$

$$BW = \frac{f_T}{A_{v(mid)}} = \frac{175 \text{ MHz}}{50} = 3.5 \text{ MHz}$$

dB/Octave

Sometimes, the gain roll-off of an amplifier is expressed in dB/octave rather than dB/decade. An octave corresponds to a doubling or halving of the frequency. For example, an increase in frequency from 100 Hz to 200 Hz is an octave. Likewise, a decrease in frequency from 100 kHz to 50 kHz is also an octave. A rate of -20 dB/decade is approximately -6 dB/octave, a rate of -40 dB/decade is approximately -12 dB/octave, and so on.

Half-Power Points

The upper and lower cutoff frequencies are sometimes called the *half-power frequencies*. This term is derived from the fact that the output power of an amplifier at its cutoff frequencies is one-half of its midrange power, as previously mentioned. This can be shown as follows, starting with the fact that the output voltage is 0.707 of its midrange value at the cutoff frequencies.

$$V_{out(f_c)} = 0.707 \ V_{out(mid)}$$

$$P_{out(f_c)} = \frac{V^2_{out(f_c)}}{R_{out}} = \frac{(0.707 \ V_{out(mid)})^2}{R_{out}}$$

$$= \frac{0.5 \ V^2_{out(mid)}}{R_{out}} = 0.5 \ P_{out(mid)}$$

SECTION REVIEW 11–5

1. What is the bandwidth of an amplifier when $f_{ch} = 25$ kHz and $f_{cl} = 100$ Hz?

2. The f_T of a certain transistor is 130 MHz. What gain can be achieved with a bandwidth of 50 MHz?

FREQUENCY RESPONSE OF FET AMPLIFIERS

In the first part of this section we will use a DE MOSFET amplifier and evaluate its frequency response. A zero-biased amplifier with capacitive coupling on the input and output is shown in Figure 11–37.

FIGURE 11–37
Zero-biased DE MOSFET amplifier.

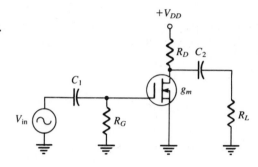

Low-Frequency Response

The midrange voltage gain of this amplifier is determined as follows.

$$A_{v(\text{mid})} = g_m R_d \qquad\qquad (11\text{–}23)$$

As you know, this is the gain at frequencies high enough so that the capacitive reactances are approximately 0. The amplifier in Figure 11–37 has only two RC networks that influence its low-frequency response. One network is formed by the input coupling capacitor C_1 and the input impedance, as shown in Figure 11–38. The other network is formed by the output coupling capacitor C_2 and the output impedance looking in at the drain.

FIGURE 11–38
Input RC network for the amplifier in Figure 11–37.

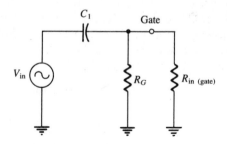

The Input RC Network

Just as in the previous case for the bipolar transistor amplifier, the reactance of the input coupling capacitor increases as the frequency decreases. When $X_{C1} = R_{\text{in}}$, the gain is down 3 dB below its midrange value. The lower critical frequency is

$$f_c = \frac{1}{2\pi R_{\text{in}} C_1}$$

The input impedance is

$$R_{in} = R_G \| R_{in(gate)}$$

where $R_{in(gate)}$ is determined from data sheet information as follows.

$$R_{in(gate)} = \left| \frac{V_{GS}}{I_{GSS}} \right|$$

So

$$f_c = \frac{1}{2\pi(R_G \| R_{in(gate)})C_1} \qquad \textbf{(11–24)}$$

The gain roll-off below f_c is 20 dB/decade, as previously shown, and the phase shift is $\theta = \arctan(X_{C1}/R_{in})$.

EXAMPLE 11–13

What is the critical frequency of the input RC network in Figure 11–39?

FIGURE 11–39

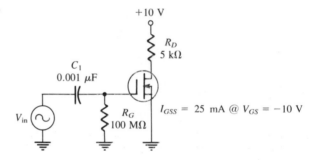

Solution

First determine R_{in}.

$$R_{in(gate)} = \left| \frac{V_{GS}}{I_{GSS}} \right| = \frac{10 \text{ V}}{25 \text{ nA}} = 400 \text{ M}\Omega$$

$$R_{in} = R_G \| R_{in(gate)} = 100 \text{ M}\Omega \| 400 \text{ M}\Omega = 80 \text{ M}\Omega$$

$$f_c = \frac{1}{2\pi R_{in} C_1} = \frac{1}{2\pi(80 \text{ M}\Omega)(0.001 \ \mu\text{F})} = 1.989 \text{ Hz}$$

The critical frequency of the input RC network of an FET amplifier is usually extremely low because of the very high input impedance.

The Output RC Network

The second RC network that affects the low-frequency response of the amplifier in Figure 11–37 is formed by the coupling capacitor C_2 and the output impedance looking in at the

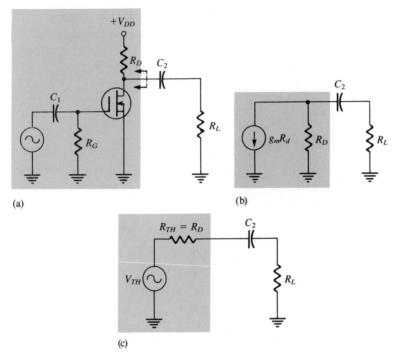

FIGURE 11–40
Development of output RC network.

drain, as shown in Figure 11–40(a). R_L is also included. As in the case of the bipolar transistor, the FET is treated as a current source, and the upper end of R_D is effectively ac ground, as shown in Figure 11–40(b). The Thevenin equivalent of the circuit to the left of C_2 is shown in Figure 11–40(c). The critical frequency for this network is determined as follows.

$$f_c = \frac{1}{2\pi(R_D + R_L)C_2} \tag{11–25}$$

The effect of the output RC network on the amplifier's gain below the midrange is similar to that of the input RC network. The network with the *higher* critical frequency dominates because it is the one that first causes the gain to roll off as the frequency drops below its midrange values. The phase shift in the output RC network is

$$\theta = \arctan\left(\frac{X_{C_2}}{R_D + R_L}\right) \tag{11–26}$$

Again, at the critical frequency, the phase angle is 45° and approaches 90° as the frequency approaches 0. However, starting at the critical frequency, the phase angle decreases from 45° and becomes very small as the frequency goes higher.

EXAMPLE 11–14

Determine the total low-frequency response of the FET amplifier in Figure 11–41. Assume that the load is another identical amplifier with the same R_{in}. $I_{GSS} = 100$ nA at $V_{GS} = -12$ V.

FIGURE 11–41

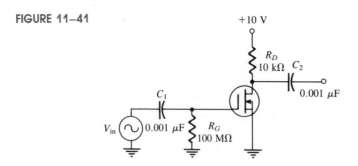

Solution

First, we find the critical frequency for the input RC network.

$$R_{in(gate)} = \left| \frac{V_{GS}}{I_{GSS}} \right| = \frac{12 \text{ V}}{100 \text{ nA}} = 120 \text{ M}\Omega$$

$$R_{in} = R_G \| R_{in(gate)} = 100 \text{ M}\Omega \| 120 \text{ M}\Omega = 54.55 \text{ M}\Omega$$

$$f_{c(input)} = \frac{1}{2\pi R_{in} C_1} = \frac{1}{2\pi (54.55 \text{ M}\Omega)(0.001 \ \mu\text{F})} = 2.92 \text{ Hz}$$

The output RC network has a critical frequency of

$$f_{c(output)} = \frac{1}{2\pi (R_D + R_L) C_2}$$

$$= \frac{1}{2\pi (54.56 \text{ M}\Omega)(0.001 \ \mu\text{F})}$$

$$= 2.92 \text{ Hz}$$

The approach to the high-frequency analysis of an FET amplifier is very similar to that of a bipolar amplifier. The basic differences are the specifications of the internal FET capacitances and the determination of the input impedance.

Figure 11–42(a) shows a JFET common-source amplifier that we will use to illustrate high-frequency analysis. An equivalent circuit for high-frequency analysis of the amplifier is shown in part (b). Notice that the coupling and bypass capacitors are assumed to have 0 reactances and are thus treated as effective shorts. The internal capacitances C_{gs} and C_{gd} appear in the equivalent circuit because their reactances are significant at high frequencies and thus influence the amplifier's response.

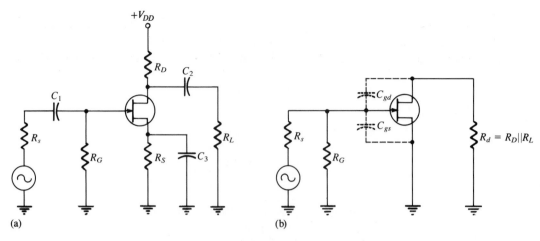

FIGURE 11–42
JFET amplifier and its high-frequency equivalent circuit.

Getting the Values of C_{gs} and C_{gd}

FET data sheets do not normally provide values for C_{gs} and C_{gd} directly. Instead, two other values are usually specified because they are easier to measure. These are C_{iss}, the *input capacitance,* and C_{rss}, the *reverse transfer capacitance.* Because of the manufacturer's method of measurement, the following relationships allow us to determine the capacitor values needed for analysis.

$$C_{gd} = C_{rss} \qquad\qquad (11\text{–}27)$$

$$C_{gs} = C_{iss} - C_{rss} \qquad\qquad (11\text{–}28)$$

EXAMPLE 11–15
The data sheet for a 2N3823 JFET gives $C_{iss} = 6$ pF and $C_{rss} = 2$ pF. Determine C_{gs} and C_{gd}.

Solution

$$C_{gd} = C_{rss} = 2 \text{ pF}$$
$$C_{gs} = C_{iss} - C_{rss} = 6 \text{ pF} - 2 \text{ pF} = 4 \text{ pF}$$

Using Miller's Theorem

Miller's theorem is applied in the same way for high-frequency FET analysis as was done for bipolar analysis, using the midrange gain. Looking in from the signal source in Figure 11–42(b), C_{gd} effectively appears in the Miller input capacitance as follows.

$$C_{\text{in(Miller)}} = C_{gd}(A_v + 1) \qquad\qquad (11\text{–}29)$$

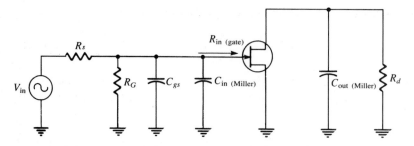

FIGURE 11–43
High-frequency equivalent circuit after applying Miller's theorem.

C_{gs} simply appears as a capacitance to ac ground in parallel with $C_{in(Miller)}$, as shown in Figure 11–43. Looking in at the drain, C_{gd} effectively appears in the Miller output capacitance from drain to ground in parallel with R_d, as shown in Figure 11–43.

$$C_{out(Miller)} = C_{gd}\left(\frac{A_v + 1}{A_v}\right) \tag{11–30}$$

These two Miller capacitances create a high-frequency input RC network and a high-frequency output RC network. Both act as lag networks (low-pass filters).

The Input RC Network

The high-frequency input network forms a low-pass type of filter and is shown in Figure 11–44(a). Because both R_G and the input impedance at the gate of FETs are extremely high, the controlling resistance for the input network is the resistance of the signal source as long as $R_s \ll R_{in}$. This is because R_s appears in parallel with R_{in} when Thevenin's theorem is applied. The simplified input RC network appears in part (b). The critical frequency is

$$f_c = \frac{1}{2\pi R_s C_T} \tag{11–31}$$

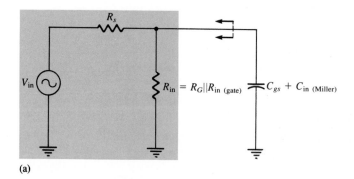

(a)

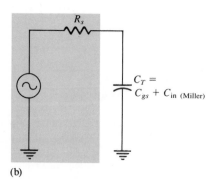

(b)

FIGURE 11–44
High-frequency input RC network.

where $C_T = C_{gs} + C_{in(Miller)}$. The input network produces a phase shift of

$$\theta = \arctan\left(\frac{R_s}{X_{C_T}}\right) \qquad (11\text{–}32)$$

The effect of the input RC network is to reduce the midrange gain of the amplifier by 3 dB at the critical frequency and to cause the gain to decrease at -20 dB/decade beyond f_c.

EXAMPLE 11–16

Find the critical frequency of the input RC network for the amplifier in Figure 11–45. $C_{iss} = 8$ pF and $C_{rss} = 3$ pF. $g_m = 6500$ μS.

FIGURE 11–45

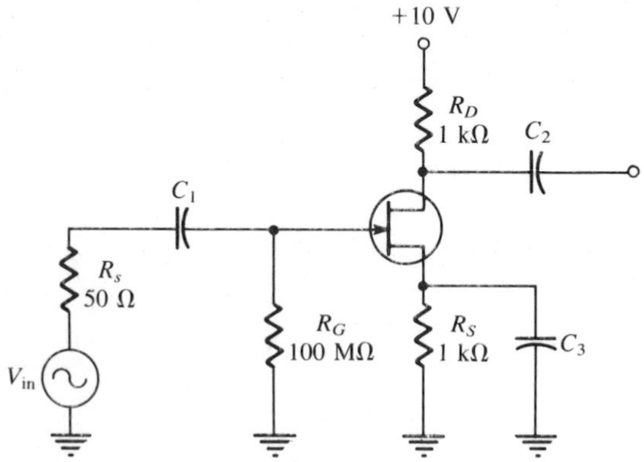

Solution

$$C_{gd} = C_{rss} = 3 \text{ pF}$$
$$C_{gs} = C_{iss} - C_{rss} = 8 \text{ pF} - 3 \text{ pF} = 5 \text{ pF}$$

The input RC network is derived as follows.

$$A_v = g_m R_d = (6500 \ \mu\text{S})(1 \ \text{k}\Omega) = 6.5$$
$$C_{in(Miller)} = C_{gd}(A_v + 1) = (3 \text{ pF})(7.5) = 22.5 \text{ pF}$$

The total input capacitance is

$$C_T = C_{gs} + C_{in(Miller)} = 5 \text{ pF} + 22.5 \text{ pF} = 27.5 \text{ pF}$$

The critical frequency is

$$f_c = \frac{1}{2\pi R_s C_T} = \frac{1}{2\pi(50 \ \Omega)(27.5 \text{ pF})} = 115.75 \text{ MHz}$$

The Output RC Network

The high-frequency output RC network is formed by the Miller output capacitance and the output impedance looking in at the drain, as shown in Figure 11–46(a). As in the case of

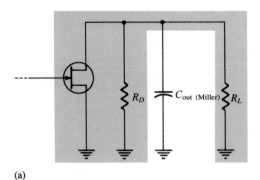 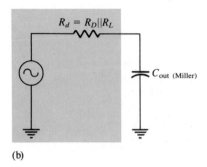

(a)   (b)

FIGURE 11–46
High-frequency output RC network.

the bipolar transistor, the FET is treated as a current source. Application of Thevenin's theorem produces an equivalent output RC network consisting of R_D in parallel with R_L and an equivalent output capacitance as follows.

$$C_{out(Miller)} = C_{gd}\left(\frac{A_v + 1}{A_v}\right)$$

This equivalent output network is shown in Figure 11–44(b). The critical frequency of the output RC lag network is

$$f_c = \frac{1}{2\pi R_d C_{out(Miller)}} \tag{11–33}$$

The output network produces a phase shift of

$$\theta = \arctan\left(\frac{R_d}{X_{C_{out(Miller)}}}\right) \tag{11–34}$$

EXAMPLE 11–17

Determine the critical frequency of the output RC network for the amplifier in Figure 11–45. What is the phase shift introduced by this network at the critical frequency?

Solution
Since R_L is very large compared to R_D, it can be neglected, and the equivalent output resistance is

$$R_d = R_D = 1 \text{ k}\Omega$$

The equivalent output capacitance is

$$C_{out(Miller)} = C_{gd}\left(\frac{A_v + 1}{A_v}\right) = 3 \text{ pF}\left(\frac{7.5}{6.5}\right) = 3.46 \text{ pF}$$

Therefore, the critical frequency is

$$f_c = \frac{1}{2\pi R_d C_{out(Miller)}} = \frac{1}{2\pi (1 \text{ k}\Omega)(3.46 \text{ pF})} = 46 \text{ MHz}$$

The phase angle is always 45° at f_c for one RC network. At high frequencies, the output network causes a phase lag.

In Example 11–16, the critical frequency of the input RC network was found to be 115.75 MHz. Therefore, the critical frequency for the output network is dominant since it is the lesser of the two.

---

1. What is the amount of phase shift contributed by an input network when $X_C = 0.5\, R_{in}$?
2. What is f_c when $R_D = 1.5$ kΩ, $R_L = 5$ kΩ, and $C_2 = 0.002$ μF?
3. What are the capacitances that are usually specified on an FET data sheet?
4. If $C_{gs} = 4$ pF and $C_{gd} = 3$ pF, what is the total input capacitance of an FET amplifier whose voltage gain is 25?

FREQUENCY RESPONSE MEASUREMENT TECHNIQUES

In this section, two basic methods of measuring the frequency response of an amplifier are covered. We will concentrate on determining the two dominant critical frequencies and thus obtain the bandwidth.

11–7

Frequency Measurement

Figure 11–47 shows an amplifier connected with a sine wave signal source and frequency meter on the input and an ac voltmeter on the output. The signal frequency is set at a midrange value, and the signal level is set at a value to establish an output signal reference level, as in Figure 11–48(a). This output reference level should be set at a convenient value within the linear range of the amplifier; for example, 100 mV, 1 V, 10 V, and so on. In this case, we use 1 V.

FIGURE 11–47
Test set-up for measuring cutoff frequencies.

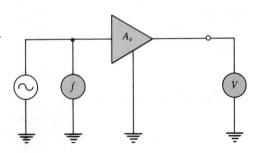

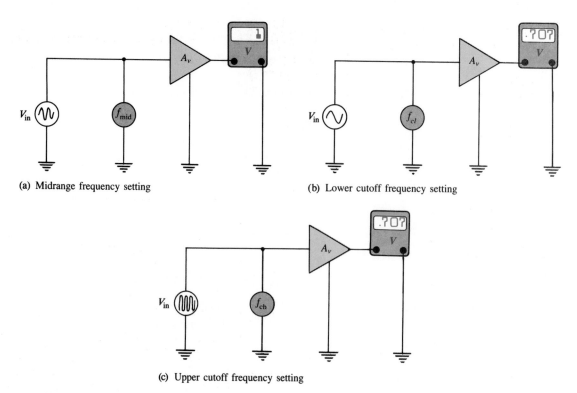

(a) Midrange frequency setting   (b) Lower cutoff frequency setting

(c) Upper cutoff frequency setting

FIGURE 11–48
Test procedure for measuring cutoff frequencies.

Next, the frequency of the signal source is lowered until the ac voltmeter indicates an output level of 70.7 percent of the reference level. If a dB scale is used, the reference can be set at 0 dB, and then the 70.7 percent point is at -3 dB. At this point, the frequency of the signal source is measured with a frequency meter and the resulting value is f_{cl}, as shown in Figure 11–48(b).

Next, the signal frequency is raised back to midrange and increased until the output signal level again drops to 70.7 percent of the reference level, as shown in Figure 11–48(c). The frequency at this point is f_{ch}. From these frequency measurements, the bandwidth is derived by the formula $BW = f_{ch} - f_{cl}$.

Step Response

Another method of determining f_{cl} and f_{ch} is *step response* measurement. Rather than using a sinusoidal signal source as in the previous method, a step input is used. The test set-up for this procedure is shown in Figure 11–49. A pulse generator provides the input, and an oscilloscope is the output measurement instrument.

When a step input is applied, as shown in Figure 11–50(a), the amplifier's high-frequency RC networks prevent the output from responding immediately to the step input

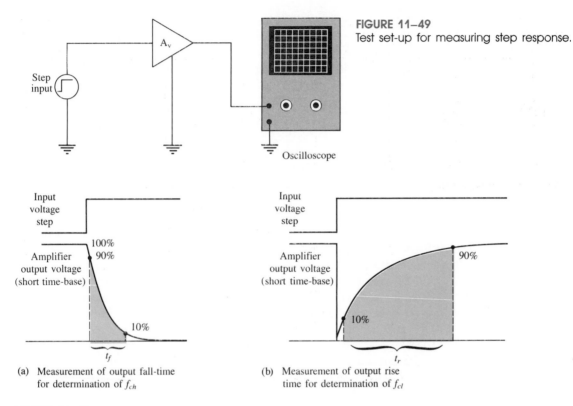

FIGURE 11–49
Test set-up for measuring step response.

Oscilloscope

(a) Measurement of output fall-time
for determination of f_{ch}

(b) Measurement of output rise
time for determination of f_{cl}

FIGURE 11–50
Step responses of capacitively coupled amplifier.

due to the internal capacitors. As a result, the output voltage has a fall-time (t_f) associated with it, as shown. The fall-time is measured from the 90 percent amplitude point to the 10 percent amplitude point, as indicated. Once this measurement is made on the oscilloscope, the upper cutoff frequency can be calculated.

$$f_{ch} = \frac{0.35}{t_f} \qquad \textbf{(11–35)}$$

Next, the oscilloscope's time base is adjusted so that a longer time interval can be measured, and the input is adjusted to a step voltage of long duration. The amplifier's low-frequency RC networks will cause the output voltage to appear, as in Figure 11–50(b), due to the charging of the coupling and bypass capacitors. The rise-time (t_r) of the charge curve is then measured from its 10 percent point to its 90 percent point, as indicated, and the lower cutoff frequency calculated as

$$f_{cl} = \frac{0.35}{t_r} \qquad \textbf{(11–36)}$$

1. In a step response measurement, the output of an amplifier has a fall-time of 5 ns. What is the upper cutoff frequency?

2. The fall-time is measured between what two points on the voltage transition?

A SYSTEM APPLICATION

11–8

As mentioned in the introduction to this chapter, AM and FM superhet principles are used in television receivers. Other than having different frequency requirements, the receiver circuits in a television are basically the same as those in a radio. We will highlight some of the aspects of frequency response in a TV receiver. A complete study of the operation of television receivers is left for the electronic communications course.

There are twelve channels allotted for transmission of VHF (very high frequency) television signals (channels 2 through 13) and eight channels for UHF (ultra high frequency). Each channel has a bandwidth of 6 MHz in which the video, audio, and synchronizing signals are carried. Thus the signals for each channel range from a lower frequency to a frequency that is 6 MHz above the lowest. Television frequency allocations are listed in Table 11–2. The frequency band for Channel 2, for example, is from 54 MHz to 60 MHz, and for Channel 3, from 60 MHz to 66 MHz, and so on.

The RF amplifier in a TV receiver such as that shown in the partial block diagram of Figure 11–51(a) must respond to a 6 MHz range of frequencies around the carrier frequency for which it is tuned when a specific channel is selected. The frequency response curve for a television signal is shown in Figure 11–51(b). Notice that the picture (video) carrier is 1.25 MHz above the lowest frequency and the sound (audio) carrier is 5.75 MHz above. Thus, the frequency band for Channel 2, for example, is from 54 MHz to 60 MHz, and for Channel 3, it is 60 MHz to 66 MHz, and so on.

The mixer output that contains both AM video and FM sound goes to the video IF amplifier section where most of the system amplification and selectivity are provided. The standard IF frequencies for television are 45.75 MHz for the picture carrier and 41.25 MHz for the sound carrier. Even though the RF sound carrier is above the RF

TABLE 11–2
TV channel frequencies

Lower VHF band		Upper VHF band		UHF band	
Channel	Lowest frequency in MHz	Channel	Lowest frequency in MHz	Channel	Lowest frequency in MHz
2	54	7	174	14	470
3	60	8	180	24	530
4	66	9	186	34	590
5	76	10	192	44	650
6	82	11	198	54	710
		12	204	64	770
		13	210	74	830
				83	884

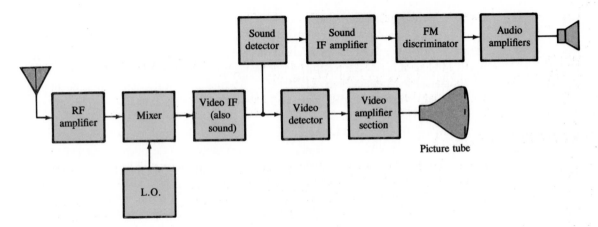

(a) Partial block diagram of a TV receiver

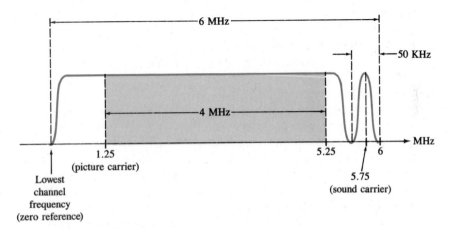

(b) Frequency curves of transmitted TV signal referenced to 0 Hz (the lowest channel frequency)

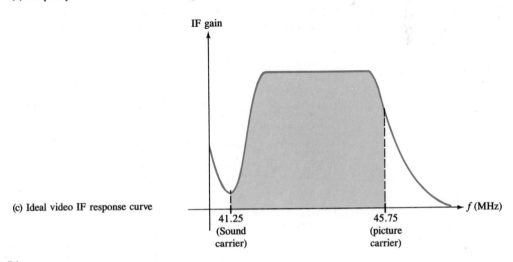

(c) Ideal video IF response curve

FIGURE 11–51
Block diagram of a television receiver and frequency response curves.

picture carrier when the signal is received, this relationship is reversed by the action of the mixer. For example, the RF picture carrier frequency for Channel 5 is 77.25 MHz and the RF sound carrier frequency is 81.25 MHz. Mixing the local oscillator frequency of 123 MHz with the incoming RF generates difference frequencies of 41.25 MHz for sound and 45.75 MHz for picture.

An ideal frequency response curve for the video IF amplifiers is shown in Figure 11–51(c). The IF sound carrier is attenuated much more than the other frequencies in the IF band to minimize the interference effects that the sound would have on the picture. Notice the steepness of the response curve near the sound carrier. The IF amplifiers must have a bandwidth of approximately 4 MHz. Since the output of the video detector is the original video modulating signal and contains frequencies from 0 to 4 MHz, the video amplifier must be capable of responding to this range of frequencies. The IF output of the separate sound detector is an FM signal with a carrier frequency of 4.5 MHz.

FORMULAS

Miller's theorem and decibels

(11–1) $C_{in(Miller)} = C(A_v + 1)$ Miller input capacitance, where $C = C_{bc}$ or C_{gd}

(11–2) $C_{out(Miller)} = C\left(\dfrac{A_v + 1}{A_v}\right)$ Miller output capacitance, where $C = C_{bc}$ or C_{gd}

(11–3) A_p (dB) $= 10 \log(A_p)$ dB power gain

(11–4) A_v (dB) $= 20 \log A_v$ dB voltage gain

Low-frequency analysis

(11–5) $A_{v(mid)} = \dfrac{R_c}{r_e}$ Midrange gain (bipolar)

(11–6) $V_{out} = \left(\dfrac{R_{in}}{\sqrt{R_{in}^2 + X_{C1}^2}}\right) V_{in}$ RC lead network

(11–7) $f_c = \dfrac{1}{2\pi R_{in} C_1}$ Critical low frequency (input)

(11–8) $\theta = \arctan\left(\dfrac{X_{C1}}{R_{in}}\right)$ Phase angle, input network

(11–9) $f_c = \dfrac{1}{2\pi(R_C + R_L)C_2}$ Critical low frequency (output)

(11–10) $\theta = \arctan\left(\dfrac{X_{C2}}{R_C + R_L}\right)$ Phase angle, output network

(11–11) $R_{out} = \dfrac{R_{TH}}{\beta} + r_e$ Output impedance (emitter)

(11–12) $f_c = \dfrac{1}{2\pi[(r_e + R_{TH}/\beta)\|R_E]C_3}$ Critical low frequency (bypass)

High-frequency analysis

(11–13) $C_{in(Miller)} = C_{bc}(A_v + 1)$ Miller input capacitance

(11–14) $C_{out(Miller)} = C_{bc}\left(\dfrac{A_v + 1}{A_v}\right)$ Miller output capacitance

(11–15) $X_{C_T} = R_s\|R_1\|R_2\|\beta r_e$ Condition at f_c

(11–16) $f_c = \dfrac{1}{2\pi(R_s\|R_1\|R_2\|\beta r_e)C_T}$ Critical high frequency (input)

(11–17) $\theta = \arctan\left(\dfrac{X_{C_T}}{R_s\|R_1\|R_2\|\beta r_e}\right)$ Phase angle, input network

(11–18) $C_{out(Miller)} \cong C_{bc}$ When $A_v > 10$

(11–19) $f_c = \dfrac{1}{2\pi R_c C_{out(Miller)}}$ Critical high frequency (output)

(11–20) $\theta = \arctan\left(\dfrac{R_c}{X_{C_{out(Miller)}}}\right)$ Phase angle, output network

Total response

(11–21) $BW = f_{ch} - f_{cl}$ Bandwidth

(11–22) $f_T = A_{v(mid)}BW$ Gain-bandwidth product

FET frequency response

(11–23) $A_{v(mid)} = g_m R_d$ Midrange gain

(11–24) $f_c = \dfrac{1}{2\pi(R_G\|R_{in(gate)})C_1}$ Critical low frequency (input)

(11–25) $f_c = \dfrac{1}{2\pi(R_D + R_L)C_2}$ Critical low frequency (output)

(11–26) $\theta = \arctan\left(\dfrac{X_{C_2}}{R_D + R_L}\right)$ Phase angle, output network

(11–27) $C_{gd} = C_{rss}$ Gate-to-drain capacitance

(11–28) $C_{gs} = C_{iss} - C_{rss}$ Gate-to-source capacitance

(11–29) $C_{in(Miller)} = C_{gd}(A_v + 1)$ Miller input capacitance

(11–30) $C_{out(Miller)} = C_{gd}\left(\dfrac{A_v + 1}{A_v}\right)$ Miller output capacitance

(11–31) $f_c = \dfrac{1}{2\pi R_s C_T}$ Critical high frequency (input)

(11–32) $\theta = \arctan\left(\dfrac{R_s}{X_{C_T}}\right)$ Phase angle, input network

(11–33) $f_c = \dfrac{1}{2\pi R_d C_{out(Miller)}}$ Critical high frequency (output)

(11–34) $\theta = \arctan\left(\dfrac{R_d}{X_{C_{out(Miller)}}}\right)$ Phase angle, output network

Measurement techniques

(11–35) $f_{ch} = \dfrac{0.35}{t_f}$ Upper cutoff frequency

(11–36) $f_{cl} = \dfrac{0.35}{t_r}$ Lower cutoff frequency

SUMMARY

☐ The coupling and bypass capacitors of an amplifier affect the low-frequency response.

☐ The internal transistor capacitances affect the high-frequency response.

☐ Critical frequencies are values of frequency at which the RC networks start to cause a reduction in voltage gain.

☐ Each RC network causes the gain to drop at a rate of 20 dB/decade.

☐ For the low-frequency networks, the *highest* critical frequency value is the *dominant* critical frequency.

☐ For the high-frequency networks, the *lowest* critical frequency value is the *dominant* critical frequency.

☐ A decade of frequency change is a ten-times change (increase or decrease).

☐ An octave of frequency change is a two-times change (increase or decrease).

☐ The bandwidth of an amplifier is the range of frequencies between the lower cutoff frequency and the upper cutoff frequency.

☐ The gain-bandwidth product is a transistor parameter that is a constant and equal to the unity-gain frequency.

SELF-TEST

1. In the amplifier of Figure 11–52, list the capacitors that affect the low-frequency response of the amplifier and those that affect the high-frequency response.

FIGURE 11–52

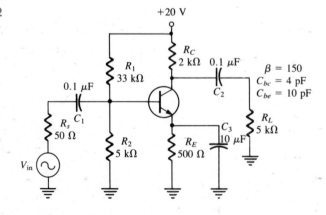

2. Determine the Miller input capacitance in Figure 11–52.

3. Determine the Miller output capacitance in Figure 11–52.

4. Express the midrange gain in dB for the amplifier in Figure 11–52.

5. Determine the low-frequency break points in Figure 11–52.

6. Determine the high-frequency break points in Figure 11–52.

7. What are the upper and lower cutoff frequencies in Figure 11–52?

8. Calculate the bandwidth of the amplifier in Figure 11–52.

9. For a certain transistor, $f_T = 75$ MHz. The transistor is to be used in an amplifier that is to have a bandwidth of 10 MHz. How much voltage gain can be realized if this transistor is used?

10. A step voltage input is applied to an amplifier, and an output fall-time of 10 ns is measured. What is the upper cutoff frequency for this amplifier?

11. If the lower cutoff frequency is negligible compared to f_{ch} obtained in problem 11, what is the amplifier's bandwidth?

Section 11–1

11–1 In a capacitively coupled amplifier, the input coupling capacitor and the output coupling capacitor form two of the networks (along with the respective resistances) that determine the low-frequency response. Assuming that the input and output impedances are the same and neglecting the bypass network, which network will first cause the gain to drop from its midrange value as the frequency is lowered?

11–2 Explain why the coupling capacitors do not have a significant effect on gain at sufficiently high signal frequencies.

11–3 List the capacitances that affect high-frequency gain in both bipolar and FET amplifiers.

Section 11–2

11–4 Determine the Miller input and output capacitances for the amplifier in Figure 11–53. Necessary data sheet information is given in the diagram.

FIGURE 11–53

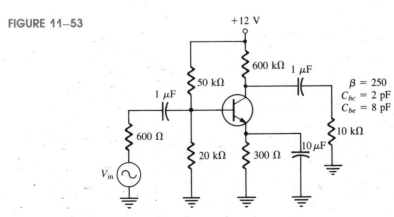

11–5 Determine the high-frequency equivalent circuit for the amplifier in Figure 11–53.

11–6 Repeat problems 11–4 and 11–5 for the amplifier in Figure 11–54.

FIGURE 11–54

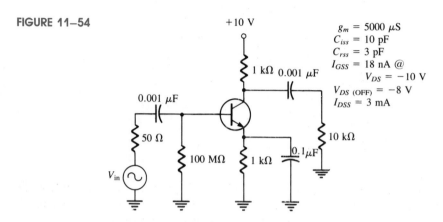

$g_m = 5000\ \mu S$
$C_{iss} = 10\ pF$
$C_{rss} = 3\ pF$
$I_{GSS} = 18\ nA$ @
$V_{DS} = -10\ V$
$V_{DS\ (OFF)} = -8\ V$
$I_{DSS} = 3\ mA$

11–7 A certain amplifier exhibits an output power of 5 W with an input power of 0.5 W. What is the power gain in dB?

11–8 If the output voltage of an amplifier is 1.2 V rms and its voltage gain is 50, what is the rms input voltage? What is the gain in dB?

11–9 The midrange voltage gain of a certain amplifier is 65. At a certain frequency beyond midrange, the gain drops to 25. What is the gain reduction in dB?

11–10 What are the dBm values corresponding to the following power values?
(a) 2 mW
(b) 1 mW
(c) 4 mW
(d) 0.25 mW

Section 11–3

11–11 Determine the critical frequencies of each RC network in Figure 11–55.

FIGURE 11–55

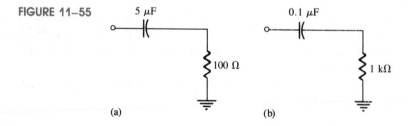

(a)

(b)

11–12 Determine the critical frequencies associated with the low-frequency response of the amplifier in Figure 11–56. Which is the dominant critical frequency? Sketch the Bode plot.

FIGURE 11–56

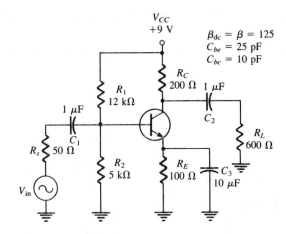

$$V_{CC}$$
$$+9 \text{ V}$$

$\beta_{dc} = \beta = 125$
$C_{be} = 25 \text{ pF}$
$C_{bc} = 10 \text{ pF}$

11–13 Determine the voltage gain of the amplifier in Figure 11–56 at one-tenth of the dominant critical frequency, at the dominant critical frequency, and at ten times the dominant critical frequency for the low-frequency reponse.

11–14 Determine the phase shift at each of the frequencies used in problem 11–13.

Section 11–4

11–15 Determine the critical frequencies associated with the high-frequency response of the amplifier in Figure 11–56. Identify the dominant critical frequency and sketch the Bode plot.

11–16 Determine the voltage gain of the amplifier in Figure 11–56 at the following frequencies: $0.1 f_c$, f_c, $10 f_c$, and $100 f_c$, where f_c is the dominant critical frequency in the high-frequency response.

Section 11–5

11–17 A particular amplifier has the following low critical frequencies: 25 Hz, 42 Hz, and 136 Hz. It also has high critical frequencies of 8 kHz and 20 kHz. Determine the upper and lower cutoff frequencies.

11–18 Determine the bandwidth of the amplifier in Figure 11–56.

11–19 $f_T = 200$ MHz is taken from the data sheet of a transistor used in a certain amplifier. If the midrange gain is determined to be 38 and if f_{cl} is low enough to be neglected compared to f_{ch}, what bandwidth would you expect? What value of f_{ch} would you expect?

11–20 If the midrange gain of a given amplifier is 50 dB and therefore 47 dB at f_{ch}, how much gain is there at $2 f_{ch}$? At $4 f_{ch}$? At $10 f_{ch}$?

Section 11–6

11–21 Determine the critical frequencies associated with the low-frequency response of the amplifier in Figure 11–57. Indicate the dominant critical frequency and draw the Bode plot.

FIGURE 11–57

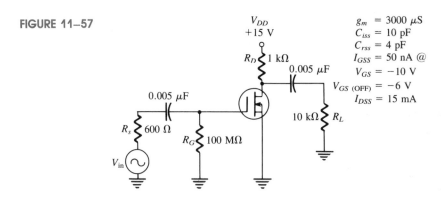

11–22 **(a)** Find the voltage gain of the amplifier in Figure 11–57 at the following frequencies: f_c, $0.1 f_c$, and $10 f_c$, where f_c is the dominant critical frequency.

(b) Find the phase shift at each frequency in part (a).

11–23 The data sheet for the FET in Figure 11–57 gives $C_{rss} = 4$ pF and $C_{iss} = 10$ pF. Determine the critical frequencies associated with the high-frequency response of the amplifier, and indicate the dominant frequency.

11–24 Determine the voltage gain in dB and the phase shift at each of the following multiples of the dominant critical frequency in Figure 11–57 for the high-frequency response: $0.1 f_c$, f_c, $10 f_c$, and $100 f_c$.

Section 11–7

11–25 In a step response test of a certain amplifier, $t_f = 20$ ns and $t_r = 1$ ms. Determine f_{cl} and f_{ch}.

11–26 Suppose you are measuring the frequency response of an amplifier with a signal source and an ac voltmeter. Assume that the signal level and frequency are set such that the voltmeter indicates an output voltage level of 5 V rms in the midrange of the amplifier's response. If you wish to determine the upper cutoff frequency, indicate what you would do and what voltmeter indication you would look for.

ANSWERS TO SECTION REVIEWS

Section 11–1

1. The coupling capacitors and the bypass capacitor.

2. Internal capacitances.

Section 11–2

1. 4.

2. 13.98 dB.

3. 1 mW.

Section 11–3

1. f_{c2}.

2. 47 dB.

3. −20 dB.

Section 11–4

1. Internal capacitance.

2. 324 pF.

3. Input network.

Section 11–5

1. 24.9 kHz.

2. 2.6.

Section 11–6

1. 26.57°.

2. 12.24.

3. C_{iss} and C_{rss}.

4. 82 pF.

Section 11–7

1. 70 MHz.

2. 90 percent point and 10 percent point.

Example 11–2: **(a)** 61.58 dB; **(b)** 16.99 dB; **(c)** 101.9 dB

Example 11–3: **(a)** 50 V; **(b)** 6.25 V; **(c)** 1.5625 V

Example 11–4: **(a)** 7.23 Hz; **(b)** 0.707; **(c)** 353.5

Example 11–5: 212.1 @ 400 Hz, 30 @ 40 Hz, 3 @ 4 Hz

Example 11–6: **(a)** 2.19 Hz; **(b)** 0.479; **(c)** 70.7

Example 11–7: 91.8 Hz

Example 11–9: 1.735 MHz, 209.4 Ω in series with 438 pF

**ANSWERS
TO
PRACTICE
EXERCISES**

Operational Amplifiers

In this chapter you will learn

☐ What an operational amplifier (op-amp) is

☐ The ideal and nonideal characteristics of an op-amp

☐ What a differential amplifier is and how it works

☐ What is meant by single-ended and differential operation of a differential amplifier

☐ The definition of *common-mode rejection ratio*

☐ What the common-mode gain is

☐ How differential amplifiers are used in an operational amplifier

☐ The meanings of op-amp data sheet parameters

☐ What negative feedback is and how it is used in amplifiers

☐ The benefits of negative feedback in op-amp performance

☐ How negative feedback affects the input impedance, output impedance, and gain of an op-amp

☐ The basic op-amp configurations of inverting, noninverting, and voltage follower

☐ How op-amps are used in a specific system application

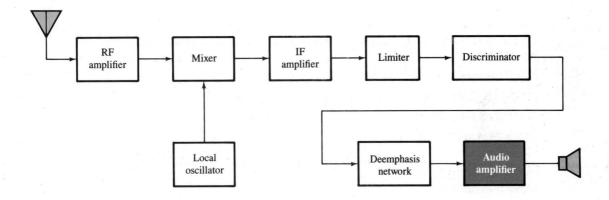

The previous chapters of this book have introduced a number of electronic devices. These devices, such as the diode and the transistor, are individually packaged and connected in a circuit with other individual devices to form a functional unit. Individually packaged devices are referred to as *discrete* components. Studying devices individually provides a fundamental understanding of their operation and how they are used in various circuit configurations.

In this chapter we will move into the area of *integrated circuits,* where many transistors, diodes, resistors, and capacitors are fabricated on a single silicon chip and packaged in a single case to form a functional circuit. In the study of integrated circuits, the entire functional circuit is treated as a device. That is, we will be concerned with what the circuit does more from an *external* viewpoint than from an internal, component-level viewpoint.

The *operational amplifier* is the most versatile and widely used of the linear integrated circuits. In many receiver systems, for example, the op-amp can be applied as an audio amplifier instead of using a discrete transistor amplifier. Typically, the op-amp is used as the first audio stage driving a discrete transistor push-pull power amplifier.

INTRODUCTION TO OPERATIONAL AMPLIFIERS

12–1

Early operational amplifiers (op-amps) were used primarily to perform *mathematical operations* such as addition, subtraction, integration, and differentiation, hence the term *operational*. These early devices were constructed with vacuum tubes and worked with high voltages. Today's op-amps are *linear integrated circuits* that use relatively low supply voltages and are reliable and inexpensive.

Symbol and Terminals

The standard op-amp symbol is shown in Figure 12–1(a). It has *two input terminals,* called the *inverting input* (−) and the *noninverting input* (+), and one output terminal. The typical op-amp operates with two dc supply voltages, one positive and the other negative, as shown in part (b). Usually these dc voltage terminals are left off the schematic symbol for simplicity but are always understood to be there.

FIGURE 12–1
Op-amp symbols and packages.

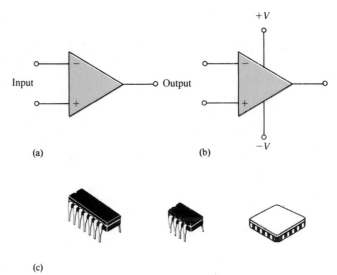

(a)   (b)

(c)

The Ideal Op-Amp

In order to get a concept of what an op-amp is, we will consider its *ideal* characteristics. A practical op-amp, of course, falls short of these ideal standards, but it is much easier to understand and analyze the device from an ideal point of view.

First, the ideal op-amp has *infinite voltage gain* and *infinite bandwidth*. Also, it has an *infinite input impedance* (open), so that it does not draw any power from the driving source. Finally, it has a *0 output impedance*. These characteristics are illustrated in Figure 12–2. The input voltage V_{in} appears between the two input terminals, and the output voltage is $A_v V_{in}$, as indicated by the internal voltage source symbol. The concept of

FIGURE 12–2
Ideal op-amp representation.

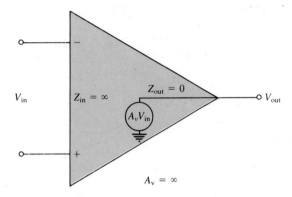

infinite input impedance is a particularly valuable analysis tool for the various op-amp configurations covered later.

The Practical Op-Amp

Although modern integrated circuit op-amps approach parameter values that can be treated as ideal in many cases, the ideal device has not been and probably will not be developed even though improvements continue to be made. Any device has limitations, and the integrated circuit op-amp is no exception. Op-amps have both voltage and current limitations. Peak-to-peak output voltage, for example, is usually limited to slightly less than the two supply voltages. Output current is also limited by internal restrictions such as power dissipation and component ratings. Characteristics of a practical op-amp are high voltage gain, high input impedance, low output impedance, and wide bandwidth. These are illustrated in Figure 12–3.

FIGURE 12–3
Practical op-amp representation.

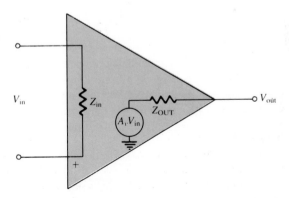

1. Sketch the symbol for an op-amp.
2. Describe some of the characteristics of a *practical* op-amp.

SECTION REVIEW 12–1

THE DIFFERENTIAL AMPLIFIER

12–2

The op-amp, in its basic form, typically consists of two or more *differential amplifier* stages. Because the differential amplifier (diff-amp) is fundamental to the op-amp's internal operation, it is useful to spend some time in acquiring a basic understanding of this type of circuit. A basic differential amplifier circuit is shown in Figure 12–4(a) and its block symbol in part (b). The diff-amp stages that make up part of the op-amp provide *high voltage gain* and *common-mode rejection* (to be defined).

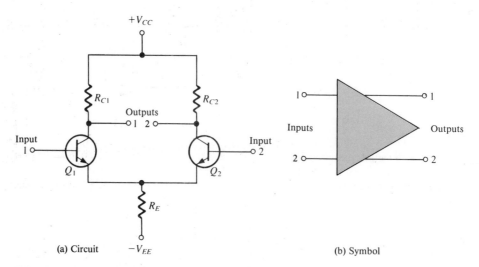

(a) Circuit $-V_{EE}$ (b) Symbol

FIGURE 12–4
Basic differential amplifier.

Basic Operation

Although an op-amp typically has more than one differential amplifier stage, we will use a single diff-amp to illustrate the basic operation. The following discussion is in relation to Figure 12–5 and consists of a basic dc analysis of the diff-amp's operation. First, when both inputs are grounded (0 V), the emitters are at -0.7 V, as indicated in Figure 12–5(a). It is assumed that the transistors are identically matched by careful process control during manufacturing so that their dc emitter currents are the same with no input signal.

$$I_{E_1} = I_{E_2}$$

Since both emitter currents combine through R_E,

$$I_{E_1} = I_{E_2} = \frac{I_{R_E}}{2} \qquad (12\text{–}1)$$

where

$$I_{R_E} = \frac{V_E - V_{EE}}{R_E} \qquad (12\text{–}2)$$

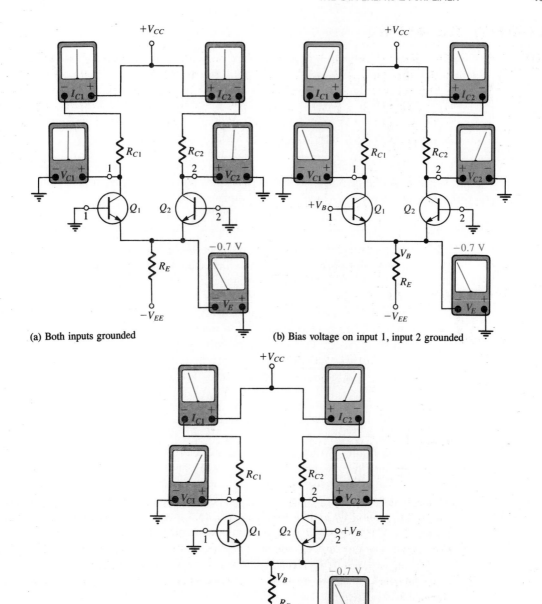

(a) Both inputs grounded

(b) Bias voltage on input 1, input 2 grounded

(c) Bias voltage on input 2, input 1 grounded

FIGURE 12–5
Basic operation of a differential amplifier (ground is zero volts).

Based on the approximation that $I_C \cong I_E$, it can be stated that

$$I_{C_1} = I_{C_2} \cong \frac{I_{R_E}}{2} \qquad (12\text{–}3)$$

Since both collector currents and both collector resistors are equal (when the input voltage is 0),

$$V_{C_1} = V_{C_2} = V_{CC} - I_{C_1}R_{C_1} \qquad (12\text{–}4)$$

This condition is illustrated in Figure 12–5(a). Next, input 2 is left grounded, and a positive bias voltage is applied to input 1, as shown in part (b). The positive voltage on the base of Q_1 increases I_{C_1} and raises the emitter voltage to

$$V_E = V_B - 0.7 \text{ V} \qquad (12\text{–}5)$$

This action *reduces* the forward bias (V_{BE}) of Q_2 because its base is held at 0 volts (ground), thus causing I_{C_2} to *decrease* as indicated in part (b) of the diagram. The net result is that the increase in I_{C_1} causes a decrease in V_{C_1}, and the decrease in I_{C_2} causes an increase in V_{C_2}, as shown. Finally, input 1 is grounded and a positive bias voltage is applied to input 2, as shown in Figure 12–5(c).

The positive bias voltage causes Q_2 to conduct more, thus increasing I_{C_2}. Also, the emitter voltage is raised. This reduces the forward bias of Q_1, since its base is held at ground, and causes I_{C_1} to decrease. The result is that the increase in I_{C_2} produces a decrease in V_{C_2}, and the decrease in I_{C_1} causes V_{C_1} to increase, as shown.

Modes of Signal Operation

Single-Ended Input.   When a diff-amp is operated in this mode, one input is grounded and the signal voltage is applied only to the other input, as shown in Figure 12–6. In the case where the signal voltage is applied to input 1 as in part (a), an inverted, amplified signal voltage appears at output 1 as shown. Also, a signal voltage appears in-phase at the emitter of Q_1. Since the emitters of Q_1 and Q_2 are common, this emitter signal becomes an input to Q_2 which functions as a common-base amplifier. The signal is amplified by Q_2 and appears, noninverted, at output 2. This action is illustrated in part (a).

In the case where the signal is applied to input 2 with input 1 grounded, as in part (b), an inverted, amplified signal voltage appears at output 2. In this situation, Q_1 acts as a common-base amplifier, and a noninverted, amplified signal appears at output 1. This action is illustrated in part (b) of the figure.

Differential Input.   In this mode, two *opposite-polarity* (out-of-phase) signals are applied to the inputs, as shown in Figure 12–7(a) (p. 454). This type of operation is also referred to as *double-ended*. Each input affects the outputs, as you will see in the following discussion. Figure 12–7(b) shows the output signals due to the signal on input 1 acting alone as a single-ended input. Figure 12–7(c) shows the output signals due to the signal on input 2 acting alone as a single-ended input. Notice, in parts (b) and (c), that the signals on output 1 are of the same polarity. The same is also true for output 2. By superimposing

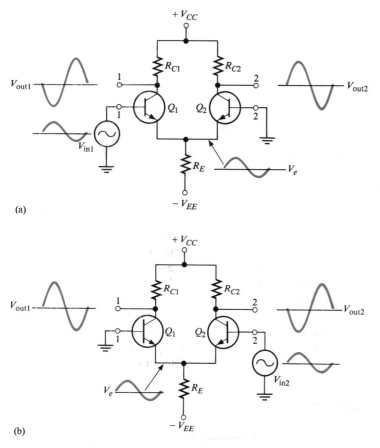

(a)

(b)

FIGURE 12–6
Single-ended input operation of a differential amplifier.

both output 1 signals and both output 2 signals, we get the total differential operation, as pictured in Figure 12–7(d).

Common-Mode Input.   One of the most important aspects of the operation of a differential amplifier can be seen by considering the case where two signal voltages of the *same phase, frequency, and amplitude* are applied to the two inputs, as shown in Figure 12–8(a) (p. 455). Again, by considering each input signal as acting alone, the basic operation can be understood. Figure 12–8(b) shows the output signals due to the signal on only input 1, and part (c) shows the output signals due to the signal on only input 2. Notice that the component signals on output 1 are of the opposite polarity, and so are the ones on output 2. When these are superimposed, they cancel, resulting in a 0 output voltage, as shown in Figure 12–8(d).

This action is called *common-mode rejection*. Its importance lies in the situation where an *unwanted* signal appears commonly on both diff-amp inputs. Common-mode rejection means that this unwanted signal will not appear on the outputs to distort the

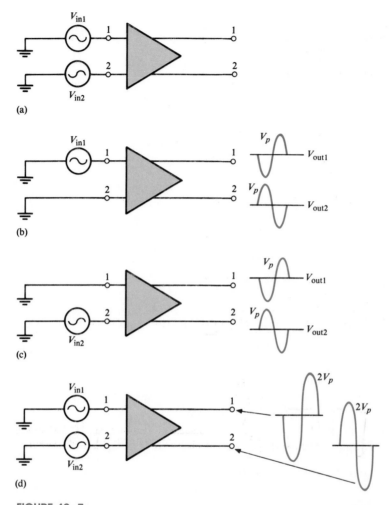

FIGURE 12–7

Differential operation of a differential amplifier: (a) Differential inputs. (b) Outputs due to V_{in_1}. (c) Outputs due to V_{in_2}. (d) Total outputs due to differential inputs.

desired signal. Common-mode signals (noise) generally are the result of the pick-up of radiated energy on the input lines, from adjacent lines, or the 60 Hz power line, or other sources.

      In summary, *desired* signals appear on only one input or with opposite polarities on both input lines. These desired signals are amplified and appear on the outputs as previously discussed. Unwanted signals (noise) appearing with the same polarity on both input lines are essentially cancelled by the diff-amp and do not appear on the outputs. The *measure* of an amplifier's ability to reject common-mode signals is a parameter called the *common-mode rejection ratio* (CMRR) and is discussed below.

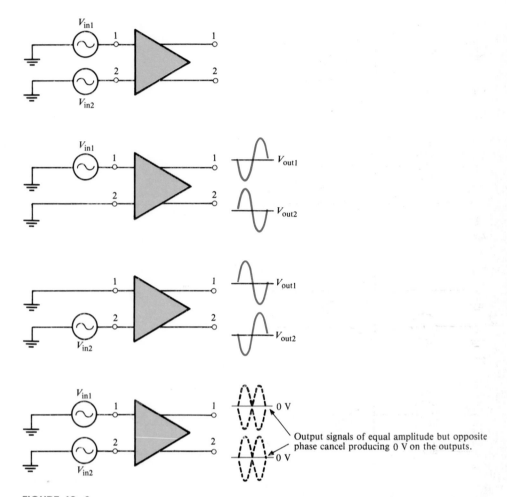

FIGURE 12–8
Common-mode operation of a differential amplifier. (a) Common-mode inputs. (b) Outputs due to V_{in_1}. (c) Outputs due to V_{in_2}. (d) Outputs cancel when common-mode signals are applied.

Common-Mode Gain

Ideally, a differential amplifier provides a very high gain for desired signals (single-ended or differential), and 0 gain for common-mode signals. Practical diff-amps, however, do exhibit a very small common-mode gain (usually much less than 1), while providing a high differential voltage gain (usually several thousand). The higher the differential gain with respect to the common-mode gain, the better the performance of the diff-amp in terms of rejection of common-mode signals. This suggests that a good measure of the diff-amp's performance in rejecting unwanted common-mode signals is the ratio of the differential gain $A_{v(d)}$ to the common-mode gain, A_{cm}. This ratio is called the *common-mode rejection ratio*, CMRR.

$$\text{CMRR} = \frac{A_{v(d)}}{A_{cm}} \qquad \text{(12–6)}$$

The higher the CMRR, the better, as you can see. A very high value of CMRR means that the differential gain $A_{v(d)}$ is high and the common-mode gain A_{cm} is low. The CMRR is often expressed in decibels (dB) as

$$\text{CMRR} = 20 \log \frac{A_{v(d)}}{A_{cm}} \qquad \text{(12–7)}$$

EXAMPLE 12–1

A certain differential amplifier has a differential voltage gain of 2000 and a common-mode gain of 0.2. Determine the CMRR and express it in dB.

Solution

$$A_{v(d)} = 2000, \text{ and } A_{cm} = 0.2$$

$$\text{CMRR} = \frac{A_{v(d)}}{A_{cm}} = \frac{2000}{0.2} = 10,000$$

In dB, CMRR = 20 log(10,000) = 80 dB.

Practice Exercise 12–1
Determine the CMRR and express it in dB for an amplifier with a differential voltage gain of 8500 and a common-mode gain of 0.25.

A CMRR of 10,000, for example, means that the desired input signal (differential) is amplified 10,000 times more than the unwanted noise (common-mode). So, as an example, if the amplitudes of the differential input signal and the common-mode noise are equal, the desired signal will appear on the output 10,000 times greater in amplitude than the noise. Thus, the noise or interference has been essentially eliminated. Example 12–2 should help reinforce the idea of *common-mode rejection* and the general signal operation of the differential amplifier.

EXAMPLE 12–2

The differential amplifier shown in Figure 12–9 has a differential voltage gain of 2500 and a CMRR of 30,000. In part (a) of the figure, a single-ended input signal of 500 μV rms is applied. At the same time a 1 V, 60 Hz interference signal appears on both inputs as a result of radiated pick-up from the ac power system. In part (b) of the figure, differential input signals of 500 μV rms each are applied to the inputs. The common-mode interference is the same as in part (a).

(a) Determine the common-mode gain.
(b) Express the CMRR in dB.
(c) Determine the rms output signal for parts (a) and (b) of the figure.
(d) Determine the rms interference voltage on the output.

FIGURE 12–9

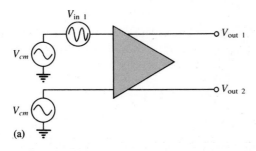

(a)

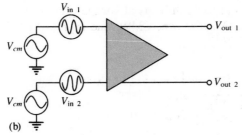

(b)

Solution

(a) $\text{CMRR} = \dfrac{A_{v(d)}}{A_{cm}}$

$A_{cm} = \dfrac{A_{v(d)}}{\text{CMRR}} = \dfrac{2500}{30,000} = 0.083$

(b) $\text{CMRR} = 20 \log(30,000) = 89.5 \text{ dB}$

(c) In Figure 12–9(a), the differential input voltage is the *difference* between the voltage on input 1 and that on input 2. Since input 2 is grounded, its voltage is zero. Therefore,

$$V_{\text{in(diff)}} = V_{\text{in}_1} - V_{\text{in}_2}$$
$$= 500 \ \mu\text{V} - 0 \text{ V}$$
$$= 500 \ \mu\text{V}$$

The output signal voltage in this case is taken at ouput 1.

$$V_{\text{out}_1} = A_{v(d)}V_{\text{in(diff)}}$$
$$= (2500)(500 \ \mu\text{V})$$
$$= 1.25 \text{ V rms}$$

In Figure 12–9(b), the differential input voltage is the difference between the two opposite-polarity, 500 μV signals.

$$V_{\text{in(diff)}} = V_{\text{in}_1} - V_{\text{in}_2}$$
$$= 500 \ \mu\text{V} - (-500 \ \mu\text{V})$$
$$= 1000 \ \mu\text{V}$$
$$= 1 \ \text{mV}$$

The output signal voltage is

$$V_{\text{out}_1} = A_{v(d)} V_{\text{in(diff)}}$$
$$= (2500)(1 \ \text{mV})$$
$$= 2.5 \ \text{V rms}$$

This shows that a differential input (two opposite-polarity signals) results in a gain that is double that for a single-ended input.

(d) The common-mode input is 1 V rms. The common-mode gain A_{cm} is 0.083. The interference voltage on the output is therefore

$$A_{cm} = \frac{V_{\text{out(cm)}}}{V_{\text{in(cm)}}}$$

$$V_{\text{out(cm)}} = A_{cm} V_{\text{in(cm)}}$$
$$= (0.083)(1 \ \text{V})$$
$$= 0.083 \ \text{V}$$

Practice Exercise 12–2
The amplifier in Figure 12–9 has a differential voltage gain of 4200 and a CMRR of 25,000. For the same single-ended and differential input signals as described in the example: **(a)** Find A_{cm}. **(b)** Express the CMRR in dB. **(c)** Determine the rms output signal for parts (a) and (b) of the figure. **(d)** Determine the rms interference voltage appearing on the output.

Simple Op-Amp Arrangement

Figure 12–10 shows two differential amplifier (diff-amp) stages and an emitter-follower connected to form a simple op-amp. The first stage can be used with a single-ended or a differential input. The differential outputs of the first stage are directly coupled into the differential inputs of the second stage. The output of the second stage is single-ended to drive an emitter-follower to achieve a relatively low output impedance. Both differential stages together provide a high voltage gain and a high CMRR.

SECTION REVIEW 12–2

1. Distinguish between differential and single-ended inputs.
2. Define *common-mode rejection.*

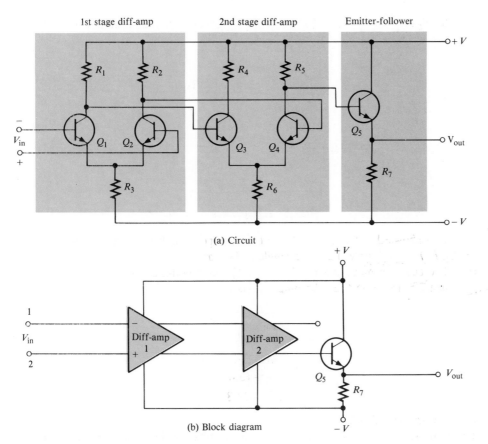

(a) Circuit

(b) Block diagram

FIGURE 12–10
Internal circuitry of a simple op-amp.

OP-AMP DATA SHEET PARAMETERS

12–3

In this section, several important operational amplifier parameters are defined, and four popular integrated circuit op-amps are compared in terms of their parameter values.

Input Offset Voltage

The ideal op-amp produces 0 volts out for 0 volts in. In a practical op-amp, however, a small dc voltage appears at the output when no differential input voltage is applied. Its primary cause is a slight mismatch of the base-to-emitter voltages of the differential input stage, as illustrated in Figure 12–11(a). The output voltage of the differential input stage is expressed as

$$V_{OUT} = I_{C_2}R_C - I_{C_1}R_C \qquad \textbf{(12–8)}$$

A small difference in the base-to-emitter voltages of Q_1 and Q_2 causes a small difference in the collector currents. This results in a nonzero value of V_{OUT}. (The collector resistors are equal.) As specified on an op-amp data sheet, the *input offset voltage V_{OS} is*

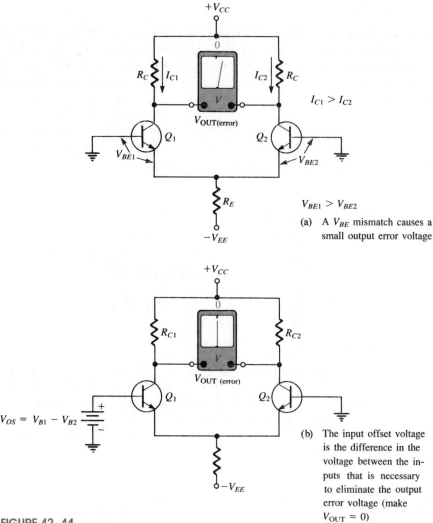

(a) A V_{BE} mismatch causes a small output error voltage

(b) The input offset voltage is the difference in the voltage between the inputs that is necessary to eliminate the output error voltage (make $V_{OUT} = 0$)

FIGURE 12–11
Input offset voltage, V_{OS}.

the differential dc voltage required between the inputs to force the differential output to 0 volts, as demonstrated in Figure 12–11(b). Typical values of input offset voltage are in the range of 2 mV or less. In the ideal case, it is 0.

Input Offset Voltage Drift

The input offset voltage drift is a parameter related to V_{OS} that specifies how much change occurs in the input offset voltage for each degree change in temperature. Typical values range anywhere from about 5 μV per degree Celsius to about 50 μV per degree Celsius. Usually, an op-amp with a higher nominal value of input offset voltage exhibits a higher drift.

Input Bias Current

You have seen that the input terminals of a bipolar differential amplifier are the transistor bases and, therefore, the input currents are the base currents. *The input bias current is the dc current required by the inputs of the amplifier to properly operate the first stage.* By definition, the input bias current is the *average* of both input currents and is calculated as follows:

$$I_{BIAS} = \frac{I_1 + I_2}{2} \qquad \text{(12–9)}$$

The concept of input bias current is illustrated in Figure 12–12.

FIGURE 12–12
Input bias current is the average of the two op-amp input currents.

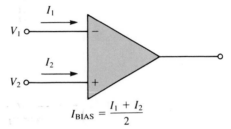

Input Impedance

There are two basic ways of specifying the input impedance of an op-amp. The *differential input impedance* is the total resistance between the inverting and the noninverting inputs. This is illustrated in Figure 12–13(a). Differential impedance is measured by determining the change in bias current for a given change in differential input voltage. The *common-mode input impedance* is measured from the common inputs to ground and is depicted in Figure 12–13(b).

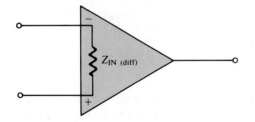

(a) Differential input impedance

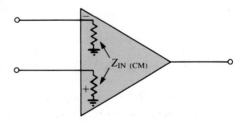

(b) Common-mode input impedance

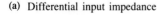

FIGURE 12–13
Op-amp input impedance.

Input Offset Current

Ideally, the two input bias currents are equal, and thus their difference is 0. In a practical op-amp, however, the bias currents are not exactly equal. *The input offset current is the difference of the input bias currents,* expressed as

$$I_{OS} = |I_1 - I_2|$$ (12–10)

Actual magnitudes of offset current are usually at least an order of magnitude (ten times) less than the bias current. In many applications the offset current can be neglected. However, high-gain, high-input impedance amplifiers should have as little I_{OS} as possible, because the difference in currents through large input resistances develops a substantial offset voltage, as shown in Figure 12–14.

The offset voltage developed by the input offset current is

$$V_{OS} = I_1 R_{in} - I_2 R_{in}$$
$$= (I_1 - I_2) R_{in}$$
$$V_{OS} = I_{OS} R_{in}$$ (12–11)

The error created by I_{OS} is amplified by the gain A_v of the op-amp and appears in the output as

$$V_{OUT(error)} = A_v I_{OS} R_{in}$$ (12–12)

The *change* in offset current with temperature is often an important consideration. Values of temperature coefficient in the range of 0.5 nA per degree Celsius are common.

FIGURE 12–14
Effect of input offset current.

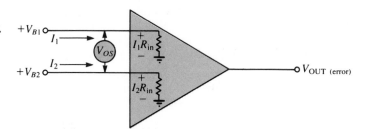

Output Impedance

This parameter is the resistance viewed from the output terminal of the op-amp, as indicated in Figure 12–15.

FIGURE 12–15
Op-amp output impedance.

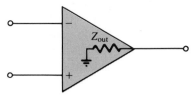

Common-Mode Range

All op-amps have limitations on the range of voltages over which they will operate. The *common-mode range* is the range of input voltages which, when applied to both inputs, will not cause clipping or other output distortion. Many op-amps have common-mode ranges of ± 10 V with dc supply voltages of ± 15 V.

Open-Loop Voltage Gain

This is the gain of the op-amp without any external feedback from output to input. A good op-amp has a very high open-loop gain; 50,000 to 200,000 is typical.

Common-Mode Rejection Ratio

The *common-mode rejection ratio* (CMRR), as discussed in conjunction with the diff-amp, is a measure of an op-amp's ability to reject common-mode signals. An infinite value of CMRR means that the output is 0 when the same signal is applied to both inputs (common-mode).

An infinite CMRR is never achieved in practice, but a good op-amp does have a very high value of CMRR. As previously mentioned, common-mode signals are unde-sired interference voltages such as 60 Hz power-supply ripple and noise voltages due to pick-up of radiated energy. A high CMRR enables the op-amp to virtually eliminate these interference signals from the output.

The accepted definition of CMRR for an op-amp is the *open-loop gain* (A_{ol}) divided by the common-mode gain.

$$\text{CMRR} = \frac{A_{ol}}{A_{cm}} \qquad\qquad (12\text{--}13)$$

It is commonly expressed in decibels as follows:

$$\text{CMRR} = 20 \log \frac{A_{ol}}{A_{cm}} \qquad\qquad (12\text{--}14)$$

EXAMPLE 12–3

A certain op-amp has an open-loop gain of 100,000 and a common-mode gain of 0.25. Determine the CMRR and express it in dB.

Solution

$$\text{CMRR} = \frac{A_{ol}}{A_{cm}} = \frac{100,000}{0.25} = 400,000$$

$$\text{CMRR} = 20 \log(400,000) = 112 \text{ dB}$$

Slew Rate

The maximum rate of change of the output voltage in response to a step input voltage is the *slew rate* of an op-amp. The slew rate is dependent upon the frequency response of the amplifier stages within the op-amp. Slew rate is measured with an op-amp connected as shown in Figure 12–16(a). This particular op-amp connection is a unity-gain, noninvert-ing configuration which will be discussed later. It gives a *worst-case* (slowest) slew rate.

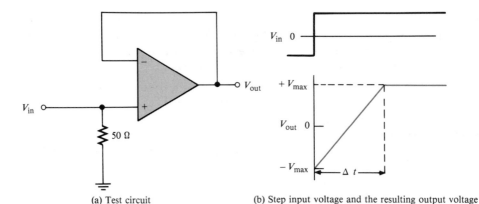

(a) Test circuit

(b) Step input voltage and the resulting output voltage

FIGURE 12–16
Slew rate measurement.

A pulse is applied to the input as shown, and the ideal output voltage is measured as indicated in Figure 12–16(b). The width of the input pulse must be sufficient to allow the output to "slew" from its lower limit to its upper limit, as shown. As you can see, a certain time interval, Δt, is required for the output voltage to go from its lower limit $-V_{max}$ to its upper limit $+V_{max}$, once the input step is applied. The slew rate is expressed as

$$\text{Slew rate} = \frac{\Delta V_{out}}{\Delta t} \qquad (12\text{–}15)$$

where $\Delta V_{out} = +V_{max} - (-V_{max})$. The unit of slew rate is volts per microsecond (V/μs).

EXAMPLE 12–4

The output voltage of a certain op-amp appears as shown in Figure 12–17 in response to a step input. Determine the slew rate.

FIGURE 12–17

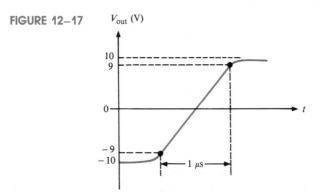

Solution

The output goes from the lower to the upper limit in 1 μs. Since this is not an ideal response, the limits are taken at the 90 percent points, as indicated. So, the upper limit is +9 V and the lower limit is −9 V. The slew rate is

$$\frac{\Delta V}{\Delta t} = \frac{+9 \text{ V} - (-9 \text{ V})}{1 \text{ } \mu s} = 18 \text{ V}/\mu s$$

Practice Exercise 12–4

When a pulse is applied to an op-amp, the output voltage goes from −8 V to +7 V in 0.75 μs. What is the slew rate?

Frequency Response

The internal amplifier stages that make up an op-amp have voltage gains limited by junction capacitances, as discussed in Chapter 11. Although the differential amplifiers used in op-amps are somewhat different from the basic amplifiers discussed, the same principles apply. An op-amp has no internal coupling capacitors, however; therefore the low-frequency response extends down to dc. Frequency-related characteristics will be discussed in the next chapter.

Comparison of Op-Amp Parameters

Table 12–1 provides a comparison of values of some of the parameters just described for four integrated circuit op-amps. Any values not listed were not given on the manufacturer's data sheet. All values are typical at 25°C.

TABLE 12–1

Op-amp type	μA741C	LM101A	LM108	LM218
Input offset voltage	1 mV	1 mV	0.7 mV	2 mV
Input bias current	80 nA	120 nA	0.8 nA	120 nA
Input offset current	20 nA	40 nA	0.05 nA	6 nA
Input impedance	2 MΩ	800 kΩ	70 MΩ	3 MΩ
Output impedance	75 Ω	—	—	—
Open-loop gain	200,000	160,000	300,000	200,000
Slew rate	0.5 V/μs	—	—	70 V/μs
CMRR	90 dB	90 dB	100 dB	100 dB

Other Features

Most available op-amps have two important features: short-circuit protection and no latch-up. Short-circuit protection keeps the circuit from being damaged if the output becomes shorted, and the no latch-up feature prevents the op-amp from hanging up in one output state (high- or low-voltage level) under certain input conditions.

---

1. List ten or more op-amp parameters.
2. List two parameters, not including frequency response, that are frequency dependent.

NEGATIVE FEEDBACK

12–4

Negative feedback is one of the most useful concepts in electronic circuits, particularly in op-amp applications. Negative feedback is the *process whereby a portion of the output voltage of an amplifier is returned to the input with a phase angle that opposes (or subtracts from) the input signal.* This is illustrated in Figure 12–18. The inverting input effectively makes the feedback signal 180° out-of-phase with the input signal.

FIGURE 12–18
Illustration of negative feedback.

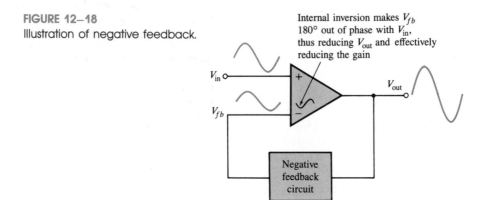

Internal inversion makes V_{fb} 180° out of phase with V_{in}, thus reducing V_{out} and effectively reducing the gain

V_{in}

V_{fb}

V_{out}

Negative feedback circuit

Why Use Negative Feedback?

As you have seen, the inherent open-loop gain of a typical op-amp is very high (usually greater than 100,000). Therefore, an extremely small input voltage drives the op-amp into its saturated output states. In fact, even the input offset voltage of the op-amp can drive it into saturation. For example, assume $V_{in} = 1$ mV and $A_{ol} = 100,000$. Then, $V_{in}A_{ol} = (1$ mV$)(100,000) = 100$ V. Since the output level of an op-amp can never reach 100 V, it is driven deep into saturation and the output is limited to its maximum output levels, as illustrated in Figure 12–19 for both a positive and a negative input voltage of 1 mV.

The usefulness of an op-amp operated in this manner is severely restricted and is generally limited to comparator applications (to be studied later). With negative feedback, the overall voltage gain (A_{cl}) can be reduced and controlled so that the op-amp can function as a linear amplifier. In addition to providing a controlled, stable voltage gain, negative feedback also provides for control of the input and output impedances and amplifier bandwidth. (We will cover these topics in detail later.) Table 12–2 summarizes the general effects of negative feedback on op-amp performance.

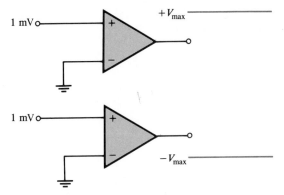

FIGURE 12–19
Without negative feedback, a small input voltage drives the op-amp into saturation.

TABLE 12–2

	Voltage gain	*Input Z*	*Output Z*	*Bandwidth*
Without negative feedback	A_{ol} is too high for linear amplifier applications	Relatively high (see Table 12–1)	Relatively low (see Table 12–1)	Relatively narrow
With negative feedback	A_{cl} is set by the feedback circuit to desired value	Can be increased or reduced to a desired value depending on type of circuit	Can be reduced to a desired value	Significantly wider

---

1. List the benefits of negative feedback in an op-amp circuit.
2. Explain why it is necessary to reduce the gain of an op-amp from its open-loop value.

SECTION REVIEW 12–4

OP-AMPS WITH NEGATIVE FEEDBACK

12–5

In this section we will discuss several basic ways in which an op-amp can be connected using *negative feedback* to stabilize the gain and increase frequency response. As mentioned, the extremely high open-loop gain of an op-amp creates an unstable situation because a small noise voltage on the input can be amplified to a point where the amplifier is driven out of its linear region. Also, unwanted oscillations can occur. In addition, the open-loop gain parameter of an op-amp can vary greatly from one device to the next.

Negative feedback takes a portion of the output and applies it back *out-of-phase* with the input, creating an effective reduction in gain. This closed-loop gain is usually much less than the open-loop gain and independent of it.

Noninverting Amplifier

An op-amp connected as a *noninverting* amplifier with a controlled amount of voltage gain is shown in Figure 12–20. The input signal is applied to the noninverting input. The

FIGURE 12–20
Noninverting amplifier.

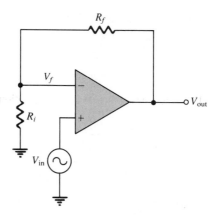

output is applied back to the inverting input through the feedback network formed by R_i and R_f. This creates *negative feedback* as follows. R_i and R_f form a voltage-divider network which reduces the output V_{out} and connects the reduced voltage V_f to the inverting input. The feedback voltage is expressed as

$$V_f = \left(\frac{R_i}{R_i + R_f}\right)V_{out} \tag{12–16}$$

The difference of the input voltage V_{in} and the feedback voltage V_f is the *differential input* to the op-amp, as shown in Figure 12–21. This differential voltage is amplified by the open-loop gain of the op-amp (A_{ol}) and produces an output voltage expressed as

$$V_{out} = A_{ol}(V_{in} - V_f) \tag{12–17}$$

Letting $R_i/(R_i + R_f) = B$ and then substituting BV_{out} for V_f in equation (12–17), we get the following algebraic steps:

$$V_{out} = A_{ol}(V_{in} - BV_{out})$$
$$V_{out} = A_{ol}V_{in} - A_{ol}BV_{out}$$
$$V_{out} + A_{ol}BV_{out} = A_{ol}V_{in}$$
$$V_{out}(1 + A_{ol}B) = A_{ol}V_{in}$$

FIGURE 12–21
Differential input.

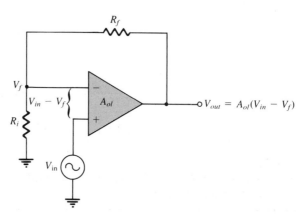

Since the total voltage gain of the amplifier in Figure 12–20 is V_{out}/V_{in}, it can be expressed as

$$\frac{V_{out}}{V_{in}} = \frac{A_{ol}}{1 + A_{ol}B} \qquad (12\text{--}18)$$

The product $A_{ol}B$ is usually much greater than 1, so equation (12–18) simplifies to

$$\frac{V_{out}}{V_{in}} = \frac{A_{ol}}{A_{ol}B}$$

Since

$$A_{cl(NI)} = \frac{V_{out}}{V_{in}}$$

$$A_{cl(NI)} = \frac{1}{B} = \frac{R_i + R_f}{R_i} = 1 + \frac{R_f}{R_i} \qquad (12\text{--}19)$$

Equation (12–19) shows that the *closed-loop gain* $A_{cl(NI)}$ of the noninverting (NI) amplifier is the reciprocal of the attenuation (B) of the feedback network (voltage-divider). It is interesting to note that the closed-loop gain is not at all dependent on the op-amp's open-loop gain under the condition $A_{ol}B \gg 1$. The closed-loop gain can be set by selecting values of R_i and R_f.

EXAMPLE 12–5

Determine the gain of the amplifier in Figure 12–22. The open-loop voltage gain is 100,000.

FIGURE 12–22

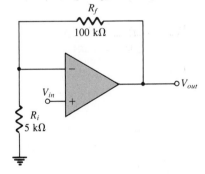

Solution
This is a noninverting op-amp configuration. Therefore, the closed-loop gain is

$$A_{cl(NI)} = 1 + \frac{R_f}{R_i} = 1 + \frac{100 \text{ k}\Omega}{5 \text{ k}\Omega} = 21$$

Practice Exercise 12–5
If the open-loop gain of the amplifier in Figure 12–22 is 150,000 and R_f is increased to 150 kΩ, determine the closed-loop gain.

Voltage Follower

The voltage-follower configuration is a special case of the noninverting amplifier where all of the output voltage is fed back to the inverting input, as shown in Figure 12–23. As you can see, the straight feedback connection has a voltage gain of approximately 1. The closed-loop voltage gain of a noninverting amplifier is $1/B$ as previously derived. Since $B = 1$, the closed-loop gain of the voltage follower is

$$A_{cl(VF)} = \frac{1}{B} \cong 1 \qquad\qquad (12\text{–}20)$$

The most important features of the voltage-follower configuration are its very high input impedance and its very low output impedance. These features make it a nearly ideal buffer amplifier for interfacing high-impedance sources and low-impedance loads. This is discussed further in the next section.

FIGURE 12–23
Op-amp voltage follower.

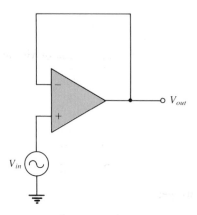

Inverting Amplifier

An op-amp connected as an *inverting* amplifier with a controlled amount of voltage gain is shown in Figure 12–24. The input signal is applied through a series input resistor R_i to the inverting input. Also, the output is fed back through R_f to the same input. The noninverting input is grounded.

FIGURE 12–24
Inverting amplifier.

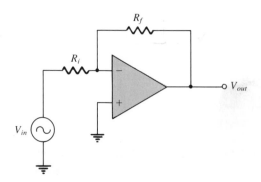

At this point, the ideal op-amp parameters mentioned earlier are very useful in simplifying the analysis of this circuit. In particular, the concept of infinite input imped-ance is of great value. An infinite input impedance implies 0 *current* to the inverting input. If there is 0 current through the input impedance, then there must be *no* voltage drop between the inverting and noninverting inputs. This means that the voltage at the inverting ($-$) input is 0 because the other input ($+$) is grounded. This 0 voltage at the inverting input terminal is referred to as *virtual ground*. This condition is illustrated in Figure 12–25(a).

Since there is no current into the inverting input, the current through R_i and the current through R_f are equal, as shown in Figure 12–25(b).

$$I_{in} = I_f$$

The voltage across R_i equals V_{in} because of virtual ground on the other side of the resistor. Therefore,

$$I_{in} = \frac{V_{in}}{R_i}$$

Also, the voltage across R_f equals $-V_{out}$ because of virtual ground, and therefore,

$$I_f = \frac{-V_{out}}{R_f}$$

Since $I_f = I_{in}$,

$$\frac{-V_{out}}{R_f} = \frac{V_{in}}{R_i}$$

Rearranging the terms, we get

$$\frac{V_{out}}{V_{in}} = -\frac{R_f}{R_i}$$

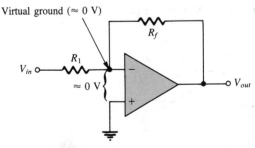

Virtual ground (≈ 0 V)

(a) Virtual ground

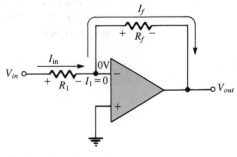

(b) $I_{in} = I_f$ and $I_1 = 0$

FIGURE 12–25
Virtual ground concept and closed-loop voltage gain development for the inverting amplifier.

Of course you recognize V_{out}/V_{in} as the overall gain of the amplifier.

$$A_{cl(I)} = -\frac{R_f}{R_i} \qquad (12\text{--}21)$$

Equation (12–21) shows that the closed-loop voltage gain $A_{cl(I)}$ of the inverting amplifier is the ratio of the feedback resistance R_f to the resistance R_i. The *closed-loop gain* is independent of the op-amp's internal *open-loop gain*. Thus the negative feedback stabilizes the voltage gain. The negative sign indicates inversion.

EXAMPLE 12–6

Given the op-amp configuration in Figure 12–26, determine the value of R_f required to produce a closed-loop voltage gain of 100.

FIGURE 12–26

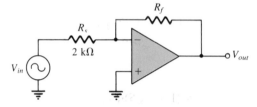

Solution

$$R_i = 2 \text{ k}\Omega, \text{ and } A_{cl(I)} = 100$$

$$A_{cl(I)} = \left| \frac{R_f}{R_i} \right|$$

$$R_f = A_{cl(I)}R_i$$
$$= (100)(2 \text{ k}\Omega)$$
$$= 200 \text{ k}\Omega$$

Practice Exercise 12–6
If R_i is changed to 2.7 kΩ in Figure 12–26, what value of R_f is required to produce a closed-loop gain of 25?

SECTION REVIEW 12–5

1. What is the main purpose of negative feedback?
2. The closed-loop voltage gain of each of the op-amp configurations discussed is dependent on the internal open-loop voltage gain of the op-amp (T or F).
3. The attenuation of the negative feedback network of a noninverting op-amp configuration is 0.02. What is the closed-loop gain of the amplifier?

EFFECTS OF NEGATIVE FEEDBACK ON OP-AMP IMPEDANCES

Input Impedance of the Noninverting Amplifier

12–6

The input impedance of this op-amp configuration is developed with the aid of Figure 12–27. For this analysis, a small differential voltage V_{diff} is assumed to exist between the two inputs, as indicated. This means that the op-amp's input impedance is not assumed to be infinite, nor the input current to be 0. The input voltage can be expressed as

$$V_{in} = V_{diff} + V_f$$

Substituting BV_{out} for V_f,

$$V_{in} = V_{diff} + BV_{out}$$

Since $V_{out} \cong A_{ol}V_{diff}$ (A_{ol} is the open-loop gain of the op-amp),

$$V_{in} = V_{diff} + A_{ol}BV_{diff}$$
$$= (1 + A_{ol}B)V_{diff}$$

Because $V_{diff} = I_{in}Z_{in}$,

$$V_{in} = (1 + A_{ol}B)I_{in}Z_{in}$$

where Z_{in} is the open-loop input impedance of the op-amp (without feedback connections).

$$\frac{V_{in}}{I_{in}} = (1 + A_{ol}B)Z_{in}$$

V_{in}/I_{in} is the overall input impedance of the closed-loop noninverting configuration.

$$Z_{in(NI)} = (1 + A_{ol}B)Z_{in} \qquad \qquad \textbf{(12–22)}$$

This equation shows that the input impedance of this amplifier configuration with negative feedback is much greater than the internal input impedance of the op-amp itself (without feedback).

FIGURE 12–27

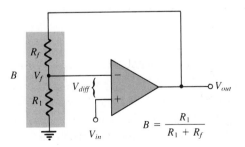

$$B = \frac{R_1}{R_1 + R_f}$$

Output Impedance of the Noninverting Amplifier

An expression for output impedance is developed with the aid of Figure 12–28. By applying Kirchhoff's law to the output circuit, we get

$$V_{\text{out}} = A_{ol}V_{\text{diff}} - Z_{\text{out}}I_{\text{out}}$$

The differential input voltage is $V_{\text{in}} - V_f$; therefore, under the assumption that $A_{ol}V_{\text{diff}} \gg Z_{\text{out}}I_{\text{out}}$, the output voltage can be expressed as

$$V_{\text{out}} \cong A_{ol}(V_{\text{in}} - V_f)$$

Substituting BV_{out} for V_f, we get

$$V_{\text{out}} \cong A_{ol}(V_{\text{in}} - BV_{\text{out}})$$

Remember, B is the attenuation of the negative feedback network. Expanding and factoring, we get

$$V_{\text{out}} \cong A_{ol}V_{\text{in}} - A_{ol}BV_{\text{out}}$$
$$A_{ol}V_{\text{in}} \cong V_{\text{out}} + A_{ol}BV_{\text{out}}$$
$$\cong (1 + A_{ol}B)V_{\text{out}}$$

Since the output impedance of the noninverting configuration is $Z_{\text{out(NI)}} = V_{\text{out}}/I_{\text{out}}$, we can substitute $I_{\text{out}}Z_{\text{out(NI)}}$ for V_{out}.

$$A_{ol}V_{\text{in}} = (1 + A_{ol}B)I_{\text{out}}Z_{\text{out(NI)}}$$

Dividing both sides of the above expression by I_{out}, we get

$$\frac{A_{ol}V_{\text{in}}}{I_{\text{out}}} = (1 + A_{ol}B)Z_{\text{out(NI)}}$$

The term on the left is the *internal output* impedance of the op-amp (Z_{out}) because, without feedback, $A_{ol}V_{\text{in}} = V_{\text{out}}$. Therefore,

$$Z_{\text{out}} = (1 + A_{ol}B)Z_{\text{out(NI)}}$$

FIGURE 12–28

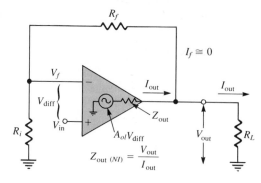

Thus,

$$Z_{out(NI)} = \frac{Z_{out}}{1 + A_{ol}B} \qquad (12\text{--}23)$$

This equation shows that the output impedance of this amplifier configuration with negative feedback is much less than the internal output impedance of the op-amp itself (without feedback).

EXAMPLE 12–7

(a) Determine the input and output impedances of the amplifier in Figure 12–29. The op-amp data sheet gives $Z_{in} = 2\ M\Omega$, $Z_{out} = 75\ \Omega$, and $A_{ol} = 200{,}000$.
(b) Find the closed-loop voltage gain.

FIGURE 12–29

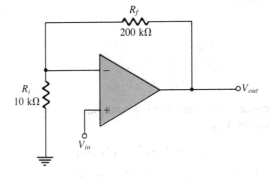

Solution

(a) The attenuation of the feedback network is

$$B = \frac{R_i}{R_i + R_f} = \frac{10\ k\Omega}{210\ k\Omega} = 0.048$$

$$
\begin{aligned}
Z_{in(NI)} &= (1 + A_{ol}B)Z_{in}\\
&= [1 + (200{,}000)(0.048)](2\ M\Omega)\\
&= (1 + 9600)(2\ M\Omega)\\
&= 19{,}202\ M\Omega
\end{aligned}
$$

$$
\begin{aligned}
Z_{out(NI)} &= \frac{Z_{out}}{1 + A_{ol}B}\\
&= \frac{75\ \Omega}{1 + 9600}\\
&= 0.0078\ \Omega
\end{aligned}
$$

(b) $\quad A_{cl(NI)} = \dfrac{1}{B} = \dfrac{1}{0.048} = 20.83$

Practice Exercise 12–7

(a) Determine the input and output impedances in Figure 12–29 for op-amp data sheet values of $Z_{in} = 3.5$ MΩ, $Z_{out} = 82$ Ω, and $A_{ol} = 135{,}000$.

(b) Find A_{cl}.

Voltage-Follower Impedances

Since the voltage follower is a special case of the noninverting configuration, the same impedance formulas are used with $B = 1$.

$$Z_{in(VF)} = (1 + A_{ol})Z_{in} \tag{12–24}$$

$$Z_{out(VF)} = \frac{Z_{out}}{1 + A_{ol}} \tag{12–25}$$

As you can see, the voltage-follower input impedance is greater for a given A_{ol} and Z_{in} than for the noninverting configuration with the voltage-divider feedback network. Also, its output impedance is much smaller.

EXAMPLE 12–8

The same op-amp in Example 12–7 is used in a voltage-follower configuration. Determine the input and output impedances.

Solution

Since $B = 1$,

$$Z_{in(VF)} = (1 + A_{ol})Z_{in}$$
$$= (1 + 200{,}000)(2 \text{ MΩ})$$
$$\cong 400{,}000 \text{ MΩ}$$

$$Z_{out(VF)} = \frac{Z_{out}}{1 + A_{ol}}$$
$$= \frac{75 \text{ Ω}}{1 + 200{,}000}$$
$$= 0.00038 \text{ Ω}$$

Notice that $Z_{in(VF)}$ is much greater than $Z_{in(NI)}$, and $Z_{out(VF)}$ is much less than $Z_{out(NI)}$ from Example 12–7.

Input Impedance of an Inverting Amplifier

The input impedance of this op-amp configuration is developed with the aid of Figure 12–30. Because both the input signal and the negative feedback are applied, through resistors, to the inverting terminal, Miller's theorem can be applied to this configuration.

FIGURE 12–30
Inverting amplifier.

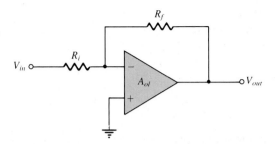

According to Miller's theorem, the effective input impedance of an amplifier with a feedback resistor from output to input as in Figure 12–30 is

$$Z_{in(Miller)} = \frac{R_f}{A_{ol} + 1} \qquad (12\text{–}26)$$

and

$$Z_{out(Miller)} = \left(\frac{A_{ol}}{A_{ol} + 1}\right)R_f \qquad (12\text{–}27)$$

Applying Miller's theorem to the circuit of Figure 12–30, we get the equivalent circuit of Figure 12–31(a). As shown in Figure 12–31(b), the Miller input impedance appears in parallel with the internal input impedance of the op-amp, and R_i appears in series with this as follows:

$$Z_{in(I)} = R_i + \frac{R_f}{A_{ol} + 1}\|Z_{in} \qquad (12\text{–}28)$$

Typically, $R_f/(A_{ol} + 1)$ is much less than the Z_{in} of an open-loop op-amp; also, $A_{ol} \gg 1$. So equation (12–28) simplifies to

$$Z_{in(I)} \cong R_i + \frac{R_f}{A_{ol}}$$

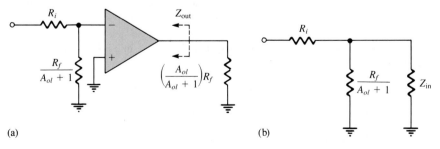

(a)   (b)

FIGURE 12–31
(a) Miller equivalent for inverting amplifier. (b) Input impedance equivalent circuit.

Since R_i appears in series with R_f/A_{ol} and if $R_i \gg R_f/A_{ol}$, $Z_{in(I)}$ reduces to

$$Z_{in(I)} \cong R_i \qquad (12\text{–}29)$$

The Miller output impedance appears in parallel with Z_{out} of the op-amp.

$$Z_{out(I)} = \left(\frac{A_{ol}}{A_{ol} + 1}\right) R_f \| Z_{out} \qquad (12\text{–}30)$$

Normally $A_{ol} \gg 1$ and $R_f \gg Z_{out}$, so $Z_{out(I)}$ simplifies to

$$Z_{out(I)} \cong Z_{out} \qquad (12\text{–}31)$$

EXAMPLE 12–9

Find the values of the input and output impedances in Figure 12–32. Also, determine the closed-loop voltage gain. The op-amp has the following parameters: $A_{ol} = 50{,}000$, $Z_{in} = 4\ M\Omega$, and $Z_{out} = 50\ \Omega$.

FIGURE 12–32

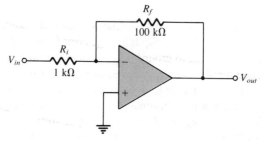

Solution

$$Z_{in(I)} \cong R_i = 1\ k\Omega$$
$$Z_{out(I)} \cong Z_{out} = 50\ \Omega$$
$$A_{cl(I)} = -\frac{R_f}{R_i} = -\frac{100\ k\Omega}{1\ k\Omega} = -100$$

Practice Exercise 12–9
Determine the input and output impedances and the closed-loop voltage gain in Figure 12–32. The op-amp parameters and circuit values are as follows: $A_{ol} = 100{,}000$, $Z_{in} = 5\ M\Omega$, $Z_{out} = 75\ \Omega$, $R_i = 560\ \Omega$, and $R_f = 82\ k\Omega$.

SECTION REVIEW 12–6

1. How does the input impedance of a noninverting amplifier configuration compare to the input impedance of the op-amp itself?

2. Connecting an op-amp in a voltage-follower configuration (increases, decreases) the input impedance.

3. Given that $R_f = 100\ k\Omega$, $R_i = 2\ k\Omega$, $A_{ol} = 120{,}000$, $Z_{in} = 2\ M\Omega$, and $Z_{out} = 60\ \Omega$, determine $Z_{in(I)}$ and $Z_{out(I)}$ for an inverting amplifier configuration.

BIAS CURRENT AND OFFSET VOLTAGE COMPENSATION

Up until now, the op-amp has been treated as an ideal device in many of our discussions. However, since it is not an ideal device, certain "flaws" in the op-amp must be recognized because of their effects on its operation. Transistors within the op-amp must be biased so that they have the correct values of base and collector current and collector-to-emitter voltages. The ideal op-amp has no input current at its terminals, but in fact, the practical op-amp has small input bias currents. Also, small internal imbalances in the transistors effectively produce a small offset voltage between the inputs. These nonideal parameters were described in section 12–3.

Effect of an Input Bias Current

Figure 12–33 is an *inverting amplifier with 0 input voltage*. The current through R_i is 0 because the input voltage is 0 and the voltage at the inverting (−) terminal is 0. The small input current I_1 is furnished from the output terminal through R_f. I_1 creates a voltage drop across R_f, as indicated. The positive side of R_f is the output terminal, and therefore, the output error voltage is I_1R_f when it should be 0.

Figure 12–34 is a *voltage follower* with 0 input voltage and a source resistance R_s. In this case, an input current I_2 creates an output voltage error (a path exists for I_2 through the negative voltage supply and back to ground). I_2 produces a drop across R_s, as shown. The voltage at the inverting input terminal decreases to $-I_2R_s$ because the negative feedback tends to maintain a differential voltage of 0, as indicated. Since the inverting terminal is connected directly to the output terminal, the output error voltage is $-I_2R_s$.

Figure 12–35 is a *noninverting amplifier* with 0 input voltage. The voltage at the inverting terminal is also 0, as indicated. The input current I_1 produces a voltage drop across R_f and thus creates an output error voltage of I_1R_f, just as with the inverting amplifier.

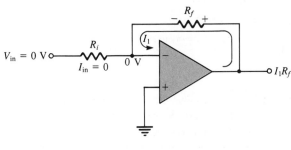

FIGURE 12–33

Input bias current creates output error voltage (I_1R_f) in inverting amplifier.

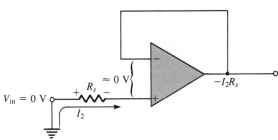

FIGURE 12–34

Input bias current creates output error voltage in voltage follower.

FIGURE 12–35

Input bias current creates output error voltage in noninverting amplifier.

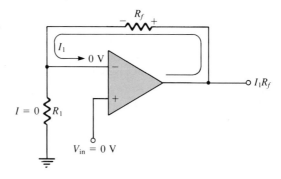

Bias Current Compensation in a Voltage Follower

The output error voltage due to bias currents in a voltage follower can be sufficiently reduced by adding a resistor equal to R_s in the feedback path, as shown in Figure 12–36. The voltage drop created by I_1 across the added resistor subtracts from the $-I_2R_s$ output error voltage. If $I_1 = I_2$, then the output voltage is 0. Usually I_1 does not quite equal I_2; but even in this case, the output error voltage is reduced as follows, because I_{OS} is less than I_2.

$$V_{OUT(error)} = (|I_1 - I_2|)R_s$$
$$V_{OUT(error)} = I_{OS}R_s \qquad\qquad (12\text{–}32)$$

where I_{OS} is the input offset current.

FIGURE 12–36

Bias current compensation in a voltage follower.

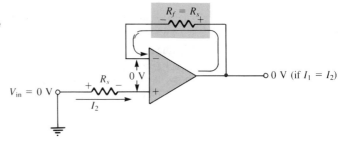

Bias Current Compensation in Other Op-Amp Configurations

To compensate for the effect of bias current in the noninverting amplifier, a resistor R_c is added, as shown in Figure 12–37(a). The compensating resistor value equals the parallel combination of R_i and R_f. The input current I_2 creates a voltage drop across R_c that offsets the voltage across the R_i-R_f combination, thus sufficiently reducing the output error voltage. The inverting amplifier is similarly compensated, as shown in Figure 12–37(b).

Use of a BIFET Op-Amp to Eliminate the Need for Bias Current Compensation

A BIFET op-amp uses both bipolar junction transistors and JFETs in its internal circuitry. The JFETs are used as the input devices to achieve a higher input impedance than is

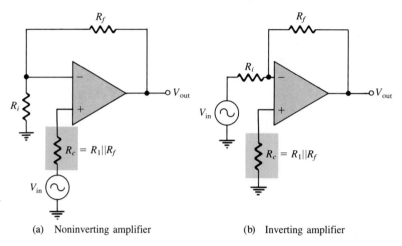

(a)   Noninverting amplifier   (b)   Inverting amplifier

FIGURE 12–37
Bias current compensation.

possible with standard bipolar amplifiers. Because of their very high input impedance, BIFETs typically have input bias currents that are much smaller than in bipolar op-amps, thus reducing or eliminating the need for bias current compensation.

Effect of Input Offset Voltage

The output voltage of an op-amp should be 0 when the differential input is 0. However, there is always a small output error voltage present whose value typically ranges from microvolts to millivolts. This is due to unavoidable imbalances within the internal op-amp transistors aside from the bias currents previously discussed. In a negative feedback configuration, the input offset voltage V_{IO} can be visualized as an equivalent small dc voltage source, as illustrated in Figure 12–38. The output error voltage due to the input offset voltage is

$$V_{OUT(error)} = V_{IO} \qquad \qquad (12\text{–}33)$$

FIGURE 12–38
Input offset voltage equivalent.

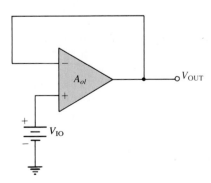

Input Offset Voltage Compensation

Most integrated circuit op-amps provide a means of compensating for offset voltage. This is usually done by connecting an external potentiometer to designated pins on the IC package, as illustrated in Figure 12–39(a) and (b) for a μA741 op-amp. The two terminals are labelled *offset null*. The potentiometer is simply adjusted until the output voltage reads 0, as shown in part (c).

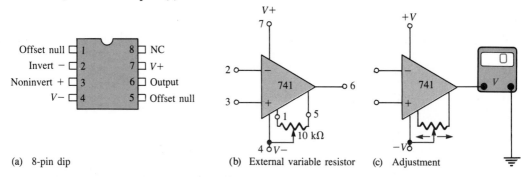

(a) 8-pin dip   (b) External variable resistor   (c) Adjustment

FIGURE 12–39
Input offset voltage-compensated μA741.

SECTION REVIEW 12–7

1. Name two sources of dc output error voltages.
2. How do you compensate for bias current in a voltage follower?

A SYSTEM APPLICATION

12–8

An op-amp can be used as part of the audio amplifier section of an AM or FM receiver, as shown in Figure 12–40. Capacitor C_1 provides a form of *feed-forward* compensation (to

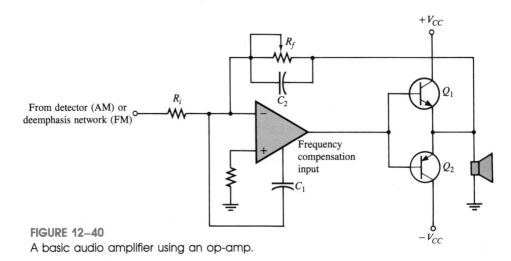

FIGURE 12–40
A basic audio amplifier using an op-amp.

be discussed in the next chapter), which improves the high-frequency response of the op-amp. Capacitor C_2 across R_f insures stability. As you know, R_f and R_i set the amplifier's voltage gain. R_f is variable for volume control. The output of the op-amp drives a pair of discrete transistors, Q_1 and Q_2, to achieve the final power amplification to drive the speaker. This amplifier will respond to frequencies down to dc, and for most audio applications, a high-frequency limit of at least 15 kHz is necessary.

Differential amplifiers FORMULAS

(12–1)	$I_{E_1} = I_{E_2} = \dfrac{I_{R_E}}{2}$	Diff-amp emitter current
(12–2)	$I_{R_E} = \dfrac{V_E - V_{EE}}{R_E}$	Combined emitter current
(12–3)	$I_{C_1} = I_{C_2} \cong \dfrac{I_{R_E}}{2}$	Diff-amp collector current
(12–4)	$V_{C_1} = V_{C_2} = V_{CC} - I_{C_1}R_{C_1}$	Diff-amp collector voltage
(12–5)	$V_E = V_B - 0.7\ \text{V}$	Diff-amp emitter voltage
(12–6)	$\text{CMRR} = \dfrac{A_{v(d)}}{A_{cm}}$	Common-mode rejection ratio (diff-amp)
(12–7)	$\text{CMRR} = 20 \log \dfrac{A_{v(d)}}{A_{cm}}$	Common-mode rejection ratio (dB)

Op-amp parameters

(12–8)	$V_{\text{OUT}} = I_{C_2}R_C - I_{C_1}R_C$	Differential output
(12–9)	$I_{\text{BIAS}} = \dfrac{I_1 + I_2}{2}$	Input bias current
(12–10)	$I_{\text{OS}} = |I_1 - I_2|$	Input offset current
(12–11)	$V_{\text{OS}} = I_{\text{OS}}R_{\text{in}}$	Offset voltage
(12–12)	$V_{\text{OUT(error)}} = A_v I_{\text{OS}}R_{\text{in}}$	Output error voltage
(12–13)	$\text{CMRR} = \dfrac{A_{ol}}{A_{cm}}$	Common-mode rejection ratio (op-amp)
(12–14)	$\text{CMRR} = 20 \log \dfrac{A_{ol}}{A_{cm}}$	Common-mode rejection ratio (dB)
(12–15)	$\text{Slew rate} = \dfrac{\Delta V_{\text{out}}}{\Delta t}$	Slew rate

Op-amp configurations

(12–16)	$V_f = \left(\dfrac{R_i}{R_i + R_f}\right)V_{\text{out}}$	Feedback voltage (noninverting)
(12–17)	$V_{\text{out}} = A_{ol}(V_{\text{in}} - V_f)$	Output voltage (noninverting)

(12–18) $\dfrac{V_{\text{out}}}{V_{\text{in}}} = \dfrac{A_{ol}}{1 + A_{ol}B}$ Voltage gain (noninverting)

(12–19) $A_{cl(\text{NI})} = \dfrac{1}{B} = 1 + \dfrac{R_f}{R_i}$ Voltage gain (noninverting)

(12–20) $A_{cl(\text{VF})} = \dfrac{1}{B} \cong 1$ Voltage gain (voltage follower)

(12–21) $A_{cl(\text{I})} = -\dfrac{R_f}{R_i}$ Voltage gain (inverting)

Op-amp impedances

(12–22) $Z_{\text{in(NI)}} = (1 + A_{ol}B)Z_{\text{in}}$ Input impedance (noninverting)

(12–23) $Z_{\text{out(NI)}} = \dfrac{Z_{\text{out}}}{1 + A_{ol}B}$ Output impedance (noninverting)

(12–24) $Z_{\text{in(VF)}} = (1 + A_{ol})Z_{\text{in}}$ Input impedance (voltage follower)

(12–25) $Z_{\text{out(VF)}} = \dfrac{Z_{\text{out}}}{1 + A_{ol}}$ Output impedance (voltage follower)

(12–26) $Z_{\text{in(Miller)}} = \dfrac{R_f}{A_{ol} + 1}$ Miller input impedance (inverting)

(12–27) $Z_{\text{out(Miller)}} = \left(\dfrac{A_{ol}}{A_{ol} + 1}\right)R_f$ Miller output impedance (inverting)

(12–28) $Z_{\text{in(I)}} = R_i + \dfrac{R_f}{A_{ol} + 1}\|Z_{\text{in}}$ Input impedance (inverting)

(12–29) $Z_{\text{in(I)}} \cong R_i$ Input impedance (inverting)

(12–30) $Z_{\text{out(I)}} = \left(\dfrac{A_{ol}}{A_{ol} + 1}\right)R_f\|Z_{\text{out}}$ Output impedance (inverting)

(12–31) $Z_{\text{out(I)}} \cong Z_{\text{out}}$ Output impedance (inverting)

Error voltage

(12–32) $V_{\text{OUT(error)}} = I_{\text{OS}}R_s$ Output error voltage

(12–33) $V_{\text{OUT(error)}} = V_{\text{IO}}$ Output error voltage (input offset)

SUMMARY

☐ The basic op-amp has three terminals not including power and ground: inverting input (−), noninverting input (+), and output.

☐ Most op-amps require both a positive and a negative dc supply voltage.

☐ The ideal (perfect) op-amp has infinite input impedance, 0 output impedance, infinite open-loop voltage gain, infinite bandwidth, and infinite CMRR.

☐ A good practical op-amp has high input impedance, low output impedance, high open-loop voltage gain, and wide bandwidth.

☐ A differential amplifier is normally used for the input stage of an op-amp.

☐ A differential input voltage appears between the inverting and noninverting inputs of a differential amplifier.

☐ A single-ended input voltage appears between one input and ground (with the other inputs grounded).

☐ A differential output voltage appears between two output terminals of a diff-amp.

☐ A single-ended output voltage appears between the output and ground of a diff-amp.

☐ Common-mode occurs when equal in-phase voltages are applied to both input terminals.

☐ Input offset voltage produces an output error voltage (with no input voltage).

☐ Input bias current also produces an output error voltage (with no input voltage).

☐ Input offset current is the difference between the two bias currents.

☐ Open-loop voltage gain is the gain of the op-amp with no external feedback connections.

☐ The common-mode rejection ratio (CMRR) is a measure of an op-amp's ability to reject common-mode inputs.

☐ Slew rate is the rate in volts per microsecond that the output voltage of an op-amp can change in response to a step input.

☐ There are three basic op-amp configurations: inverting, noninverting, and voltage follower.

☐ All op-amp configurations (except comparators) employ negative feedback. Negative feedback occurs when a portion of the output voltage is connected back to the inverting input such that it subtracts from the input voltage, thus reducing the voltage gain but increasing the stability and bandwidth.

☐ A noninverting amplifier configuration has a higher input impedance and a lower output impedance than the op-amp itself (without feedback).

☐ An inverting amplifier configuration has an input impedance approximately equal to the input resistor R_i and an output impedance approximately equal to the output impedance of the op-amp itself.

☐ The voltage follower has the highest input impedance and the lowest output impedance of the three configurations.

☐ All practical op-amps have small input bias currents and input offset voltages that produce small output error voltages.

☐ The input bias current effect can be compensated for with external resistors as described.

☐ The input offset voltage can be compensated for with an external potentiometer between the two *offset* null pins provided on the IC op-amp package and as recommended by the manufacturer.

---

1. Name all the terminals on an integrated op-amp.   **SELF-TEST**

2. Which of the following characteristics do not *necessarily* apply to an op-amp?
   (a) High gain
   (b) Low power
   (c) High input impedance
   (d) Low output impedance

(e) High CMRR

(f) dc isolation

3. Identify the quantity being measured by each meter in Figure 12–41.

FIGURE 12–41

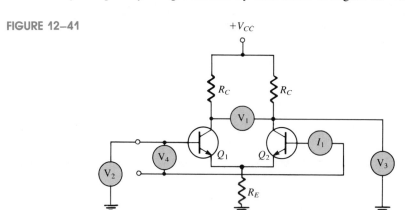

4. Describe how a differential amplifier basically differs from an operational amplifier.

5. A differential amplifier stage has collector resistors of 5 kΩ each. If $I_{C_1} = 1.35$ mA and $I_{C_2} = 1.29$ mA, what is the differential output voltage?

6. A certain op-amp has a bias current into its inverting input of 50 μA and a bias current into its noninverting input of 49.3 μA. What is the input offset current?

7. Explain the difference between common-mode input impedance and the differential input impedance.

8. (a) Ideally, what value of CMRR is desirable?

    (b) One particular op-amp has a CMRR of 80 dB and another has a CMRR of 100 dB. Which would you probably use, all other factors being equal?

9. The output voltage of a particular op-amp increases 8 V in 12 μs in response to a step voltage on the input. Determine the slew rate.

10. Identify each of the op-amp configurations in Figure 12–42.

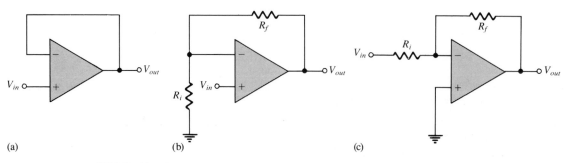

(a)   (b)   (c)

FIGURE 12–42

Op-Amp Frequency Response, Stability, and Compensation

In this chapter you will learn

☐ The difference between *open-loop* gain and *closed-loop* gain of an op-amp

☐ How an op-amp responds to frequency under both open-loop and closed-loop conditions

☐ How negative feedback affects bandwidth of an op-amp in a closed-loop configuration

☐ The meaning of the *gain-bandwidth product*

☐ What happens when positive feedback occurs in an op-amp

☐ The definition of *phase margin*

☐ The meaning of *stability* and what factors affect the stability of an op-amp

☐ How compensation is used to insure stability in an op-amp

☐ Where op-amps can be used in a specific system application

Chapter 13

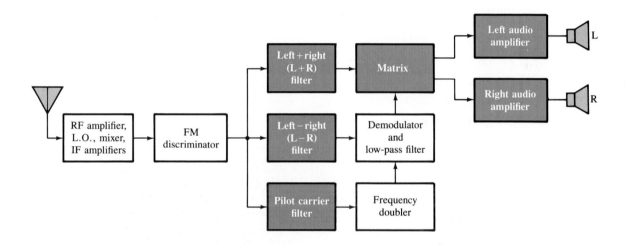

In this chapter we will more closely examine frequency response, bandwidth, phase shift, and other frequency-related parameters mentioned in the previous chapter. We also referred to the effects of negative feedback on amplifier *stability*. You will learn more about stability criteria in this chapter and how to *compensate* op-amps to obtain stable operation.

Many applications of the op-amp depend not only on its gain, but also on its frequency-related performance, as you will see in chapters to follow. The block diagram of the FM stereo multiplex receiver shows specific areas where op-amps can be applied—such as the audio amplifiers (as previously discussed), filters, adders, and inverters found in this type of system.

BASIC RELATIONSHIPS

13–1

The last chapter demonstrated how closed-loop voltage gains of op-amp configurations are determined, and the basic distinction between open-loop gain and closed-loop gain was established. Because of the importance of these two different types of gain, the definitions are restated as follows.

Open-Loop Gain

The open-loop gain of an op-amp is the *internal* voltage gain of the device and represents the ratio of output voltage to input voltage, as indicated in Figure 13–1(a). Notice that there are no external components, so the open-loop gain is set entirely by the internal design. Open-loop gain can range up to 200,000 and is not a well-controlled parameter. Data sheets often refer to the open-loop gain as the *large-signal voltage gain*.

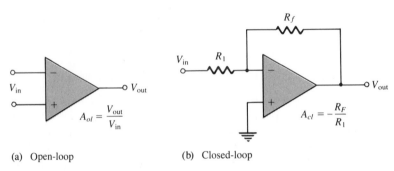

(a) Open-loop   (b) Closed-loop

FIGURE 13–1
Open-loop and closed-loop configurations.

Closed-Loop Gain

The closed-loop gain is for an entire amplifier configuration consisting of the op-amp and an external negative feedback circuit that connects the output to the inverting input. The closed-loop gain is determined by the external component values, as illustrated in Figure 13–1(b) for an inverting amplifier configuration. The closed-loop gain can be precisely controlled by external component values.

The Gain Is Frequency Dependent

In the last chapter, all of the gain expressions applied to the *midrange gain* and were considered independent of the frequency. The midrange open-loop gain of an op-amp extends from 0 frequency (dc) up to a critical frequency at which the gain is 3 dB less than the midrange value. This concept should be familiar from your study of Chapter 11. The difference now is that op-amps are *dc amplifiers* (no capacitive coupling between stages), and therefore, there is no lower critical frequency. This means that the midrange gain extends down to 0 frequency, and dc voltages are amplified the same as midrange signal frequencies.

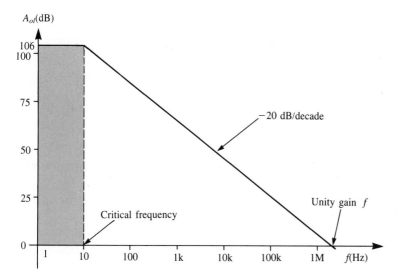

FIGURE 13–2
Ideal plot of open-loop voltage gain versus frequency for typical op-amp.

An open-loop response curve (Bode plot) for a certain op-amp is shown in Figure 13–2. Most op-amp data sheets show this type of curve or specify the midrange open-loop gain. Notice that the curve rolls off at −20 dB per decade (−6 dB per octave). The midrange gain is 200,000, which is 106 dB, and the critical (cutoff) frequency is approximately 10 Hz.

3 dB Open-Loop Bandwidth

Recall that the bandwidth of an ac amplifier is the frequency range between the points where the gain is 3 dB less than midrange, equalling the upper cutoff (critical) frequency minus the lower cutoff frequency.

$$BW = f_{c(\text{high})} - f_{c(\text{low})} \qquad (13\text{–}1)$$

Since f_{cl} for an op-amp is 0, the bandwidth is simply equal to the upper cutoff frequency.

$$BW = f_{c(\text{high})} \qquad (13\text{–}2)$$

From now on, $f_{c(\text{high})}$ will be referred to simply as f_c. Open-loop (ol) or closed-loop (cl) subscript designators will be used.

Unity-Gain Bandwidth

Notice in Figure 13–2 that the gain steadily decreases to a point where it is equal to 1 (0 dB). The value of frequency at which this unity gain occurs is the *unity-gain bandwidth*.

Gain-versus-Frequency Analysis

The RC lag (low-pass) networks within an op-amp are responsible for the roll-off in gain as the frequency increases, just as was discussed for the discrete amplifiers in Chapter 11. From basic ac circuit theory, the attenuation of an RC lag network, such as in Figure 13–3, is expressed as

$$\frac{V_{out}}{V_{in}} = \frac{X_C}{\sqrt{R^2 + X_C^2}} \qquad (13\text{–}3)$$

Dividing both the numerator and denominator to the right of the equal sign by X_C, we get

$$\frac{V_{out}}{V_{in}} = \frac{1}{\sqrt{1 + R^2/X_C^2}} \qquad (13\text{–}4)$$

The critical frequency of an RC network is

$$f_c = \frac{1}{2\pi RC}$$

Dividing both sides by f gives

$$\frac{f_c}{f} = \frac{1}{2\pi RCf} = \frac{1}{(2\pi fC)R}$$

Since $X_C = 1/(2\pi fC)$, the above expression can be written as

$$\frac{f_c}{f} = \frac{X_C}{R} \qquad (13\text{–}5)$$

Substituting this result into equation (13–4) produces the following expression for the attenuation of an RC lag network.

$$\frac{V_{out}}{V_{in}} = \frac{1}{\sqrt{1 + f^2/f_c^2}} \qquad (13\text{–}6)$$

If an op-amp is represented by a voltage gain element and a single RC lag network, as shown in Figure 13–4, then the total open-loop gain is the *product* of the midrange open-loop gain $A_{ol(mid)}$ and the attenuation of the RC network.

$$A_{ol} = \frac{A_{ol(mid)}}{\sqrt{1 + f^2/f_c^2}} \qquad (13\text{–}7)$$

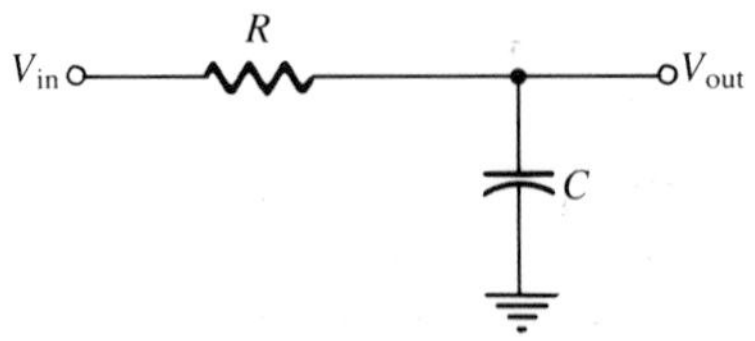

FIGURE 13–3
RC lag network.

FIGURE 13–4

Op-amp represented by gain element and internal RC network.

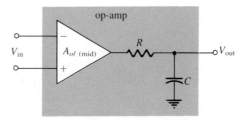

As you can see from equation (13–7), the open-loop gain equals the midrange value when the signal frequency f is much less than the critical frequency f_c, and drops off as the frequency increases. Since f_c is an inherent part of the open-loop response of an op-amp, we will refer to it as $f_{c(ol)}$.

EXAMPLE 13–1

Determine A_{ol} for the following values of f. Assume $f_{c(ol)} = 100$ Hz and $A_{ol(mid)} = 100,000$.

(a) $f = 0$ Hz
(b) $f = 10$ Hz
(c) $f = 100$ Hz
(d) $f = 1000$ Hz

Solution

(a) $A_{ol} = \dfrac{A_{ol(mid)}}{\sqrt{1 + f^2/f_{c(ol)}^2}} = \dfrac{100,000}{\sqrt{1 + 0}} = 100,000$

(b) $A_{ol} = \dfrac{100,000}{\sqrt{1 + (0.1)^2}} = 99,503$

(c) $A_{ol} = \dfrac{100,000}{\sqrt{1 + (1)^2}} = \dfrac{100,000}{\sqrt{2}} = 70,710$

(d) $A_{ol} = \dfrac{100,000}{\sqrt{1 + (10)^2}} = 9950$

This exercise has demonstrated how the open-loop gain decreases as the frequency increases above $f_{c(ol)}$.

Practice Exercise 13–1

Find A_{ol} for the following frequencies. Assume $f_{c(ol)} = 200$ Hz, $A_{ol(mid)} = 80,000$.
(a) $f = 2$ Hz; (b) $f = 10$ Hz; (c) $f = 2500$ Hz

Phase Shift

As you know, an RC network causes a propagation delay from input to output, thus creating a phase difference between the input signal and the output signal. An RC lag network such as the type inherent in an op-amp stage causes the output signal voltage to

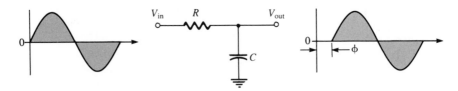

FIGURE 13–5
Output voltage lags input voltage.

lag the input, as shown in Figure 13–5. From basic ac circuit theory, the phase angle of this circuit is

$$\phi = -\arctan\left(\frac{R}{X_C}\right) \tag{13–8}$$

Substituting the relationship in equation (13–5), we get

$$\phi = -\arctan\left(\frac{f}{f_c}\right) \tag{13–9}$$

The negative sign indicates that the output *lags* the input in-phase. This equation shows that the phase angle increases with frequency and approaches $-90°$ as f becomes much greater than f_c.

EXAMPLE 13–2

Calculate the phase angle for an RC lag network for each of the following frequencies, and then plot the curve of phase angle versus frequency. Assume $f_c = 100$ Hz. (The phase angle versus frequency curve is plotted in Figure 13–6.)
(a) $f = 1$ Hz
(b) $f = 10$ Hz
(c) $f = 100$ Hz
(d) $f = 1000$ Hz
(e) $f = 10,000$ Hz

Solution

(a) $\phi = -\arctan\left(\dfrac{f}{f_c}\right) = -\arctan\left(\dfrac{1 \text{ Hz}}{100 \text{ Hz}}\right) = -0.573°$

(b) $\phi = -\arctan\left(\dfrac{10 \text{ Hz}}{100 \text{ Hz}}\right) = -5.71°$

(c) $\phi = -\arctan\left(\dfrac{100 \text{ Hz}}{100 \text{ Hz}}\right) = -45°$

(d) $\phi = -\arctan\left(\dfrac{1000 \text{ Hz}}{100 \text{ Hz}}\right) = -84.29°$

(e) $\phi = -\arctan\left(\dfrac{10,000 \text{ Hz}}{100 \text{ Hz}}\right) = -89.43°$

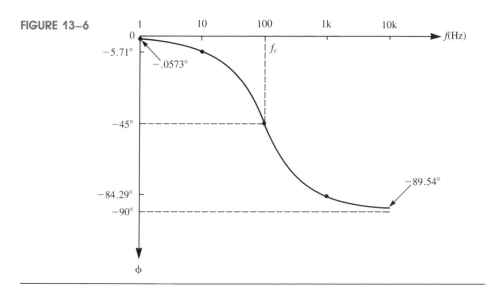

FIGURE 13-6

---

1. How do the open-loop gain and the closed-loop gain of an op-amp differ?

2. The upper cutoff frequency of a particular op-amp is 100 Hz. What is its open-loop 3 dB bandwidth?

3. Open-loop gain (increases, decreases) with frequency above the cutoff frequency.

**SECTION
REVIEW
13-1**

OP-AMP OPEN-LOOP RESPONSE

Frequency Response

13-2

In the previous section, we considered an op-amp to have a constant roll-off of −20 dB/ decade above its critical frequency. Actually, the situation is often more complex than that. A typical IC operational amplifier may consist of two or more cascaded amplifier stages. The gain of each stage is frequency dependent and rolls off at −20 dB/decade above its critical frequency. Therefore, the total response of an op-amp is a *composite* of the individual responses of the internal stages. As an example, a three-stage op-amp is represented in Figure 13-7(a), and the frequency response of each stage is shown in part (b). As you know, dB gains are added so that the total op-amp frequency response is as shown in Figure 13-7(c). Since the roll-off rates are additive, the total roll-off rate increases by −20 dB/decade (−6 dB/octave) as each critical frequency is reached.

Phase Response

In a multistage amplifier, each stage contributes to the total phase lag. As you have seen, *each* RC lag network can produce up to a −90° phase shift. Since each stage in an op-amp includes an RC lag network, a three-stage op-amp, for example, can have a *maximum* phase lag of −270°. Also, the phase lag of *each* stage is less than −45° below the critical

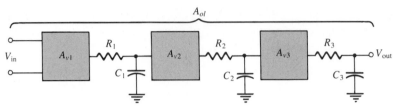

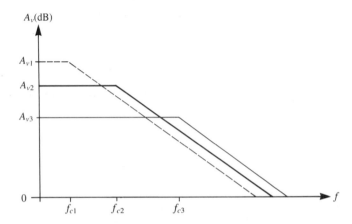

(a) Representation of an op-amp with three internal stages

(b) Individual responses

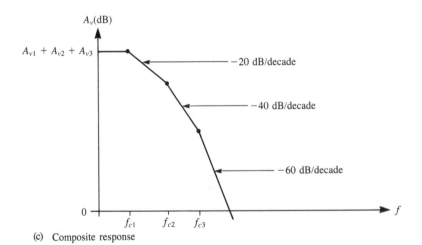

(c) Composite response

FIGURE 13–7

Op-amp open-loop frequency response.

frequency, equal to $-45°$ at the critical frequency, and greater than $-45°$ above the critical frequency. The phase lags of the stages of an op-amp are added to produce a total phase lag, according to the following formula for three stages.

$$\phi_{tot} = -\arctan\left(\frac{f}{f_{c_1}}\right) - \arctan\left(\frac{f}{f_{c_2}}\right) - \arctan\left(\frac{f}{f_{c_3}}\right) \qquad (13\text{--}10)$$

EXAMPLE 13–3

A certain op-amp has three internal amplifier stages with the following gains and critical frequencies:

Stage 1: $A_{v_1} = 40$ dB, $f_{c_1} = 2000$ Hz
Stage 2: $A_{v_2} = 32$ dB, $f_{c_2} = 40$ kHz
Stage 3: $A_{v_3} = 20$ dB, $f_{c_3} = 150$ kHz

Determine the open-loop midrange dB gain and the total phase lag when $f = f_{c_1}$.

Solution

$$A_{ol(mid)} = A_{v_1} + A_{v_2} + A_{v_3}$$
$$= 40 \text{ dB} + 32 \text{ dB} + 20 \text{ dB}$$
$$= 92 \text{ dB}$$

$$\phi_{tot} = -\arctan\left(\frac{f}{f_{c_1}}\right) - \arctan\left(\frac{f}{f_{c_2}}\right) - \arctan\left(\frac{f}{f_{c_3}}\right)$$
$$= -\arctan(1) - \arctan\left(\frac{2}{40}\right) - \arctan\left(\frac{2}{150}\right)$$
$$= -45° - 2.86° - 0.76°$$
$$= -48.62°$$

Practice Exercise 13–3
The internal stages of a two-stage amplifier have the following characteristics: $A_{v_1} = 50$ dB, $A_{v_2} = 25$ dB, $f_{c_1} = 1500$ Hz, and $f_{c_2} = 3000$ Hz. Determine the open-loop midrange gain in dB and the total phase lag when $f = f_{c_1}$.

---

1. If the individual stage gains of an op-amp are 20 dB and 30 dB, what is the total gain in dB?

2. If the individual phase lags are $-49°$ and $-5.2°$, what is the total phase lag?

OP-AMP CLOSED-LOOP RESPONSE

Op-amps are normally used with *negative feedback* for precise control of the gain and bandwidth. As previously mentioned, the open-loop gain, like the β of a transistor, varies greatly from device to device and cannot be depended upon to have a constant value. In this section you will see how feedback affects the overall frequency response of an op-amp. Recall from the previous chapter that midrange gain is *reduced* by negative feedback, as indicated by the following closed-loop gain expressions for the three configurations previously covered, where B is the feedback attenuation.

$$A_{cl(NI)} = \frac{A_{ol}}{1 + A_{ol}B} \cong \frac{1}{B}$$

$$A_{cl(I)} \cong \frac{R_f}{R_i}$$

$$A_{cl(VF)} \cong 1$$

Effect of Negative Feedback on Bandwidth

You know how negative feedback affects gain; now you will learn how it affects bandwidth. The formula for open-loop gain in equation (13–7) can be expressed in complex notation as

$$A_{ol} = \frac{A_{ol(mid)}}{1 + jf/f_{c(ol)}} \tag{13–11}$$

Substituting the above expression into the equation $A_{cl} = A_{ol}/(1 + BA_{ol})$, we get a formula for the total closed-loop gain.

$$A_{cl} = \frac{A_{ol(mid)}/(1 + jf/f_{c(ol)})}{1 + BA_{ol(mid)}/(1 + jf/f_{c(ol)})}$$

Multiplying the numerator and denominator by $1 + jf/f_{c(ol)}$ yields

$$A_{cl} = \frac{A_{ol(mid)}}{1 + BA_{ol(mid)} + jf/f_{c(ol)}}$$

Dividing the numerator and denominator by $1 + BA_{ol(mid)}$ gives

$$A_{cl} = \frac{A_{ol(mid)}/(1 + BA_{ol(mid)})}{1 + j[f/(f_{c(ol)}(1 + BA_{ol(mid)}))]}$$

The above expression is of the form of equation (13–11):

$$A_{cl} = \frac{A_{cl(mid)}}{1 + jf/f_{c(cl)}}$$

where $f_{c(cl)}$ is the closed-loop critical frequency. Thus,

$$f_{c(cl)} = f_{c(ol)}(1 + BA_{ol(mid)}) \tag{13–12}$$

This expression shows that the closed-loop critical frequency is higher than the open-loop critical frequency $f_{c(ol)}$ by the factor $1 + BA_{ol(mid)}$. Since $f_{c(cl)}$ = bandwidth for the closed-loop amplifier, the bandwidth is increased by the same factor.

$$BW_{cl} = BW_{ol}(1 + BA_{ol(mid)}) \tag{13–13}$$

EXAMPLE 13–4

A certain amplifier has an open-loop midrange gain of 150,000 and an open-loop 3 dB bandwidth of 200 Hz. The attenuation of the feedback loop is 0.002. What is the closed-loop bandwidth?

Solution

$$BW_{cl} = BW_{ol}(1 + BA_{ol(mid)})$$
$$= 200 \text{ Hz}[1 + (150{,}000)(0.002)]$$
$$= 60.2 \text{ kHz}$$

Figure 13–8 graphically illustrates the concept of closed-loop response. When the open-loop gain of an op-amp is reduced by negative feedback, the bandwidth is increased. The closed-loop gain is independent of the open-loop gain up to the point of intersection of the two gain curves. This point of intersection is the critical frequency for the closed-loop response. Notice that the closed-loop gain has the same roll-off rate as the open-loop gain, beyond the closed-loop critical frequency.

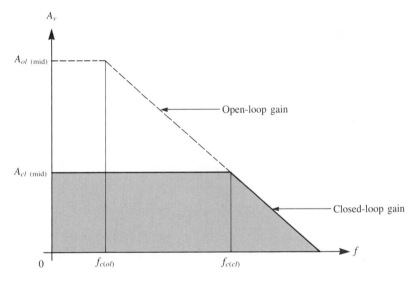

FIGURE 13–8
Closed-loop gain compared to open-loop gain.

Gain-Bandwidth Product

An increase in closed-loop gain causes a *decrease* in the bandwidth and vice versa, such that the *product* of gain and bandwidth is a *constant*. This is true as long as the roll-off rate is fixed. Letting A_{cl} stand for the gain of any of the closed-loop configurations and $f_{c(cl)}$ for the closed-loop critical frequency (also the bandwidth), then

$$A_{cl}f_{c(cl)} = A_{ol}f_{c(ol)} \qquad \qquad \textbf{(13–14)}$$

The gain-bandwidth product is always equal to the frequency at which the op-amp's open-loop gain is unity (unity-gain bandwidth).

$$A_{cl}f_{c(cl)} = \text{unity-gain bandwidth} \qquad (13\text{--}15)$$

EXAMPLE 13–5

Determine the bandwidth of each of the amplifiers in Figure 13–9. Both op-amps have an open-loop gain of 100 dB and a unity-gain bandwidth of 3 MHz.

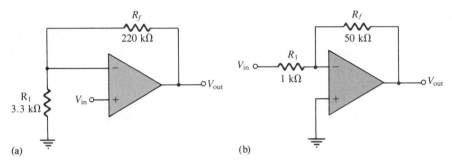

FIGURE 13–9

Solution

(a) For the noninverting amplifier in part (a) of the figure, the closed-loop gain is

$$A_{cl} \cong \frac{1}{B} = \frac{1}{R_i/(R_i + R_f)} = \frac{1}{3.3 \text{ k}\Omega/223.3 \text{ k}\Omega} = 67.67$$

Using equation (13–15) and solving for $f_{c(cl)}$, we get (where $f_{c(cl)} = BW_{cl}$):

$$f_{c(cl)} = BW_{cl} = \frac{\text{unity-gain } BW}{A_{cl}}$$

$$BW_{cl} = \frac{3 \text{ MHz}}{67.67} = 44.33 \text{ kHz}$$

(b) For the inverting amplifier in part (b) of the figure, the closed-loop gain is

$$A_{cl} = \frac{R_f}{R_i} = \frac{50 \text{ k}\Omega}{1 \text{ k}\Omega} = 50$$

The closed-loop bandwidth is

$$BW_{cl} = \frac{3 \text{ MHz}}{50} = 60 \text{ kHz}$$

Practice Exercise 13–5

Determine the bandwidth of each of the amplifiers in Figure 13–9. Both op-amps have an A_{ol} of 90 dB and a unity-gain bandwidth of 2 MHz.

1. The closed-loop gain is always less than the open-loop gain (T or F).

2. A certain op-amp is used in a feedback configuration having a gain of 30 and a bandwidth of 100 kHz. If the external resistor values are changed to increase the gain to 60, what is the new bandwidth?

3. What is the unity-gain bandwidth of the op-amp in question 2?

POSITIVE FEEDBACK AND STABILITY

Stability is a very important consideration when using op-amps. Stable operation means that the op-amp does not *oscillate* under any condition. Instability produces oscillations, which are unwanted voltage swings on the output when there is no signal present on the input, or in response to noise or transient voltages on the input.

13–4

Positive Feedback

To understand stability, instability and its causes must first be examined. As you know, with negative feedback, the signal fed back to the input is out-of-phase with the input signal, thus subtracting from it and effectively reducing the voltage gain. As long as the feedback is negative, the amplifier is stable. When the signal fed back from output to input is *in-phase* with the input signal, a *positive feedback* condition exists. That is, positive feedback occurs when *the total phase shift through the op-amp and feedback network is 360°*. 360° is equivalent to no phase shift (0°).

Loop Gain

For instability to occur, (1) there must be *positive feedback,* and (2) *the loop gain of the closed-loop amplifier must be greater than 1*. The loop gain of a closed-loop amplifier is defined to be the op-amp's open-loop gain times the attenuation of the feedback network.

$$\text{Loop gain} = A_{ol}B \qquad\qquad (13\text{–}16)$$

Phase Margin

Notice that for each amplifier configuration in Figure 13–10, the feedback loop is connected to the *inverting* input. There is an inherent phase shift of 180° because of the *inversion* between input and output. Additional phase shift (ϕ_{tot}) is produced by the RC lag networks within the amplifier. So, the total phase shift around the loop is 180° + ϕ_{tot}.

The *phase margin* is the amount of additional phase shift required to make the *total* phase shift around the loop 360°. (360° is equivalent to 0°.) It is expressed as

$$\theta_{pm} = 180° - |\phi_{tot}| \qquad\qquad (13\text{–}17)$$

If the phase margin is positive, the total phase shift is less than 360° and the amplifier is stable. If the phase margin is 0 or negative, then the amplifier is potentially unstable because the signal fed back can be in-phase with the input. As you can see from equation (13–17), when the total lag network phase shift ϕ_{tot} equals or exceeds 180°, then the phase margin is 0° or negative and an unstable condition exists.

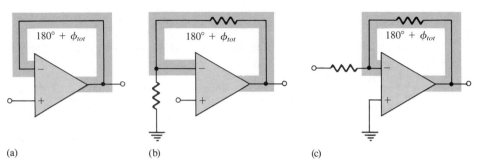

(a)   (b)   (c)

FIGURE 13–10
Feedback-loop phase shift.

Stability Analysis

Since most op-amp configurations use a loop gain greater than 1 ($A_{ol}B > 1$), the criteria for stability are based on the phase angle of the internal lag networks. As previously mentioned, operational amplifiers are composed of multiple stages, each of which has a critical frequency. For purposes of illustrating the concept of stability, we will use a three-stage op-amp with an open-loop response as shown in the Bode plot of Figure 13–11. Notice that there are three different critical frequencies, which indicates three internal RC lag networks. At the first critical frequency, f_{c_1}, the gain begins rolling off at −20 dB/decade; when the second critical frequency, f_{c_2}, is reached, the gain decreases at −40 dB/decade; and when the third critical frequency, f_{c_3}, is reached, the gain drops at −60 dB/decade.

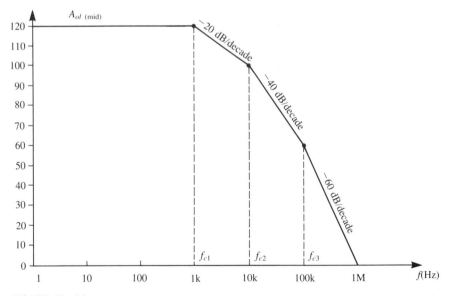

FIGURE 13–11
Bode plot of example of three-stage op-amp response.

To analyze a closed-loop amplifier for stability, the phase margin must be determined. A positive phase margin will indicate that the amplifier is stable for a given value of closed-loop gain. Three example cases will be considered in order to demonstrate the conditions for instability.

Case 1. The closed-loop gain intersects the open-loop response on the -20 dB/decade slope, as shown in Figure 13–12. For this example, the midrange closed-loop gain is 106 dB, and the closed-loop critical frequency is 5 kHz. If we assume that the amplifier is not operated out of its midrange, the *maximum* phase shift for the 106 dB amplifier occurs at the *highest* midrange frequency (in this case, 5 kHz). The total phase shift at this frequency due to the three lag networks is calculated as follows:

$$\phi_{tot} = -\arctan\left(\frac{f}{f_{c_1}}\right) - \arctan\left(\frac{f}{f_{c_2}}\right) - \arctan\left(\frac{f}{f_{c_3}}\right)$$

where $f = 5$ kHz, $f_{c_1} = 1$ kHz, $f_{c_2} = 10$ kHz, and $f_{c_3} = 100$ kHz. Therefore,

$$\phi_{tot} = -\arctan\left(\frac{5\text{ kHz}}{1\text{ kHz}}\right) - \arctan\left(\frac{5\text{ kHz}}{10\text{ kHz}}\right) - \arctan\left(\frac{5\text{ kHz}}{100\text{ kHz}}\right)$$
$$= -78.69° - 26.57° - 2.86°$$
$$= -108.12°$$

The phase margin is

$$\theta_{pm} = 180° - |\phi_{tot}| = 180° - 108.12° = +71.88°$$

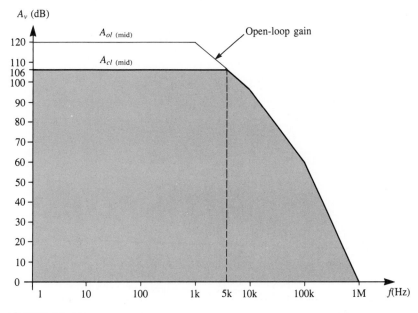

FIGURE 13–12
Case where closed-loop gain intersects open-loop gain on -20 dB/decade slope (stable operation).

θ_{pm} is positive, so *the amplifier is stable* for all frequencies in its midrange. In general, an amplifier is stable for all midrange frequencies if its closed-loop gain intersects the open-loop response curve on a -20 dB/decade slope.

Case 2.   The closed-loop gain is lowered to where it intersects the open-loop response on the -40 dB/decade slope, as shown in Figure 13–13. The midrange closed-loop gain in this case is 72 dB, and the closed-loop critical frequency is approximately 30 kHz. The total phase shift at $f = 30$ kHz due to the three lag networks is calculated as follows.

$$\phi_{tot} = -\arctan\left(\frac{30\text{ kHz}}{1\text{ kHz}}\right) - \arctan\left(\frac{30\text{ kHz}}{10\text{ kHz}}\right) - \arctan\left(\frac{30\text{ kHz}}{100\text{ kHz}}\right)$$

$$= -88.09° - 71.57° - 16.7°$$

$$= -176.36°$$

The phase margin is

$$\theta_{pm} = 180° - 176.36° = +3.64°$$

The phase margin is positive, so the amplifier is still stable for all frequencies in its midrange, but a very slight increase in frequency above f_c would cause it to oscillate. Therefore, it is *marginally stable* and very close to instability because instability occurs where $\theta_{pm} = 0°$ and, as a general rule, a minimum 45° phase margin is recommended to avoid marginal conditions.

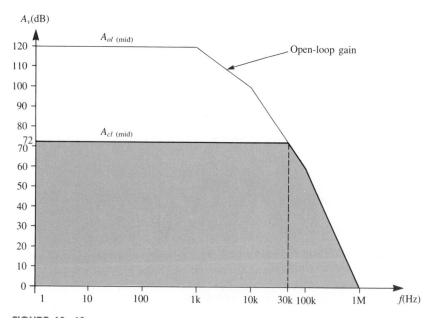

FIGURE 13–13
Case where closed-loop gain intersects open-loop gain on -40 dB/decade slope (marginally stable operation).

Case 3. The closed-loop gain is further decreased until it intersects the open-loop response on the −60 dB/decade slope, as shown in Figure 13–14. The midrange closed-loop gain in this case is 18 dB, and the closed-loop critical frequency is 500 kHz. The total phase shift at f = 500 kHz due to the three lag networks is

$$\phi_{tot} = -\arctan\left(\frac{500\text{ kHz}}{1\text{ kHz}}\right) - \arctan\left(\frac{500\text{ kHz}}{10\text{ kHz}}\right) - \arctan\left(\frac{500\text{ kHz}}{100\text{ kHz}}\right)$$

$$= -89.89° - 88.85° - 78.69° = -257.43°$$

The phase margin is

$$\theta_{pm} = 180° - 257.43° = -77.43°$$

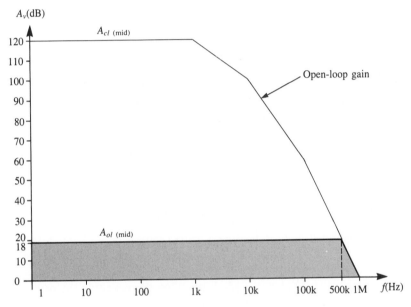

FIGURE 13–14
Case where closed-loop gain intersects open-loop gain on −60 dB/decade slope (unstable operation).

Here the phase margin is negative and *the amplifier is unstable* at the upper end of its midrange. Using the following program, you can determine the approximate midrange frequency at which instability occurs. The program computes the phase margin for each of a specified number of frequency values and determines whether the op-amp is stable or unstable. The display shows a list of frequencies with corresponding phase shifts, phase margins, and stability conditions. Instability occurs between the last two frequencies printed. Inputs required for this program are all of the critical frequencies in the open-loop response of the given op-amp, the highest frequency of operation, and the increments of frequency for which the parameters are to be computed.

```
10  CLS
20  PRINT"THIS PROGRAM COMPUTES THE PHASE MARGINS FOR EACH"
30  PRINT"FREQUENCY IN A SPECIFIED RANGE AND DETERMINES THE"
40  PRINT"MAXIMUM FREQUENCY FOR STABLE OPERATION OF THE OP AMP"
50  PRINT:PRINT:PRINT
60  INPUT "TO CONTINUE PRESS 'ENTER'";X
70  CLS
80  INPUT "NUMBER OF OPEN LOOP CRITICAL FREQUENCIES";N
90  CLS
100 FOR Y=1 TO N
110 INPUT "CRITICAL FREQUENCY IN HERTZ";FC(Y)
120 NEXT
130 INPUT "THE HIGHEST INPUT FREQUENCY";FH
140 INPUT "INCREMENTS OF FREQUENCY FROM ZERO TO THE HIGHEST";FI
150 CLS
160 PRINT"FREQUENCY","PHASE SHIFT", "PHASE MARGIN","STABILITY"
170 FOR F=0 TO FH STEP FI
180 PH=0
190 FOR Y=1 TO N
200 PH=PH-ATN(F/FC(Y))*57.29578
210 NEXT Y
220 PM=180+PH
230 IF PM<=0 THEN S$="UNSTABLE" ELSE S$="STABLE"
240 PRINT F,PH,PM,S$
250 IF PM<=0 THEN END
260 NEXT F
```

Summary of Stability Criteria

This analysis has demonstrated that an amplifier's closed-loop gain must intersect the open-loop gain curve on a -20 dB/decade slope to insure stability for all of its midrange frequencies. If the closed-loop gain is lowered to a value that intersects on a -40 dB/decade slope, then marginal stability or complete instability can occur. In the previous example, the closed-loop gain should be greater than 72 dB.

If the closed-loop gain intersects the open-loop response on a -60 dB/decade slope, definite instability will occur at some frequency within the amplifier's midrange. Therefore, to insure stability for all of the midrange frequencies, an op-amp must be operated at a closed-loop gain such that the roll-off rate beginning at its dominant critical frequency does not exceed -20 dB/decade.

**SECTION
REVIEW
13–4**

1. An amplifier can oscillate when the feedback is _____.

2. Theoretically, the lag network phase shift can be as great as _____ degrees before instability occurs. At this point the phase margin is _____.

3. To insure stability, it is desirable that the open-loop gain of an op-amp not exceed a roll-off rate of _____.

13–5

The last section demonstrated that instability can occur when an op-amp's response has roll-off rates exceeding -20 dB/decade and the op-amp is operated in a closed-loop configuration having a gain curve that intersects a higher roll-off rate portion of the open-loop response. In situations like those examined in the last section, the closed-loop voltage gain is restricted to very high values. In many applications lower values of closed-loop gain are necessary or desirable. To allow op-amps to be operated at low closed-loop gain, phase lag compensation is required.

Phase Lag Compensation

As you have seen, the cause of instability is excessive phase shift through an op-amp's internal lag networks. When these phase shifts equal or exceed 180°, the amplifier can oscillate. *Compensation* is used to either *eliminate* open-loop roll-off rates greater than -20 dB/decade or *extend* the -20 dB/decade rate to a lower gain. These concepts are illustrated in Figure 13–15.

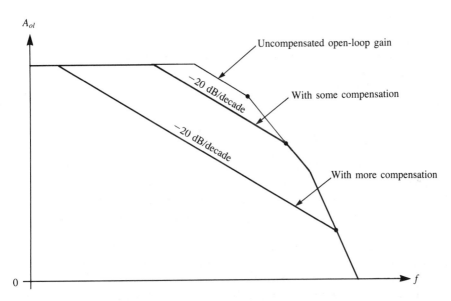

FIGURE 13–15
Bode plot illustrating effect of phase compensation on open-loop gain of typical op-amp.

Compensating Network

There are two basic methods of compensation for integrated circuit op-amps: *internal* and *external*. In either case an RC network is added. The basic compensating action is as follows. Consider first the RC network shown in Figure 13–16(a). At low frequencies

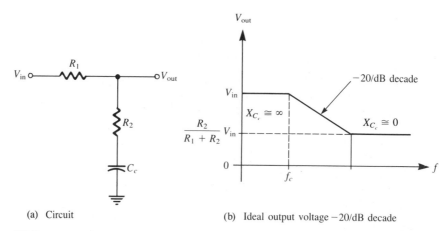

(a) Circuit

(b) Ideal output voltage −20/dB decade

FIGURE 13–16
Basic compensating network action.

where X_{C_c} is extremely large, the output voltage approximately equals the input voltage. When the frequency reaches its critical value, $f_c = 1/[2\pi(R_1 + R_2)C_c]$, the output voltage decreases at −20 dB/decade. This roll-off rate continues until $X_{C_c} \cong 0$, at which point the output voltage levels off to a value determined by R_1 and R_2, as indicated in Figure 13–16(b). This is the principle used in phase-compensating an op-amp.

To see how a compensating network changes the open-loop response of an op-amp, refer to Figure 13–17. This diagram represents a two-stage op-amp. The individual stages are within blocks along with the associated lag networks. A compensating network is shown connected at point A on the output of stage 1.

The critical frequency of the compensating network is set to a value *less* than the dominant (lowest) critical frequency of the internal lag networks. This causes the −20 dB/decade roll-off to begin at the compensating network's critical frequency. The roll-off of the compensating network continues up to the critical frequency of the domi-

FIGURE 13–17
Representation of op-amp with compensation.

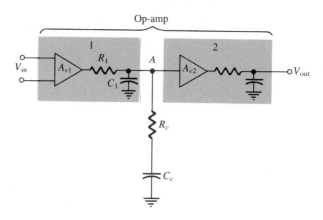

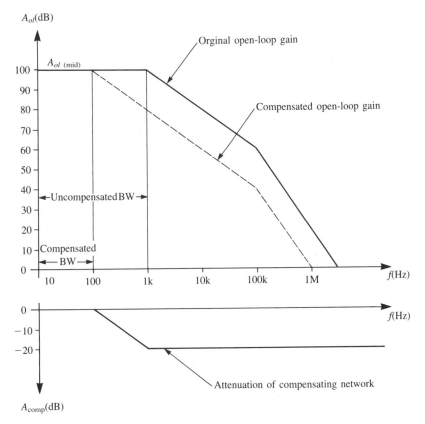

FIGURE 13–18
Example of compensated op-amp frequency response.

nant lag network. At this point, the response of the compensating network levels off, and the −20 dB/decade roll-off of the dominant lag network takes over. The net result is a shift of the open-loop response to the left, thus reducing the bandwidth, as shown in Figure 13–18. The response curve of the compensating network is shown in proper relation to the overall open-loop response.

EXAMPLE 13–6

A certain op-amp has the open-loop response in Figure 13–19. As you can see, the lowest closed-loop gain for which stability is assured is approximately 40 dB (where the closed-loop gain line still intersects the −20 dB/decade slope). In a particular application, a 20 dB closed-loop gain is required.
(a) Determine the critical frequency for the compensating network.
(b) Sketch the ideal response curve for the compensating network.
(c) Sketch the total ideal compensated open-loop response.

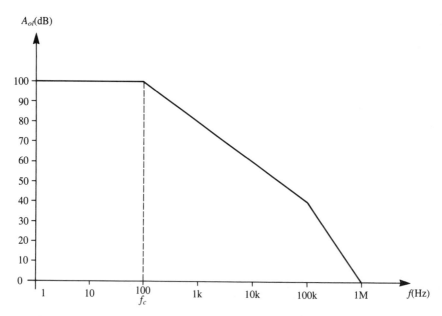

FIGURE 13–19
Original open-loop response.

Solution

(a) The gain must be dropped so that the −20 dB/decade roll-off extends down to 20 dB rather than to 40 dB. To achieve this, the midrange open-loop gain must be made to roll off a decade sooner. Therefore, the critical frequency of the compensating network must be 10 Hz.

(b) The roll-off of the compensating network must end at 100 Hz, as shown in Figure 13–20(a).

(c) The total open-loop response resulting from compensation is shown in Figure 13–20(b).

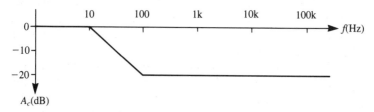

(a) Compensating network response

FIGURE 13–20

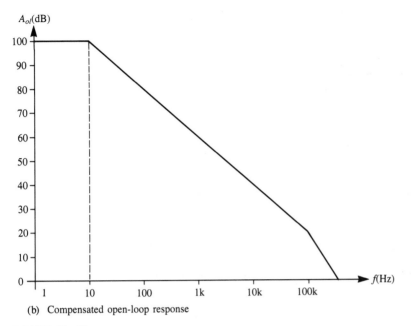

(b) Compensated open-loop response

FIGURE 13–20
(continued)

Extent of Compensation

A larger compensating capacitor will cause the open-loop roll-off to begin at a lower frequency and thus extend the −20 dB/decade roll-off to lower gain levels, as shown in Figure 13–21(a). With a sufficiently large compensating capacitor, an op-amp can be made *unconditionally stable*, as illustrated in Figure 13–21(b), where the −20 dB/decade slope is extended all the way down to unity gain. This is normally the case when *internal* compensation is provided by the manufacturer. An internally, *fully* compensated op-amp can be used for any value of closed-loop gain and remain stable. The μA741 is an example of an internally compensated device.

A disadvantage of fully compensated op-amps is that bandwidth is sacrificed; thus the slew rate is decreased. Therefore, many IC op-amps have provisions for *external* compensation. Figure 13–22 shows the package layouts of an LM101 op-amp with pins available for external compensation with a small capacitor. With provisions for external connections, just enough compensation can be used for a given application without sacrificing more performance than necessary.

FIGURE 13–21
Extent of compensation.

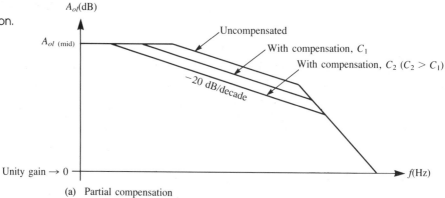

(a) Partial compensation

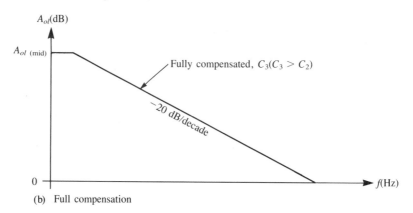

(b) Full compensation

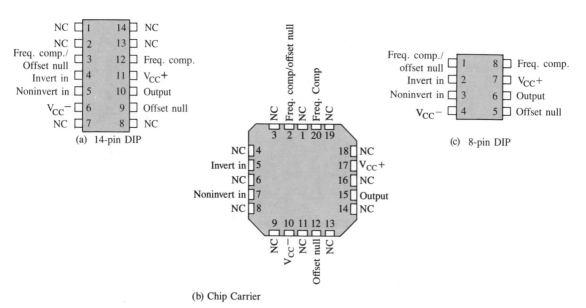

(a) 14-pin DIP

(b) Chip Carrier

(c) 8-pin DIP

FIGURE 13–22
Op-amp packages.

Single-Capacitor Compensation

As an example of compensating an IC op-amp, a capacitor C_1 is connected to pins 1 and 8 of an LM101 in an inverting amplifier configuration, as shown in Figure 13–23(a). Part (b) of the figure shows the open-loop frequency response curves for two values of C_1. The 3 pF compensating capacitor produces a unity-gain bandwidth approaching 10 MHz. Notice that the -20 dB/decade slope extends to a very low gain value. When C_1 is increased ten times to 30 pF, the bandwidth is reduced by a factor of ten. Notice that the -20 dB/decade slope now extends through unity gain.

When the op-amp is used in a closed-loop configuration, as in Figure 13–23(c), the useful frequency range depends on the compensating capacitor. For example, with a closed-loop gain of 40 dB as shown in the figure, the bandwidth is approximately 10 kHz for $C_1 = 30$ pF and increases to approximately 100 kHz when C_1 is decreased to 3 pF.

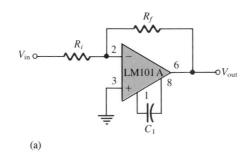

(a)

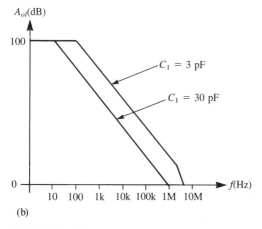

(b)

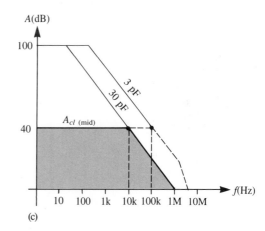

(c)

FIGURE 13–23
Example of single-capacitor compensation.

Feed-Forward Compensation

Another method of phase compensation is called *feed-forward*. This type of compensation results in less bandwidth reduction than the method previously discussed. The basic concept is to bypass the internal input stage of the op-amp at high frequencies, driving the higher-frequency second stage, as shown in Figure 13–24.

FIGURE 13–24
Feed-forward compensation showing
high-frequency bypassing of first stage.

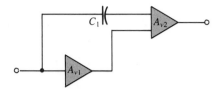

Feed-forward compensation of an LM101 is shown in Figure 13–25(a). The feed-forward capacitor C_1 is connected from the inverting input to the compensating terminal. A small capacitor is needed across R_f to insure stability. The Bode plot in Figure 13–25(b) shows the feed-forward compensated response and the standard compensated response that was discussed previously. The use of feed-forward compensation is restricted to the inverting amplifier configuration. Other compensation methods are also used. Often, recommendations are provided by the manufacturer on the data sheet.

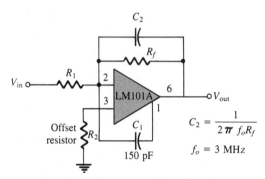

$$C_2 = \frac{1}{2\pi f_o R_f}$$

$$f_o = 3 \text{ MHz}$$

(a) Manufacturers' recommended configuration

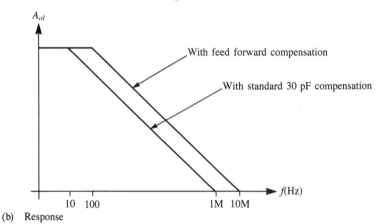

(b) Response

FIGURE 13–25
Feed-forward compensation.

1. What is the purpose of phase compensation?

2. What is the main difference between internal and external compensation?

3. To compensate an amplifier, you must sacrifice _____ .

A SYSTEM APPLICATION

13–6

Stereo systems use two separate FM signals to reproduce sound as, for example, from the left and right sides of the stage in a concert performance. When processed by a stereo receiver, the sound comes from both the left and right speakers, and the original sound distribution and direction are reproduced. When an FM stereo broadcast is received by a standard single-speaker system (monophonic), the output to the speaker is equal to the sum of the left plus the right channels (L + R) so you get the original sound without separation. When a stereo receiver is used, the full stereo effect is reproduced by the two speakers. Stereo FM signals are transmitted on an RF carrier of 88 MHz to 108 MHz with the frequency pattern shown in Figure 13–26(a). The sum of the left and right modulating signals (L + R) extends from 30 Hz to 15 kHz the same as the full modulating signal in standard monophonic FM. In addition, stereo FM consists of a signal corresponding to the left minus the right modulating signals (L − R) which extends from 23 kHz to 53 kHz. This L − R signal appears in two sidebands with a suppressed carrier at 38 kHz as indicated. A 19 kHz frequency, known as the pilot subcarrier, is also included.

FM stereo receivers are identical to standard monophonic (nonstereo) receivers up to the output of the discriminator, as indicated in Figure 13–26(b). The discriminator output in a stereo system contains the L + R signal, the L − R signal, and the pilot subcarrier. If a nonstereo FM receiver is tuned to a stereo station, the output of its discriminator also contains all the stereo frequencies, but the speaker output is not affected by any signals except the L + R. This is because the audio amplifiers and speakers in most receivers will not pass the frequencies of 19 kHz and above; but even if they do, the normal human ear cannot hear these frequencies. Thus the standard nonstereo receiver reproduces only the frequencies of the L + R signal (30 Hz to 15 kHz).

As you can see in Figure 13–26(b), the stereo FM receiver becomes quite complex after the discriminator. The three frequency components of the stereo signal are separated by the three filters (op-amps can be used for these filter circuits, as you will see in Chapter 16). The L + R signal is obtained through a low-pass filter with a response from 0 Hz to 15 kHz. The L − R signal is separated by a band-pass filter with a bandwidth from 23 kHz to 53 kHz. The pilot subcarrier is extracted by a band-pass filter with a very narrow bandwidth centered around 19 kHz. The pilot frequency is then doubled to 38 kHz (equal to the L − R suppressed carrier) and combined with the L − R signal in a nonlinear demodulator to produce sum and difference outputs which are then filtered by a low-pass circuit to obtain the 30 Hz to 15 kHz L − R audio signal. What has happened is that the 23 kHz to 53 kHz signal has been translated down to its original audio range (30 Hz to 15 kHz). At this point, the L + R and the L − R audio signals are applied to the matrix and deemphasis network, as shown in Figure 13–26(c). Within the matrix, the L + R and

FIGURE 13–26
The stereo FM system.

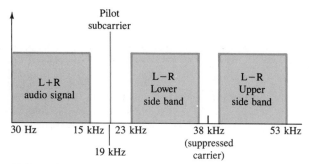

(a) Transmitted FM stereo modulating frequency bands

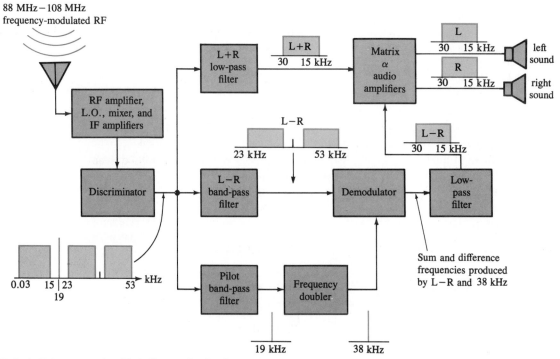

(b) Basic FM stereo receiver block diagram showing frequencies of processed signals

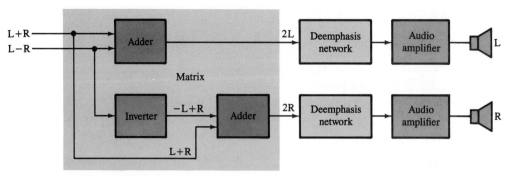

(c) Matrix, deemphasis, and audio amplifier section

L − R signals are applied to the inputs of an adder (you will see how an op-amp can be used as an adder circuit in Chapter 14) to produce an output of (L + R) + (L − R) = 2L, which is the left-side audio. Also, the L − R signal is inverted (again, an op-amp can be used as an inverting circuit) and applied to another adder along with the L + R signal to produce an output of (−L + R) + (L + R) = 2R, which is the right-side audio. The left and right audio signals then go through separate channels for deemphasis and amplification.

(13–1) $BW = f_{c(\text{high})} - f_{c(\text{low})}$ General bandwidth

(13–2) $BW = f_{c(\text{high})}$ Op-amp bandwidth

(13–3) $\dfrac{V_{\text{out}}}{V_{\text{in}}} = \dfrac{X_C}{\sqrt{R^2 + X_C^2}}$ RC attenuation, lag network

(13–4) $\dfrac{V_{\text{out}}}{V_{\text{in}}} = \dfrac{1}{\sqrt{1 + R^2/X_C^2}}$ RC attenuation, lag network

(13–5) $\dfrac{f_c}{f} = \dfrac{X_C}{R}$ Ratio of frequencies equals ratio of reactances

(13–6) $\dfrac{V_{\text{out}}}{V_{\text{in}}} = \dfrac{1}{\sqrt{1 + f^2/f_c^2}}$ RC attenuation

(13–7) $A_{ol} = \dfrac{A_{ol(\text{mid})}}{\sqrt{1 + f^2/f_c^2}}$ Open-loop gain

(13–8) $\phi = -\arctan\left(\dfrac{R}{X_C}\right)$ RC phase shift

(13–9) $\phi = -\arctan\left(\dfrac{f}{f_c}\right)$ RC phase shift

(13–10) $\phi_{tot} = -\arctan\left(\dfrac{f}{f_{c_1}}\right)$ Total phase shift

$- \arctan\left(\dfrac{f}{f_{c_2}}\right) - \arctan\left(\dfrac{f}{f_{c_3}}\right)$

(13–11) $A_{ol} = \dfrac{A_{ol(\text{mid})}}{1 + jf/f_{c(ol)}}$ Open-loop gain (complex)

(13–12) $f_{c(cl)} = f_{c(ol)}(1 + BA_{ol(\text{mid})})$ Closed-loop critical frequency

(13–13) $BW_{cl} = BW_{ol}(1 + BA_{ol(\text{mid})})$ Closed-loop bandwidth

(13–14) $A_{cl}f_{c(cl)} = A_{ol}f_{c(ol)}$ Gain-bandwidth product

(13–15) $A_{cl}f_{c(cl)} = $ unity-gain bandwidth

(13–16)   Loop gain $= A_{ol}B$

(13–17) $\theta_{pm} = 180° - |\phi_{tot}|$ Phase margin

SUMMARY

☐ *Open-loop gain* is the voltage gain of an op-amp without feedback.

☐ *Closed-loop gain* is the voltage gain of an op-amp with negative feedback.

☐ The closed-loop gain is always less than the open-loop gain.

☐ The midrange gain of an op-amp extends down to dc.

☐ The gain of an op-amp decreases as frequency increases above the critical frequency.

☐ The bandwidth of an op-amp equals the upper cutoff frequency (the lowest critical frequency in the open-loop gain).

☐ The internal RC lag networks that are inherently part of the amplifier stages cause the gain to roll off as frequency goes up.

☐ The internal RC lag networks also cause a phase shift between input and output signals.

☐ *Negative feedback* lowers the gain and increases the bandwidth.

☐ The product of gain and bandwidth is constant for a given op-amp.

☐ The gain-bandwidth product equals the frequency at which unity voltage gain occurs.

☐ *Positive feedback* occurs when the total phase shift through the op-amp (including 180° inversion) and feedback network is 0° (equivalent to 360°) or more.

☐ The *phase margin* is the amount of additional phase shift required to make the total phase shift around the loop 360°.

☐ When the closed-loop gain of an op-amp intersects the open-loop response curve on a −20 dB/decade (−6 dB/octave) slope, the amplifier is stable.

☐ When the closed-loop gain intersects the open-loop response curve on a slope greater than −20 dB/decade, the amplifier can be either marginally stable or unstable.

☐ A minimum phase margin of 45° is recommended to provide a sufficient safety factor for stable operation.

☐ A fully compensated op-amp has a −20 dB/decade roll-off all the way down to unity gain.

☐ Compensation reduces bandwidth and increases slew rate.

☐ Internally compensated op-amps such as the μA741 are available. These are usually fully compensated with a large sacrifice in bandwidth.

☐ Externally compensated op-amps such as the LM101 are available. External compensating networks can be connected to specified pins, and the compensation can be tailored to a specific application. In this way, bandwidth and slew rate are not degraded more than necessary.

SELF-TEST

1. Explain the difference between *open-loop* gain and *closed-loop* gain.

2. What is the bandwidth of an *ac amplifier* having a lower cutoff frequency of 1 kHz and an upper cutoff frequency of 10 kHz?

3. What is the bandwidth of a dc amplifier having an upper cutoff frequency of 100 kHz?

4. Calculate the attenuation of each RC network in Figure 13–27 for a frequency of 10 kHz.

FIGURE 13–27

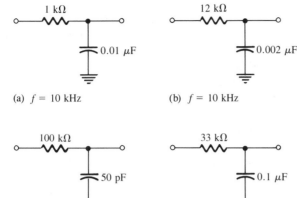

(a) $f = 10$ kHz

(b) $f = 10$ kHz

(c) $f = 10$ kHz

(d) $f = 10$ kHz

5. Determine the open-loop gain at each of the frequencies below of an op-amp having a midrange gain of 80,000 and a cutoff frequency of 500 Hz.
    (a) $f = 100$ Hz
    (b) $f = 1$ kHz
    (c) $f = 5$ kHz
    (d) $f = 20$ kHz

6. Convert each gain value in problem 5 to dB.

7. Calculate the phase shift for each circuit in Figure 13–27 at the specified frequency.

8. How much phase shift is possible in a three-stage op-amp if each stage consists of one RC lag network?

9. A certain op-amp has an open-loop midrange gain of 100,000 and an open-loop bandwidth of 150 Hz. Determine the bandwidth when the op-amp is operated in a negative feedback configuration where $B = 0.004$.

10. What is the gain-bandwidth product of an op-amp that has a gain of 1 at 800 kHz?

11. Refer to the open-loop response curve in Figure 13–28. What is the minimum closed-loop gain at which the op-amp can be operated with stability assured?

12. In a certain op-amp configuration, the total phase shift due to the internal lag networks is 120°. What is the phase margin? Is the amplifier stable?

13. At what point does feedback become positive?

14. The critical frequency of the compensating network is always (greater, less) than the cutoff frequency (dominant critical frequency) of the op-amp's open-loop response.

15. What is the advantage of external phase compensation over internal compensation?

16. If an op-amp is compensated so that it begins rolling off a decade below its uncompensated open-loop cutoff frequency, how much is the gain down when the frequency reaches the uncompensated open-loop cutoff value?

FIGURE 13–28

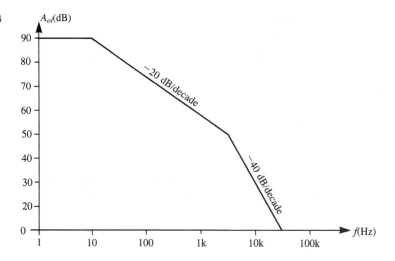

PROBLEMS

Section 13–1

13–1 The midrange open-loop gain of a certain op-amp is 120 dB. Negative feedback reduces this gain by 50 dB. What is the closed-loop gain?

13–2 The upper cutoff frequency of an op-amp's open-loop response is 200 Hz. If the midrange gain is 175,000, what is the ideal gain at 200 Hz? What is the actual gain? What is the op-amp's open-loop bandwidth?

13–3 An RC lag network has a critical frequency of 5 kHz. If the resistance value is 1 kΩ, what is X_C when $f = 3$ kHz?

13–4 Determine the attenuation of an RC lag network with $f_c = 12$ kHz for each of the following frequencies.
(a) 1 kHz
(b) 5 kHz
(c) 12 kHz
(d) 20 kHz
(e) 100 kHz

13–5 The midrange open-loop gain of a certain op-amp is 80,000. If the open-loop critical frequency is 1 kHz, what is the open-loop gain at each of the following frequencies?
(a) 100 Hz
(b) 1 kHz
(c) 10 kHz
(d) 1 MHz

13–6 Determine the phase shift through each network in Figure 13–29 at a frequency of 2 kHz.

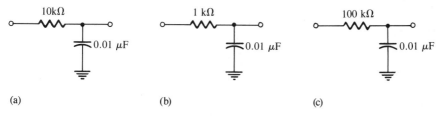

(a)   (b)   (c)

FIGURE 13–29

13–7 An RC lag network has a critical frequency of 8.5 kHz. Determine the phase for each frequency and plot a graph of its phase angle versus frequency.
(a) 100 Hz
(b) 400 Hz
(c) 850 Hz
(d) 8.5 kHz
(e) 25 kHz
(f) 85 kHz

Section 13–2

13–8 A certain op-amp has three internal amplifier stages with midrange gains of 30 dB, 40 dB, and 20 dB. Each stage also has a critical frequency associated with it as follows: f_{c_1} = 600 Hz, f_{c_2} = 50 kHz, and f_{c_3} = 200 kHz.
(a) What is the midrange open-loop gain of the op-amp, expressed in dB?
(b) What is the total phase shift through the amplifier, including inversion, when the signal frequency is 10 kHz?

13–9 What is the gain roll-off rate in problem 13–8 between the following frequencies?
(a) 0 Hz and 600 Hz
(b) 600 Hz and 50 kHz
(c) 50 kHz and 200 kHz
(d) 200 kHz and 1 MHz

Section 13–3

13–10 Determine the midrange gain in dB of each amplifier in Figure 13–30. Are these open- or closed-loop gains?

13–11 A certain amplifier has an open-loop gain in midrange of 180,000 and an open-loop cutoff frequency of 1500 Hz. If the attenuation of the feedback path is 0.015, what is the closed-loop bandwidth?

13–12 Given that $f_{c(ol)}$ = 750 Hz, A_{ol} = 89 dB, and $f_{c(cl)}$ = 5.5 kHz, determine the closed-loop gain in dB.

13–13 What is the unity-gain bandwidth in problem 13–12?

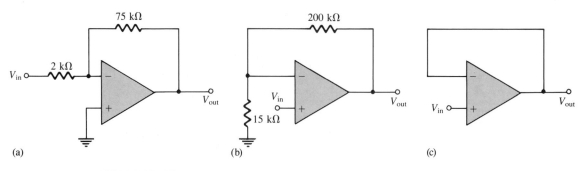

FIGURE 13–30

13–14 For each amplifier in Figure 13–31, determine the closed-loop gain and band-width. The op-amps in each circuit exhibit an open-loop gain of 125 dB and a unity-gain bandwidth of 2.8 MHz.

FIGURE 13–31

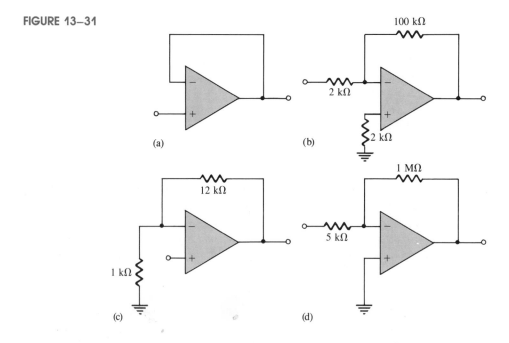

13–15 Which of the amplifiers in Figure 13–32 has the smaller bandwidth?

Section 13–4

13–16 It has been determined that the op-amp circuit in Figure 13–33 has three internal critical frequencies as follows: 1.2 kHz, 50 kHz, 250 kHz. If the midrange open-loop gain is 100 dB, is the amplifier configuration stable, marginally stable, or unstable?

FIGURE 14–33

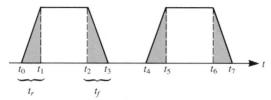

Solution

For simplicity, the rising and falling edges of the pulse are represented with constant slopes (ramps).

During the pulse rise-time, the derivative is a positive constant equal to the slope (rate of change). The same is true during the fall-time, except the constant is negative. Between the rising and falling edges, the pulse has a constant amplitude whose rate of change is, of course, 0. The derivative is therefore 0. Figure 14–34 shows the graph of the derivative of the pulse.

FIGURE 14–34

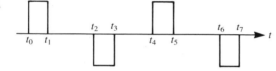

Operation of the Op-Amp Differentiator

First consider the op-amp differentiator's performance with the application of a triangular input as shown in Figure 14–35.

Again, remember that the inverting input is at virtual ground (0 V). When the triangular voltage is applied to the input, the voltage across the capacitor increases at a constant rate. Recall from circuit theory that the current in a capacitor is proportional to the rate of change of the voltage across the capacitor. Since the rate of change of the voltage ramp is constant, the current is also constant and equal to

$$I = \frac{CV_{pp}}{T/2} \qquad \textbf{(14–10)}$$

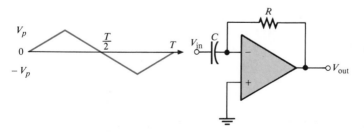

FIGURE 14–35
Differentiator with triangular input voltage.

Also, since the output voltage equals the voltage across R, and since I is constant, the output voltage is constant and equal to

$$V_{\text{out}} = \pm RC\left(\frac{V_{pp}}{T/2}\right) \tag{14-11}$$

A negative output with an amplitude given by equation (14–11) occurs during the positive slope of the triangular input due to the inversion of this op-amp. Likewise, a positive output occurs during the negative slope of the input. V_{pp} is the peak-to-peak amplitude of the input, and T is its period.

The waveform diagrams of the input and output for the circuit in Figure 14–35 are shown in Figure 14–36. Other input waveforms are also possible, of course, but the triangular input has been used to illustrate the basic operation.

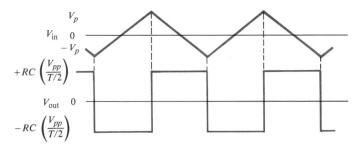

FIGURE 14–36
Input and output waveforms for the differentiator in Figure 14–35.

SECTION REVIEW 14–3
1. Sketch the circuit diagram for an op-amp integrator.
2. An integrator produces a _____ output in response to a step input.
3. A differentiator produces a _____ output in response to a ramp input.

INSTRUMENTATION AMPLIFIER

14–4
A simplified version of an instrumentation amplifier is shown in Figure 14–37. It consists of three operational amplifiers and several resistors. Integrated circuit manufacturers provide this circuitry on a single chip and package it as one device. Common characteristics are high input impedance (typically 300 MΩ), high voltage gain, and excellent CMRR (typically in excess of 100 dB).

Basic Operation

Op-amps A_1 and A_2 are noninverting amplifier stages that provide high input impedance and voltage gain. Op-amp A_3 is a unity-gain amplifier. When R_G is connected externally, as illustrated in Figure 14–38, op-amp A_1 receives the differential input signal V_{in_1} on its noninverting input and amplifies it with a gain of $1 + R_{f_1}/R_G$. Op-amp A_1 also receives the

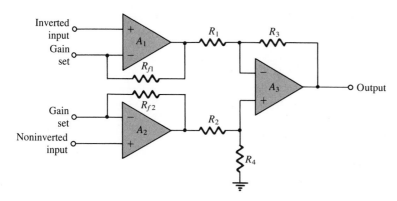

FIGURE 14–37
Basic instrumentation amplifier.

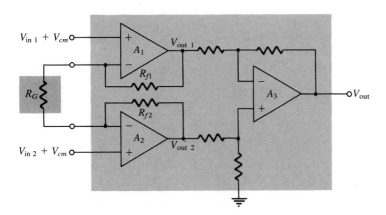

FIGURE 14–38
Instrumentation amplifier with gain-setting resistor R_G connected.

input signal V_{in_2} through op-amp A_2, R_{f_2}, and R_G. V_{in_2} appears on the inverting input of op-amp A_1 and is amplified by a gain of R_f/R_G. Also, the common-mode voltage on the noninverting input is amplified by the common-mode gain of A_1. (A_{cm} is typically less than unity.) The total output voltage of op-amp A_1 is as follows.

$$V_{\text{out}_1} = \left(1 + \frac{R_{f_1}}{R_G}\right)V_{\text{in}_1} - \left(\frac{R_{f_1}}{R_G}\right)V_{\text{in}_2} + V_{cm} \qquad \textbf{(14–12)}$$

A similar analysis can be applied to op-amp A_2, resulting in the following expression.

$$V_{\text{out}_2} = \left(1 + \frac{R_{f_2}}{R_G}\right)V_{\text{in}_2} - \left(\frac{R_{f_2}}{R_G}\right)V_{\text{in}_1} + V_{cm} \qquad \textbf{(14–13)}$$

The differential input voltage to op-amp A_3 is $V_{out_2} - V_{out_1}$.

$$V_{out_2} - V_{out_1} = \left(1 + \frac{R_{f_2}}{R_G} + \frac{R_{f_1}}{R_G}\right)V_{in_2}$$

$$- \left(\frac{R_{f_2}}{R_G} + 1 + \frac{R_{f_1}}{R_G}\right)V_{in_1} + V_{cm} - V_{cm}$$

For $R_{f_1} = R_{f_2} = R_f$:

$$V_{out_2} - V_{out_1} = \left(1 + \frac{2R_f}{R_G}\right)V_{in_2} - \left(1 + \frac{2R_f}{R_G}\right)V_{in_1} + V_{cm} - V_{cm}$$

Notice that, since the common-mode voltages (V_{cm}) are equal, they *cancel out*. Factoring gives the following result which is the differential input to op-amp A_3.

$$V_{out_2} - V_{out_1} = \left(1 + \frac{2R_f}{R_G}\right)(V_{in_2} - V_{in_1})$$

Because op-amp A_3 has unity gain, the final output of the instrumentation amplifier is $V_{out} = (1)(V_{out_2} - V_{out_1})$.

$$V_{out} = \left(1 + \frac{2R_f}{R_G}\right)(V_{in_2} - V_{in_1}) \qquad \textbf{(14–14)}$$

The closed-loop gain is

$$A_{cl} = 1 + \frac{2R_f}{R_G} \qquad \textbf{(14–15)}$$

Equation (14–15) shows that the differential gain of the instrumentation amplifier can be set by the value of R_G. R_{f_1} and R_{f_2} are normally internal to the IC chip, and their value is set by the manufacturer. As an example, for the LH0036, $R_{f_1} = R_{f_2} = 25$ kΩ.

Rearranging equation (14–15) gives an expression for calculating the value of R_G needed for a desired value of closed-loop gain if R_f is known.

$$A_{cl} = \frac{R_G + 2R_f}{R_G}$$

$$A_{cl}R_G = R_G + 2R_f$$

$$A_{cl}R_G - R_G = 2R_f$$

$$R_G(A_{cl} - 1) = 2R_f$$

$$R_G = \frac{2R_f}{A_{cl} - 1} \qquad \textbf{(14–16)}$$

EXAMPLE 14–11

Determine the value of the external gain-setting resistor R_G for an IC instrumentation amplifier having an $R_{f_1} = R_{f_2} = 25$ kΩ for a closed-loop gain of 500.

Solution

$$R_G = \frac{2R_f}{A_{cl} - 1} = \frac{50 \text{ k}\Omega}{500 - 1} \cong 100 \ \Omega$$

Practice Exercise 14–11

What value of external gain-setting resistor is required for an instrumentation amplifier having $R_{f_1} = R_{f_2} = 39$ kΩ for a closed-loop gain of 275?

Applications

The instrumentation amplifier is normally used to measure small differential signal voltages that are superimposed on a common-mode voltage often larger than the signal voltage. Applications often include a situation where a quantity is measured by a remote sensing device (transducer), and the resulting small electrical signal is sent over a long line subject to large common-mode voltages. The instrumentation amplifier at the end of the line must amplify the small signal and reject the large common-mode voltage. Figure 14–39 illustrates this.

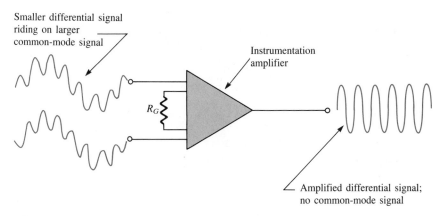

Smaller differential signal riding on larger common-mode signal

R_G

Instrumentation amplifier

Amplified differential signal; no common-mode signal

FIGURE 14–39
Instrumentation amplifier showing rejection of large common-mode signal.

1. What makes up the internal circuitry of a basic instrumentation amplifier?
2. Determine the value of the gain-setting resistor for a closed-loop gain of 800. The instrumentation amplifier has a specified $R_f = 15$ kΩ.

SECTION REVIEW 14–4

MORE OP-AMP APPLICATIONS

14–5

This section introduces a few more op-amp applications. These and the ones previously covered represent some basic uses of op-amps. The following three chapters continue the coverage of op-amp applications with a study of oscillators, filters, and voltage regulators.

Constant Current Source

The purpose of a constant-current source is to deliver a load current that remains constant when the load resistance changes. Figure 14–40 shows a basic circuit in which a stable voltage source (V_{IN}) provides a constant current (I_i) through the input resistor (R_i). Since the inverting ($-$) input of the op-amp is at virtual ground (0 V), the value of I_i is determined by V_{IN} and R_i as

$$I_i = \frac{V_{IN}}{R_i}$$

Now, since the internal input impedance of the op-amp is extremely high (ideally infinite), practically all of I_i flows through R_L, which is connected in the feedback path. Since $I_i = I_L$,

$$I_L = \frac{V_{IN}}{R_i} \tag{14–17}$$

If R_L changes, I_L remains constant as long as V_{IN} and R_i are held constant.

FIGURE 14–40
A basic constant-current source.

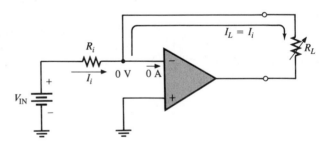

Current-to-Voltage Converter

The purpose of a current-to-voltage converter is to convert a variable input current to a proportional output voltage. A basic circuit that accomplishes this is shown in Figure 14–41(a). Since practically all of I_i flows through the feedback path, the voltage dropped across R_f is $I_i R_f$. Because the left side of R_f is at virtual ground (0 V), the output voltage equals the voltage across R_f, which is proportional to I_i.

$$V_{out} = I_i R_f \tag{14–18}$$

A specific application of this circuit is illustrated in Figure 14–41(b), where a photoconductive cell is used to sense changes in light level. As the amount of light

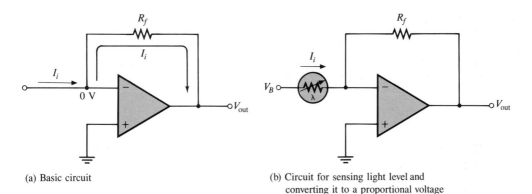

(a) Basic circuit

(b) Circuit for sensing light level and converting it to a proportional voltage

FIGURE 14–41
Current-to-voltage converter.

changes, the current through the photoconductive cell varies because of the cell's change in resistance. This change in resistance produces a proportional change in the output voltage ($\Delta V_{OUT} = \Delta I_i R_f$).

Voltage-to-Current Converter

A basic voltage-to-current converter is shown in Figure 14–42. This circuit is used in applications where it is necessary to have an output (load) current that is controlled by an input voltage.

Neglecting the input offset voltage, both inverting and noninverting input terminals of the op-amp are at the same voltage, V_{in}. Therefore, the voltage across R_1 equals V_{in}. Since negligible current flows into the inverting input, the same current that flows through R_1 also flows through R_L; thus

$$I_L = \frac{V_{in}}{R_1} \qquad (14-19)$$

FIGURE 14–42
Voltage-to-current converter.

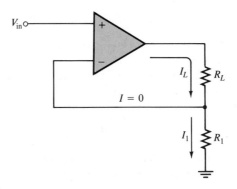

Peak Detector

An interesting application of the op-amp is in a peak detector circuit such as the one shown in Figure 14–43. In this case the op-amp is used as a comparator. The purpose of this circuit is to detect the peak of the input voltage and store that peak voltage on a capacitor. For example, this circuit can be used to detect and store the maximum value of a voltage surge; this value can then be measured at the output with a voltmeter or recording device. The basic operation is as follows. When a positive voltage is applied to the noninverting input of the op-amp through R_i, the high-level output voltage of the op-amp forward-biases the diode and charges the capacitor. The capacitor continues to charge until its voltage reaches a value equal to the input voltage and thus both op-amp inputs are at the same voltage. At this point, the op-amp comparator switches, and its output goes to the low level. The diode is now reverse-biased, and the capacitor stops charging. It has reached a voltage equal to the peak of V_{in} and will hold this voltage until the charge eventually leaks off. If a greater input peak occurs, the capacitor charges to the new peak.

FIGURE 14–43
A basic peak detector.

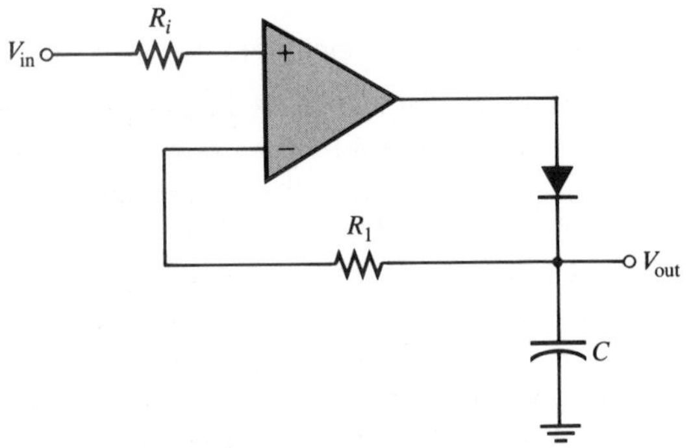

<table>
<tr><td>

SECTION
REVIEW
14–5

</td><td>

1. For the constant-current source in Figure 14–40, the input reference voltage is 6.8 V and R_i is 10 kΩ. What value of constant current does the circuit supply to a 1 kΩ load? To a 5 kΩ load?

2. What element determines the constant of proportionality that relates input current to output voltage in the current-to-voltage converter?

</td></tr>
</table>

TROUBLESHOOTING

14–6

Although integrated circuit op-amps are extremely reliable and trouble-free, failures do occur from time to time. One type of failure mode is a condition where the output is in a saturated state resulting in a constant high or constant low level, regardless of the input. Figure 14–44 illustrates this failure with a comparator circuit. External component fail-

Comparators

$(14–1) \quad V_{\text{REF}} = \dfrac{R_2}{R_1 + R_2}(+V)$ Comparator reference

$(14–2) \quad V_{\text{UTP}} = \dfrac{R_2}{R_1 + R_2}[+V_{\text{out(max)}}]$ Upper trigger point

$(14–3) \quad V_{\text{LTP}} = \dfrac{R_2}{R_1 + R_2}[-V_{\text{out(max)}}]$ Lower trigger point

$(14–4) \quad V_{\text{HYS}} = V_{\text{UTP}} - V_{\text{LTP}}$ Hysteresis voltage

Summing amplifier

$(14–5) \quad V_{\text{OUT}} = -(V_{\text{IN}_1} + V_{\text{IN}_2})$ Two-input adder

$(14–6) \quad V_{\text{OUT}} = -(V_{\text{IN}_1} + V_{\text{IN}_2} + \cdots + V_{\text{IN}_n})$ n-input adder

$(14–7) \quad V_{\text{OUT}} = -\dfrac{R_f}{R}(V_{\text{IN}_1} + V_{\text{IN}_2} + \cdots + V_{\text{IN}_n})$ Adder with gain

$(14–8) \quad V_{\text{OUT}} = -\left(\dfrac{R_f}{R_1}V_{\text{IN}_1} + \dfrac{R_f}{R_2}V_{\text{IN}_2} + \cdots + \dfrac{R_f}{R_n}V_{\text{IN}_n}\right)$ Adder with gain

Integrator and differentiator

$(14–9) \quad \dfrac{\Delta V_{\text{out}}}{\Delta t} = -\dfrac{V_{\text{IN}}}{RC}$ Integrator, rate of change

$(14–10) \quad I = \dfrac{CV_{pp}}{T/2}$ Differentiator current with triangular input

$(14–11) \quad V_{\text{out}} = \pm RC\left(\dfrac{V_{pp}}{T/2}\right)$ Differentiator output with triangular input

Instrumentation amplifier

$(14–12) \quad V_{\text{out}_1} = \left(1 + \dfrac{R_{f_1}}{R_G}\right)V_{\text{in}_1} - \left(\dfrac{R_{f_1}}{R_G}\right)V_{\text{in}_2} + V_{cm}$ Instrumentation amplifier

$(14–13) \quad V_{\text{out}_2} = \left(1 + \dfrac{R_{f_2}}{R_G}\right)V_{\text{in}_2} - \left(\dfrac{R_{f_2}}{R_G}\right)V_{\text{in}_1} + V_{cm}$ Instrumentation amplifier

(14–14) $V_{\text{out}} = \left(1 + \dfrac{2R_f}{R_G}\right)(V_{\text{in}_2} - V_{\text{in}_1})$ Instrumentation amplifier

(14–15) $A_{cl} = 1 + \dfrac{2R_f}{R_G}$ Instrumentation amplifier, closed-loop gain

(14–16) $R_G = \dfrac{2R_f}{A_{cl} - 1}$ Gain-setting resistor

Miscellaneous

(14–17) $I_L = \dfrac{V_{\text{IN}}}{R_i}$ Constant-current source

(14–18) $V_{\text{out}} = I_i R_f$ Current-to-voltage converter

(14–19) $I_L = \dfrac{V_{\text{in}}}{R_1}$ Voltage-to-current converter

---

SUMMARY

- [ ] In an op-amp comparator, when the input voltage exceeds a specified reference voltage, the output changes state.
- [ ] Hysteresis gives an op-amp noise immunity.
- [ ] A comparator switches to one state when the input reaches the UTP and back to the other state when the input drops below the LTP.
- [ ] The difference between the UTP and the LTP is the hysteresis voltage.
- [ ] Bounding limits the output amplitude of a comparator.
- [ ] The output voltage of a summing amplifier is proportional to the *sum* of the input voltages.
- [ ] An *averaging amplifier* is a summing amplifier with a closed-loop gain equal to the reciprocal of the number of inputs.
- [ ] In a *scaling adder,* a different weight can be assigned to each input, thus making the input contribute more or contribute less to the output.
- [ ] Integration is a mathematical process for determining the area under a curve.
- [ ] The integral of a step is a ramp.
- [ ] Differentiation is a mathematical process for determining the rate of change of a function.
- [ ] The derivative of a ramp is a step.
- [ ] A basic instrumentation amplifier consists of three op-amps.
- [ ] An instrumentation amplifier possesses very high input impedance, high CMRR, and high voltage gain adjustable by an external resistor.

SELF-TEST

1. Determine the output level (maximum positive or maximum negative) for each comparator in Figure 14–49.
2. Calculate the UTP and LTP in Figure 14–50. $V_{\text{out(max)}} = \pm 10$ V.
3. What is the hysteresis voltage in Figure 14–50?

FIGURE 14–49

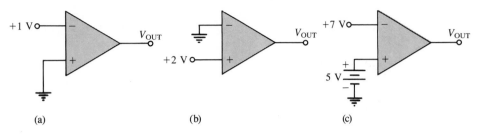

(a)   (b)   (c)

FIGURE 14–50

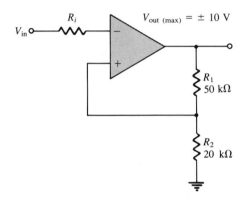

4. A 6.2 V zener diode is connected from the output to the inverting input in Figure 14–50 with the cathode at the output. What are the positive and negative output levels?

5. Determine the output voltage for each circuit in Figure 14–51.

FIGURE 14–51

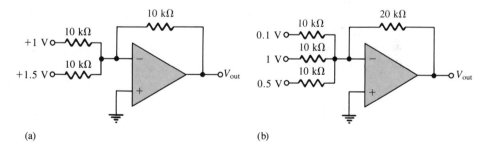

(a)   (b)

6. Determine the gain for each input in Figure 14–52.

FIGURE 14–52

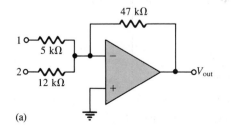

(a)

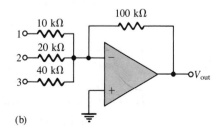

(b)

7. Plot the points representing the area under the step voltage curve in Figure 14–53 at each interval; then connect the points and define the resulting graph in relationship to the step.

FIGURE 14–53

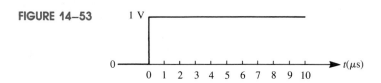

8. A voltage is plotted on a graph. Its rate of change at a certain point is 2.50 V/s. What is the derivative at this same point?

9. Identify each circuit in Figure 14–54.

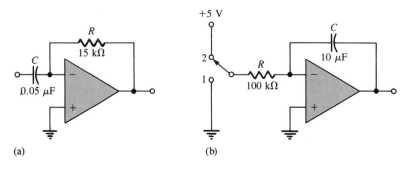

(a)   (b)

FIGURE 14–54

10. A triangular waveform with a peak-to-peak voltage of 2 V and a period of 1 ms is applied to the differentiator in Figure 14–54. What is the output voltage?

11. Beginning in position 1 in Figure 14–54(b), the switch is thrown into position 2 and held there for 10 ms, then back to position 1 for 10 ms, and so forth. Sketch the resulting output waveform. The saturated output levels of the op-amp are ±12 V.

PROBLEMS   **Section 14–1**

14–1 A certain op-amp has an open-loop gain of 80,000. The maximum saturated output levels of this particular device are ±12 V when the dc supply voltages are ±15 V. If a differential voltage of 0.15 mV rms is applied between the inputs, what is the peak-to-peak value of the output?

14–2 What is the main problem with an op-amp comparator without hysteresis?

14–3 Sketch the output voltage waveform for each circuit in Figure 14–55 with respect to the input. Show voltage levels.

14–4 Determine the hysteresis voltage for each comparator in Figure 14–56. The maximum output levels are ±11 V.

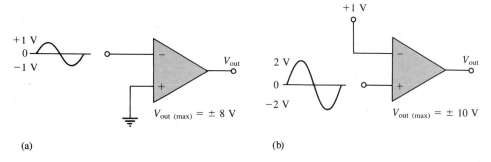

(a)   (b)

FIGURE 14–55

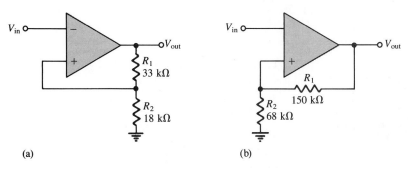

(a)   (b)

FIGURE 14–56

14–5 Determine the output voltage waveform in Figure 14–57.

FIGURE 14–57

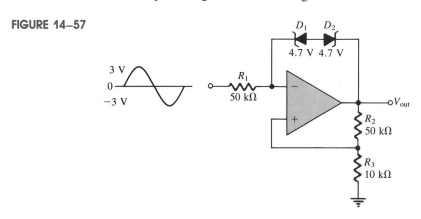

Section 14–2

14–6 Refer to Figure 14–58. Determine the following:
   (a) V_{R_1} and V_{R_2}
   (b) Current through R_f
   (c) V_{OUT}

FIGURE 14–58

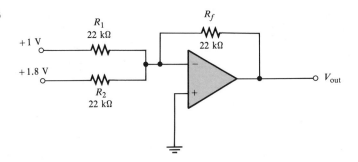

14–7 Find the value of R_f necessary to produce an output that is five times the sum of the inputs in Figure 14–58.

14–8 Design a summing amplifier that will average eight input voltages. Use input resistances of 10 kΩ each.

14–9 Find the output voltage when the input voltages shown in Figure 14–59 are applied to the scaling adder. What is the current through R_f?

FIGURE 14–59

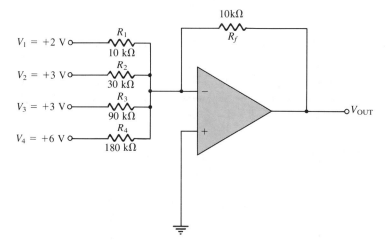

14–10 Determine the values of the input resistors required in a six-input scaling adder so that the lowest weighted input is 1 and each successive input has a weight *twice* the previous one. Use $R_f = 100$ kΩ.

Section 14–3

14–11 Determine the rate of change of the output voltage in response to the step input to the integrator in Figure 14–60.

14–12 A triangular waveform is applied to the input of the circuit in Figure 14–61 as shown. Determine what the output should be and sketch its waveform in relation to the input.

FIGURE 14–60

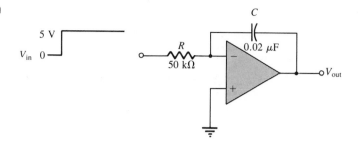

FIGURE 14–61

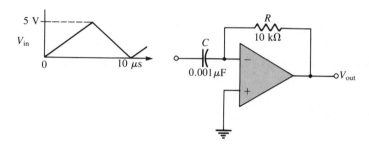

14–13 What is the magnitude of the capacitor current in problem 14–12?

Section 14–4

14–14 Refer to the graph in Figure 14–62, which shows closed-loop gain versus gain-setting resistor for a certain instrumentation amplifier. Determine the value of R_G for a gain of 200.

FIGURE 14–62

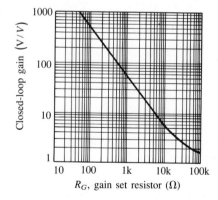

14–15 An instrumentation amplifier is shown in Figure 14–63. If the internal R_f is 25 kΩ, what is the rms output voltage?

14–16 What is the closed-loop voltage gain of the circuit in Figure 14–63?

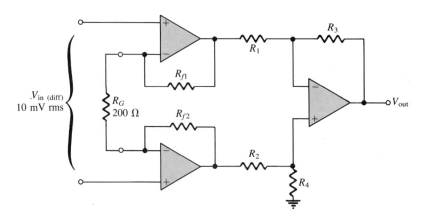

FIGURE 14–63

Section 14–5

14–17 Determine the load current in each circuit of Figure 14–64.

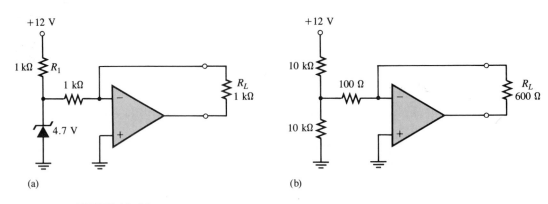

(a)   (b)

FIGURE 14–64

14–18 Devise a circuit for remotely sensing temperature and producing a proportional voltage that can then be converted to digital form for display. A thermistor can be used as the temperature-sensing element.

Section 14–6

14–19 The waveforms given in Figure 14–65(a) are observed at the indicated points in Figure 14–65(b). Is the circuit operating properly? If not, what is a likely fault?

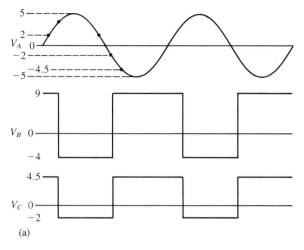

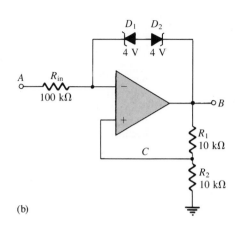

(a)   (b)

FIGURE 14–65

Section 14–1

1. 1.36 V, −3.43 V.

2. True.

3. Bounding limits the amplitude to a specified level.

Section 14–2

1. The terminal of the op-amp where the input resistors are commonly connected.

2. ⅕.

3. 20 kΩ.

Section 14–3

1. See Figure 14–66.

FIGURE 14–66

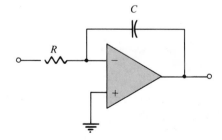

2. Ramp.

3. Step.

Section 14–4

1. Three op-amps and several resistors.

2. $37.5 \; \Omega$

Section 14–5

1. 0.68 mA, 0.68 mA

Section 14–6

1. Stays at one output level.

2. Replace suspected components one at a time.

ANSWERS TO PRACTICE EXERCISES

Example 14–1: 1.96 V

Example 14–2: ± 3.83 V

Example 14–3: The output changes when V_{in} reaches ± 1.23 V. The two output voltage levels are ± 4 V.

Example 14–5: -5.73 V

Example 14–7: Weights: 0.45, 0.12, 0.18; $V_{OUT} = 3.03$ V

Example 14–11: 285 Ω

Oscillators

In this chapter you will learn

☐ What an oscillator is and how it works
☐ The basic principles of positive feedback
☐ The basic operation of various types of sinusoidal oscillators that use RC networks in the feedback loop, including Wien-bridge, phase-shift, and twin-T
☐ How to assure that an oscillator will start up properly
☐ The basic operation of various types of sinusoidal oscillators that use LC networks and crystals in the feedback loop, including Colpitts, Clapp, Hartley, and Armstrong
☐ The basic operation of various types of oscillators that produce square, triangular, and sawtooth waveforms
☐ How a voltage-controlled oscillator (VCO) works
☐ How to use the 555 timer IC in oscillator applications
☐ How a phase-locked loop works
☐ How oscillators and phase-locked loops are used in a system application

Chapter 15

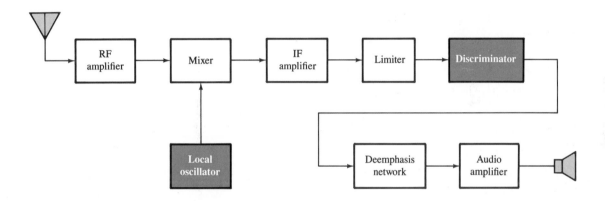

Oscillators are circuits that generate an output signal without an input signal. Different types of oscillators produce sine waves, pulses, square waves, triangular waves, and sawtooth waves. This chapter introduces several types of basic oscillator circuits, both discrete and integrated. Oscillators are based on the principle of *positive feedback,* where a portion of the output signal is fed back to the input in a way that causes it to reinforce itself and thus sustain a continuous output signal.

Oscillators are widely applied in most communications systems as well as in digital systems to generate required frequencies and timing signals. As you have already seen, oscillators are used in superheterodyne receivers to produce a variable frequency that is mixed with the RF to obtain the IF. An oscillator is also used in the discriminator section of some FM receivers as part of a *phase-locked loop,* which extracts the audio from the frequency-modulated IF signal.

DEFINITION OF AN OSCILLATOR

15–1

An oscillator is a circuit that produces a repetitive waveform on its output with only the dc supply voltage as an input. A repetitive input signal is not required. The output voltage can be either sinusoidal or nonsinusoidal, depending on the type of oscillator. This basic concept is illustrated in Figure 15–1.

Essentially, an oscillator converts electrical energy in the form of dc to electrical energy in the form of ac. A basic oscillator consists of an amplifier for gain (either discrete transistor or op-amp) and a *positive feedback* circuit that produces phase shift and provides attenuation, as shown in Figure 15–2.

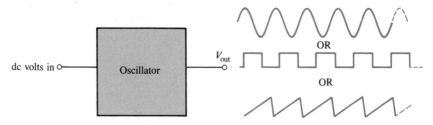

FIGURE 15–1
The basic oscillator concept showing three possible types of output waveforms.

FIGURE 15–2
Basic elements of an oscillator.

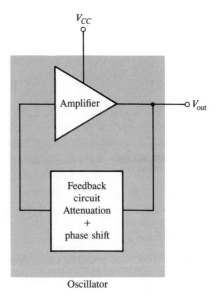

SECTION REVIEW 15–1

1. What is an oscillator?

2. What type of feedback does an oscillator use?

OSCILLATOR PRINCIPLES

15–2

With the exception of the relaxation oscillator, which we will talk about later in this chapter, oscillator operation is based on the principle of *positive feedback*. We will examine this concept and look at the general conditions required for oscillation.

Positive Feedback

Positive feedback is characterized by the condition wherein a portion of the output voltage of an amplifier is fed back to the input with no *net* phase shift, resulting in a reinforcement of the output signal. This basic idea is illustrated in Figure 15–3. As you can see, the in-phase feedback voltage is amplified to produce the output voltage, which in turn produces the feedback voltage. That is, a *loop* is created in which the signal sustains itself and a continuous sine wave output is produced. This phenomenon is called *oscillation*.

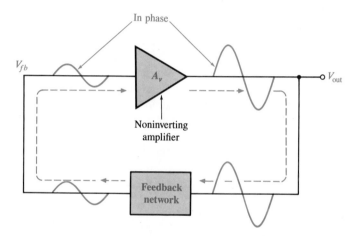

FIGURE 15–3
Positive feedback produces oscillation.

Conditions for Oscillation

Two conditions are required for a sustained state of oscillation:

1. The phase shift around the feedback loop must be 0°
2. The voltage gain around the closed feedback loop must equal 1 (unity)

The voltage gain around the closed feedback loop (A_{cl}) is the product of the amplifier gain (A_v) and the attenuation of the feedback circuit (B).

$$A_{cl} = A_v B$$

For example, if the amplifier has a gain of 100, the feedback circuit must have an attenuation of 0.01 to make the loop gain equal to 1 ($A_v B = 100 \times 0.01 = 1$). These conditions for oscillation are illustrated in Figure 15–4.

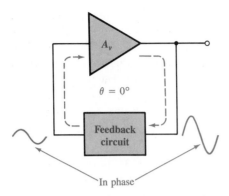

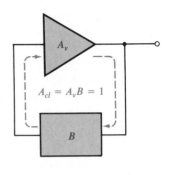

(a) The phase shift around the loop is 0°   (b) The closed loop gain is 1

FIGURE 15–4
Conditions for oscillation.

Start-Up Conditions

So far, we have seen what it takes for an oscillator to produce a continuous sine wave output. Now we examine the requirements for the oscillation to start when the dc supply voltage is turned on. As you have seen, the unity-gain condition must be met for oscillation to be *sustained*. For oscillation to *begin*, the voltage gain around the positive feedback loop must be *greater than 1* so that the amplitude of the output can build up to a desired level. The gain must then decrease to 1 so that the output stays at the desired level. (Several ways to achieve this reduction in gain after start-up are discussed later.) The conditions for both starting and sustaining oscillation are illustrated in Figure 15–5.

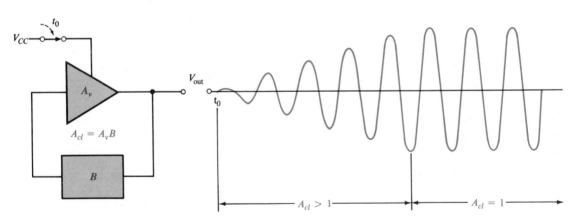

FIGURE 15–5
When oscillation starts at t_0, the condition $A_{cl} > 1$ causes the sinusoidal output voltage amplitude to build up to a desired level, where A_{cl} decreases to 1 and maintains the desired amplitude.

A question that normally arises is this: If the oscillator is off (no dc voltage) and there is no output voltage, how does a feedback signal originate to start the positive feedback build-up process? Initially, a small positive feedback voltage develops from thermally produced broad-band noise in the resistors or other components or from turn-on transients. The feedback circuit permits only a voltage with a frequency equal to the selected oscillation frequency to appear in-phase on the amplifier's input. This initial feedback voltage is amplified and continually reinforced, resulting in a build-up of the output voltage as previously discussed.

1. Discuss the conditions required for a circuit to oscillate.
2. Define positive feedback.

SECTION
REVIEW
15–2

OSCILLATORS WITH RC FEEDBACK CIRCUITS

This section introduces three types of oscillator circuits that produce sinusoidal outputs: the *Wien-bridge oscillator, the phase-shift oscillator,* and the *twin-T oscillator.* Generally, RC oscillators are used for frequencies up to about 1 MHz. The Wien-bridge is by far the most widely used type of RC oscillator for this range of frequencies.

15–3

The Wien-Bridge Oscillator

One type of sine wave oscillator is the *Wien-bridge* oscillator. A fundamental part of the Wien-bridge oscillator is a *lead-lag* network like that shown in Figure 15–6(a). R_1 and C_1 together form the *lag* portion of the network; R_2 and C_2 form the *lead* portion. The operation of this circuit is as follows. At lower frequencies, the lead network dominates due to the high reactance of C_2. As the frequency increases, X_{C_2} decreases, thus allowing the output voltage to increase. At some specified frequency, the response of the lag network takes over, and the decreasing value of X_{C_1} causes the output voltage to decrease.

So, we have a response curve like that shown in Figure 15–6(b) where the output voltage peaks at a frequency f_r. At this point, the attenuation of the network is $\frac{1}{3}$, as

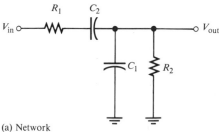

(a) Network

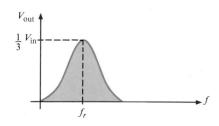

(b) Response curve

FIGURE 15–6
Lead-lag network.

shown in the following derivation ($R_1 = R_2 = R$, and $X_{C_1} = X_{C_2} = X$):

$$\frac{V_{out}}{V_{in}} = \frac{R(-jX)/(R-jX)}{(R-jX) + R(-jX)/(R-jX)}$$

$$= \frac{R(-jX)}{(R-jX)^2 - jRX}$$

Multiplying the numerator and denominator by j,

$$\frac{V_{out}}{V_{in}} = \frac{RX}{j(R-jX)^2 + RX}$$

$$= \frac{RX}{RX + j(R^2 - j2RX - X^2)}$$

$$= \frac{RX}{RX + jR^2 + 2RX - jX^2}$$

$$= \frac{RX}{3RX + j(R^2 - X^2)}$$

For a $0°$ phase angle there can be no j term. Recall from complex numbers in ac theory that a *nonzero* angle is associated with a complex number having a j term. Therefore, at f_r the j term is 0.

$$R^2 - X^2 = 0$$

Thus,

$$\frac{V_{out}}{V_{in}} = \frac{RX}{3RX}$$

Cancelling, we get

$$\frac{V_{out}}{V_{in}} = \frac{1}{3} \tag{15-1}$$

From the fact that $R^2 - X^2 = 0$, a formula for f_r can be developed.

$$R^2 = X^2$$

Therefore

$$R = X$$

$$R = \frac{1}{2\pi f_r C}$$

$$f_r = \frac{1}{2\pi RC} \tag{15-2}$$

To summarize, the lead-lag network has a *resonant frequency f_r* at which the *phase shift through the network is $0°$ and the attenuation is $\frac{1}{3}$.* Below f_r, the lead network dominates

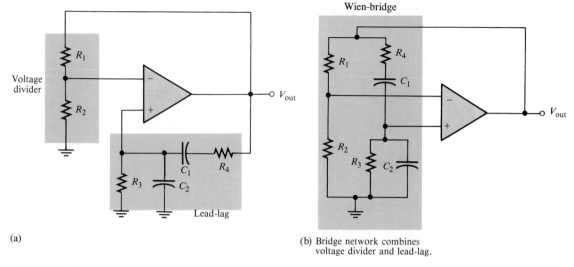

FIGURE 15–7
Two ways to draw the schematic of a Wien-bridge oscillator.

and the output leads the input. Above f_r, the lag network dominates and the output lags the input.

The lead-lag network is used in the *positive* feedback loop of an op-amp, as shown in Figure 15–7(a). A voltage divider is used in the *negative* feedback loop.

Basic Circuit

This oscillator circuit can be viewed as a *noninverting* amplifier configuration with the input signal fed back from the output through the lead-lag network. Recall that the *closed-loop* gain of the amplifier is determined by the voltage divider.

$$A_{cl} = \frac{1}{B} = \frac{1}{R_2/(R_1 + R_2)} = \frac{R_1 + R_2}{R_2}$$

The circuit is redrawn in Figure 15–7(b) to show that the op-amp is connected *across* the Wien-bridge. One leg of the bridge is the lead-lag network, and the other is the voltage divider.

The Positive Feedback Conditions for Oscillation

As you know, for the circuit to produce a sustained sine wave output (oscillate), the phase shift around the positive feedback loop must be 0° and the gain around the loop must be at least unity (1). The 0° phase shift condition is met when the frequency is f_r, because the phase shift through the lead-lag network is 0° and there is no inversion from the noninverting input (+) of the op-amp to the output. This is shown in Figure 15–8(a).

The unity-gain condition in the feedback loop is met when

$$A_{cl} = 3$$

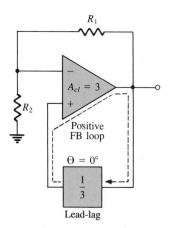
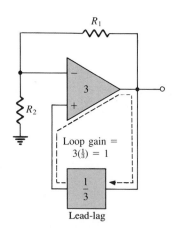

(a) The phase shift around the loop is 0°.   (b) The voltage gain around the loop is 1.

FIGURE 15–8
Conditions for oscillation.

This offsets the ⅓ attenuation of the lead-lag network, thus making the gain around the positive feedback loop equal to 1, as depicted in Figure 15–8(b). To achieve a closed-loop gain of 3,

$$R_1 = 2R_2$$

Then

$$A_{cl} = \frac{R_1 + R_2}{R_2} = \frac{2R_2 + R_2}{R_2} = \frac{3R_2}{R_2} = 3$$

Start-Up Conditions

Initially, the closed-loop gain of the amplifier must be more than 1 ($A_{cl} > 3$) until the output signal builds up to a desired level. The gain must then decrease to 1 so that the output signal stays at the desired level. This is illustrated in Figure 15–9.

The circuit in Figure 15–10 illustrates a basic method for achieving the condition described above. Notice that the voltage-divider network has been modified to include an additional resistor R_3 in parallel with a back-to-back zener diode arrangement. When dc power is first applied, both zener diodes appear as opens. This places R_3 in series with R_1, thus increasing the closed-loop gain as follows ($R_1 = 2R_2$).

$$A_{cl} = \frac{R_1 + R_2 + R_3}{R_2} = \frac{3R_2 + R_3}{R_2} = 3 + \frac{R_3}{R_2}$$

Initially, a small positive feedback signal develops from noise or turn-on transients. The lead-lag network permits only a signal with a frequency equal to f_r to appear *in-phase* on the noninverting input. This feedback signal is amplified and continually reinforced, resulting in a build-up of the output voltage. When the output signal reaches the zener

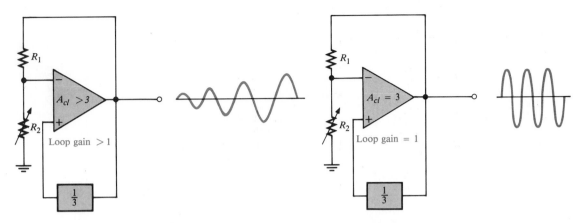

(a) Loop gain greater than 1 causes output to build up.

(b) Loop gain of 1 causes a sustained constant output.

FIGURE 15–9
Oscillator start-up conditions.

FIGURE 15–10
Self-starting Wien-bridge oscillator using
back-to-back zener diodes.

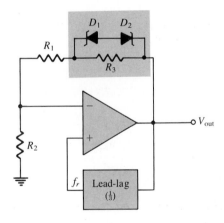

breakdown voltage, the zeners conduct and effectively short out R_3. This lowers the amplifier's closed-loop gain to 3. At this point the output signal levels off and the oscillation is sustained. (Incidentally, the frequency of oscillation can be adjusted by using gang-tuned capacitors in the lead-lag network.)

Another method sometimes used to insure self-starting employs a tungsten lamp in the voltage divider, as shown in Figure 15–11. When the power is first turned on, the resistance of the lamp is lower than its nominal value. This keeps the negative feedback small and makes the closed-loop gain of the amplifier greater than 3. As the output voltage builds up, the voltage across the tungsten lamp—and thus its current—increases. As a result, the lamp resistance increases until it reaches a value equal to one-half the feedback resistance. At this point the closed-loop gain is 3, and the output is sustained at a constant level.

FIGURE 15–11

Self-starting Wien-bridge oscillator using a tungsten lamp in the negative feedback loop.

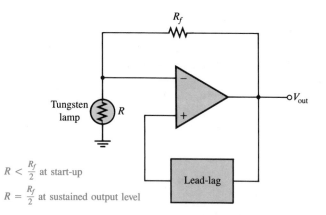

$R < \dfrac{R_f}{2}$ at start-up

$R = \dfrac{R_f}{2}$ at sustained output level

EXAMPLE 15–1

Determine the frequency of the Wien-bridge oscillator in Figure 15–12. Also verify that oscillations will start and then continue when the output signal reaches 5.4 V.

FIGURE 15–12

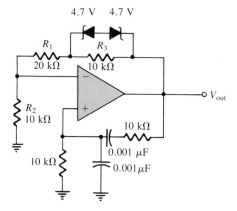

Solution

The frequency is

$$f_r = \frac{1}{2\pi RC} = \frac{1}{2\pi (10 \text{ k}\Omega)(0.001 \ \mu\text{F})} = 15.92 \text{ kHz}$$

Initially the closed-loop gain is

$$A_{cl} = \frac{R_1 + R_2 + R_3}{R_2} = \frac{40 \text{ k}\Omega}{10 \text{ k}\Omega} = 4$$

Since $A_{cl} > 3$, the start-up condition is met.

When the output reaches 5.4 V (4.7 V + 0.7 V), the zeners conduct (their forward resistance is assumed small, compared to 10 kΩ), and the closed-loop gain is reached. Thus oscillation is sustained.

$$A_{cl} = \frac{R_1 + R_2}{R_2} = \frac{30 \text{ k}\Omega}{10 \text{ k}\Omega} = 3$$

The Phase-Shift Oscillator

One type of *phase-shift* oscillator is shown in Figure 15–13. Each of the three RC networks in the feedback loop can provide a maximum phase shift of 90°. Oscillation occurs at the frequency where the total phase shift through the three RC networks is 180°. The inversion of the op-amp, itself, provides the additional 180° to meet the requirement for oscillation.

The attenuation B of the three-section RC feedback network is

$$B = \frac{1}{29} \tag{15–3}$$

The derivation of this unusual result is given in Appendix B. To meet the greater-than-unity loop gain requirement, the closed-loop voltage gain of the op-amp must be greater than 29 (set by R_f and R_i). The frequency of oscillation is also derived in Appendix B and stated in the following equation.

$$f_r = \frac{1}{2\pi\sqrt{6}RC} \tag{15–4}$$

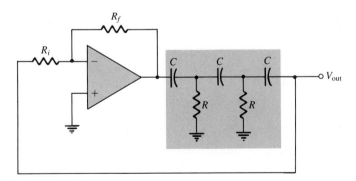

FIGURE 15–13
Op-amp phase-shift oscillator.

EXAMPLE 15–2

(a) For $R_i = 5\ k\Omega$, determine the value of R_f necessary for the circuit in Figure 15–14 to operate as an oscillator.

(b) Determine the frequency of oscillation.

Solution

(a) $A_{cl} = 29$, and $A_{cl} = \dfrac{R_f}{R_i}$

$$\frac{R_f}{R_i} = 29$$

$$R_f = 29R_i = 29(10\ k\Omega) = 290\ k\Omega$$

(b) $f_r = \dfrac{1}{2\pi\sqrt{6}RC} = \dfrac{1}{2\pi\sqrt{6}(10\ k\Omega)(0.001\ \mu F)} \cong 6.5\ kHz$

Practice Exercise 15–2

(a) If R_i in Figure 15–14 is changed to 8.2 $k\Omega$, what value must R_f be for oscillation?

(b) What is f_r?

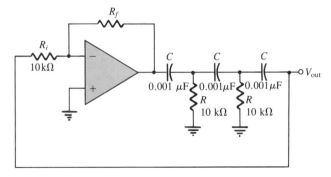

FIGURE 15–14

Twin-T Oscillator

Another type of RC oscillator is called the *twin-T* because of the two T-type RC filters used in the negative feedback loop, as shown in Figure 15–15(a). One of the twin-T filters has a low-pass response and the other has a high-pass response. The combined parallel filters produce a band-stop or notch response with a center frequency equal to the desired frequency of oscillation, f_r, as shown in Figure 15–15(b).

Oscillation cannot occur at frequencies above or below f_r because of the negative feedback through the filters. At f_r, however, there is negligible negative feedback and, thus, the positive feedback through the voltage divider (R_1 and R_2) allows the circuit to oscillate. Self-starting can be achieved by using a tungsten lamp in the place of R_1.

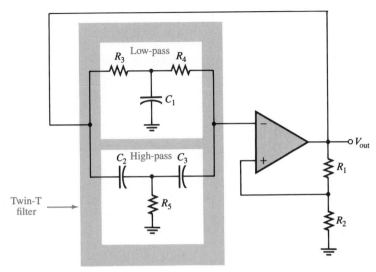

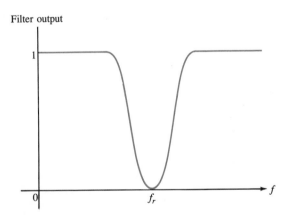

(a) Oscillator circuit

(b) Twin-T filter's frequency response curve

FIGURE 15–15
Twin-T oscillator and twin-T filter response.

**SECTION
REVIEW
15–3**

1. There are two feedback loops in the Wien-bridge oscillator. State the purpose of each.

2. A certain lead-lag network has $R_1 = R_2$ and $C_1 = C_2$. An input voltage of 5 V rms is applied. Its frequency equals the resonant frequency of the network. What is the rms output voltage?

3. Why must the phase shift through the RC feedback circuit in a phase-shift oscillator equal 180°?

OSCILLATORS WITH LC FEEDBACK CIRCUITS

15–4

Although the RC oscillators, particularly the Wien-bridge, are generally suitable for frequencies up to about 1 MHz, LC feedback elements are normally used in oscillators that require higher frequencies of oscillation. Also, because of the frequency limitation (lower unity-gain frequency) of most op-amps, discrete transistors are often employed as the gain element in LC oscillators. This section introduces several types of resonant LC oscillators: the *Colpitts, Clapp, Hartley, Armstrong,* and *crystal-controlled* oscillators.

The Colpitts Oscillator

One basic type of resonant circuit oscillator is the Colpitts, named after its inventor—as are most of the others we cover here. As shown in Figure 15–16, this type of oscillator uses an LC circuit in the feedback loop to provide the necessary phase shift and to act as a resonant filter that passes only the desired frequency of oscillation.

The approximate frequency of oscillation is the resonant frequency of the LC circuit and is established by the values of C_1, C_2, and L according to this familiar formula:

$$f_r \cong \frac{1}{2\pi\sqrt{LC_T}} \tag{15–5}$$

Because the capacitors effectively appear in series around the tank circuit, the total capacitance is

$$C_T = \frac{C_1 C_2}{C_1 + C_2} \tag{15–6}$$

Conditions for Oscillation and Start-Up. The attenuation, B, of the resonant feedback circuit in the Colpitts oscillator is basically determined by the values of C_1 and C_2.

FIGURE 15–16
A basic Colpitts oscillator.

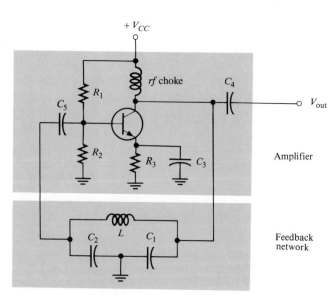

FIGURE 15–17
The attenuation of the tank circuit is the output of the tank (V_f) divided by the input to the tank (V_{out}). $B = V_f/V_{out} = C_1/C_2$. For $A_v B > 1$, A_v must be greater than C_2/C_1.

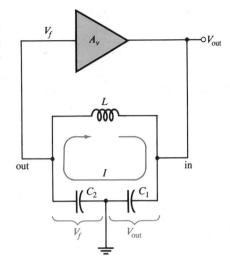

Figure 15–17 shows that the circulating tank current flows through both C_1 and C_2 (they are effectively in series). The voltage developed across C_1 is the oscillator's output voltage (V_{out}) and the voltage developed across C_2 is the feedback voltage (V_f), as indicated. The expression for the attenuation is

$$B = \frac{V_f}{V_{out}} \cong \frac{IX_{C2}}{IX_{C1}} = \frac{X_{C2}}{X_{C1}} = \frac{1/2\pi f_r C_2}{1/2\pi f_r C_1}$$

Cancelling the $2\pi f_r$ terms gives

$$B = \frac{C_1}{C_2} \qquad\qquad (15\text{–}7)$$

As you know, a condition for oscillation is $A_v B = 1$. Since $B = C_1/C_2$,

$$A_v = \frac{C_2}{C_1} \qquad\qquad (15\text{–}8)$$

where A_v is the voltage gain of the transistor amplifier. With this condition met, $A_v B = (C_2/C_1)(C_1/C_2) = 1$. Actually, for the oscillator to be self-starting, $A_v B$ must be greater than 1 ($A_v B > 1$). Therefore, the voltage gain must be made slightly greater than C_2/C_1.

$$A_v > \frac{C_2}{C_1} \qquad\qquad (15\text{–}9)$$

Loading the Feedback Circuit Affects the Frequency of Oscillation. As indicated in Figure 15–18, the input impedance of the transistor amplifier acts as a load on the resonant feedback circuit and reduces the Q of the circuit. Recall from your study of resonance

FIGURE 15–18

R_{in} of the transistor amplifier loads the feedback circuit and lowers its Q, thus lowering the resonant frequency.

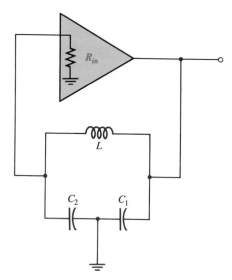

that the resonant frequency of a parallel resonant circuit depends on the Q, according to the following formula:

$$f_r = \frac{1}{2\pi\sqrt{LC_T}}\sqrt{\frac{Q^2}{Q^2 + 1}} \qquad (15\text{–}10)$$

As a rule of thumb, for a Q greater than 10, the frequency is approximately $1/2\pi\sqrt{LC_T}$, as stated in Equation (15–5). When Q is less than 10, however, f_r is reduced significantly.

An FET can be used in place of a bipolar transistor, as shown in Figure 15–19, to minimize the loading effect of the transistor's input impedance. Recall that FETs have much higher input impedances than do bipolar transistors. Also, when an external load is connected to the oscillator output, as shown in Figure 15–20(a), f_r may decrease, again because of a reduction in Q. This happens if the load resistance is too small. In some cases, one way to eliminate the effects of a load resistance is by transformer coupling as indicated in Figure 15–20(b).

EXAMPLE 15–3

(a) Determine the frequency of oscillation for the oscillator in Figure 15–21 (p. 598). Assume there is negligible loading on the feedback circuit and its Q is greater than 10.

(b) Find f_r if the oscillator is loaded to a point where the Q drops to 8.

FIGURE 15–19
A basic FET Colpitts oscillator.

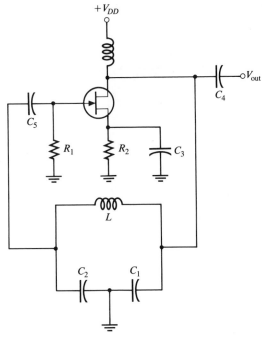

FIGURE 15–20
Oscillator loads.

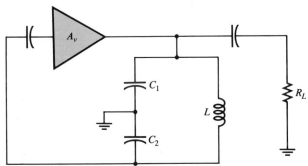

(a) Load capacitively coupled to oscillator output
    can reduce circuit Q and f_r

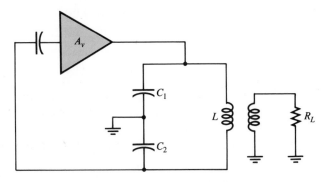

(b) Transformer coupling of load can reduce loading
    effect by impedance transformation

FIGURE 15–21

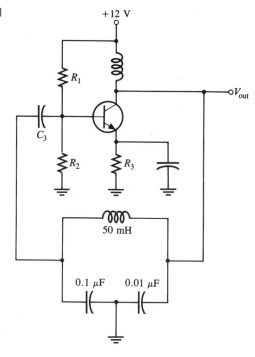

Solution

(a) $C_T = \dfrac{C_1 C_2}{C_1 + C_2} = \dfrac{(0.01 \ \mu\text{F})(0.1 \ \mu\text{F})}{0.11 \ \mu\text{F}} = 0.091 \ \mu\text{F}$

$f_r \cong \dfrac{1}{2\pi\sqrt{LC_T}} = \dfrac{1}{2\pi\sqrt{(50 \ \text{mH})(0.091 \ \mu\text{F})}} = 2359 \ \text{Hz}$

(b) $f_r = \dfrac{1}{2\pi\sqrt{LC_T}} \sqrt{\dfrac{Q^2}{Q^2 + 1}} = (2359 \ \text{Hz})(0.9923) = 2341 \ \text{Hz}$

The Clapp Oscillator

The Clapp oscillator is a variation of the Colpitts. The basic difference is an additional capacitor in series with the inductor in the resonant feedback circuit, as shown in Figure 15–22. C_3 is the added capacitor. Since C_3 is in series with C_1 and C_2, the total capacitance around the tank circuit is

$$C_T = \dfrac{1}{1/C_1 + 1/C_2 + 1/C_3} \qquad (15\text{–}11)$$

and the approximate frequency of oscillation $(Q > 10)$ is $f_r \cong 1/2\pi\sqrt{LC_T}$.

If C_3 is much smaller than C_1 and C_2, the resonant frequency is controlled almost entirely by C_3 $(f_r \cong 1/2\pi\sqrt{LC_3})$. Since C_1 and C_2 are both connected to ground at one end, the junction capacitance of the transistor and other stray capacitances appear in

FIGURE 15–22
A basic Clapp oscillator.

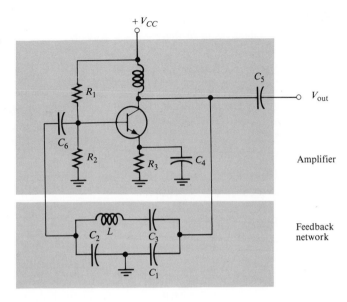

parallel with C_1 and C_2 to ground, altering their effective values. C_3 is not affected, however, and thus provides a more accurate and stable frequency of oscillation.

The Hartley Oscillator

The Hartley oscillator is similar to the Colpitts except that the feedback network consists of two series inductors and a parallel capacitor as shown in Figure 15–23.

In this circuit, the frequency of oscillation for $Q > 10$ is

$$f_r \cong \frac{1}{2\pi\sqrt{L_T C}}$$

(15–12)

FIGURE 15–23
A basic Hartley oscillator.

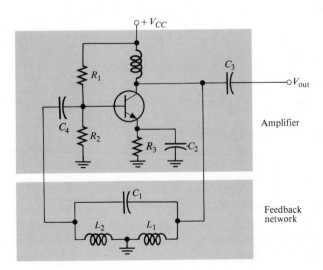

where $L_T = L_1 + L_2$. The inductors act in a role similar to C_1 and C_2 in the Colpitts to determine the attenuation of the feedback network.

$$B \cong \frac{L_2}{L_1} \qquad \text{(15–13)}$$

To assure start-up of oscillation, A_v must be greater than $1/B$.

$$A_v > \frac{L_1}{L_2} \qquad \text{(15–14)}$$

Loading of the tank circuit has the same effect in the Hartley as in the Colpitts; that is, the Q is decreased and thus f_r decreases.

The Armstrong Oscillator

This type of LC oscillator uses transformer coupling to feed back a portion of the signal voltage, as shown in Figure 15–24. It is sometimes called a "tickler" oscillator in reference to the transformer secondary or "tickler coil" that provides the feedback to keep the oscillation going. The Armstrong is less common than the Colpitts, Clapp, and Hartley, mainly because of the disadvantage of transformer size and cost. The frequency of oscillation is set by the inductance of the primary winding (L_p) in parallel with C_1 ($f_r = 1/2\pi\sqrt{L_pC_1}$).

FIGURE 15–24
A basic Armstrong oscillator.

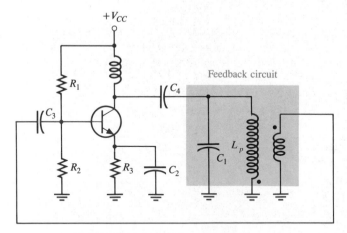

Crystal-Controlled Oscillators

The most stable and accurate type of oscillators use a *piezoelectric crystal* in the feedback loop to control the frequency.

The Piezoelectric Effect. Quartz is one type of crystalline substance found in nature that exhibits a property called the *piezoelectric effect*. When a changing mechanical stress is applied across the crystal to cause it to vibrate, a voltage develops at the frequency of the mechanical vibration. Conversely, when an ac voltage is applied across the crystal, it

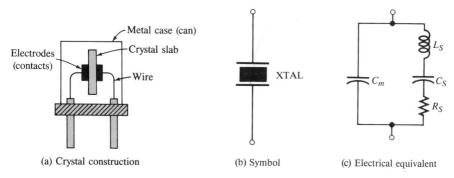

(a) Crystal construction   (b) Symbol   (c) Electrical equivalent

FIGURE 15–25
Crystals in electronics applications.

vibrates at the frequency of the applied voltage. The greatest vibration occurs at the crystal's natural resonant frequency which is determined by the physical dimensions and by the way the crystal is cut.

Crystals used in electronic applications are typically mounted between two electrodes in a protective "can" as shown in Figure 15–25(a). A schematic symbol is shown in part (b) and an equivalent LC circuit for the crystal appears in part (c). As you can see, the crystal's equivalent circuit is a series-parallel LC network and can operate in either series resonance or parallel resonance. At the series resonant frequency, the inductive reactance is cancelled by the reactance of C_S. The remaining series resistance, R_S, determines the impedance of the crystal. Parallel resonance occurs when the inductive reactance and the reactance of the parallel capacitance, C_m, are equal. The parallel resonant frequency is usually at least 1 kHz higher than the series resonant frequency. A great advantage of the crystal is that it exhibits a very high Q (Qs of several thousand are typical).

Modes of Oscillation in the Crystal. Piezoelectric crystals can oscillate in either of two modes: *fundamental* and *overtone*. The fundamental frequency of a crystal is the lowest frequency at which it is naturally resonant. The fundamental frequency depends on the crystal's mechanical dimensions, type of cut, and other factors, and is *inversely proportional to the thickness of the crystal slab*. Because a slab of crystal cannot be cut too thin without fracturing, there is an upper limit on the fundamental frequency. For most crystals, this upper limit is less than 20 MHz. For higher frequencies, the crystal must be operated in the overtone mode. Overtones are approximate integer multiples of the fundamental frequency. The overtone frequencies are usually, but not always, odd multiples (3, 5, 7, . . .) of the fundamental.

An oscillator that uses a crystal as a series resonant tank circuit is shown in Figure 15–26(a). The impedance of the crystal is *minimum* at the series resonant frequency, thus providing maximum feedback. The series capacitor, C_c, is used to "fine tune" the oscillator frequency by "pulling" the resonant frequency of the crystal slightly up or down. A modified Colpitts configuration is shown in Figure 15–26(b) with a crystal acting as a parallel resonant tank circuit. The impedance of the crystal is *maximum* at parallel

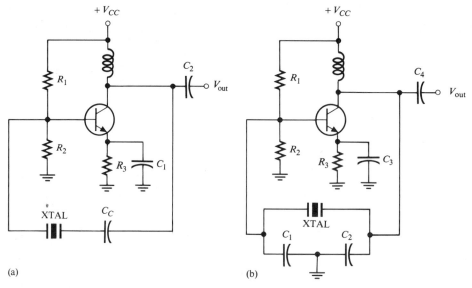

FIGURE 15–26
Basic crystal oscillators.

resonance, thus developing the maximum voltage across the capacitors. The voltage across C_2 is fed back to the input.

SECTION REVIEW 15–4

1. What is the basic difference between the Colpitts and the Hartley oscillators?
2. What is the advantage of an FET amplifier in a Colpitts or Hartley oscillator?
3. How can you distinguish a Colpitts oscillator from a Clapp oscillator?

NONSINUSOIDAL OSCILLATORS

15–5

In this section, several types of oscillator circuits that produce square, triangular, or sawtooth waveforms are discussed. Some of these types of oscillators are frequently referred to as *generators* and *multivibrators* depending on the particular circuit implementation.

A Triangular Wave Oscillator

The op-amp integrator covered in the last chapter can be used as the basis for a triangular wave generator. The basic idea is illustrated in Figure 15–27(a) where a dual-polarity, switched input is used. We use the switch only to introduce the concept; it is not a practical way to implement this circuit. When the switch is in position 1, the negative voltage is applied and the output is a positive-going ramp. When the switch is thrown into position 2, a negative-going ramp is produced. If the switch is thrown back and forth at fixed intervals, the output is a triangular wave consisting of alternating positive-going and negative-going ramps, as shown in Figure 15–27(b).

FIGURE 15-27
Basic triangular wave generator.

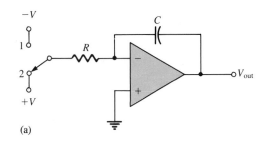

(a)

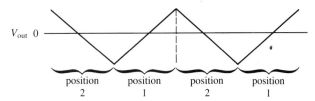

(b) Output voltage as the switch is thrown
back and forth at regular intervals

A Practical Circuit

One practical implementation of a triangular wave generator utilizes an op-amp comparator to perform the switching function, as shown in Figure 15–28. The operation is as
follows. To begin, assume that the output voltage of the comparator is at its maximum
negative level. This output is connected to the inverting input of the integrator through R_1,
producing a positive-going ramp on the output of the integrator. When the ramp voltage
reaches the UTP, the comparator switches to its maximum positive level. This positive
level causes the integrator ramp to change to a negative-going direction. The ramp continues in this direction until the LTP of the comparator is reached. At this point, the comparator output switches back to the maximum negative level and the cycle repeats. This
action is illustrated in Figure 15–29.

Since the comparator produces a square-wave output, the circuit can be used as both
a triangular wave generator and a square-wave generator. Devices of this type are

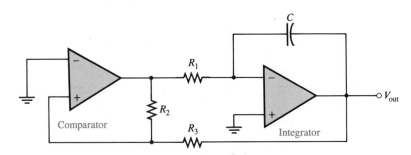

FIGURE 15-28
A triangular wave generator using two op-amps.

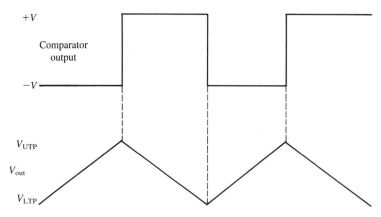

FIGURE 15–29
Waveforms for the circuit in Figure 15–28.

commonly known as *function generators* because they produce more than one output function. The output amplitude of the square wave is set by the output swing of the comparator, and the resistors R_2 and R_3 set the amplitude of the triangular output. The frequency of both waveforms depends on the R_1C time constant as well as the amplitude-setting resistors, R_2 and R_3. By varying R_1, the frequency of oscillation can be adjusted without changing the output amplitude.

$$f = \frac{1}{R_1C}\left(\frac{R_2}{R_3}\right) \tag{15–15}$$

EXAMPLE 15–4

Determine the frequency of the circuit in Figure 15–30. To what value must R_1 be changed to make the frequency 20 kHz?

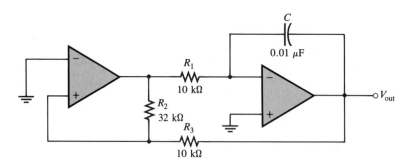

FIGURE 15–30

Solution

$$f = \frac{1}{R_1C}\left(\frac{R_2}{R_3}\right) = \left[\frac{1}{(10 \text{ k}\Omega)(0.01 \text{ }\mu\text{F})}\right]\left(\frac{32 \text{ k}\Omega}{10 \text{ k}\Omega}\right) = 32 \text{ kHz}$$

To make $f = 20$ kHz:

$$R_1 = \frac{1}{fC}\left(\frac{R_2}{R_3}\right) = \left[\frac{1}{(20 \text{ kHz})(0.01 \ \mu\text{F})}\right]\left(\frac{32 \text{ k}\Omega}{10 \text{ k}\Omega}\right) = 16 \text{ k}\Omega$$

A Voltage-Controlled Sawtooth Oscillator (VCO)

The VCO is an oscillator whose frequency can be changed by a variable dc control voltage. VCOs can be either sinusoidal or nonsinusoidal. One way to build a voltage-controlled *sawtooth* oscillator is with an op-amp integrator that uses a switching device (PUT) in parallel with the feedback capacitor to terminate each ramp at a prescribed level and effectively "reset" the circuit. Figure 15–31(a) shows the implementation.

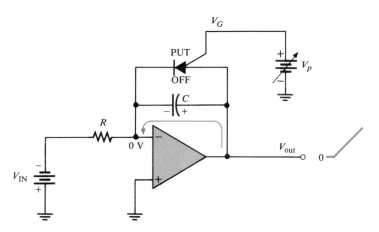

(a) Initially, the capacitor charges, the output ramp begins, and the PUT is off.

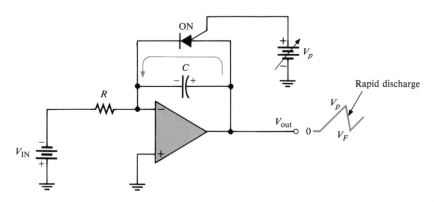

(b) The capacitor rapidly discharges when the PUT momentarily turns on.

FIGURE 15–31
Voltage-controlled sawtooth oscillator operation.

The PUT is a programmable unijunction transistor with an anode, a cathode, and a gate terminal. The gate is always biased positively with respect to the cathode. When the anode voltage exceeds the gate voltage by approximately 0.7 V, the PUT turns on and acts as a forward-biased diode. When the anode voltage falls below this level, the PUT turns off. Also, the current must be above the *holding* value to maintain conduction. (This device is covered in detail in Chapter 18.)

The operation of the sawtooth generator begins when the negative dc input voltage, $-V_{IN}$, produces a positive-going ramp on the output. During the time that the ramp is increasing, the circuit acts as a regular integrator. The PUT triggers on when the output ramp (at the anode) exceeds the gate voltage by 0.7 V. The gate is set to the approximate desired sawtooth peak voltage. When the PUT turns on, the capacitor rapidly discharges, as shown in Figure 15–31(b). The capacitor does not discharge completely to 0 because of the PUT's forward voltage, V_F. Discharge continues until the PUT current falls below the holding value. At this point, the PUT turns off and the capacitor begins to charge again, thus generating a new output ramp. The cycle continually repeats, and the resulting output is a repetitive sawtooth waveform, as shown. The sawtooth amplitude and period can be adjusted by varying the PUT gate voltage.

The frequency is determined by the RC time constant of the integrator and the peak voltage set by the PUT. Recall that the charging rate of the capacitor is V_{IN}/RC. The time it takes the capacitor to charge from V_F to V_p is the period of the sawtooth (neglecting the rapid discharge time):

$$T = \frac{V_p - V_F}{|V_{IN}|/RC} \tag{15–16}$$

From $f = 1/T$, we get

$$f = \frac{|V_{IN}|}{RC}\left(\frac{1}{V_p - V_F}\right) \tag{15–17}$$

EXAMPLE 15–5

(a) Find the amplitude and frequency of the sawtooth output in Figure 15–32. Assume that the forward PUT voltage V_F is approximately 1 V.

(b) Sketch the output waveform.

Solution

(a) First, find the gate voltage in order to establish the approximate voltage at which the PUT turns on:

$$V_G = \frac{R_4}{R_3 + R_4}(+V) = \frac{10 \text{ k}\Omega}{20 \text{ k}\Omega}(15 \text{ V}) = 7.5 \text{ V}$$

This voltage sets the approximate maximum peak value of the sawtooth output (neglecting the 0.7 V):

$$V_p \cong 7.5 \text{ V}$$

FIGURE 15-32

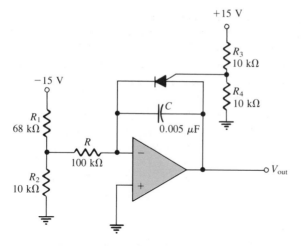

The minimum peak value (low point) is

$$V_F \cong 1 \text{ V}$$

The period is determined as follows:

$$V_{IN} = \frac{R_2}{R_1 + R_2}(-V) = \frac{10 \text{ k}\Omega}{78 \text{ k}\Omega}(-15 \text{ V}) = -1.92 \text{ V}$$

$$T = \frac{V_p - V_F}{|V_{IN}|/RC} = \frac{7.5 \text{ V} - 1 \text{ V}}{1.92 \text{ V}/(100 \text{ k}\Omega)(0.005 \text{ } \mu\text{F})} = 1.69 \text{ ms}$$

$$f = \frac{1}{1.69 \text{ ms}} \cong 592 \text{ Hz}$$

(b) The output waveform is shown in Figure 15–33.

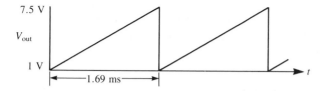

FIGURE 15-33
Output of the circuit in Figure 15–32.

A Square-Wave Oscillator

The basic square-wave generator shown in Figure 15–34 is a type of *relaxation oscillator* because its operation is based on the charging and discharging of a capacitor. Notice that the op-amp's inverting input is the capacitor voltage and the noninverting input is a portion of the output fed back through resistors R_2 and R_3. When the circuit is first turned on, the capacitor is uncharged, and thus the inverting input is at 0 V. This makes the

FIGURE 15–34
A square-wave oscillator.

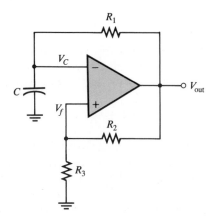

output a positive maximum, and the capacitor begins to charge toward V_{out} through R_1. When the capacitor voltage reaches a value equal to the feedback voltage on the non-inverting input, the op-amp switches to the maximum negative state. At this point the capacitor begins to discharge from $+V_f$ toward $-V_f$. When the capacitor voltage reaches $-V_f$, the op-amp switches back to the maximum positive state. This action continues to repeat, as shown in Figure 15–35, and a square-wave output voltage is obtained.

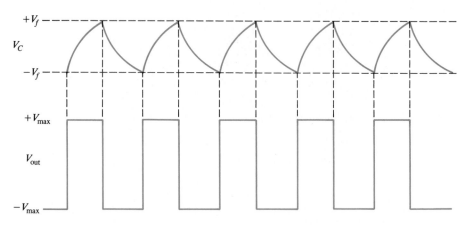

FIGURE 15–35
Waveforms for the square-wave oscillator.

SECTION REVIEW 15–5

1. What is a VCO and, basically, what does it do?
2. Upon what principle does a relaxation oscillator operate?

THE 555 TIMER AS AN OSCILLATOR

The 555 timer is a versatile integrated circuit with many applications. In this section, you will see how the 555 is configured as an *astable* or *free-running multivibrator,* which is essentially a square-wave oscillator. We will also discuss the use of the 555 timer as a voltage-controlled oscillator (VCO). The 555 timer consists basically of *two comparators,* a *flip-flop,* a *discharge transistor,* and a *resistive voltage divider,* as shown in Figure 15–36.

The *flip-flop* (bistable multivibrator) is a digital device that is perhaps unfamiliar to you at this point unless you already have taken a digital fundamentals course. Briefly, it is a *two-state* device whose output can be at either a high voltage level (set) or a low voltage level (reset). The state of the output can be changed with proper input signals.

The resistive voltage divider is used to set the voltage comparator levels. All three resistors are of equal value; therefore, the upper comparator has a reference of $\frac{2}{3} V_{CC}$,

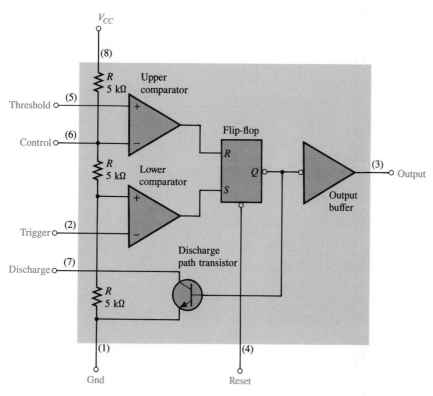

FIGURE 15–36
Internal diagram of a 555 integrated circuit timer. (Dual in-line package pin numbers are in parentheses.)

and the lower comparator has a reference of $\frac{1}{3}$ V_{CC}. The comparators' outputs control the state of the flip-flop. When the *trigger* voltage goes below $\frac{1}{3}$ V_{CC}, the flip-flop sets and the output jumps to its high level. The *threshold* input is normally connected to an external RC timing network. When the external capacitor voltage exceeds $\frac{2}{3}$ V_{CC}, the upper comparator resets the flip-flop, which in turn switches the output back to its low level. When the device output is low, the discharge transistor Q_d is turned on and provides a path for rapid discharge of the external timing capacitor. This basic operation allows the timer to be configured with external components as an *oscillator,* a *one-shot,* or a *time-delay element*.

Astable Operation

A 555 timer connected to operate in the astable mode as a free-running nonsinusoidal oscillator is shown in Figure 15–37. Notice that the threshold input (THRESH) is now connected to the trigger input (TRIG). The external components R_1, R_2, and C_{ext} form the timing network that sets the frequency of oscillation. The 0.01 μF capacitor connected to the control (CONT) input is strictly for decoupling and has no effect on the operation; in some cases it can be left off.

    Initially, when the power is turned on, the capacitor C_{ext} is uncharged and thus the trigger voltage (2) is at 0 V. This causes the output of the lower comparator to be high and the output of the upper comparator to be low, forcing the output of the flip-flop, and thus the base of Q_d, low and keeping the transistor off. Now, C_{ext} begins charging through R_1

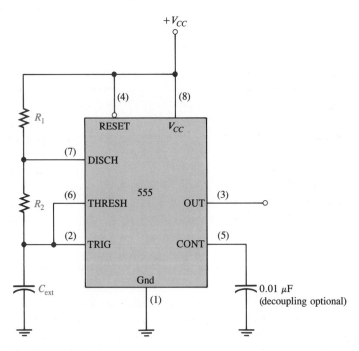

FIGURE 15–37
The 555 timer connected as an astable multivibrator.

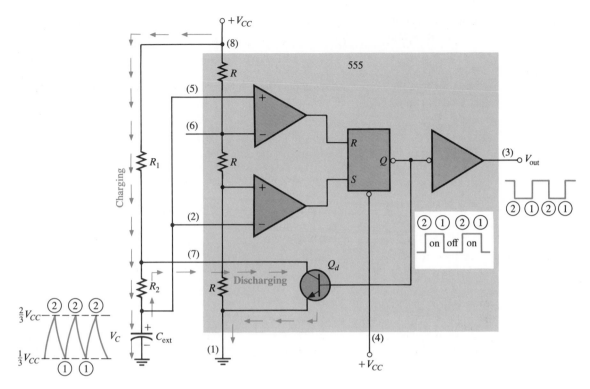

FIGURE 15–38
Operation of the 555 timer in the astable mode.

and R_2 as indicated in Figure 15–38. When the capacitor voltage reaches $\frac{1}{3} V_{CC}$, the lower comparator switches to its low output state, and when the capacitor voltage reaches $\frac{2}{3} V_{CC}$, the upper comparator switches to its high output state. This resets the flip-flop, causes the base of Q_d to go high, and turns on the transistor. This sequence creates a discharge path for the capacitor through R_2 and the transistor, as indicated. The capacitor now begins to discharge, causing the upper comparator to go low. At the point where the capacitor discharges down to $\frac{1}{3} V_{CC}$, the lower comparator switches high, setting the flip-flop, which makes the base of Q_d low and turns off the transistor. Another charging cycle begins, and the entire process repeats. The result is a rectangular wave output whose duty cycle depends on the values of R_1 and R_2. The frequency of oscillation is given by the formula

$$f = \frac{1.44}{(R_1 + 2R_2)C_{ext}} \tag{15–18}$$

By selecting R_1 and R_2, the duty cycle of the output can be adjusted. Since C_{ext} charges through $R_1 + R_2$ and discharges only through R_2, duty cycles approaching a *minimum* of 50 percent can be achieved if $R_2 \gg R_1$ so that the charging and discharging times are approximately equal. An expression to calculate the duty cycle is developed as follows.

The time that the output is high is how long it takes C_{ext} to charge from $\frac{1}{3}\,V_{CC}$ to $\frac{2}{3}\,V_{CC}$. It is expressed as

$$t_H = 0.693(R_1 + R_2)C_{ext}$$

The time that the output is low is how long it takes C_{ext} to discharge from $\frac{2}{3}\,V_{CC}$ to $\frac{1}{3}\,V_{CC}$. It is expressed as

$$t_L = 0.693R_2C_{ext}$$

The *period* of the output waveform is the sum of t_H and t_L.

$$T = t_H + t_L = 0.693(R_1 + 2R_2)C_{ext}$$

This is the reciprocal of f in equation (15–18). Finally, the duty cycle is

$$\text{Duty cycle} = \frac{t_H}{T} = \frac{t_H}{t_H + t_L}$$

$$\text{Duty cycle} = \frac{R_1 + R_2}{R_1 + 2R_2} \times 100\% \qquad \textbf{(15–19)}$$

To achieve duty cycles of less than 50 percent, the circuit in Figure 15–37 can be modified so that C_{ext} charges through only R_1 and discharges through R_2. This is achieved with a diode D_1 placed as shown in Figure 15–39. The duty cycle can be made less than

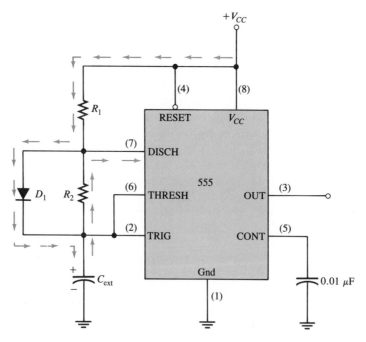

FIGURE 15–39
The addition of diode D_1 allows the duty cycle of the output to be adjusted to less than 50 percent by making $R_1 < R_2$.

50 percent by making R_1 less than R_2. Under this condition, the expression for the duty cycle is

$$\text{Duty cycle} = \frac{R_1}{R_1 + R_2} \times 100\% \qquad (15\text{--}20)$$

EXAMPLE 15–6

A 555 timer configured to run in the astable mode (oscillator) is shown in Figure 15–40. Determine the frequency of the output and the duty cycle.

FIGURE 15–40

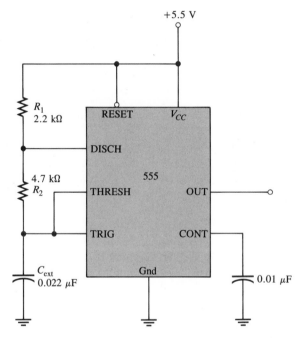

Solution

$$f = \frac{1.44}{(R_1 + 2R_2)C_{\text{ext}}} = \frac{1.44}{(2.2 \text{ k}\Omega + 9.4 \text{ k}\Omega)0.022 \text{ } \mu\text{F}} = 6.48 \text{ kHz}$$

$$\text{Duty cycle} = \frac{R_1 + R_2}{R_1 + 2R_2} \times 100\% = \frac{2.2 \text{ k}\Omega + 4.7 \text{ k}\Omega}{2.2 \text{ k}\Omega + 9.4 \text{ k}\Omega} \times 100\% = 59.5\%$$

Practice Exercise 15–6

Determine the duty cycle in Figure 15–40 if a diode is connected across R_2 as indicated in Figure 15–39.

Operation as a Voltage-Controlled Oscillator (VCO)

A 555 timer can be configured to operate as a VCO by using the same external connections as for astable operation, with the exception that a variable control voltage is applied to the CONT input (pin 5), as indicated in Figure 15–41.

FIGURE 15–41

The 555 timer connected as a voltage-controlled oscillator (VCO). Note the variable control voltage input on pin 5.

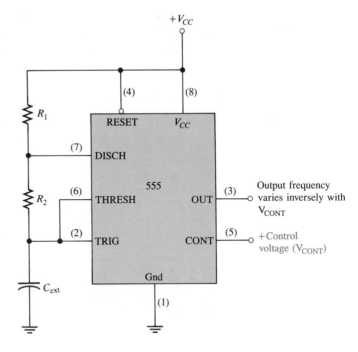

As shown in Figure 15–42, the control voltage (V_{CONT}) changes the threshold values of $\frac{1}{3} V_{CC}$ and $\frac{2}{3} V_{CC}$ for the internal comparators. With the control voltage, the upper value is V_{CONT} and the lower value is $\frac{1}{2} V_{CONT}$, as you can see by examining the internal diagram of the 555 timer. When the control voltage is varied, the output frequency also varies. An increase in V_{CONT} increases the charging and discharging time of the external capacitor and causes the frequency to decrease. A decrease in V_{CONT} decreases charging and discharging time of the capacitor and causes the frequency to increase.

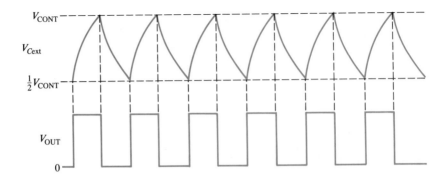

FIGURE 15–42

The VCO output frequency varies inversely with V_{CONT} because the charging and discharging time of C_{ext} is directly dependent on the control voltage.

An interesting application of the VCO is in *phase-locked loops,* which are used in various types of communications receivers to track variations in the frequency of incoming signals. We will cover the basic operation of a phase-locked loop in the next section.

SECTION REVIEW 15–6

1. Name the five basic elements in a 555 timer IC.
2. When the 555 timer is configured as an astable multivibrator, how is the duty cycle determined?

THE PHASE-LOCKED LOOP

15–7

The phase-locked loop (PLL) is an electronic feedback circuit consisting of a *phase detector,* a *low-pass filter,* and a *voltage-controlled oscillator* (VCO). It is capable of locking onto or synchronizing with an incoming signal. When the phase changes, indicating that the incoming frequency is changing, the phase detector's output voltage (error voltage) increases or decreases just enough to keep the oscillator frequency the same as the incoming frequency. A basic block diagram of the PLL is shown in Figure 15–43. Phase-locked loops are used in a wide variety of communication system applications, including TV receivers, FM demodulation, modems, telemetry, and tone decoders.

Basic Operation

Using Figure 15–43 as reference, the general operation of the PLL is as follows. When there is no input signal, the error voltage is 0 and the frequency f_o of the VCO is called the *free-running* or *center* frequency. When an input signal is applied, the phase detector compares the phase and frequency of the input signal with the VCO frequency and produces error voltage V_e. This error voltage is proportional to the phase and frequency difference of the incoming frequency and the VCO frequency. The error voltage contains

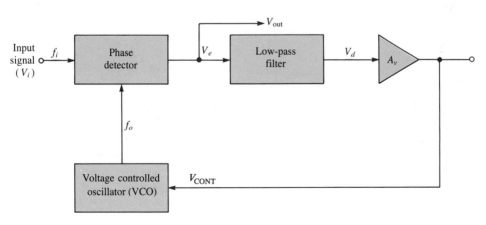

FIGURE 15–43
Basic phase-locked loop block diagram.

components that are the *sum* and *difference* of the two compared frequencies. The low-pass filter passes only the difference frequency V_d, which is the lower of the two components. This signal is amplified and fed back to the VCO as a control voltage V_{CONT}. The control voltage forces the VCO frequency to change in a direction that *reduces* the difference between the incoming frequency f_i and the VCO frequency f_o. When f_i and f_o are sufficiently close in value, the feedback action of the PLL causes the VCO to *lock* onto the incoming signal. Once the VCO locks, its frequency is the same as the input frequency with a slight difference in phase. This phase difference ϕ is necessary to keep the PLL in the lock condition.

Lock Range.   Once the PLL is locked, it can *track* frequency changes in the incoming signal. The range of frequencies over which the PLL can maintain lock with the incoming signal is called the *lock* or *tracking* range. The lock range is usually expressed as a percentage of the VCO frequency.

Capture Range.   The range of frequencies over which the PLL can *acquire* lock with an input signal is called the *capture* range. The parameter is normally expressed in percentage of f_o, also.

Sum and Difference Frequencies.   The phase detector operates as a *multiplier* circuit to produce the sum and difference of the input frequency f_i and the VCO frequency f_o. This action can best be described mathematically as follows. Recall from ac circuit theory that a sine wave voltage can be expressed as

$$v = V_p \sin 2\pi ft$$

where V_p is the peak value, f is the frequency, and t is the time. Using this basic expression, the input signal voltage and the VCO voltage can be written as

$$v_i = V_{ip} \sin 2\pi f_i t$$
$$v_o = V_{op} \sin 2\pi f_o t$$

When these two signals are multiplied in the phase detector, we get a product at the output as follows.

$$V_{out} = V_{ip} V_{op} (\sin 2\pi f_i t)(\sin 2\pi f_o t) \qquad \textbf{(15–21)}$$

Substituting the trigonometric identity

$$(\sin A)(\sin B) = \frac{1}{2}[\cos(A - B) - \cos(A + B)]$$

into equation (15–20), we get

$$V_{out} = \frac{V_{ip} V_{op}}{2}[\cos(2\pi f_i t - 2\pi f_o t) - \cos(2\pi f_i t + 2\pi f_o t)]$$

$$V_{out} = \frac{V_{ip} V_{op}}{2} \cos 2\pi(f_i - f_o)t - \frac{V_{ip} V_{op}}{2} \cos 2\pi(f_i + f_o)t \qquad \textbf{(15–22)}$$

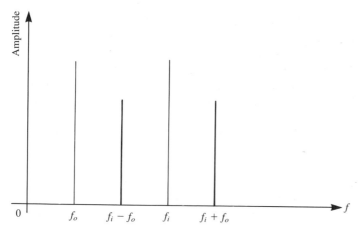

FIGURE 15–44
Frequency spectrum of the phase detector.

You can see in equation (15–22) that V_{out} of the phase detector consists of a difference frequency component $(f_i - f_o)$ and a sum frequency component $(f_i + f_o)$. This concept is illustrated in Figure 15–44 with frequency spectrum graphs. Each vertical line represents a specific signal frequency, and the height is its amplitude.

EXAMPLE 15–7

A 10 kHz signal f_i and an 8 kHz signal f_o are applied to a phase detector. Determine the sum and difference frequencies.

Solution

$$f_i - f_o = 10 \text{ kHz} - 8 \text{ kHz} = 2 \text{ kHz}$$
$$f_i + f_o = 10 \text{ kHz} + 8 \text{ kHz} = 18 \text{ kHz}$$

When the PLL Is in Lock

When the PLL is in a lock condition, the VCO frequency equals the input frequency $(f_o = f_i)$. Thus the difference frequency is $f_o - f_i = 0$. A 0 frequency indicates a *dc* component. The low-pass filter removes the sum frequency $(f_i + f_o)$ and passes the dc (0-frequency) component, which is amplified and fed back to the VCO. *When the PLL is in lock, the difference frequency component is always dc and is always passed by the filter, so the lock range is independent of the bandwidth of the low-pass filter.* This is illustrated in Figure 15–45.

When the PLL Is Not in Lock

In the case where the PLL has not yet locked onto the incoming frequency, the phase detector still produces sum and difference frequencies. In this case, however, the difference frequency may lie outside the pass band of the low-pass filter and will not be fed

FIGURE 15–45
When the PLL is in lock, the difference frequency is 0 and passes through the filter. The lock range is independent of the filter bandwidth.

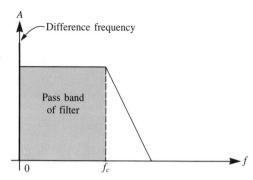

back to the VCO, as illustrated in Figure 15–46. The VCO will remain at its center frequency (free-running) as long as this condition exists.

As the input frequency approaches the VCO frequency, the difference component produced by the phase detector decreases and eventually falls into the pass band of the filter and drives the VCO toward the incoming frequency. When the VCO reaches the incoming frequency the PLL locks on it.

FIGURE 15–46
When the PLL is not in lock, the difference frequency can be outside the pass band.

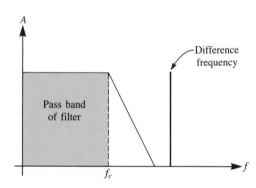

SECTION REVIEW 15–7

1. List the basic blocks in a phase-locked loop.
2. State the general purpose of a phase-locked loop.

A SYSTEM APPLICATION

15–8 The Local Oscillator

As you have learned in previous chapters, superheterodyne receivers require local oscillators (LOs). These oscillators produce frequencies that are above the incoming RF carrier frequency. In the case of AM, the LO frequency is 455 kHz above the RF and in FM it is 10.7 MHz above. Local oscillators must be variable in frequency so they can be adjusted to "track" the frequency of the RF. The LO in an AM receiver must be tuned over a frequency range of 995 kHz to 2095 kHz. The LO in an FM receiver must be tuned from 98.7 MHz to 118.7 MHz. Tuning can be accomplished with a variable air capacitor or with a varactor diode. Use of varactor diodes in tuned circuits is called *electronic tuning*.

Typically, discrete transistor circuits are used for the "front end" of a receiver (RF amplifier, local oscillator, and mixer). Local oscillators are found in many variations of the basic configurations introduced in this chapter. The Hartley oscillator, or variations of it, are commonly used for the LO function.

The Phase-Locked Loop as an FM Discriminator

Among many other applications, the phase-locked loop has been increasingly employed as a method of FM demodulation. Other, more conventional types of FM discriminators depend on intricate coil adjustments, which the phase-locked loop eliminates.

The input to the PLL discriminator is the frequency-modulated 10.7 MHz IF. As the frequency of the incoming modulated IF changes, the PLL produces an output proportional to the change in frequency, which is actually the audio signal. The output of the discriminator is taken from the low-pass filter and amplifier portion of the PLL, as indicated in Figure 15–47 (p. 620). This audio output signal is also fed back to the VCO control input, causing the VCO to match the FM input to the phase detector. A drift in the IF frequency caused by a drift in the LO makes the PLL automatically readjust itself. No tuned circuits are necessary. In a conventional FM discriminator, any frequency drift results in a distorted output. Complete phase-locked loops are available in IC form, where the detector, local oscillator, filter, and amplifier are contained in a single IC package. The 560 is one common type of PLL.

FORMULAS

(**15–1**) $\dfrac{V_{\text{out}}}{V_{\text{in}}} = \dfrac{1}{3}$ Wien-bridge positive feedback attenuation

(**15–2**) $f_r = \dfrac{1}{2\pi RC}$ Wien-bridge frequency

(**15–3**) $B = \dfrac{1}{29}$ Phase-shift feedback attenuation

(**15–4**) $f_r = \dfrac{1}{2\pi \sqrt{6}\,RC}$ Phase-shift oscillator frequency

(**15–5**) $f_r \cong \dfrac{1}{2\pi \sqrt{LC_T}}$ Colpitts approximate resonant frequency

(**15–6**) $C_T = \dfrac{C_1 C_2}{C_1 + C_2}$ Colpitts feedback capacitance

(**15–7**) $B = \dfrac{C_1}{C_2}$ Colpitts feedback attenuation

(**15–8**) $A_v = \dfrac{C_2}{C_1}$ Colpitts amplifier gain

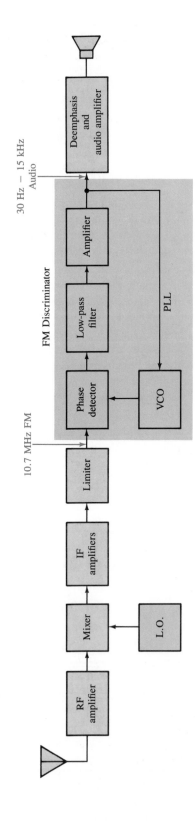

FIGURE 15–47
An FM superheterodyne receiver with a PLL discriminator.

Active Filters

In this chapter you will learn

☐ The general shape and characteristics of low-pass, high-pass, band-pass, and band-stop filter responses

☐ How to recognize Butterworth, Chebyshev, and Bessel response characteristics

☐ How the *damping factor* in a filter determines the response characteristic

☐ The meaning of the term *pole* in filter terminology

☐ How filter cutoff frequencies are determined

☐ How the roll-off rate and the number of poles are related

☐ How to achieve higher roll-offs by cascading

☐ How to implement the Butterworth response in a filter

☐ How to analyze Sallen and Key low-pass and high-pass filters

☐ How to analyze multiple-feedback and state variable band-pass and band-stop filters

☐ Two basic methods for measuring filter responses

☐ How filters are used in a specific system application

Chapter 16

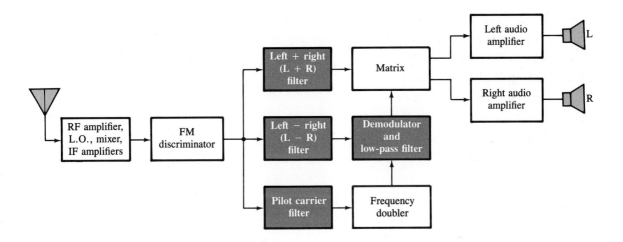

We introduced power-supply filters in Chapter 3. In this chapter, we will discuss active filters used for signal processing. Filters are circuits capable of passing input signals with desired frequencies through to the output while rejecting signals with other frequencies. This property is called *selectivity*.

Active filters use transistors or op-amps to provide gain and impedance characteristics and passive networks to provide the frequency selection. In terms of their general response, there are four basic categories of filters: *low-pass, high-pass, band-pass,* and *band-stop*. An excellent example of the use of filters in a system application is the FM superheterodyne stereo receiver. Filters play an important role in the separation of the left and right channel audio signals, as indicated in the block diagram. This particular application uses both low-pass and band-pass filters, not only for their selectivity but also their voltage gain.

BASIC FILTER RESPONSE CURVES

16–1

Filters are usually categorized by the manner in which the output voltage varies with the frequency of the input voltage. The categories are *low-pass, high-pass, band-pass,* and *band-stop.* We will examine each of these general responses.

Low-Pass Response

The *pass band* of the basic low-pass filter is defined to be from 0 Hz (dc) up to the cutoff frequency, f_c, at which the output voltage is 70.7 percent of the pass band voltage, as indicated in Figure 16–1(a). The *ideal* pass band, shown by the shaded region within the dashed lines, has an instantaneous roll-off at f_c. The bandwidth of this filter is equal to f_c.

$$BW = f_c$$

Although the ideal response is not achievable in practice, roll-off rates of -20 dB/decade and higher are obtainable. Figure 16–1(b) illustrates ideal low-pass filter response curves with several roll-off rates. The -20 dB/decade rate is obtained with a single RC network consisting of one resistor and one capacitor. The higher roll-off rates require additional RC networks. Each network is called a *pole.*

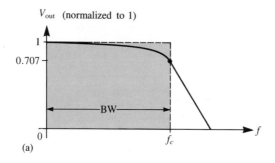

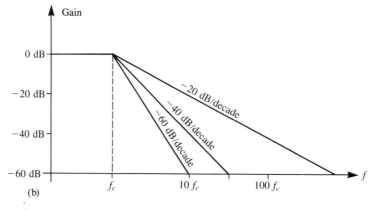

FIGURE 16–1
Low-pass responses.

As explained in Chapter 11, the cutoff frequency of the RC low-pass filter occurs when $X_C = R$, where

$$f_c = \frac{1}{2\pi RC} \qquad\qquad \textbf{(16–1)}$$

High-Pass Response

A high-pass response is one that significantly attenuates all frequencies below f_c and passes all frequencies above f_c. The cutoff frequency is, of course, the frequency at which the output voltage is 70.7 percent of the pass band voltage, as shown in Figure 16–2(a). The ideal response, shown by the shaded region within the dashed lines, has an instantaneous drop at f_c, which, of course, is not achievable. Roll-off rates of 20 dB/decade/pole are realizable. Figure 16–2(b) illustrates high-pass responses with several roll-off rates.

As with the RC low-pass filter, the high-pass cutoff frequency corresponds to the value where $X_C = R$ and is calculated with the formula $f_c = 1/2\pi RC$. The response of a high-pass filter extends from f_c up to a frequency that is determined by the limitations of the active element (transistor or op-amp) used.

FIGURE 16–2
High-pass responses.

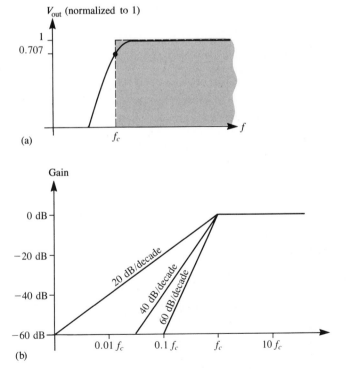

(a)

(b)

Band-Pass Response

A band-pass filter passes all signals lying within a band between a lower- and an upper-frequency limit and essentially rejects all other frequencies that are outside this specified

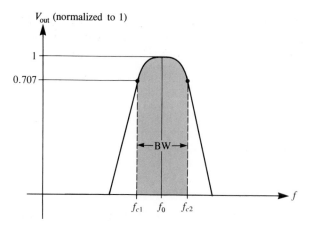

FIGURE 16–3
General band-pass response curve.

band. A generalized band-pass response curve is shown in Figure 16–3. The *bandwidth* (*BW*) is defined as the difference between the upper cutoff frequency (f_{c2}) and the lower cutoff frequency (f_{c1}).

$$BW = f_{c2} - f_{c1} \qquad (16\text{–}2)$$

The cutoff frequencies are, of course, the points at which the response curve is 70.7 percent of its maximum. Recall from Chapter 12 that these cutoff frequencies are also called *3 dB frequencies*. The frequency about which the pass band is centered is called the *center frequency, f_0*, defined as the geometric mean of the cutoff frequencies:

$$f_0 = \sqrt{f_{c1}f_{c2}} \qquad (16\text{–}3)$$

Quality Factor.   The *quality factor* (*Q*) of a band-pass filter is the ratio of the center frequency to the bandwidth.

$$Q = \frac{f_0}{BW} \qquad (16\text{–}4)$$

The value of Q is an indication of the *selectivity* of a band-pass filter. The higher the value of Q, the narrower the bandwidth and the better the selectivity for a given value of f_0. Band-pass filters are sometimes classified as narrow-band ($Q > 10$) or wide-band ($Q < 10$). The Q can also be expressed in terms of the *damping factor* (*DF*) of the filter as

$$Q = \frac{1}{DF} \qquad (16\text{–}5)$$

We will study the damping factor later in the chapter.

EXAMPLE 16–1

A certain band-pass filter has a center frequency of 15 kHz and a bandwidth of
1 kHz. Determine the Q and classify the filter as narrow-band or wide-band.

Solution

$$Q = \frac{f_0}{BW} = \frac{15 \text{ kHz}}{1 \text{ kHz}} = 15$$

Because $Q > 10$, this is a narrow-band filter.

Band-Stop Response

Another category of active filter is the band-stop, also known as *notch, band-reject,* or
band-elimination filters. You can think of the operation as opposite to that of the band-
pass filter because frequencies within a certain bandwidth are rejected and frequencies
outside the bandwidth are passed. A general response curve for a band-stop filter is shown
in Figure 16–4. Notice that the bandwidth is the band of frequencies between the 3 dB
points, just as in the case of the band-pass response.

FIGURE 16–4
General band-stop response.

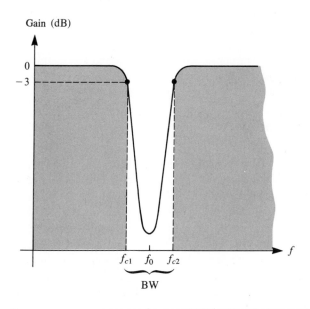

1. The bandwidth of a low-pass filter is determined by _____ .
2. What limits the bandwidth of an active high-pass filter?
3. How are the Q and the bandwidth of a band-pass filter related? Explain how the
   selectivity is affected by the Q of a filter.

**SECTION
REVIEW
16–1**

FILTER RESPONSE CHARACTERISTICS

16–2

Each type of response (low-pass, high-pass, band-pass, or band-stop) can be tailored by circuit component values to have either a *Butterworth, Chebyshev,* or *Bessel* characteristic. Each of these characteristics is identified by the shape of the response curve, and each has an advantage in certain applications.

The Butterworth Characteristic

The Butterworth characteristic provides a very flat amplitude response in the pass band and a roll-off rate of 20 dB/decade/pole. The phase response is not linear, however, and the phase shift (thus, time delay) of signals passing through the filter varies nonlinearly with frequency. Therefore, a pulse applied to a filter with a Butterworth response will cause overshoots on the output, because each frequency component of the pulse's rising and falling edges experiences a different time delay. Filters with the Butterworth response are normally used when all frequencies in the pass band must have the same gain. The Butterworth response is often referred to as a *maximally flat* response.

The Chebyshev Characteristic

Filters with the Chebyshev response characteristic are useful when a rapid roll-off is required because it provides a roll-off rate greater than 20 dB/decade/pole. This is a greater rate than that of the Butterworth, so filters can be implemented with the Chebyshev response with fewer poles and less complex circuitry for a given roll-off rate. This type of filter response is characterized by overshoot or ripples in the pass band (depending on the number of poles) and an even less linear phase response than the Butterworth.

The Bessel Characteristic

The Bessel response exhibits a linear phase characteristic, meaning that the phase shift increases linearly with frequency. The result is almost no overshoot on the output with a pulse input. For this reason, filters with the Bessel response are used for filtering pulse waveforms without distorting the shape of the waveform.

Butterworth, Chebyshev, or Bessel response characteristics can be realized with most active filter circuit configurations by proper selection of certain component values, as we will see later. A general comparison of the three response characteristics for a low-pass response curve is shown in Figure 16–5. High-pass and band-pass filters can also be designed to have any one of the characteristics.

The Damping Factor

As mentioned, an active filter can be designed to have either a Butterworth, Chebyshev, or Bessel response characteristic regardless of whether it is a low-pass, high-pass, band-pass, or band-stop type. The *damping factor (DF)* of an active filter circuit determines which response characteristic the filter exhibits. To explain the basic concept, a general-

A_v

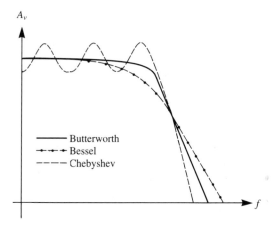

——— Butterworth
—•—•— Bessel
– – – – Chebyshev

ized active filter is shown in Figure 16–6. It includes an amplifier, a negative feedback circuit, and a filter section. The amplifier and feedback are connected in a noninverting configuration. The damping factor is determined by the negative feedback circuit and is defined by the following equation:

$$DF = 2 - \frac{R_1}{R_2} \qquad (16\text{–}6)$$

The value of the damping factor required to produce a desired response characteristic depends on the *order* (number of poles) of the filter. To achieve a second-order Butterworth response, for example, the damping factor must be 1.414. To implement this damping factor, the feedback resistor ratio must be

$$\frac{R_1}{R_2} = 2 - DF = 2 - 1.414 = 0.586$$

FIGURE 16–6
General diagram of an active filter.

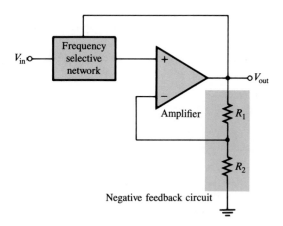

This ratio gives the closed-loop gain of the noninverting filter amplifier, $A_{cl(NI)}$, a value of 1.586, derived as follows:

$$A_{cl(NI)} = \frac{1}{B} = \frac{R_1 + R_2}{R_2} = \frac{R_1}{R_2} + 1 = 0.586 + 1 = 1.586$$

EXAMPLE 16-2

If resistor R_2 in the feedback circuit of an active two-pole filter of the type in Figure 16–6 is 10 kΩ, what value must R_1 be to obtain a maximally flat Butterworth response?

Solution

$$\frac{R_1}{R_2} = 0.586$$

$$R_1 = 0.586R_2 = 0.586(10 \text{ k}\Omega) = 5860 \ \Omega$$

Using the nearest standard 5 percent value of 5600 Ω will get very close to the ideal Butterworth response.

Cutoff Frequency and Roll-Off Rate

The cutoff frequency is determined by the values of the resistor and capacitors in the RC network, as shown in Figure 16–6. For a single-pole (first-order) filter, as shown in Figure 16–7, the cutoff frequency is

$$f_c = \frac{1}{2\pi RC}$$

Although we show a low-pass configuration, the same formula is used for the f_c of a single-pole high-pass filter. The number of poles determines the roll-off rate of the filter. A Butterworth response produces 20 dB/decade/pole. So, a first-order (one-pole) filter has a roll-off of 20 dB/decade; a second-order (two-pole) filter has a roll-off rate of 40 dB/decade; a third-order (three-pole) filter has a roll-off rate of 60 dB/decade; and so on.

Generally, to obtain a filter with three poles or more, one-pole or two-pole filters are *cascaded*, as shown in Figure 16–8. To obtain a third-order filter, for example, we

FIGURE 16-7
First-order (one-pole) low-pass filter.

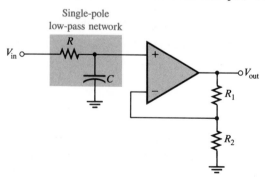

Single-pole
low-pass network

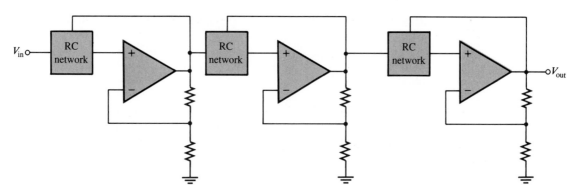

FIGURE 16–8
The number of filter poles is increased by cascading.

cascade a second-order and a first-order filter; to obtain a fourth-order filter, we cascade two second-order filters; and so on. Each filter in a cascaded arrangement is called a *stage* or *section*.

Because of its maximally flat response, the Butterworth characteristic is the most widely used. Therefore, we will limit our coverage to the Butterworth response to illustrate basic filter concepts. Table 16–1 lists the roll-off rates, damping factors, and R_1/R_2 ratios for up to sixth-order Butterworth filters.

TABLE 16–1
Values for the Butterworth response

Order	Roll-off dB/decade	1st stage			2nd stage			3rd stage		
		Poles	DF	R_1/R_2	Poles	DF	R_1/R_2	Poles	DF	R_1/R_2
1	20	1	Optional							
2	40	2	1.414	0.586						
3	60	2	1.00	1	1	1.00	1			
4	80	2	1.848	0.152	2	0.765	1.235			
5	100	2	1.00	1	2	1.618	0.382	1	0.618	1.382
6	120	2	1.932	0.068	2	1.414	0.586	2	0.518	1.482

SECTION REVIEW 16–2

1. Explain how Butterworth, Chebyshev, and Bessel responses differ.
2. What determines the response characteristic of a filter?
3. Name the basic parts of an active filter.

ACTIVE LOW-PASS FILTERS

16–3

Filters that use op-amps as the active element provide several advantages over passive filters (*R, L,* and *C* elements only). The op-amp provides gain, so that the signal is not attenuated as it passes through the filter. The high input impedance of the op-amp prevents excessive loading of the driving source, and the low output impedance of the op-amp prevents the filter from being affected by the load that it is driving. Active filters are also easy to adjust over a wide frequency range without altering the desired response.

A Single-Pole Filter

Figure 16–9(a) shows an active filter with a single low-pass RC network that provides a roll-off of -20 dB/decade above the cutoff frequency, as indicated by the response curve in Figure 16–9(b). The cutoff frequency of the single-pole filter is $f_c = 1/2\pi RC$. The op-amp in this filter is connected as a noninverting amplifier with the closed-loop voltage gain in the pass band set by the values of R_1 and R_2 ($A_{cl} = R_1/R_2 + 1$).

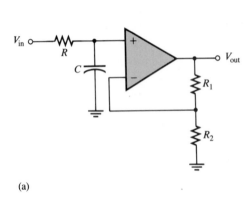

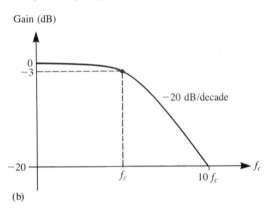

(a)   (b)

FIGURE 16–9
Single-pole active low-pass filter and response curve.

The Sallen and Key Low-Pass Filter

The Sallen and Key is one of the most common configurations for a second-order (two-pole) filter. It is also known as a VCVS (voltage-controlled voltage source) filter. A low-pass version of the Sallen and Key filter is shown in Figure 16–10. Notice that there are two low-pass RC networks that provide a roll-off of -40 dB/decade above the cutoff frequency (assuming a Butterworth characteristic). One RC network consists of R_A and

FIGURE 16–10
Basic Sallen and Key second-order low-pass filter.

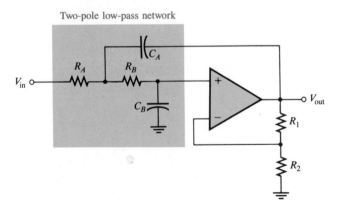

Two-pole low-pass network

C_A, and the second network consists of R_B and C_B. A unique feature is the capacitor C_A that provides feedback for shaping the response near the edge of the pass band. The cutoff frequency for the second-order Sallen and Key filter is

$$f_c = \frac{1}{2\pi\sqrt{R_A R_B C_A C_B}} \qquad (16\text{–}7)$$

For simplicity, the component values can be made equal, so that $R_A = R_B = R$ and $C_A = C_B = C$. In this case, the expression for the cutoff frequency simplifies to $f_c = 1/2\pi RC$.

As in the single-pole filter, the op-amp in the second-order Sallen and Key filter acts as a noninverting amplifier with the negative feedback provided by the R_1/R_2 network. As you have learned, the damping factor is set by the values of R_1 and R_2, thus making the filter response either Butterworth, Chebyshev, or Bessel. For example, from Table 16–1, the R_1/R_2 ratio must be 0.586 to produce the damping factor of 1.414 required for a second-order Butterworth response.

EXAMPLE 16–3

Determine the cutoff frequency of the low-pass filter in Figure 16–11, and set the value of R_1 for an approximate Butterworth response.

FIGURE 16–11

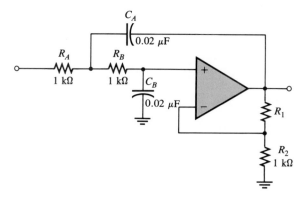

Solution
Since $R_A = R_B = 1 \text{ k}\Omega$ and $C_A = C_B = 0.02 \ \mu\text{F}$,

$$f_c = \frac{1}{2\pi RC} = \frac{1}{2\pi(1 \text{ k}\Omega)(0.02 \ \mu\text{F})} = 7.958 \text{ kHz}$$

For a Butterworth response, $R_1/R_2 = 0.586$.

$$R_1 = 0.586R_2 = 0.586(1 \text{ k}\Omega) = 586 \ \Omega$$

Select a standard value as near as possible to this calculated value.

Practice Exercise 16–3
Determine f_c for Figure 16–11 if $R_A = R_B = R_2 = 2.2 \text{ k}\Omega$ and $C_A = C_B = 0.01 \ \mu\text{F}$. Also determine the value of R_1 for a Butterworth response.

Cascading Low-Pass Filters to Achieve a Higher Roll-Off Rate

A three-pole filter is required to get a third-order low-pass response (-60 dB/decade). This is done by cascading a two-pole low-pass filter and a single-pole low-pass filter, as shown in Figure 16–12(a). Part (b) shows a four-pole configuration obtained by cascading two two-pole filters.

EXAMPLE 16–4

For the four-pole filter in Figure 16–12(b), determine the capacitance values required to produce a cutoff frequency of 2680 Hz if all the resistors are 1.8 kΩ. Also select values for the feedback resistors to get a Butterworth response.

Solution

Both stages must have the same f_c. Assuming equal-value capacitors:

$$f_c = \frac{1}{2\pi RC}$$

$$C = \frac{1}{2\pi R f_c} = \frac{1}{2\pi(1.8 \text{ k}\Omega)(2680 \text{ Hz})} = 0.032 \ \mu\text{F}$$

$$C_{A1} = C_{B1} = C_{A2} = C_{B2} = 0.032 \ \mu\text{F}$$

Select $R_2 = R_4 = 1.8$ kΩ for simplicity. Refer to Table 16–1.

First Stage:

$$DF = 1.848, \quad \frac{R_1}{R_2} = 0.152$$

$$R_1 = 0.152 R_2 = 0.152(1800 \ \Omega) = 273.6 \ \Omega$$

Choose $R_1 = 270 \ \Omega$.

Second Stage:

$$DF = 0.765, \quad \frac{R_3}{R_4} = 1.235$$

$$R_3 = 1.235 R_4 = 1.235(1800 \ \Omega) = 2.223 \text{ k}\Omega$$

Choose $R_3 = 2.2$ kΩ.

Practice Exercise 16–4

For the filter in Figure 16–12(b), determine the capacitance values for $f_c = 1$ kHz if all the filter resistors are 680 Ω. Also specify the values for the feedback resistors to produce a Butterworth response.

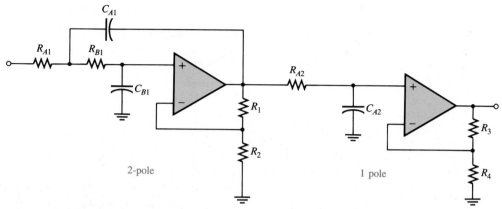

(a) Third-order

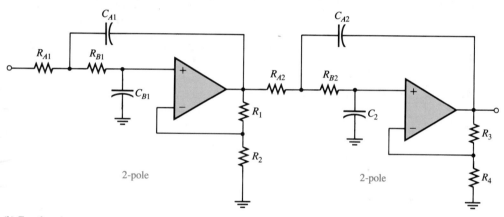

(b) Fourth-order

FIGURE 16–12
Cascaded low-pass filters.

1. How many poles does a second-order low-pass filter have? How many resistors and how many capacitors are used in the frequency-selective network?
2. Why is the damping factor of a filter important?
3. What is the primary purpose of cascading low-pass filters?

SECTION
REVIEW
16–3

ACTIVE HIGH-PASS FILTERS

16–4

A Single-Pole Filter

A high-pass active filter with a 20 dB/decade roll-off is shown in Figure 16–13(a). Notice that the input circuit is a single high-pass RC network. The negative feedback circuit is the same as for the low-pass filters previously discussed. The high-pass response curve is shown in part (b).

Ideally, a high-pass filter passes all frequencies above f_c without limit, as indicated in Figure 16–14(a), although in practice, this is not the case. As you have learned, all op-amps inherently have internal RC networks that limit the amplifier's response at high frequencies. Therefore, there is an upper-frequency limit on the high-pass filter's response which, in effect, makes it a band-pass filter with a very wide bandwidth. In the majority of applications, the internal high-frequency limitation is so much greater than that of the filter's f_c that the limitation can be neglected. In some applications, discrete transistors are used for the gain element to increase the high-frequency limitation beyond that realizable with available op-amps.

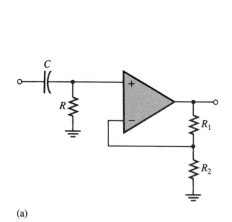

(a)

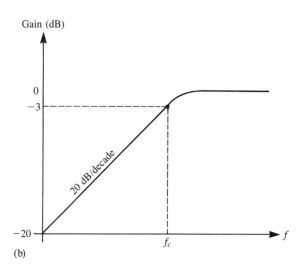

(b)

FIGURE 16–13
Single-pole active high-pass filter and response curve.

The Sallen and Key High-Pass Filter

A high-pass second-order Sallen and Key configuration is shown in Figure 16–15. The components R_A, C_A, R_B, and C_B form the two-pole frequency-selective network. Notice that the positions of the resistors and capacitors in the frequency-selective network are opposite to those in the low-pass configuration. As with the other filters, the response characteristic can be optimized by proper selection of the feedback resistors, R_1 and R_2.

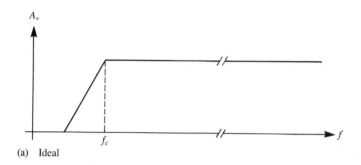

(a)   Ideal

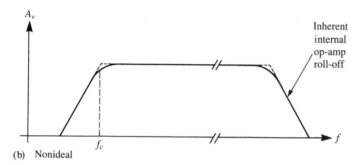

(b)   Nonideal

FIGURE 16–14
High-pass filter response.

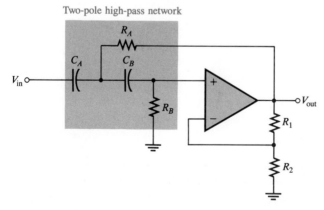

FIGURE 16–15
Basic Sallen and Key second-order high-pass filter.

EXAMPLE 16–5

Choose values for the Sallen and Key high-pass filter in Figure 16–15 to implement an equal-value second-order Butterworth response with a cutoff frequency of approximately 10 kHz.

Solution

1. Start by selecting a value for R_A and R_B (R_1 or R_2 can also be the same value as R_A and R_B for simplicity).

$$R = R_A = R_B = R_2 = 3.3 \text{ k}\Omega \quad \text{(an arbitrary selection)}$$

2. Calculate the capacitance value from $f_c = 1/2\pi RC$.

$$C = C_A = C_B = \frac{1}{2\pi R f_c} = \frac{1}{2\pi(3.3 \text{ k}\Omega)(10 \text{ kHz})} = 0.004 \text{ } \mu\text{F}$$

3. For a Butterworth response, the damping factor must be 1.414 and $R_1/R_2 = 0.586$.

$$R_1 = 0.586 R_2 = 0.586 \text{ } (3.3 \text{ k}\Omega) = 1.93 \text{ k}\Omega$$

If we had let $R_1 = 3.3 \text{ k}\Omega$, then

$$R_2 = \frac{R_1}{0.586} = \frac{3.3 \text{ k}\Omega}{0.586} = 5.63 \text{ k}\Omega$$

Either way, an approximate Butterworth response is realized by choosing the nearest standard value.

Practice Exercise 16–5

Select values for all the components in the high-pass filter of Figure 16–15 to obtain an $f_c = 300$ Hz. Use equal-value components and optimize for a Butterworth response.

Cascading High-Pass Filters

As with the low-pass configuration, first- and second-order high-pass filters can be cascaded to provide three or more poles and thereby create faster roll-off rates. Figure 16–16 shows a six-pole high-pass filter consisting of three two-pole stages. With this configuration optimized for a Butterworth response, a roll-off of 120 dB/decade is achieved.

SECTION REVIEW 16–4

1. How does a high-pass Sallen and Key filter differ from the low-pass configuration?
2. To increase the cutoff frequency of a high-pass filter, would you increase or decrease the resistor values?
3. If three two-pole high-pass filters and one single-pole high-pass filter are cascaded, what is the resulting roll-off?

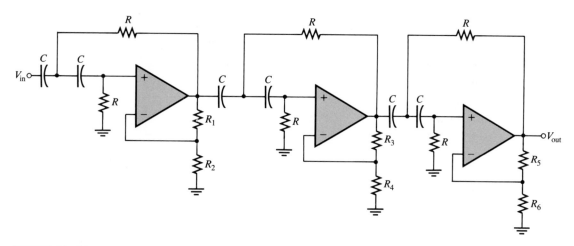

FIGURE 16–16
Sixth-order high-pass filter.

As mentioned, band-pass filters pass all frequencies bounded by a lower- and an upper-frequency limit and reject all others lying outside this specified band. A band-pass response can be thought of as the overlapping of a low-frequency response curve and a high-frequency response curve.

Cascaded Low-Pass and High-Pass Filters Achieve a Band-Pass Response

One way to implement a band-pass filter is a cascaded arrangement of a high-pass filter and a low-pass filter, as shown in Figure 16–17(a), as long as the cutoff frequencies are sufficiently separated. Each of the filters shown is a two-pole Sallen and Key Butterworth configuration so that the roll-off rates are ±40 dB/decade, indicated in the composite response curve of Figure 16–17(b). The cutoff frequency of each filter is chosen so that the response curves overlap sufficiently, as indicated. The cutoff frequency of the high-pass filter must be sufficiently lower than that of the low-pass stage.

The lower frequency f_{c1} of the pass band is the cutoff frequency of the high-pass filter. The upper frequency f_{c2} is the cutoff frequency of the low-pass filter. Ideally, as discussed earlier, the center frequency f_0 of the pass band is the geometric mean of f_{c1} and f_{c2}. The following formulas express the three frequencies of the band-pass filter in Figure 16–17.

$$f_{c1} = \frac{1}{2\pi\sqrt{R_{A1}R_{B1}C_{A1}C_{B1}}}$$

$$f_{c2} = \frac{1}{2\pi\sqrt{R_{A2}R_{B2}C_{A2}C_{B2}}}$$

$$f_0 = \sqrt{f_{c1}f_{c2}}$$

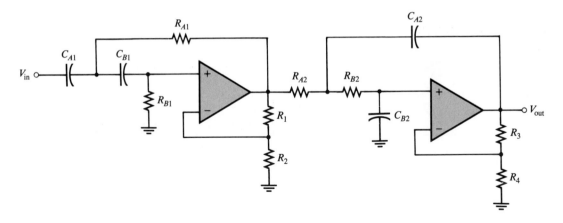

(a)   Two-pole high-pass   Two-pole low-pass

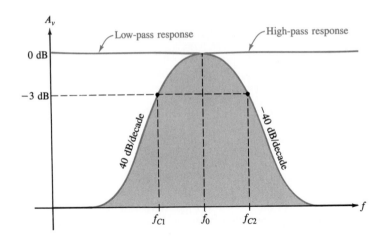

(b)

FIGURE 16–17

Band-pass filter formed by cascading a two-pole high-pass and a two-pole low-pass filter (it does not matter in which order the filters are cascaded).

Of course, if equal-value components are used in implementing each filter, the cutoff frequency equations simplify to the form $f_c = 1/2\pi RC$.

Multiple-Feedback Band-Pass Filter

Another type of filter configuration, shown in Figure 16–18, is a *multiple-feedback* band-pass filter. The two feedback paths are through R_2 and C_1. Components R_1 and C_1 provide the low-pass response, and R_2 and C_2 provide the high-pass response. The maximum gain, A_0, occurs at the center frequency. Q values of less than 10 are typical in this type of filter. An expression for the center frequency is developed as follows, recognizing that R_1

FIGURE 16–18

Multiple-feedback band-pass filter.

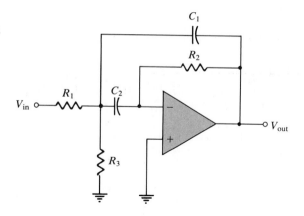

and R_3 appear in parallel as viewed from the C_1 feedback path (with the V_{in} source replaced by a short).

$$f_0 = \frac{1}{2\pi\sqrt{(R_1\|R_3)R_2C_1C_2}}$$

Making $C_1 = C_2 = C$ yields

$$f_0 = \frac{1}{2\pi\sqrt{(R_1\|R_3)R_2C^2}}$$

$$= \frac{1}{2\pi C\sqrt{(R_1\|R_3)R_2}}$$

$$= \frac{1}{2\pi C}\sqrt{\frac{1}{R_2(R_1\|R_3)}}$$

$$= \frac{1}{2\pi C}\sqrt{\frac{1}{R_2}\frac{1}{R_1R_3/(R_1 + R_3)}}$$

$$f_0 = \frac{1}{2\pi C}\sqrt{\frac{R_1 + R_3}{R_1R_2R_3}} \qquad\qquad \textbf{(16–8)}$$

A convenient value for the capacitors is chosen, then the three resistor values are calculated based on the desired values for f_0, BW, and A_0. As you know, the Q can be determined from the relation $Q = f_0/BW$, and the resistors are found using the following formulas (stated without derivation).

$$R_1 = \frac{Q}{2\pi f_0 C A_0}$$

$$R_2 = \frac{Q}{\pi f_0 C}$$

$$R_3 = \frac{Q}{2\pi f_0 C(2Q^2 - A_0)}$$

To develop a gain expression, we solve for Q in the first two equations above:

$$Q = 2\pi f_0 A_0 C R_1$$
$$Q = \pi f_0 C R_2$$

Then,

$$2\pi f_0 A_0 C R_1 = \pi f_0 C R_2$$

Cancelling, we get

$$2A_0 R_1 = R_2$$

$$A_0 = \frac{R_2}{2R_1} \qquad (16\text{--}9)$$

In order for the denominator of the equation $R_3 = Q/2\pi f_0 C(2Q^2 - A_0)$ to be positive, $A_0 < 2Q^2$, which imposes a limitation on the gain.

EXAMPLE 16–6

Determine the center frequency, maximum gain, and bandwidth for the filter in Figure 16–19.

FIGURE 16–19

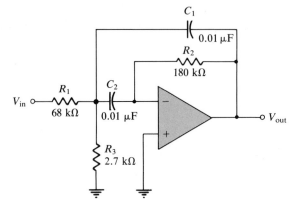

Solution

$$f_0 = \frac{1}{2\pi C}\sqrt{\frac{R_1 + R_3}{R_1 R_2 R_3}}$$

$$= \frac{1}{2\pi(0.01\ \mu F)}\sqrt{\frac{68\ k\Omega + 2.7\ k\Omega}{(68\ k\Omega)(180\ k\Omega)(2.7\ k\Omega)}}$$

$$= 736\ \text{Hz}$$

$$A_0 = \frac{R_2}{2R_1} = \frac{180\ k\Omega}{2(68\ k\Omega)} = 1.32$$

$$Q = \pi f_0 C R_2$$
$$= \pi (736 \text{ Hz})(0.01 \ \mu\text{F})(180 \text{ k}\Omega)$$
$$= 4.16$$
$$BW = \frac{f_0}{Q} = \frac{736 \text{ Hz}}{4.16} = 176.9 \text{ Hz}$$

State Variable Filters

The *state variable* or *universal active filter* is widely used for band-pass applications. As shown in Figure 16–20, it consists of a summing amplifier and two op-amp integrators (which act as single-pole low-pass filters) that are combined in a cascaded arrangement to form a second-order filter. Although used primarily as a band-pass filter, the state variable configuration also provides low-pass (LP) and high-pass (HP) outputs. The center frequency is set by the RC networks in both integrators. When used as a band-pass filter, the cutoff frequencies of the integrators are usually made equal, thus setting the center frequency of the pass band.

Basic Operation.   At input frequencies below f_c, the input signal passes through the summing amplifier and integrators and is fed back $180°$ out-of-phase. Thus, the feedback signal and input signal cancel for all frequencies below approximately f_c. As the low-pass response of the integrators rolls off, the feedback signal diminishes, thus allowing the input to pass through to the band-pass output. Above f_c, the low-pass response disappears,

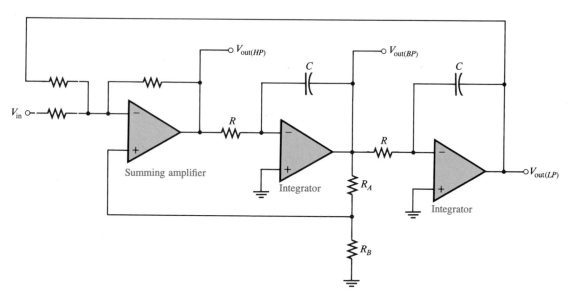

FIGURE 16–20
State variable filter.

FIGURE 16–21
General state variable response curves.

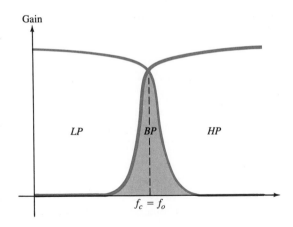

thus preventing the input signal from passing through the integrators. As a result, the band-pass output peaks sharply at f_c, as indicated in Figure 16–21. Stable Qs up to 100 can be obtained with this type of filter. The Q is set by the feedback resistors R_A and R_B according to the following equation.

$$Q = \frac{1}{3}\left(\frac{R_A}{R_B} + 1\right)$$ (16–10)

The state variable filter cannot be optimized for low-pass, high-pass, and band-pass performance simultaneously for this reason: To optimize for a low-pass or a high-pass Butterworth response, DF must equal 1.414. Since $Q = 1/DF$, a Q of 0.707 will result. Such a low Q provides a very poor band-pass response (large BW). For optimization as a band-pass filter, the Q must be set high.

EXAMPLE 16–7

Determine the center frequency, Q, and BW for the band-pass output of the state variable filter in Figure 16–22.

Solution
For each integrator:

$$f_c = \frac{1}{2\pi RC} = \frac{1}{2\pi(1 \text{ k}\Omega)(0.022 \text{ }\mu\text{F})} = 7.23 \text{ kHz}$$

The center frequency is approximately equal to the cutoff frequencies of the integrators:

$$f_0 = f_c = 7.23 \text{ kHz}$$

$$Q = \frac{1}{3}\left(\frac{R_A}{R_B} + 1\right) = \frac{1}{3}\left(\frac{100 \text{ k}\Omega}{1 \text{ k}\Omega} + 1\right) = 33.67$$

$$BW = \frac{f_0}{Q} = \frac{7.23 \text{ kHz}}{33.67} = 214.7 \text{ Hz}$$

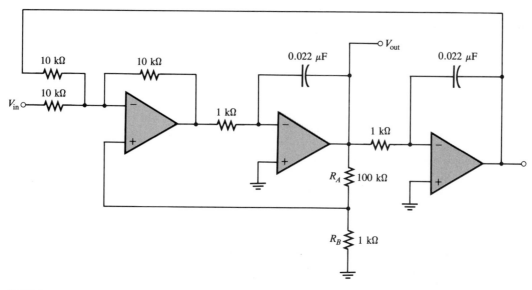

FIGURE 16–22

Practice Exercise 16–7
Determine f_0, Q, and BW for the filter in Figure 16–21 if $R_1 = R_2 = R_B = 330 \ \Omega$ with all other component values the same as shown on the schematic.

---

1. What determines *selectivity* in a band-pass filter?
2. One filter has a $Q = 5$ and another has a $Q = 25$. Which has the narrower bandwidth?
3. List the elements that make up a state variable filter.

ACTIVE BAND-STOP FILTERS

16–6

Band-stop filters reject a specified band of frequencies and pass all others. The response is opposite to that of a band-pass filter.

Multiple-Feedback Band-Stop Filter

Figure 16–23 shows a multiple-feedback band-stop filter. Notice that this configuration is similar to the band-pass version except that R_3 has been omitted and R_A and R_B have been added.

State Variable Band-Stop Filter

Summing the low-pass and the high-pass responses of the state variable filter covered in the last section creates a band-stop response as shown in Figure 16–24. One important

FIGURE 16–23
Multiple feedback band-stop filter.

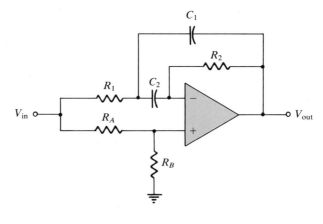

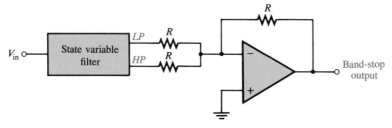

FIGURE 16–24
State variable band-stop filter.

application of this filter is minimizing the 60 Hz "hum" in audio systems by setting the center frequency to 60 Hz.

EXAMPLE 16–8

Verify that the band-stop filter in Figure 16–25 has a center frequency of 60 Hz, and optimize it for a Q of 30.

Solution

f_0 equals the f_c of the integrator stages.

$$f_0 = \frac{1}{2\pi RC} = \frac{1}{2\pi(12 \text{ k}\Omega)(0.22 \text{ }\mu\text{F})} = 60 \text{ Hz}$$

We obtain a $Q = 30$ by choosing R_B and then calculating R_A.

$$Q = \frac{1}{3}\left(\frac{R_A}{R_B} + 1\right)$$

$$R_A = (3Q - 1)R_B$$

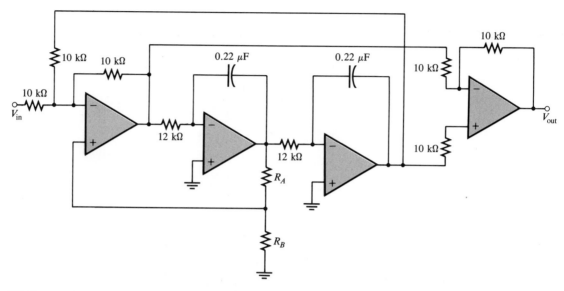

FIGURE 16–25

Choose $R_B = 1\ k\Omega$. Then

$$R_A = [3(30) - 1]1\ k\Omega = 89\ k\Omega$$

---

1. How does a band-stop response differ from a band-pass response?
2. How is a state variable band-pass filter converted to a band-stop filter?

SECTION
REVIEW
16–6

TESTING FOR FILTER RESPONSE

We will discuss two methods of determining a filter's response by measurement: *discrete point measurement* and *swept frequency measurement*.

16–7

Discrete Point Measurement

Figure 16–26 shows an arrangement for taking filter output voltage measurements at discrete values of input frequency using common laboratory instruments. The general procedure is as follows:

1. Set the amplitude of the sine wave generator to a desired voltage level.
2. Set the frequency of the sine wave generator to a value well below the expected cutoff frequency of the filter under test. For a low-pass filter, set the frequency

FIGURE 16–26
FIGURE 16–26
Test set-up for discrete point measurement of the filter response. (Readings are arbitrary and for display only.)

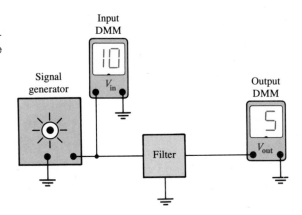

as near as possible to 0 Hz. For a band-pass filter, set the frequency well below the expected lower cutoff frequency.

3. Increase the frequency in predetermined steps sufficient to allow enough data points for an accurate response curve.
4. Maintain a constant input voltage amplitude while varying the frequency.
5. Record the output voltage at each value of frequency.
6. After recording a sufficient number of points, plot a graph of output voltage versus frequency.

Swept Frequency Measurement

The swept frequency method requires more elaborate test equipment than does the discrete point method, but is much more efficient and can result in a more accurate response curve. A general test set-up is shown in Figure 16–27 using a *swept frequency generator* and a *spectrum analyzer*.

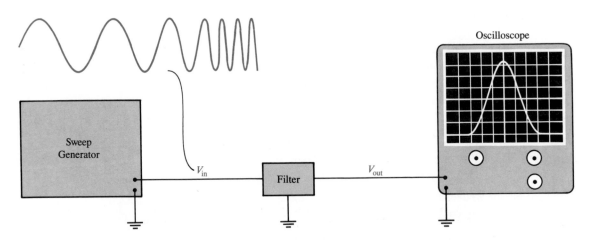

FIGURE 16–27
Test set-up for swept frequency measurement of the filter response.

The swept frequency generator produces a constant amplitude output signal whose frequency increases linearly between two preset limits, as indicated in the figure. The spectrum analyzer is essentially an elaborate oscilloscope that can be calibrated for a desired *frequency span/division* rather than for the usual *time/division* setting. Therefore, as the input frequency to the filter sweeps through a preselected range, the response curve is traced out on the screen of the spectrum analyzer. An actual swept frequency test set-up using typical equipment is shown in Figure 16–28.

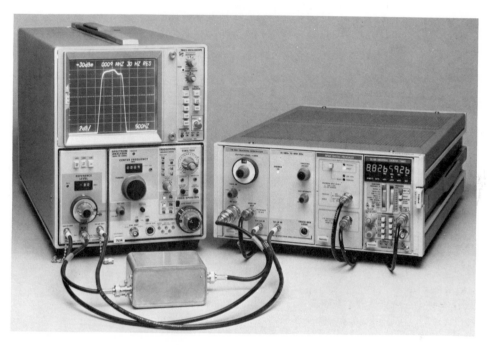

FIGURE 16–28
Courtesy of Tektronix, Inc.

---

1. What is the purpose of the two tests discussed in this section?
2. Name one advantage and one disadvantage of each test method.

**SECTION
REVIEW
16–7**

A SYSTEM APPLICATION

The filters in an FM stereo receiver separate into three signals the output of the discriminator, which contains frequency components from 30 Hz to 15 kHz (left + right), 23 kHz to 53 kHz (left − right), and 19 kHz (pilot subcarrier), as indicated in Figure 16–29.

The L + R signal is passed by the low-pass filters with a frequency response of 0 to 15 kHz, the L − R signal is passed by the band-pass filter with a response of 23 kHz to 53 kHz, and the pilot subcarrier is passed by the 19 kHz band-pass filter. After passing through the filter, the pilot carrier frequency doubles to 38 kHz and combines in the

16–8

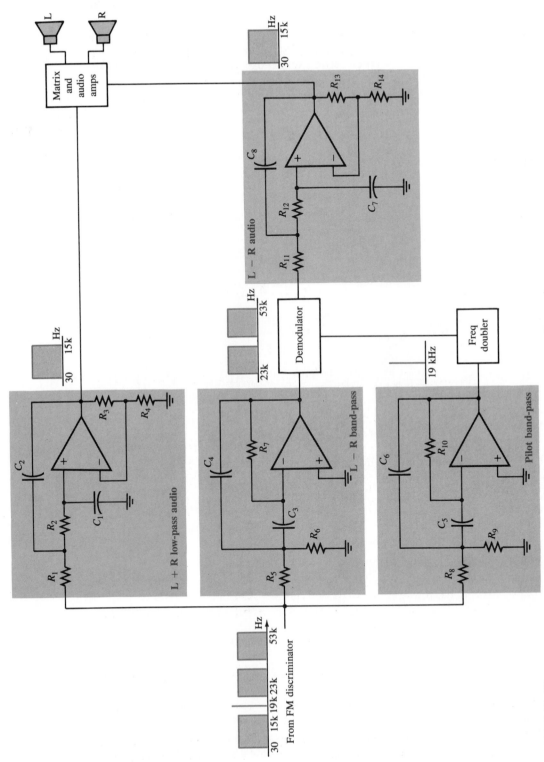

FIGURE 16–29

An example of FM stereo receiver filter circuits.

demodulator with the 23 kHz to 53 kHz L − R signal to produce sum and difference frequencies. All these frequencies except the 30 Hz to 15 kHz L − R audio are filtered out by the low-pass filter at the demodulator output. One possible implementation of the FM stereo filter system using op-amps is shown in Figure 16–29.

FORMULAS

$$(16\text{--}1) \quad f_c = \frac{1}{2\pi RC} \qquad\qquad \text{Filter cutoff frequency}$$

$$(16\text{--}2) \quad BW = f_{c2} - f_{c1} \qquad\qquad \text{Filter bandwidth}$$

$$(16\text{--}3) \quad f_0 = \sqrt{f_{c1}f_{c2}} \qquad\qquad \text{Center frequency of a band-pass filter}$$

$$(16\text{--}4) \quad Q = \frac{f_0}{BW} \qquad\qquad \text{Quality factor of a band-pass filter}$$

$$(16\text{--}5) \quad Q = \frac{1}{DF} \qquad\qquad Q \text{ in terms of damping factor}$$

$$(16\text{--}6) \quad DF = 2 - \frac{R_1}{R_2} \qquad\qquad \text{Damping factor}$$

$$(16\text{--}7) \quad f_c = \frac{1}{2\pi\sqrt{R_A R_B C_A C_B}} \qquad\qquad \text{Cutoff frequency for a second-order Sallen and Key filter}$$

$$(16\text{--}8) \quad f_0 = \frac{1}{2\pi C}\sqrt{\frac{R_1 + R_3}{R_1 R_2 R_3}} \qquad\qquad \text{Center frequency of a multiple-feedback filter}$$

$$(16\text{--}9) \quad A_0 = \frac{R_2}{2R_1} \qquad\qquad \text{Gain of a multiple-feedback filter}$$

$$(16\text{--}10) \quad Q = \frac{1}{3}\left(\frac{R_A}{R_B} + 1\right) \qquad\qquad Q \text{ of a state variable filter}$$

SUMMARY

☐ The bandwidth in a low-pass filter equals the cutoff frequency, because the response extends to 0 Hz.

☐ The bandwidth in a high-pass filter extends above the cutoff frequency and is limited only by the inherent frequency limitation of the active circuit.

☐ A band-pass filter passes all frequencies within a band between a lower and an upper cutoff frequency and rejects all others outside this band.

☐ The bandwidth of a band-pass filter is the difference between the upper cutoff frequency and the lower cutoff frequency.

☐ A band-stop filter rejects all frequencies within a specified band and passes all those outside this band.

☐ Filters with the Butterworth response characteristic have a very flat response in the pass band and exhibit a roll-off of 20 dB/decade/pole.

☐ These filters are used when all the frequencies in the pass band must have the same gain.

☐ Filters with the Chebyshev characteristic have ripples or overshoot in the pass band and exhibit a faster roll-off per pole than filters with the Butterworth characteristic.

□ Filters with the Bessel characteristic are used for filtering pulse waveforms. Their linear phase characteristic results in minimal waveshape distortion. The roll-off rate per pole is slower than for the Butterworth.

□ In filter terminology, a single RC network is called a *pole*.

□ Each pole in a Butterworth filter causes the output to roll off at a rate of 20 dB/decade.

□ The quality factor Q of a band-pass filter determines the filter's selectivity. The higher the Q, the narrower the bandwidth and the better the selectivity.

□ The damping factor determines the filter response characteristic (Butterworth, Chebyshev, or Bessel).

SELF-TEST

1. Identify each type of response (low-pass, high-pass, band-pass, or band-stop) in Figure 16–30.

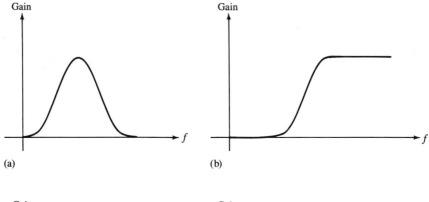

(a)   (b)

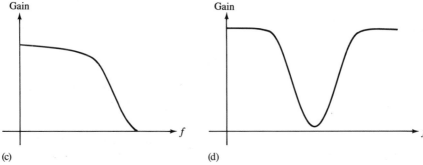

(c)   (d)

FIGURE 16–30

2. Identify each type of filter (low-pass, high-pass, band-pass, or band-stop) in Figure 16–31, and determine the order of each filter.

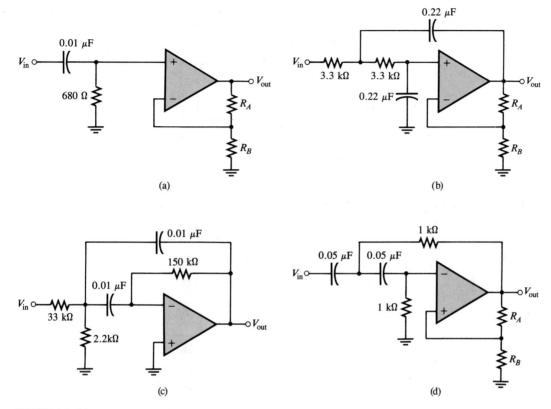

FIGURE 16–31

3. Determine the cutoff frequency or frequencies for each filter in Figure 16–31 and the center frequency, where applicable.

4. Specify the roll-off rate of each filter in Figure 16–31.

5. Determine if the filters in Figure 16–32 are optimized for the Butterworth response.

6. Response curves for high-pass second-order filters are shown in Figure 16–33. Identify each as Butterworth, Chebyshev, or Bessel.

7. Using a block diagram format, show how to implement the following roll-off rates using single-pole and two-pole low-pass filters with Butterworth responses.
   (a) −40 dB/decade
   (b) −20 dB/decade
   (c) −60 dB/decade
   (d) −100 dB/decade
   (e) −120 dB/decade

FIGURE 16–32

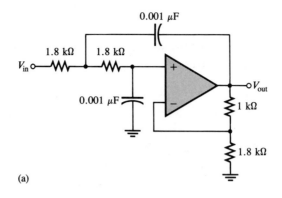

(a)

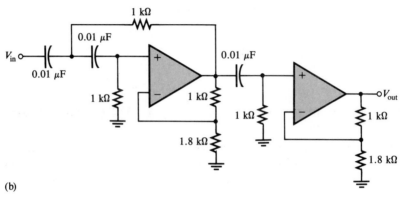

(b)

FIGURE 16–33

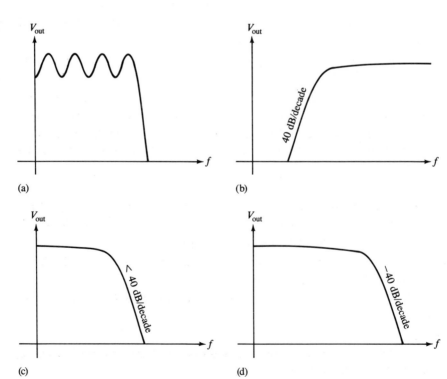

(a)

(b)

(c)

(d)

8. For the filter in Figure 16–34: **(a)** How would you increase the cutoff frequency? **(b)** How would you increase the gain?

FIGURE 16–34

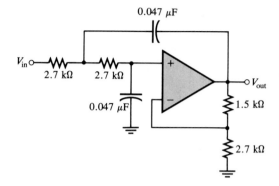

Section 16–1

16–1 A certain low-pass filter has a cutoff frequency of 800 Hz. What is its bandwidth?

16–2 A single-pole high-pass filter has a frequency-selective circuit with $R = 2.2$ kΩ and $C = 0.0015$ μF. What is the cutoff frequency? Can you determine the bandwidth from the available information?

16–3 What is the roll-off rate of the filter described in problem 16–2?

16–4 What is the bandwidth of a band-pass filter whose cutoff frequencies are 3.2 kHz and 3.9 kHz? What is the Q of this filter?

16–5 What is the center frequency of a filter with a Q of 15 and a bandwidth of 1 kHz?

Section 16–2

16–6 What is the damping factor in each active filter shown in Figure 16–35? Which filters are approximately optimized for a Butterworth response characteristic?

16–7 For the filters in Figure 16–35 that do not have a Butterworth response, specify the changes necessary to convert them to Butterworth responses. (Use nearest standard values.)

Section 16–3

16–8 Is the four-pole filter in Figure 16–36 a low-pass or a high-pass type? Is it approximately optimized for a Butterworth response? What is the roll-off rate?

16–9 Determine the cutoff frequency in Figure 16–36.

16–10 Without changing the response curve, adjust the component values in the filter of Figure 16–36 to make it an equal-value filter.

16–11 Modify the filter in Figure 16–36 to increase the roll-off rate to −120 dB/decade.

Section 16–4

16–12 Convert the filter in problem 16–10 to a low-pass with the same cutoff frequency and response characteristic.

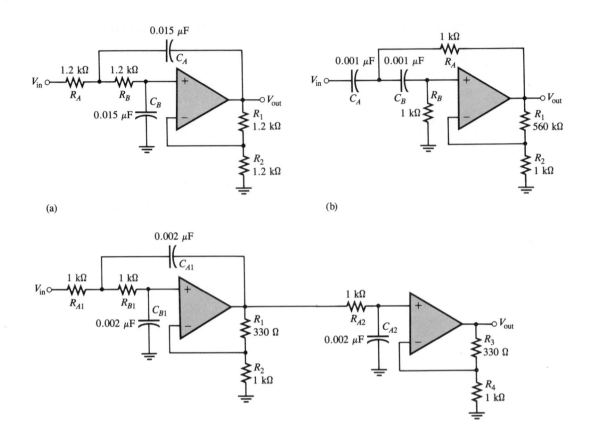

(a)

(b)

(c)

FIGURE 16–35

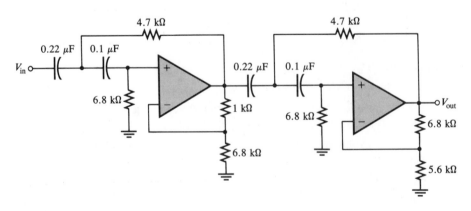

FIGURE 16–36

16–13 Make the necessary circuit modification to reduce by half the cutoff frequency in problem 16–12.

Section 16–5

16–14 Identify each band-pass filter configuration in Figure 16–37.

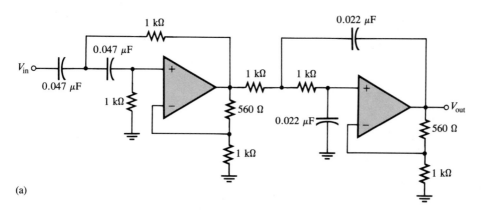

FIGURE 16–37

(a)

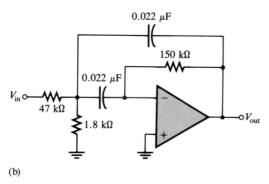

(b)

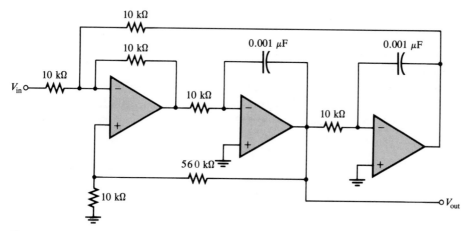

(c)

16–15 Determine the center frequency and bandwidth for each filter in Figure 16–37.

16–16 Optimize the state variable filter in Figure 16–38 for $Q = 50$. What bandwidth is achieved?

Section 16–6

16–17 Show how to make a notch (band-stop) filter using the basic circuit in Figure 16–38.

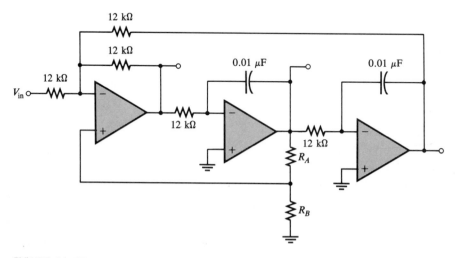

FIGURE 16–38

16–18 Modify the band-stop filter in problem 16–17 for a center frequency of 120 Hz.

ANSWERS TO SECTION REVIEWS

Section 16–1

1. Its cutoff frequency.
2. The inherent frequency limitation of the op-amp.
3. Q and BW are inversely related. The higher the Q, the better the selectivity, and vice versa.

Section 16–2

1. Butterworth is very flat in the pass band and has a 20 dB/decade/pole roll-off. Chebyshev has ripples in the pass band and has greater than 20 dB/decade/pole roll-off. Bessel has a linear phase characteristic and less than 20 dB/decade/pole roll-off.
2. The damping factor.
3. Frequency-selection network, gain element, and negative feedback circuit.

Section 16–3
1. Two poles; two resistors and two capacitors.
2. The damping factor sets the response characteristic.
3. Increase the roll-off rate.

Section 16–4
1. The positions of the Rs and Cs in the frequency-selection circuit are opposite.
2. Decrease.
3. 60 dB/decade.

Section 16–5
1. Q.
2. $Q = 25$.
3. Summing amplifier and two integrators.

Section 16–6
1. A band-stop rejects frequencies within the stop band. A band-pass passes frequencies within the pass band.
2. The low-pass and high-pass outputs are summed.

Section 16–7
1. To check the response of a filter.
2. Discrete point measurement—tedious and less complete, simpler equipment; swept frequency measurement—uses more expensive equipment, more efficient, can be more accurate and complete.

Example 16–3: 7.234 kHz, 1.29 kΩ

Example 16–4: $C_{A1} = C_{A2} = C_{B1} = C_{B2} = 0.234\ \mu F$, $R_2 = R_4 = 680\ \Omega$, $R_1 = 103\ \Omega$, $R_3 = 840\ \Omega$

Example 16–5: $R_A = R_B = R_2 = 10\ k\Omega$, $C_A = C_B = 0.053\ \mu F$, $R_1 = 586\ \Omega$

Example 16–7: $f_0 = 21.922$ kHz, $Q = 101$, $BW = 217$ Hz

ANSWERS TO PRACTICE EXERCISES

Voltage Regulators

In this chapter you will learn

- ☐ The principles of voltage regulation
- ☐ The difference between line regulation and load regulation
- ☐ How a series voltage regulator works
- ☐ What overload protection is and how it works
- ☐ How a shunt voltage regulator works
- ☐ How a switching regulator works
- ☐ Various types of IC voltage regulators
- ☐ Some specific applications of voltage regulators
- ☐ The meaning of fold-back current limiting
- ☐ How an external pass transistor increases the power-handling capability of a regulator
- ☐ How voltage regulators are used in a specific system application

Chapter 17

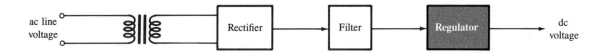

The purpose of a voltage regulator is to provide a constant dc output voltage that is practically independent of the input voltage, output load current, and temperature. The input voltage to a regulator is usually a rectified and filtered dc voltage derived from a 60 Hz ac line (as discussed in Chapter 3), or in many cases, a dc voltage from a battery.

Most voltage regulators contain four basic elements: a reference source, an error detector, a sampling element, and a control device, and sometimes protection circuitry. Typically, the reference source is a fixed voltage, the error detector is an op-amp, the sampling element is a voltage divider, the control device is a transistor, and the protection is provided by some type of current limiting.

Chapter 4 presented the basic concepts of voltage regulation and the zener diode regulator. This chapter introduces integrated circuit regulators. All electronic systems require a dc power supply to provide the dc voltages and the current to operate the circuits. As you saw earlier, the dc power supply is an important part of the superheterodyne receiver because amplifiers, oscillators, active filters, and other circuits must be biased with a dc voltage. The voltage regulator portion of a dc power supply maintains a constant dc voltage when changes in line voltage occur or when circuit demands create changes in the supply current. IC voltage regulators are common in many system applications.

VOLTAGE REGULATION

17–1

The two basic aspects of voltage regulation are *line regulation* and *load regulation*.

Line Regulation

When the dc input (line) voltage changes, the voltage regulator must maintain a nearly constant output voltage, as illustrated in Figure 17–1.

Line regulation can be defined as the percentage change in the output voltage for a given change in the input (line) voltage. It is usually expressed in units of %/V. For example, a line regulation of 0.05%/V means that the output voltage changes 0.05 percent when the input voltage increases or decreases by 1 volt. Line regulation can be calculated using the following formula (Δ means ''a change in''):

$$\text{Line regulation} = \frac{(\Delta V_{\text{OUT}}/V_{\text{OUT}})100\%}{\Delta V_{\text{IN}}} \qquad (17\text{--}1)$$

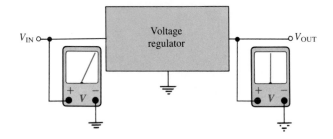

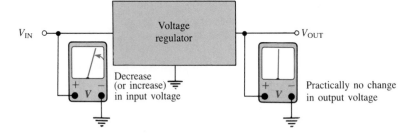

FIGURE 17–1

Line regulation: A change in input (line) voltage does not significantly affect the output voltage of a regulator (within certain limits).

EXAMPLE 17–1

When the input to a particular voltage regulator decreases by 5 V, the output decreases by 0.25 V. The nominal output is 15 V. Determine the line regulation in %/V.

Solution

The line regulation is

$$\frac{(\Delta V_{\text{OUT}}/V_{\text{OUT}})100\%}{\Delta V_{\text{IN}}} = \frac{(0.25 \text{ V}/15 \text{ V})100\%}{5 \text{ V}} = 0.333\%/\text{V}$$

Practice Exercise 17–1

The input of a certain regulator increases by 3.5 V. As a result, the output voltage increases by 0.42 V. The nominal output is 20 V. Determine the regulation in %/V.

Load Regulation

When the amount of current through a load changes due to a varying load resistance, the voltage regulator must maintain a nearly constant output voltage across the load, as illustrated in Figure 17–2.

Load regulation can be defined as the percentage change in output voltage for a given change in load current. It can be expressed as a percentage change in output voltage from no-load (NL) to full-load (FL) as follows.

$$\text{Load regulation} = \frac{(V_{\text{NL}} - V_{\text{FL}})100\%}{V_{\text{FL}}} \qquad \textbf{(17–2)}$$

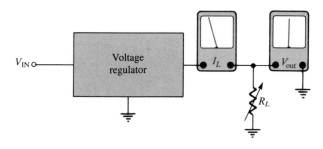

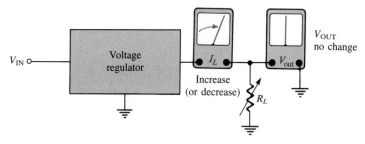

FIGURE 17–2

Load regulation: A change in load current has practically no effect on the output voltage of a regulator (within certain limits).

Alternately, the load regulation can be expressed as a percentage change in output voltage for *each* mA change in load current. For example, a load regulation of 0.01%/mA means that the output voltage changes 0.01 percent when the load current increases or decreases 1 mA.

EXAMPLE 17–2

A certain voltage regulator has a 12 V output when there is no load ($I_L = 0$). When there is a full-load current of 10 mA, the output voltage is 11.95 V. Express the voltage regulation as a percentage change from no-load to full-load and also as a percentage change for each mA change in load current.

Solution
The no-load output voltage is

$$V_{NL} = 12 \text{ V}$$

The full-load output voltage is

$$V_{FL} = 11.95 \text{ V}$$

The load regulation is

$$\left(\frac{V_{NL} - V_{FL}}{V_{FL}}\right)100\% = \left(\frac{12 \text{ V} - 11.95 \text{ V}}{11.95 \text{ V}}\right)100\%$$
$$= 0.418\%$$

The load regulation can also be expressed as

$$\frac{0.418\%}{10 \text{ mA}} = 0.0418\%/\text{mA}$$

where the change in load current from no-load to full-load is 10 mA.

Practice Exercise 17–2
A regulator has a no-load output voltage of 18 V and a full-load output of 17.85 V at a load current of 50 mA. Determine the voltage regulation as a percentage change from no-load to full-load and also as a percentage change for each mA change in load current.

SECTION REVIEW 17–1

1. Define *line regulation*.
2. Define *load regulation*.

BASIC SERIES REGULATOR

17–2

The two fundamental classes of voltage regulators are *linear* regulators and *switching* regulators. Both of these are available in integrated circuit form. There are two basic types of linear regulator. One is the *series regulator* and the other is the *shunt regulator*. In this section, we will look at the series regulator. The shunt and switching regulators are covered in the next two sections.

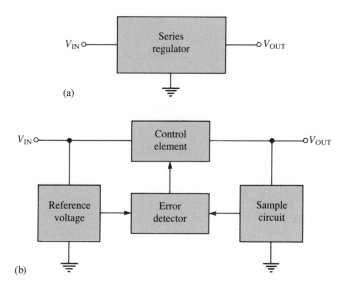

(a)

(b)

FIGURE 17–3
Simple series voltage regulator block diagram.

A simple representation of a series type of linear regulator is shown in Figure 17–3(a), and the basic components are shown in the block diagram in part (b). Notice that the control element is in *series* with the load between input and output. The output sample circuit senses a change in the output voltage. The error detector compares the sample voltage with a reference voltage and causes the control element to compensate in order to maintain a constant output voltage.

Regulating Action

A basic op-amp series regulator circuit is shown in Figure 17–4. The operation is as follows. The resistive voltage divider formed by R_2 and R_3 senses any change in the output voltage. When the output tries to decrease because of a decrease in V_{IN} or because of an increase in I_L, a *proportional* voltage decrease is applied to the op-amp's inverting input by the voltage divider. Since the zener diode holds the other op-amp input at a nearly *fixed* reference voltage V_{REF}, a small difference voltage (error voltage) is developed across the op-amp's inputs. This difference voltage is amplified, and the op-amp's output

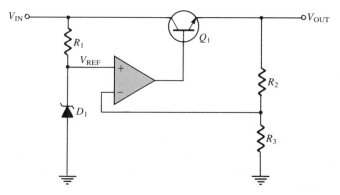

FIGURE 17–4
Basic op-amp series regulator.

voltage increases. This increase is applied to the base of Q_1, causing the emitter voltage V_{OUT} to increase until the voltage to the inverting input again equals the reference (zener) voltage. This action offsets the attempted decrease in output voltage, thus keeping it nearly constant. Q_1 is a power transistor and is often used with a heat sink because it must handle all of the load current.

The opposite action occurs when the output tries to increase. Figure 17–5 illustrates the regulating action of the circuit. The op-amp in Figure 17–4 is actually connected as a noninverting amplifier where the reference voltage V_{REF} is the input at the noninverting terminal, and the R_2/R_3 voltage divider forms the negative feedback network. The closed-loop voltage gain is

$$A_{cl} = 1 + \frac{R_2}{R_3} \qquad \textbf{(17–3)}$$

Therefore, the regulated output voltage (neglecting the base-emitter voltage of Q_1) is

$$V_{OUT} \cong \left(1 + \frac{R_2}{R_3}\right) V_{REF} \qquad \textbf{(17–4)}$$

From this analysis you can see that the output voltage is determined by the zener voltage and the resistors R_2 and R_3. It is relatively independent of the input voltage, and therefore, regulation is achieved (as long as the input voltage and load current are within specified limits).

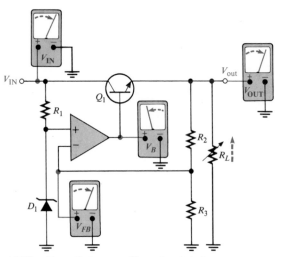

(a) When V_{IN} or R_L increases, V_{OUT} attempts to increase. The feedback voltage, V_{FB}, also attempts to increase, and, as a result, the op-amp's output voltage, V_B, applied to the base of the control transistor, attempts to decrease, thus compensating for the attempted increase in V_{OUT}.

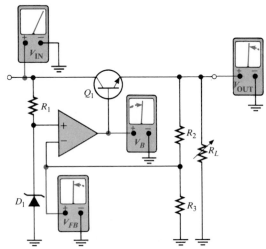

(b) When V_{IN} (or R_L) stabilizes at its new higher value, the voltages are at their original values, thus keeping V_{OUT} constant as result of the negative feedback.

FIGURE 17–5

Illustration of series regulator action that keeps V_{OUT} constant when V_{IN} or R_L changes.

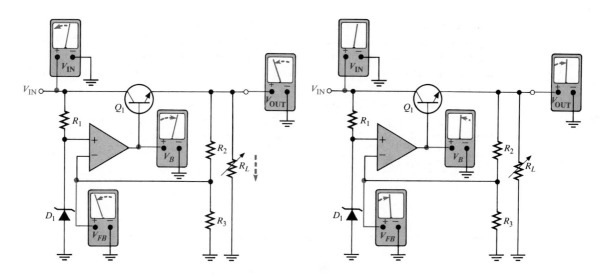

(c) When V_{IN} or R_L decreases, V_{OUT} attempts to decrease, V_{FB} also attempts to decrease, and as a result, V_B attempts to increase, thus compensating for the attempted decrease in V_{OUT}.

(d) When V_{IN} (or R_L) stabilizes at its new lower value, the voltages are at their original values, thus keeping V_{OUT} constant as a result of the negative feedback.

FIGURE 17–5
(continued)

EXAMPLE 17–3

Determine the output voltage for the regulator in Figure 17–6.

FIGURE 17–6

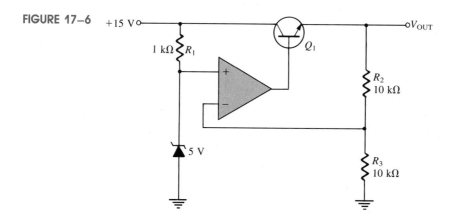

Solution

$$V_{REF} = 5 \text{ V}$$

$$V_{OUT} = \left(1 + \frac{R_2}{R_3}\right)V_{REF} = \left(1 + \frac{10 \text{ k}\Omega}{10 \text{ k}\Omega}\right)5 \text{ V} = (2)5 \text{ V} = 10 \text{ V}$$

Practice Exercise 17–3

The following changes are made in the circuit in Figure 17–6: $V_Z = 3.3$ V, $R_1 = 1.8$ kΩ, $R_2 = 22$ kΩ, and $R_3 = 18$ kΩ. What is the output voltage?

Short-Circuit or Overload Protection

If an excessive amount of load current is drawn, the series-pass transistor can be quickly damaged or destroyed. Most regulators employ some type of excess current protection in the form of a current-limiting mechanism. Figure 17–7 shows one method of current limiting to prevent overloads called *constant current limiting*. The current-limiting circuit consists of transistor Q_2 and resistor R_4.

The load current through R_4 creates a voltage from base to emitter of Q_2. When I_L reaches a predetermined maximum value, the voltage drop across R_4 is sufficient to forward-bias the base-emitter junction of Q_2, thus causing it to conduct. Enough Q_1 base current is diverted into the collector of Q_2 so that I_L is limited to its maximum value $I_{L(\max)}$. Since the base-to-emitter voltage of Q_2 cannot exceed about 0.7 V for a silicon transistor, the voltage across R_4 is held to this value, and the load current is limited to

$$I_{L(\max)} = \frac{0.7 \text{ V}}{R_4} \qquad\qquad (17\text{–}5)$$

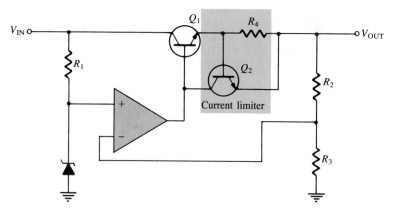

FIGURE 17–7
Series regulator with constant current limiting.

EXAMPLE 17–4

Determine the maximum current that the regulator in Figure 17–8 can provide to a load.

Solution

$$I_{L(\max)} = \frac{0.7 \text{ V}}{R_4} = \frac{0.7 \text{ V}}{1 \ \Omega} = 0.7 \text{ A}$$

FIGURE 17–8

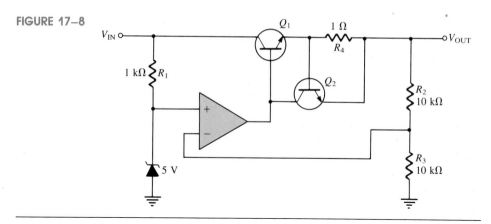

1. List the basic components in a series regulator.
2. A certain series regulator has an output voltage of 8 V. If the op-amp's closed-loop gain is 4, what is the value of the reference voltage?

BASIC SHUNT REGULATOR

17–3

As you have seen, the control element in the series regulator is the series-pass transistor. A simple representation of a shunt type of linear regulator is shown in Figure 17–9(a), and the basic components are shown in the block diagram in part (b).

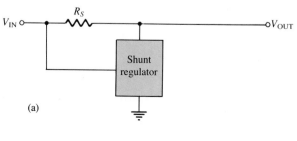

(a)

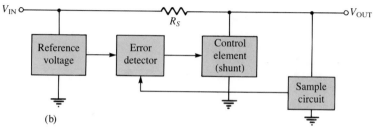

(b)

FIGURE 17–9
Simple shunt regulator block diagrams.

FIGURE 17–10

Basic op-amp shunt regulator.

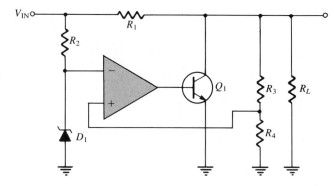

In the basic shunt regulator, the control element is a series resistor (R_1) and a transistor Q_1 in parallel with the load, as shown in Figure 17–10. The operation of the circuit is similar to that of the series regulator, except that regulation is achieved by controlling the current through the parallel transistor Q_1.

When the output voltage tries to decrease due to a change in input voltage, load current, or temperature, the attempted decrease is sensed by R_3 and R_4 and applied to the op-amp's noninverting input. The resulting difference voltage reduces the op-amp's output, driving Q_1 less, thus reducing its collector current (shunt current) and increasing its effective collector-to-emitter resistance r_{ce}. Since r_{ce} acts as a voltage divider with R_1, this action offsets the attempted decrease in V_{OUT} and maintains it at an almost constant level. The opposite action occurs when the output tries to increase. This regulating action of the shunt element is illustrated in Figure 17–11.

With I_L and V_{OUT} constant, a change in the input voltage produces a change in shunt current (I_S) as follows (Δ means "a change in"):

$$\Delta I_S = \frac{\Delta V_{IN}}{R_1} \tag{17–6}$$

With a constant V_{IN} and V_{OUT}, a change in load current causes an opposite change in shunt current:

$$\Delta I_S = -\Delta I_L \tag{17–7}$$

This formula says that if I_L increases, I_S decreases, and vice versa. The shunt regulator is less efficient than the series type but offers inherent short-circuit protection. If the output is shorted ($V_{OUT} = 0$), the load current is limited by the series resistor R_1 to a maximum value as follows ($I_S = 0$).

$$I_{L(\text{max})} = \frac{V_{IN}}{R_1} \tag{17–8}$$

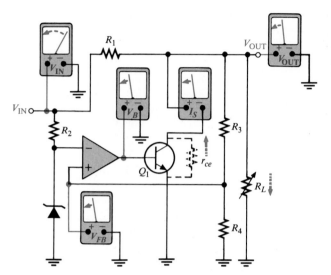

(a) Initial response to a decrease in V_{IN} or R_L

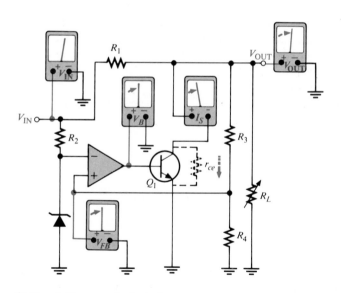

(b) V_{OUT} held constant by the feedback action.

EXAMPLE 17–5

In Figure 17–12, what power rating must R_1 have if the maximum input voltage is 12.5 V?

FIGURE 17–12

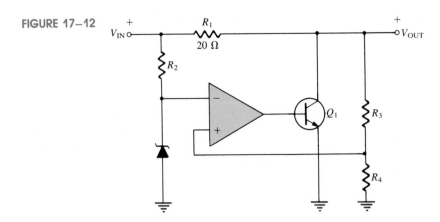

Solution
The worst-case power dissipation in R_1 occurs when the output is short-circuited. $V_{OUT} = 0$, and when $V_{IN} = 12.5$ V, the voltage dropped across R_1 is $V_{IN} - V_{OUT} = 12.5$ V. The power dissipation in R_1 is

$$P_{R1} = \frac{V_{R1}^2}{R_1} = \frac{(12.5 \text{ V})^2}{20 \text{ }\Omega} = 7.8 \text{ W}$$

Therefore, a resistor of at least 10 W should be used.

Practice Exercise 17–5
In Figure 17–12, R_1 is changed to 33 Ω. What must be the power rating of R_1 if the maximum input voltage is 24 V?

SECTION REVIEW 17–3

1. How does the control element in a shunt regulator differ from that in a series regulator?

2. Name one advantage of a shunt regulator over a series type. Name a disadvantage.

BASIC SWITCHING REGULATOR

17–4

The two types of linear regulators discussed in the preceding sections have control elements (transistors) that are *conducting* all the time, with the amount of conduction varied as demanded by changes in the output voltage or current.

The *switching regulator* is different; the control element operates as a *switch*. A greater efficiency can be realized with this type of voltage regulator than with the linear types. There are three basic configurations of switching regulators: *step-down, step-up,* and *inverting.*

Step-Down Configuration

In the step-down configuration, the output voltage is always *less* than the input voltage. A basic step-down switching regulator is shown in Figure 17–13(a), and its simplified equivalent in part (b). Transistor Q_1 is used to switch the input voltage at a duty cycle that is based on the regulator's load requirement. The LC filter is then used to average the switched voltage. Since Q_1 is either *on* (saturated) or *off,* the power lost in the control element is relatively small. Therefore, the switching regulator is useful primarily in higher power applications or in applications where efficiency is of utmost concern.

The on and off intervals of Q_1 are shown in the waveform of Figure 17–14(a). The capacitor charges during the on-time (t_{on}) and discharges during the off-time (t_{off}). When

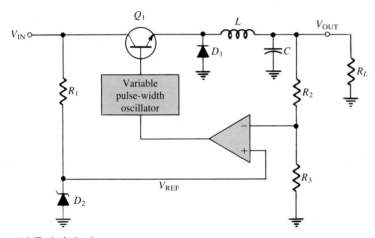

(a) Typical circuit

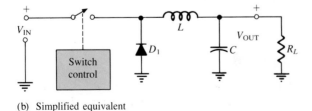

(b) Simplified equivalent

FIGURE 17–13
Step-down switching regulator.

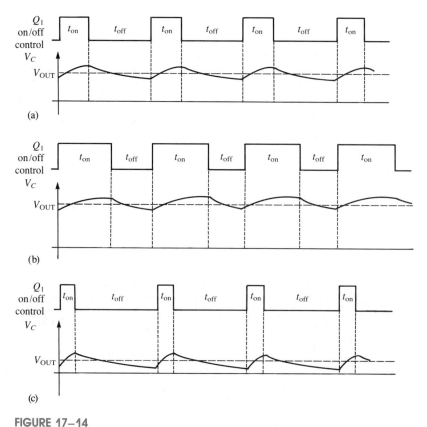

FIGURE 17–14

Switching regulator waveforms. The V_C waveform is for no inductive filtering. L and C smooth V_C to a nearly constant level, as indicated by the dashed line for V_{OUT}.

the on-time is increased relative to the off-time, the capacitor charges more, thus increasing the output voltage, as indicated in part (b). When the on-time is decreased relative to the off-time, the capacitor discharges more, thus decreasing the output voltage, as in part (c). Therefore, by adjusting the duty cycle $t_{on}/(t_{on} + t_{off})$ of Q_1, the output voltage can be varied. The inductor further smooths the fluctuations of the output voltage caused by the charging and discharging action.

The regulating action is as follows and is illustrated in Figure 17–15. When V_{OUT} tries to decrease, the on-time of Q_1 is increased, causing an additional charge on C to offset the attempted decrease. When V_{OUT} tries to increase, the on-time of Q_1 is decreased, causing C to discharge enough to offset the attempted increase.

The output voltage is expressed as

$$V_{OUT} = \left(\frac{t_{on}}{T}\right)V_{IN} \qquad (17\text{–}9)$$

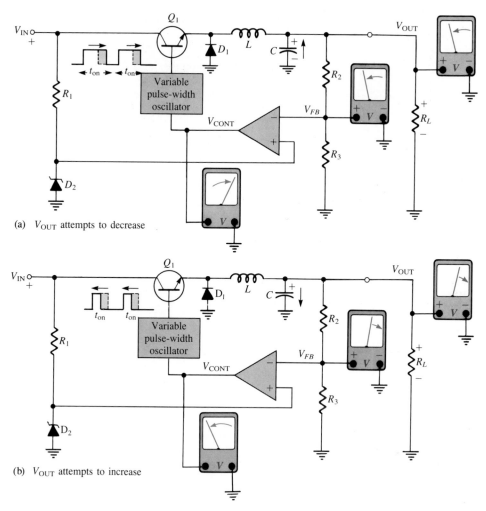

(a) V_{OUT} attempts to decrease

(b) V_{OUT} attempts to increase

FIGURE 17–15
Regulating action of step-down switching regulator.

T is the *period* of the on-off cycle of Q_1 and is related to the frequency by $T = 1/f$. The period is the sum of the on-time and the off-time.

$$T = t_{on} + t_{off} \qquad \textbf{(17–10)}$$

The ratio t_{on}/T is called the *duty cycle*.

Step-Up Configuration

A basic step-up type of switching regulator is shown in Figure 17–16. When Q_1 turns on, voltage across L increases instantaneously to $V_{IN} - V_{CE(sat)}$, and the inductor's magnetic

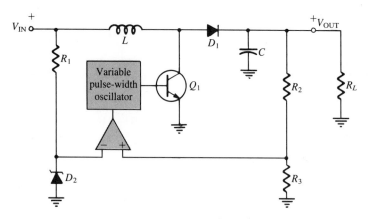

FIGURE 17–16
Step-up switching regulator.

field expands quickly, as indicated in Figure 17–17(a). During the on-time (t_{on}) of Q_1, the V_L decreases from its initial maximum, as shown. The longer Q_1 is on, the smaller V_L becomes. When Q_1 turns off, the inductor's magnetic field collapses and its polarity reverses so that its voltage adds to V_{IN}, thus producing an output voltage greater than the input, as indicated in Figure 17–17(b). During the off-time (t_{off}) of Q_1, the diode is forward-biased, allowing the capacitor to charge. The variations in the output voltage due to the charging and discharging action are sufficiently smoothed by the filtering action of L and C.

The shorter the on-time of Q_1, the greater the inductor voltage is, and thus the greater the output voltage is (greater V_L adds to V_{IN}). The longer the on-time of Q_1, the smaller are the inductor voltage and the output voltage (small V_L adds to V_{IN}). When V_{OUT} tries to decrease because of increasing load or decreasing input voltage, t_{on} decreases and the attempted decrease in V_{OUT} is offset. When V_{OUT} tries to increase, t_{on} increases and the attempted increase in V_{OUT} is offset. This regulating action is illustrated in Figure 17–18. As you can see, the output voltage is inversely related to the duty cycle of Q_1 and can be expressed as follows.

$$V_{OUT} = \left(\frac{T}{t_{on}}\right)V_{IN} \qquad (17\text{–}11)$$

where $T = t_{on} + t_{off}$.

Voltage-Inverter Configuration

A third type of switching regulator produces an output voltage that is opposite in polarity to the input. A basic diagram is shown in Figure 17–19.

When Q_1 turns on, the inductor voltage jumps to $V_{IN} - V_{CE(sat)}$ and the magnetic field rapidly expands, as shown in Figure 17–20(a). While Q_1 is on, the diode is reverse-biased and the inductor voltage decreases from its initial maximum. When Q_1 turns off,

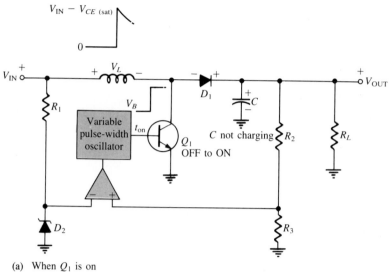

(a) When Q_1 is on

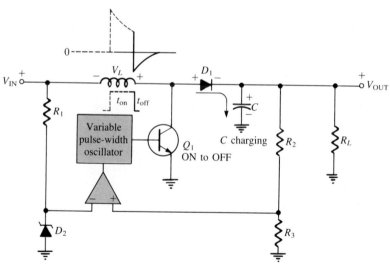

(b) When Q_1 turns off

FIGURE 17–17
Step-up action of switching regulator.

the magnetic field collapses and the inductor's polarity reverses, as shown in part (b). This forward-biases the diode, charges C, and produces a negative output voltage, as indicated. The repetitive on-off action of Q_1 produces a repetitive charging and discharging that is smoothed by the LC filter action.

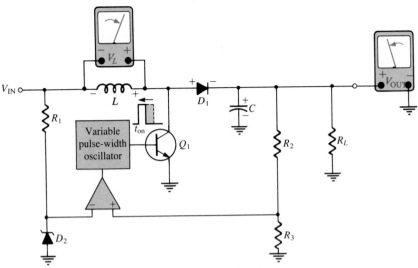

(a) When V_{OUT} tries to decrease, t_{on} decreases, causing V_L to
    increase. This compensates for the attempted decrease in V_{OUT}.

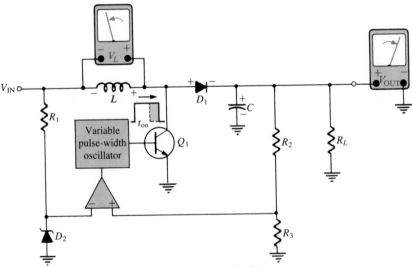

(b) When V_{OUT} tries to increase, t_{on} increases, causing V_L to
    decrease. This compensates for the attempted increase in V_{OUT}.

FIGURE 17–18
Regulating action.

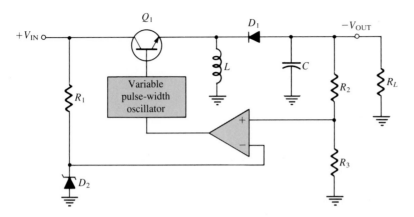

FIGURE 17–19
Inverting switching regulator.

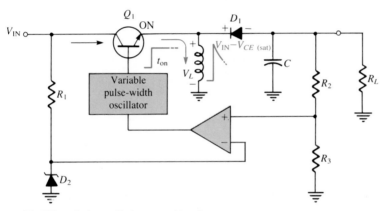

(a) When Q_1 is on, D_1 is reverse-biased

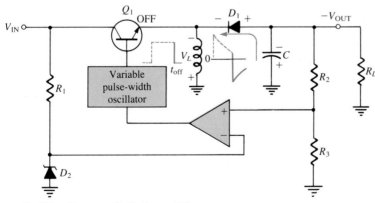

(b) When Q_1 turns off, D_1 forward biases

FIGURE 17–20
Inverting action of switching regulator.

As with the step-up regulator, the less time Q_1 is on, the greater the output voltage is, and vice versa. This regulating action is illustrated in Figure 17–21. Switching regulator efficiencies can be greater than 90 percent.

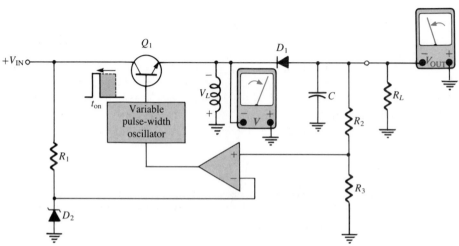

(a) When $-V_{OUT}$ tries to decrease, t_{on} decreases, causing V_L to increase. This compensates for the attempted decrease in $-V_{OUT}$.

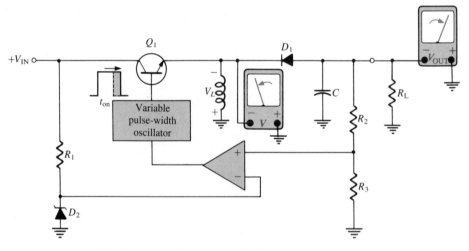

(b) When $-V_{OUT}$ tries to increase, t_{on} increases, causing V_L to decrease. This compensates for the attempted increase in $-V_{OUT}$.

FIGURE 17–21
Regulating action.

1. List three types of switching regulators.
2. What is the primary advantage of switching regulators over linear regulators?
3. Changes in output voltage are compensated for by adjusting the _____ of the switching transistor.

IC VOLTAGE REGULATORS

17–5

In the previous sections, the basic voltage regulator configurations were examined. Several types of integrated circuit voltage regulators are available, many of which can be connected to operate in any of the basic configurations. In this section some common types are introduced, along with examples of specific devices and their features. Applications and use of some of these devices are covered in the next section.

Precision Voltage Regulators

This type of IC regulator provides very precise regulation of the output voltage for both line and load variations. Generally, these devices also provide for continuously adjustable output voltages within a specified range, current limiting, and remote shutdown (to be discussed later).

The μA723 Regulator.   This particular device exhibits the following list of features, and its packaging configurations are shown in Figure 17–22(a), with an equivalent simplified schematic in part (b).

1. Typical line regulation of 0.02 percent change in V_{OUT} over a range of input voltages from 12 V to 40 V.
2. Typical load regulation of 0.03 percent change in V_{OUT} over a range of load currents from 1 mA to 50 mA.
3. The output voltage is continuously adjustable from 2 V to 37 V.
4. Current-limiting provisions.
5. Output current up to 65 mA (typical) without an external pass transistor.

Three-Terminal Positive and Negative Voltage Regulators

Generally, the three-terminal regulators provide a fixed output voltage with no external components required. Most have internal current limiting and can provide fairly high load currents (up to 1.5 A) without an external pass transistor. Normally, this type of regulator is available with a selection of several fixed output voltage values.

The μA7800 Regulators.   These devices are examples of three-terminal *positive* voltage regulators. Three-terminal regulators are generally available in three types of packages, as

FIGURE 17–22

μA723 IC voltage regulator.

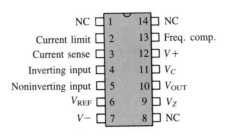

NC	1	14 NC
Current limit	2	13 Freq. comp.
Current sense	3	12 $V+$
Inverting input	4	11 V_C
Noninverting input	5	10 V_{OUT}
V_{REF}	6	9 V_Z
$V-$	7	8 NC

(a) Dual-in-line package

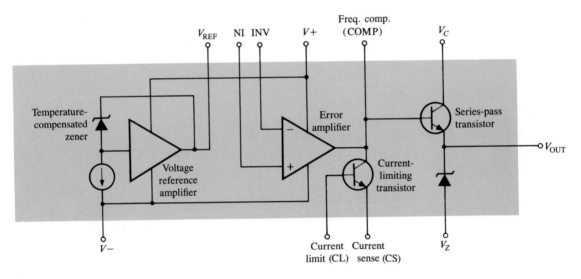

(b)

shown in Figure 17–23. They are available with fixed output voltage values of 5 V, 6 V, 8 V, 10 V, 12 V, 15 V, 18 V, and 24 V.

The LM340 5 V Regulator. This device is another example of a three-terminal regulator. It is designed specifically for a +5 V output.

SECTION REVIEW 17–5

1. Name two basic categories of IC voltage regulators.
2. Name the terminals of a three-terminal regulator.

APPLICATIONS OF IC REGULATORS

17–6

In this section, the precision voltage regulator used to illustrate application concepts is assumed to be of the μA723 type. The three-terminal devices are assumed to be similar to the μA7800 series. For specific applications refer to manufacturer's data sheets on the device you are using.

(Top view)

Input [1   8] Output
NC [2   7] NC
NC [3   6] Common
NC [4   5] NC

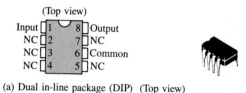

(a) Dual in-line package (DIP) (Top view)

(Top view)

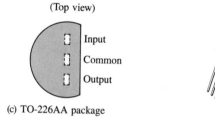

Output
Input
Common

The input terminal is in
electrical contact with
the mounting base

(b) TO-220 package

(Top view)

Input

Common

Output

(c) TO-226AA package

FIGURE 17–23
μA7800 packages.

Basic Positive Regulator Configuration

A basic regulator configuration is shown in Figure 17–24 in block form. The current-limiting sense resistor R_S is connected from the base to emitter of Q_2 (C_L to C_S). The

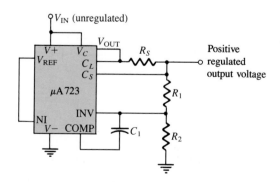

FIGURE 17–24
μA723 voltage regulator configured as a basic current-limiting positive regulator.

current at which limiting occurs is determined by

$$I_{L(\text{max})} = \frac{0.7 \text{ V}}{R_S}$$

where 0.7 V is the V_{BE} of Q_2, and $I_{L(\text{max})}$ is the maximum desired output (load) current. The value of R_S is selected based on a specified $I_{L(\text{max})}$.

The resistors R_1 and R_2 are the negative feedback resistors that determine the voltage gain of the regulator's internal op-amp. The output voltage is

$$V_{\text{OUT}} = \left(1 + \frac{R_1}{R_2}\right)V_{\text{REF}} \qquad \text{(17--12)}$$

EXAMPLE 17–6

Determine the output voltage for the μA723 regulator in Figure 17–25 if $V_{\text{REF}} = 1.6$ V. If the output is shorted, to what value is the current limited?

FIGURE 17–25

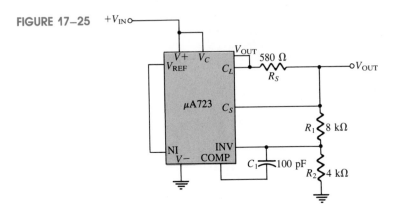

Solution

$$V_{\text{OUT}} = \left(1 + \frac{R_1}{R_2}\right)V_{\text{REF}} = \left(1 + \frac{8 \text{ k}\Omega}{4 \text{ k}\Omega}\right)1.6 \text{ V} = 4.8 \text{ V}$$

$$I_{L(\text{max})} = \frac{0.7 \text{ V}}{R_S} = \frac{0.7 \text{ V}}{580 \text{ }\Omega} = 1.2 \text{ mA}$$

Regulator with Fold-Back Current Limiting

In the previous current-limiting technique, the current is restricted to a maximum *constant* value.

Fold-back current limiting is a method used particularly in high-current regulators whereby the output current under overload conditions drops to a value well below the peak load current capability to prevent excessive power dissipation.

Basic Idea. The basic concept of fold-back current limiting is as follows, with reference to Figure 17–26. The circuit is similar to the constant current-limiting arrangement discussed in section 17–2 (Figure 17–7), with the exception of resistors R_5 and R_6. The voltage drop developed across R_4 by the load current must not only overcome the base-emitter voltage required to turn on Q_2, but it must overcome the voltage across R_5. That is, the voltage across R_4 must be

$$V_{R_4} = V_{R_5} + V_{BE}$$

In an overload or short circuit condition the load current increases to a value $I_{L(\max)}$ that is sufficient to cause Q_2 to conduct. At this point the current can increase no further. The decrease in output voltage results in a proportional decrease in the voltage across R_5; thus less current through R_4 is required to maintain the forward-biased condition of Q_1. So, as V_{OUT} decreases, I_L decreases, as shown in the graph of Figure 17–27.

The advantage of this technique is that the regulator is allowed to operate with peak load current up to $I_{L(\max)}$; but when the output becomes shorted, the current drops to a lower value to prevent overheating of the device.

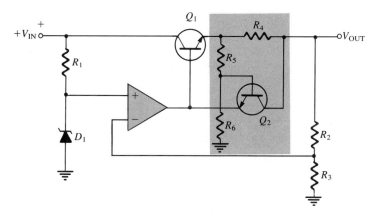

FIGURE 17–26
Series regulator with fold-back current limiting.

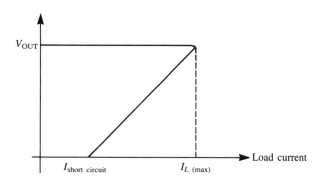

FIGURE 17–27
Fold-back current limiting (output voltage versus load current).

Fold-Back Current Limiting in the μA723.   Figure 17–28 shows the μA723 regulator configured for fold-back current limiting. Because this device has internal circuitry for current limiting, only three external resistors—R_3, R_4, and R_5—are required. A capacitor across R_3 is sometimes necessary for stability purposes.

FIGURE 17–28
μA723 regulator configured for fold-back current limiting.

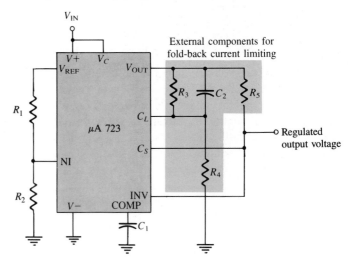

External Pass Transistor Circuits

The basic idea of using an external pass transistor to extend the usable output current range of a voltage regulator is illustrated in Figure 17–29. The internal series pass transistor is connected in a Darlington configuration with the external transistor. This increases the current gain and permits the regulator to provide a higher load current than its internal capability allows. The external pass transistor is usually a power transistor. It must be able to handle a maximum power of

$$P_{\max} = I_{L(\max)}(V_{IN} - V_{OUT}) \qquad \textbf{(17–13)}$$

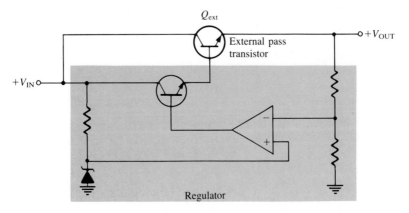

FIGURE 17–29
Regulator with external pass transistor.

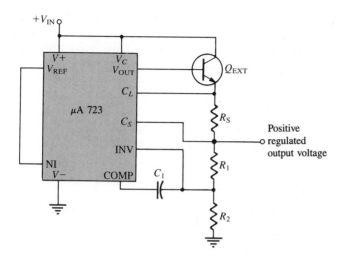

FIGURE 17–30
μA723 regulator with external pass transistor.

An IC voltage regulator with an npn external pass transistor is shown in Figure 17–30. This configuration also includes R_S for constant current limiting.

Negative Regulators

A basic negative voltage regulator circuit is shown in Figure 17–31. Q_1 acts as the external pass transistor with the negative input voltage applied to the collector, as shown. Notice that V_{REF} connects to the inverting input. The output voltage is determined as follows.

$$V_{OUT} = -\left(1 + \frac{R_1}{R_2}\right)V_{REF} \qquad (17\text{–}14)$$

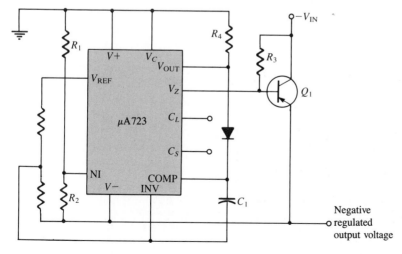

FIGURE 17–31
Basic negative regulator.

Dual Tracking Regulators

Many negative regulator applications occur in symmetrical dual-polarity supplies, such as ±15 V used by most IC op-amps. In these applications the negative supply must follow (track) changes in the positive supply voltage even during a shutdown condition. Figure 17–32 shows a dual tracking regulator configuration with two μA723 ICs. It consists of a basic positive regulator (VR_1) and a negative regulator (VR_2) like those previously discussed. Notice that the reference voltage at the inverting input (INV) of VR_2 is derived from the output voltage of VR_1 (between R_3 and R_4). As this voltage changes, so does the negative output voltage.

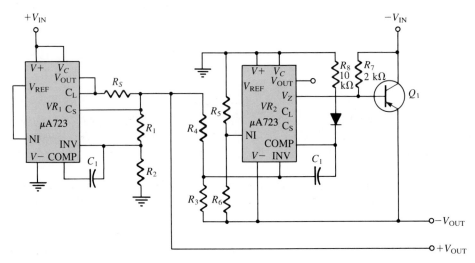

FIGURE 17–32
Dual tracking regulator.

Three-Terminal Regulator Applications

A simple three-terminal regulator, such as the μA7800 series, can be configured in many ways for various requirements. In this particular type of voltage regulator, the last two digits indicate the output voltage. For example, a μA7805 is a 5 V regulator, a μA7812 is a 12 V device, and so on. A few typical applications are introduced here. Figure 17–33 shows a simple fixed-voltage regulator. Notice that the common terminal is grounded. The input capacitor is required when the regulator is not located close to the power-supply filter.

FIGURE 17–33
Basic fixed-voltage regulator.

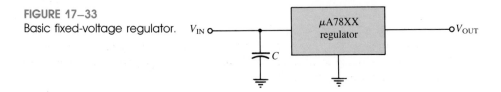

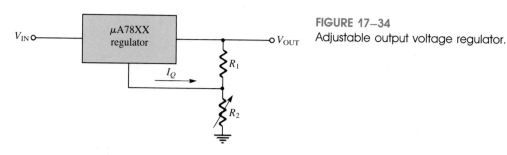

FIGURE 17–34
Adjustable output voltage regulator.

Figure 17–34 shows a circuit for increasing the output voltage above the specified value. The extra voltage is developed across R_2 by the quiescent current I_Q out of the common terminal as indicated.

$$V_{OUT} = \left(1 + \frac{R_2}{R_1}\right)V_{OUT(REG)} + I_Q R_2 \qquad (17\text{–}15)$$

where $V_{OUT(REG)}$ is the specified output voltage of the three-terminal regulator. R_2 can be an adjustable resistor to provide for an adjustable output voltage.

Figure 17–35 illustrates a three-terminal regulator with an external bypass transistor for handling currents in excess of the output capability of the basic regulator. The value of R_1 determines the point at which the transistor Q_1 begins to conduct and bypass the regulator.

The previous application examples are only a few of the many possible configurations for IC voltage regulators and serve to illustrate the versatility of these devices.

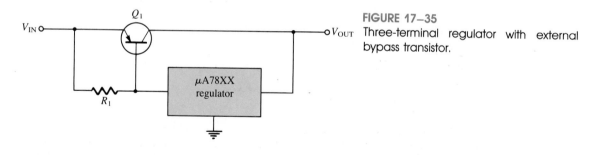

FIGURE 17–35
Three-terminal regulator with external bypass transistor.

---

1. Name two types of current limiting in voltage regulators.
2. What is the purpose of an external pass transistor?

**SECTION
REVIEW
17–6**

A SYSTEM APPLICATION

17–7

The power supply is an integral part of any electronic system. Most systems require incorporation of some sort of voltage regulation in the power-supply portion of the system. The regulator in many applications is simply a zener diode, but other applications call for more precise voltage regulation.

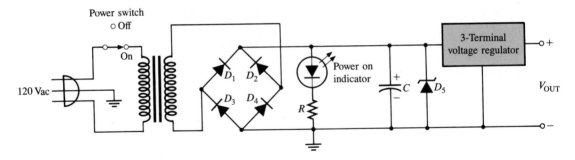

FIGURE 17–36
A basic regulated power supply.

Figure 17–36 shows a simple dc power supply with an IC three-terminal regulator. The ac line voltage, which can be turned on or off with the switch, is transformer-coupled to the full-wave bridge rectifier. Diode bridge rectifiers are available in a single package and are quite common. The LED serves as a power-on indicator. The capacitor filters the output of the rectifier and the zener diode serves, when necessary, as a preregulator to insure that the voltage does not exceed the input rating of the three-terminal regulator. The dc output voltage is set by the regulator and depends on the type of device. The 7800 series three-terminal regulator, for example, is available in a selection of fixed output voltages from 2.6 V to 24 V. Some systems may require several dc voltages, which can be achieved by distributing the rectified and filtered output to the required number of regulators.

FORMULAS

Voltage regulation

(17–1)   Line regulation $= \dfrac{(\Delta V_{OUT}/V_{OUT})100\%}{\Delta V_{IN}}$

(17–2)   Load regulation $= \dfrac{(V_{NL} - V_{FL})100\%}{V_{FL}}$

Basic series regulator

(17–3) $A_{cl} = 1 + \dfrac{R_2}{R_3}$ Closed-loop voltage gain

(17–4) $V_{OUT} \cong \left(1 + \dfrac{R_2}{R_3}\right)V_{REF}$ Regulator output

(17–5) $I_{L(max)} = \dfrac{0.7 \text{ V}}{R_4}$ For constant current limiting

Basic shunt regulator

(17–6) $\Delta I_S = \dfrac{\Delta V_{IN}}{R_1}$ Change in shunt current

(17–7) $\Delta I_S = -\Delta I_L$ Change in shunt current

(17–8) $I_{L(\text{max})} = \dfrac{V_{\text{IN}}}{R_1}$ For shunt regulator

Basic switching regulators

(17–9) $V_{\text{OUT}} = \left(\dfrac{t_{\text{on}}}{T}\right)V_{\text{IN}}$ For step-down switching regulator

(17–10) $T = t_{\text{on}} + t_{\text{off}}$ Switching period

(17–11) $V_{\text{OUT}} = \left(\dfrac{T}{t_{\text{on}}}\right)V_{\text{IN}}$ For step-up switching regulator

IC voltage regulators

(17–12) $V_{\text{OUT}} = \left(1 + \dfrac{R_1}{R_2}\right)V_{\text{REF}}$ IC regulator output

(17–13) $P_{\text{max}} = I_{L(\text{max})}(V_{\text{IN}} - V_{\text{OUT}})$ For external bypass

(17–14) $V_{\text{OUT}} = -\left(1 + \dfrac{R_1}{R_2}\right)V_{\text{REF}}$ For negative regulator

(17–15) $V_{\text{OUT}} = \left(1 + \dfrac{R_2}{R_1}\right)V_{\text{OUT(REG)}}$ Three-terminal, increased voltage
$\qquad\qquad\quad + I_Q R_2$

☐ Line regulation is the percentage change in output voltage for a given change in input voltage.　　**SUMMARY**

☐ Load regulation is the percentage change in output voltage for a given change in load current from a no-load condition to a full-load condition.

☐ A basic voltage regulator consists of a reference source, an error detector, a sampling element, and a control device. Protection circuitry is also found in most regulators.

☐ There are two basic types of *linear* regulators: series and shunt.

☐ In a *series* regulator, the control element is a transistor in series with the load.

☐ In a *shunt* regulator, the control element is a transistor (or zener) in parallel with the load.

☐ There are three basic types of switching regulators: step-down, step-up, and inverting.

☐ Switching regulators are more efficient than linear regulators.

☐ Two categories of IC voltage regulators are *precision* and *three-terminal*.

☐ IC voltage regulators can be configured for many different applications using external components.

---

1. Label the functional blocks for the voltage regulator in Figure 17–37.　　**SELF-TEST**
2. Describe the difference between line regulation and load regulation.

FIGURE 17–37

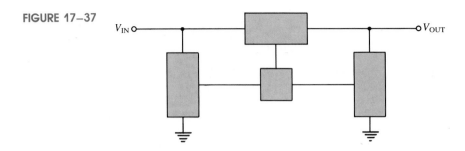

3. Identify each type of regulator in Figure 17–38.

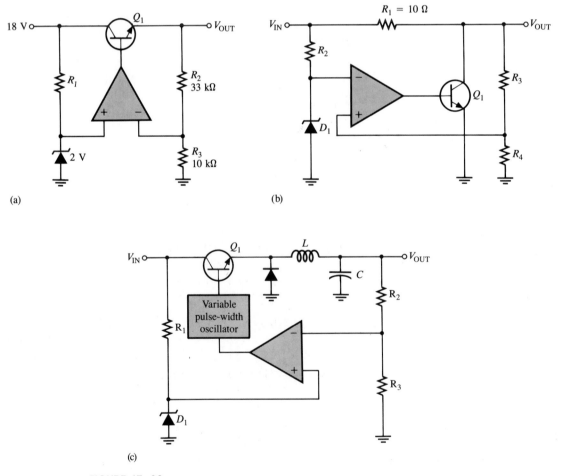

FIGURE 17–38

4. Determine the output voltage for the regulator in Figure 17–38(a).

5. Show how you would modify Figure 17–38(a) to include current limiting.

6. What is the maximum load current in Figure 17–38(b) when $V_{IN} = 12$ V?

7. What is the purpose of Q_1 in Figure 17–38(c)?

8. Determine the output voltage in Figure 17–38(c) if Q_1 is on for 1 ms and off for 3 ms with $V_{IN} = 20$ V.

9. (a) What type of device is the $\mu A723$?
   (b) What type of devices are the $\mu A7805$ and the LM109?

10. Determine V_{OUT} in Figure 17–39.

FIGURE 17–39

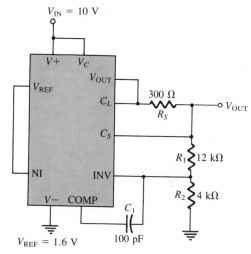

11. Explain the difference between constant current limiting and fold-back current limiting.

12. What must the power rating of a series pass transistor be when $I_{L(max)} = 1$ A, $V_{IN} = 30$ V, and $V_{OUT} = 15$ V?

13. Describe the basic function of a dual tracking regulator.

Section 17–1

17–1 The nominal output voltage of a certain regulator is 8 V. The output changes 2 mV when the input voltage goes from 12 V to 18 V. Determine the line regulation and express it as a percentage change over the entire range of V_{IN}.

17–2 Express the line regulation found in problem 17–1 in units of %/V.

17–3 A certain regulator has a no-load output voltage of 10 V and a full-load output voltage of 9.90 V. What is the percent load regulation?

17–4 In problem 17–3, if the full-load current is 250 mA, express the load regulation in %/mA.

Section 17–2

17–5   Determine the output voltage for the series regulator in Figure 17–40.

FIGURE 17–40

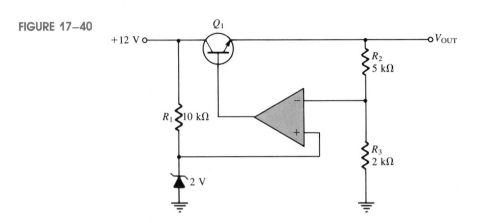

17–6   If R_3 in Figure 17–40 is doubled, what happens to the output voltage?

17–7   If the zener voltage is 2.7 V instead of 2 V in Figure 17–40, what is the output voltage?

17–8   A series voltage regulator with constant current limiting is shown in Figure 17–41. Determine the value of R_4 if the load current is to be limited to a maximum value of 250 mA. What power rating must R_4 have?

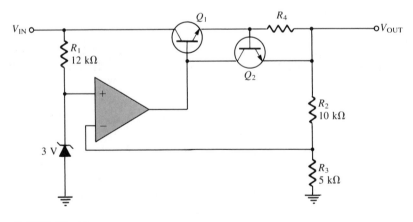

FIGURE 17–41

17–9   If the R_4 determined in problem 17–8 is halved, what is the maximum load current?

Section 17–3

17–10 In the shunt regulator of Figure 17–42, when the load current increases, does Q_1 conduct more or less? Why?

FIGURE 17–42

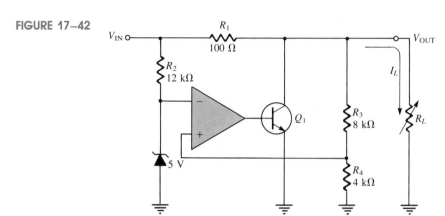

17–11 Assume I_L remains constant and V_{IN} changes by 1 V in Figure 17–42. What is the change in the collector current of Q_1?

17–12 With a constant input voltage of 17 V, the load resistance in Figure 17–42 is varied from 1 kΩ to 1.2 kΩ. Neglecting any change in output voltage, how much does the shunt current through Q_1 change?

17–13 If the maximum allowable input voltage in Figure 17–42 is 25 V, what is the maximum possible output current when the output is short-circuited? What power rating should R_1 have?

Section 17–4

17–14 A basic switching regulator is shown in Figure 17–43. If the switching frequency of the transistor is 100 Hz with an off-time of 6 ms, what is the output voltage?

FIGURE 17–43

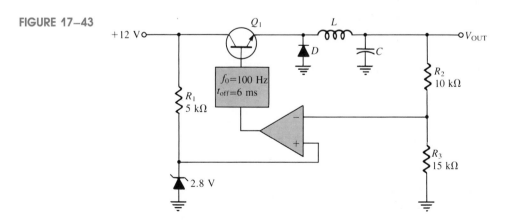

17–15   What is the duty cycle of the transistor in problem 17–14?

17–16   Determine the output voltage for the switching regulator in Figure 17–44 when the duty cycle is 40 percent.

FIGURE 17–44

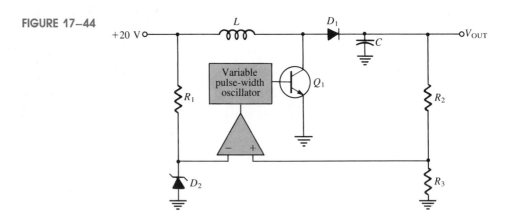

17–17   If the on-time of Q_1 in Figure 17–44 is decreased, does the output voltage increase or decrease?

Section 17–5

17–18   Associate each of the following specific voltage regulator devices with one of the classifications: three-terminal positive, three-terminal negative, precision.
  (a) μA7806
  (b) μA723
  (c) LM340

17–19   Indicate the function of each pin on the μA723 package in Figure 17–45.

FIGURE 17–45

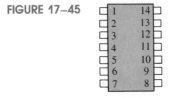

Section 17–6

17–20   Determine the output voltage for the regulator in Figure 17–46. Also, find the short circuit output current. $V_{REF} = 1.55$ V.

17–21   What value must R_1 be in order to increase the output voltage by 1.5 times in Figure 17–46?

17–22   Determine the *maximum* power dissipation for the external pass transistor Q_1 in Figure 17–47. $V_{REF} = 1.55$ V.

17–23   Identify each type of IC regulator configuration shown in Figure 17–48.

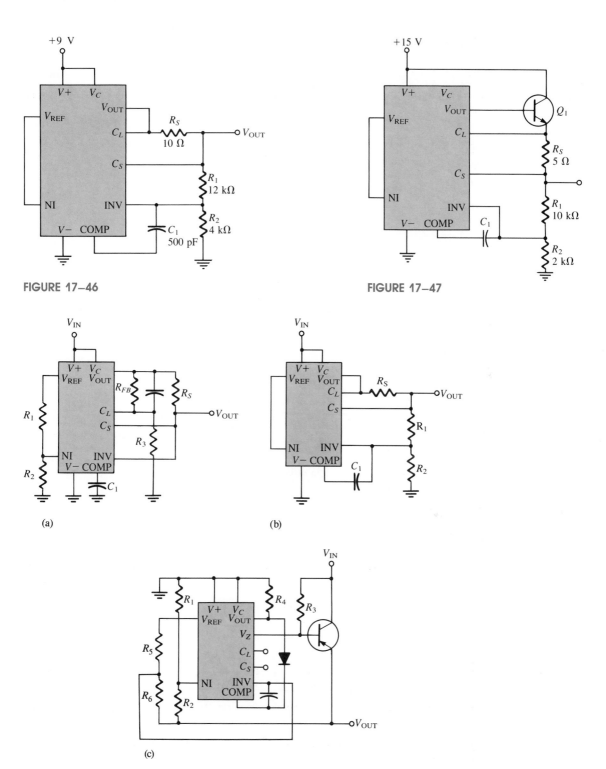

FIGURE 17–46

FIGURE 17–47

(a)

(b)

(c)

FIGURE 17–48

ANSWERS TO SECTION REVIEWS

Section 17–1
1. The percentage change in the output voltage for a given change in input voltage.
2. The percentage change in output voltage for a given change in load current.

Section 17–2
1. Control element, error detector, sampling element, reference source.
2. 2 V.

Section 17–3
1. In a shunt regulator, the control element is in parallel with the load rather than in series.
2. A shunt regulator has inherent current limiting, but is less efficient than a series regulator.

Section 17–4
1. Step-down, step-up, inverting.
2. Higher efficiency.
3. Duty cycle.

Section 17–5
1. Precision, three-terminal.
2. Input, output, common.

Section 17–6
1. Constant, fold-back.
2. Increase output current capability.

ANSWERS TO PRACTICE EXERCISES

Example 17–1:   0.6%/V
Example 17–2:   0.84%, 0.0168%/mA
Example 17–3:   7.33 V
Example 17–5:   17.45 W

Thyristors and Unijunction Transistors

In this chapter you will learn

☐ The basic construction and operation of a Shockley diode

☐ How a silicon-controlled rectifier (SCR) works and how it can be applied

☐ How a silicon-controlled switch (SCS) works and how it differs from an SCR

☐ The fundamentals of diacs and triacs and how these devices are applied

☐ The basic construction and operation of a unijunction transistor (UJT)

☐ How a UJT is used in a relaxation oscillator and in a heater control

☐ How a PUT (programmable UJT) operates and its basic applications

☐ How a thyristor device is used in a specific system application

Chapter 18

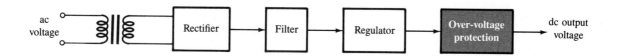

This chapter covers several important devices. First we will discuss a family of devices constructed of four layers of semiconductor material. These four-layer (pnpn) devices are classified as *thyristors* and include the *Shockley diode, silicon-controlled rectifier* (SCR), *silicon-controlled switch* (SCS), *diacs,* and *triacs*. Other important devices covered are the *unijunction transistor* (UJT) and the *programmable unijunction transistor* (PUT). These various types of thyristors share certain characteristics. They act as open circuits capable of withstanding a certain rated voltage until triggered. When triggered on, they become low-resistance current paths and remain so, even after the trigger is removed, until the current is reduced to a certain level or until they are triggered off, depending on the type of device.

Thyristors can be used to control the amount of ac power to a load and are used in lamp dimmers, motor speed control, ignition systems, and charging circuits, to name a few. UJTs and PUTs are used as trigger devices for thyristors and in oscillators and timing circuits. A good example of a thyristor system application is the SCR over-voltage protection circuit used in some dc power supplies. In this particular application, the SCR shuts down the power supply if the output voltage exceeds a specified limit, thus preventing damage to the rest of the system.

THE SHOCKLEY DIODE

18–1

The Shockley diode is a two-terminal thyristor (four-layer device). The basic construction consists of four semiconductor layers forming a pnpn structure, as shown in Figure 18–1(a). The schematic symbol is shown in part (b).

The pnpn structure can be represented by an *equivalent* circuit consisting of a pnp transistor and an npn transistor, as shown in Figure 18–2(a). The upper pnp layers form Q_1 and the lower npn layers form Q_2, with the two middle layers shared by both equivalent transistors. Notice that the base-emitter junction of Q_1 corresponds to pn junction 1 in Figure 18–1, the base-emitter junction of Q_2 corresponds to pn junction 3, and the base-collector junctions of both Q_1 and Q_2 correspond to pn junction 2.

FIGURE 18–1
Shockley diode.

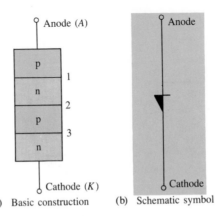

(a) Basic construction   (b) Schematic symbol

FIGURE 18–2
Shockley diode equivalent circuit.

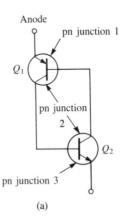

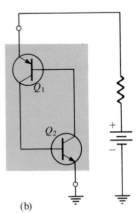

(a)   (b)

Basic Operation

When a positive bias voltage is applied to the anode with respect to the cathode, as shown in Figure 18–2(b), the base-emitter junctions of Q_1 and Q_2 (pn junctions 1 and 3 in Figure 18–1[a]) are *forward*-biased, and the common base-collector junction (pn junction 2 in Figure 18–1[a]) is *reverse*-biased. Therefore, both equivalent transistors are in the linear region.

FIGURE 18–3
Currents in a basic Shockley diode equivalent circuit.

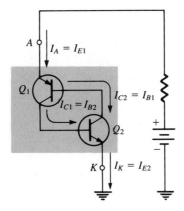

At low values of forward-bias voltage, an expression for the anode current is developed as follows, using the standard transistor relationships and Figure 18–3. In this analysis, the component of leakage current, I_{CBO}, is taken into account.

$$I_{B_1} = I_{E_1} - I_{C_1} - I_{CBO_1}$$

Since $I_C = \alpha_{dc}I_E$,

$$I_{B_1} = I_{E_1} - \alpha_{dc}I_{E_1} - I_{CBO_1}$$
$$= (1 - \alpha_{dc_1})I_{E_1} - I_{CBO_1}$$

The anode current, I_A, is the same as I_{E_1}, and therefore

$$I_{B_1} = (1 - \alpha_{dc_1})I_A - I_{CBO_1} \qquad \text{(18–1)}$$
$$I_{C_2} = \alpha_{dc_2}I_{E_2} + I_{CBO_2}$$

The cathode current, I_K, is the same as I_{E_2}, and therefore

$$I_{C_2} = \alpha_{dc_2}I_K + I_{CBO_2} \qquad \text{(18–2)}$$

Since $I_{C_2} = I_{B_1}$,

$$\alpha_{dc_2}I_K + I_{CBO_2} = (1 - \alpha_{dc_1})I_A - I_{CBO_1}$$

I_A and I_K are equal. Substituting I_A for I_K in the above equation and solving for I_A, we get

$$\alpha_{dc_2}I_A + I_{CBO_2} = (1 - \alpha_{dc_1})I_A - I_{CBO_1}$$
$$\alpha_{dc_2}I_A - (1 - \alpha_{dc_1})I_A = -I_{CBO_1} - I_{CBO_2}$$
$$I_A[(1 - \alpha_{dc_1}) - \alpha_{dc_2}] = I_{CBO_1} + I_{CBO_2}$$
$$I_A = \frac{I_{CBO_1} + I_{CBO_2}}{1 - (\alpha_{dc_1} + \alpha_{dc_2})} \qquad \text{(18–3)}$$

At low current levels, the transistor alpha (α_{dc}) is very small. Therefore, at low bias levels, there is very little anode current in the Shockley diode as equation (18–3) demonstrates, and thus it is in the *off* state or *forward-blocking region*.

EXAMPLE 18–1

A certain Shockley diode is biased in the forward-blocking region with an anode-to-cathode voltage of 20 V. Under this bias condition, $\alpha_{dc_1} = 0.35$ and $\alpha_{dc_2} = 0.45$. The leakage currents are 100 nA. Determine the anode current and the forward resistance of the diode.

Solution

$$I_A = \frac{I_{CBO_1} + I_{CBO_2}}{1 - (\alpha_{dc_1} + \alpha_{dc_2})}$$

$$= \frac{200 \text{ nA}}{1 - 0.8} = \frac{200 \text{ nA}}{0.2} = 1 \ \mu A$$

This is the forward current when the device is off, but forward-biased with $V_{AK} = +20$ V. The forward resistance is therefore

$$R_{AK} = \frac{V_{AK}}{I_A} = \frac{20 \text{ V}}{1 \ \mu A} = 20 \text{ M}\Omega$$

Forward-Breakover Voltage

At this time, the operation of the Shockley diode may seem very strange because it is forward-biased, yet acts essentially as an open switch. As previously mentioned, there is a region of forward bias, called the *forward-blocking region,* in which the device has a very high forward resistance (ideally an open). This region exists from $V_{AK} = 0$ up to a value of V_{AK} called the *forward-breakover* voltage, $V_{BR(F)}$. This is indicated on the Shockley diode characteristic curve in Figure 18–4.

As V_{AK} is increased, the anode current I_A gradually increases, as shown on the graph. As I_A increases, so do α_{dc_1} and α_{dc_2}. At the point where $\alpha_{dc_1} + \alpha_{dc_2} = 1$, the denominator in equation (18–3) becomes 0, and I_A would become infinitely large, if not limited by an external series resistor.

FIGURE 18–4
Shockley diode characteristic curve.

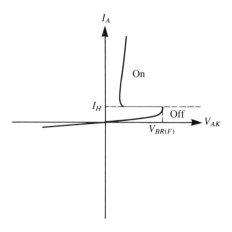

At this point, $V_{AK} = V_{BR(F)}$, and the internal transistor structures become saturated. When this happens, the forward voltage drop V_{AK} suddenly decreases to a low value approximately equal to $V_{BE} + V_{CE(sat)}$ as I_A increases, and the Shockley diode enters the forward-conduction region as indicated in Figure 18–4. Now, the device is in the *on* state and acts as a closed switch. The on/off states of the Shockley diode are illustrated in Figure 18–5.

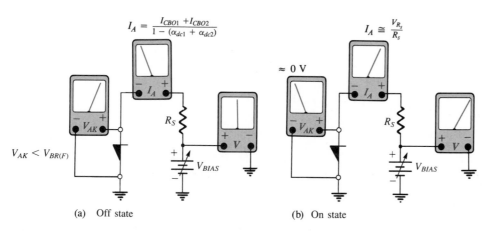

(a)   Off state   (b)   On state

FIGURE 18–5
On/off states of Shockley diode.

EXAMPLE 18–2

(a)   Determine the value of anode current in Figure 18–6(a), when the device is off and dc alphas are 0.4. The leakage currents are 0.08 μA each.

(b)   Determine the value of anode current in Figure 18–6(b) when the device is on. $V_{BR(F)} = 40$ V. Assume $V_{BE} = 0.7$ V and $V_{CE(sat)} = 0.1$ V for the internal transistor structure.

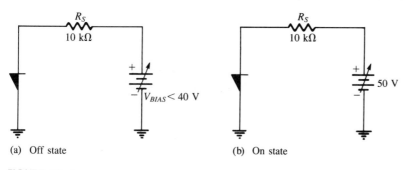

(a) Off state   (b) On state

FIGURE 18–6

Solution

(a) $I_A = \dfrac{I_{CBO_1} + I_{CBO_2}}{1 - (\alpha_{dc_1} + \alpha_{dc_2})} = \dfrac{0.16 \ \mu A}{1 - 0.8} = 0.8 \ \mu A$

(b) The voltage at the anode is

$$V_{BE} + V_{CE(sat)} = 0.7 \text{ V} + 0.1 \text{ V} = 0.8 \text{ V}$$

The voltage across R_S is

$$V_{R_S} = V_{BIAS} - V_A = 50 \text{ V} - 0.8 \text{ V} = 49.2 \text{ V}$$

The anode current is

$$I_A = \frac{V_{R_S}}{R_S} = \frac{49.2 \text{ V}}{10 \text{ k}\Omega} = 4.92 \text{ mA}$$

Holding Current

Once the Shockley diode is conducting (in the on state), it will continue to conduct until the anode current is reduced below a specified level, called the *holding current*, I_H. This parameter is *also* indicated on the characteristic curve in Figure 18–4. *When I_A falls below I_H, the device rapidly switches back to the off state and enters the forward-blocking region.*

Switching Current

The value of the anode current at the point where the device switches from the forward-blocking region (off) to the forward-conduction region (on) is called the *switching current*, I_S. This value of current is always less than the holding current, I_H, which is indicated on the characteristic curve of Figure 18–4.

An Application

The circuit in Figure 18–7(a) is a *relaxation oscillator*. The operation is as follows. When the switch is closed, the capacitor charges through R until its voltage reaches the forward-

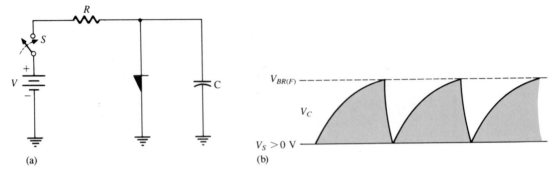

FIGURE 18–7
Shockley diode relaxation oscillator.

breakover voltage of the Shockley diode. At this point the diode switches into conduction, and the capacitor rapidly discharges through the diode. Discharging continues until the current through the diode falls below the holding value. At this point, the diode switches back to the off state, and the capacitor begins to charge again. The result of this action is a voltage waveform across C like that shown in Figure 18–7(b).

SECTION REVIEW 18–1

1. Sketch the symbol and transistor equivalent for the Shockley diode.
2. A certain Shockley diode has a $V_{BR(F)} = 50$ V. If the anode-to-cathode voltage is 30 V, is the device off or on?

SILICON-CONTROLLED RECTIFIER (SCR)

18–2

The SCR is another four-layer pnpn device similar to the Shockley diode except with three terminals: *anode, cathode,* and *gate*. The basic structure is shown in Figure 18–8(a) and the schematic symbol in part (b). Typical SCR packages and construction are shown in part (c). Like the Shockley diode, the SCR has two possible states of operation. In the *off*

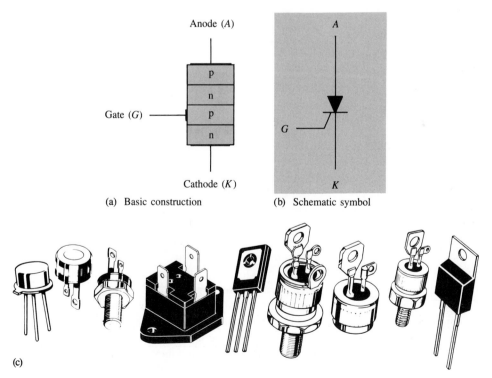

(a) Basic construction   (b) Schematic symbol

(c)

FIGURE 18–8
Silicon-controlled rectifier (SCR).

state, it acts ideally as an open circuit between the anode and the cathode; actually, rather than an open, there is a very high resistance. In the *on* state, the SCR acts ideally as a short from the anode to the cathode; actually there is a small *on* (forward) resistance. The SCR is used in many applications, including motor controls, time delay circuits, heater controls, phase controls, and relay controls, to name a few.

SCR Equivalent Circuit

As was done with the Shockley diode, the SCR operation can best be understood by thinking of its internal pnpn structure as a two-transistor arrangement, as shown in Figure 18–9. This structure is like that of the Shockley diode except for the gate connection. The upper pnp layers act as a transistor, Q_1, and the lower npn layers act as a transistor, Q_2. Again, notice that the two middle layers are ''shared.''

FIGURE 18–9
SCR equivalent circuit.

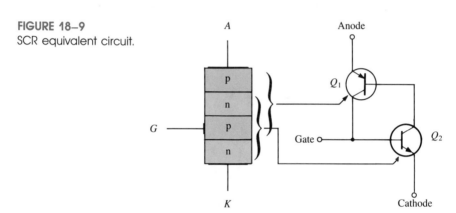

Turning the SCR On

When the gate current I_G is 0, as shown in Figure 18–10(a), the device acts as a Shockley diode in the *off* state. In this state, the very high resistance between the anode and cathode can be approximated by an open switch, as indicated. When a positive pulse of current (trigger) is applied to the gate, both transistors turn on (the anode must be more positive than the cathode). This action is shown in Figure 18–10(b). I_{B_2} turns on Q_2, providing a path for I_{B_1} into the Q_2 collector, thus turning on Q_1. The collector current of Q_1 provides additional base current for Q_2 so that Q_2 stays in conduction after the trigger pulse is removed from the gate. By this regenerative action, Q_2 sustains the saturated conduction of Q_1 by providing a path for I_{B_1}; in turn, Q_1 sustains the saturated conduction of Q_2 by providing I_{B_2}. Thus, the device stays on (latches) once it is triggered on, as shown in Figure 18–10(c). In this state, the very low resistance between the anode and cathode can be approximated by a closed switch, as indicated.

  Like the Shockley diode, an SCR can also be turned on *without* gate triggering by increasing anode-to-cathode voltage to a value exceeding the *forward-breakover* voltage $V_{BR(F)}$, as shown on the characteristic curve in Figure 18–11(a). The forward-breakover voltage decreases as I_G is increased above 0, as shown by the set of curves in Figure 18–11(b). Eventually, a value of I_G is reached at which the SCR turns on at a very low

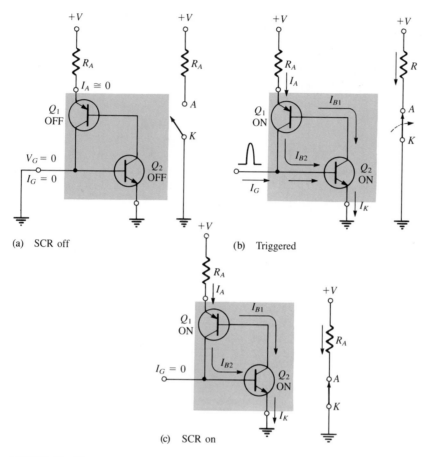

(a) SCR off

(b) Triggered

(c) SCR on

FIGURE 18–10
SCR turn-on process with switch equivalents.

anode-to-cathode voltage. So, as you can see, the gate current controls the value of forward voltage $V_{BR(F)}$ required for turn-on.

Although anode-to-cathode voltages in excess of $V_{BR(F)}$ will not damage the device if current is limited, this situation should be avoided because the normal control of the SCR is lost. It should always be triggered on only with a pulse at the gate.

Turning the SCR Off

When the gate returns to 0 after the trigger pulse is removed, the SCR *cannot* turn off; it stays in the forward-conduction region. The anode current must drop below the value of the *holding current* I_H in order for turn-off to occur. The holding current is indicated in Figure 18–11.

There are two basic methods for turning off an SCR: *anode current interruption* and *forced commutation*. The anode current can be *interrupted* by either a momentary series or parallel switching arrangement, as shown in Figure 18–12. The series switch in part (a)

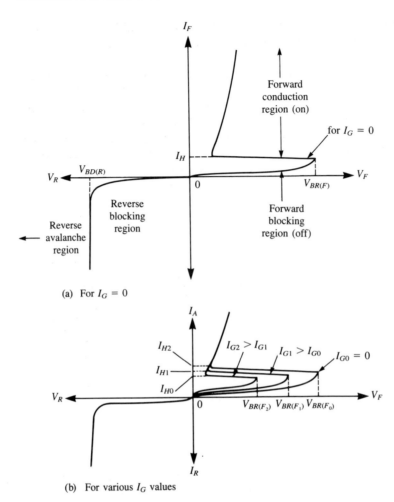

(a) For $I_G = 0$

(b) For various I_G values

FIGURE 18–11
SCR characteristic curves.

FIGURE 18–12
SCR turn-off by anode current interruption.

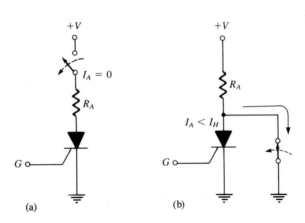

simply reduces the anode current to 0 and causes the SCR to turn off. The parallel switch in part (b) routes part of the total current away from the SCR, thereby reducing the anode current to a value less than I_H.

The *forced commutation* method basically requires momentarily forcing current through the SCR in the direction opposite to the forward conduction so that the net forward current is reduced below the holding value. The basic circuit, as shown in Figure 18–13, consists of a switch (normally a transistor switch) and a battery in parallel with the SCR. While the SCR is conducting, the switch is open, as shown in part (a). To turn off the SCR, the switch is closed, placing the battery across the SCR and forcing current through it opposite to the forward current, as shown in part (b). Typically, turn-off times for SCRs range from a few microseconds up to about 30 μs.

FIGURE 18–13
SCR turn-off by forced commutation.

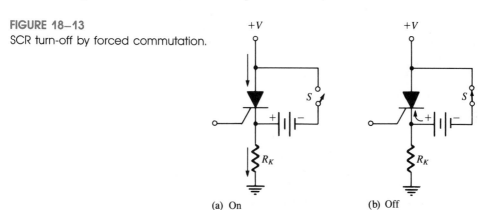

(a) On   (b) Off

SCR Characteristics and Ratings

Several of the most important SCR characteristics and ratings are defined as follows. Use the curve in Figure 18–11 for reference where appropriate.

Forward-Breakover Voltage, $V_{BR(F)}$.   This is the voltage at which the SCR enters the forward-conduction region. The value of $V_{BR(F)}$ is maximum when $I_G = 0$ and is designated $V_{BR(F_0)}$. When the gate current is increased, $V_{BR(F)}$ decreases and is designated $V_{BR(F_1)}$, $V_{BR(F_2)}$, and so on, for increasing steps in gate current (I_{G_1}, I_{G_2}, and so on).

Holding Current, I_H.   This is the value of anode current below which the SCR switches from the forward-conduction region to the forward-blocking region. The value increases with decreasing values of I_G and is maximum for $I_G = 0$.

Gate Trigger Current, I_{GT}.   This is the value of gate current necessary to switch the SCR from the forward-blocking region to the forward-conduction region under specified conditions.

Average Forward Current, $I_{F(AV)}$.   This is the maximum continuous anode current (dc) that the device can withstand in the conduction state under specified conditions.

Forward-Conduction Region.   This region corresponds to the *on* condition of the SCR where there is forward current from anode to cathode through the very low resistance (approximate short) of the SCR.

Forward- and Reverse-Blocking Regions.   These regions correspond to the *off* condition of the SCR where the forward current from anode to cathode is *blocked* by the effective open circuit of the SCR.

Reverse-Breakdown Voltage, $V_{BD(R)}$.   This parameter specifies the value of reverse voltage from cathode to anode at which the device breaks into the avalanche region and begins to conduct heavily (the same as in a pn diode).

SECTION REVIEW 18–2

1. What is an SCR?
2. Name the SCR terminals.
3. How can an SCR be turned on (made to conduct)? How can one be turned off?

SCR APPLICATIONS

18–3   The SCR has many uses in the areas of power control and switching applications. A few of the basic applications are described in this section.

On-Off Control of Current

Figure 18–14 shows an SCR circuit that permits current to be switched to a load by the momentary closure of switch S_1 and removed from the load by the momentary closure of switch S_2.

Assuming the SCR is initially off, momentary closure of S_1 provides a pulse of current into the gate, thus triggering the SCR on so that it conducts current through R_L. The SCR remains in conduction even after the momentary contact of S_1 is removed. When S_2 is momentarily closed, current is shunted around the SCR, thus reducing its anode current below the *holding* value I_H. This turns the SCR off and thus reduces the load current to 0.

FIGURE 18–14
On-off SCR control circuit.

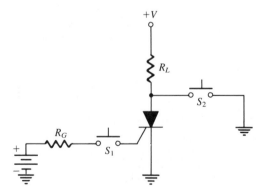

Half-Wave Power Control

A common application of SCRs is in the control of ac power for lamp dimmers, electric heaters, and electric motors.

A half-wave, variable-resistance, phase-control circuit is shown in Figure 18–15; 120 V ac are applied across terminals A and B; R_L represents the resistance of the load (for example, a heating element or lamp filament). R_1 is a current-limiting resistor, and potentiometer R_2 sets the trigger level for the SCR. By adjusting R_2, the SCR can be made to trigger at any point on the positive half-cycle of the ac waveform between 0° and 90°, as shown in Figure 18–16.

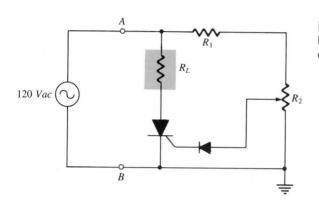

FIGURE 18–15
Half-wave, variable-resistance, phase-control circuit.

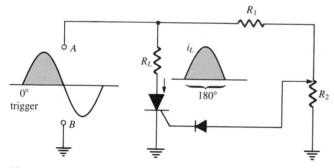

(a)   180° conduction

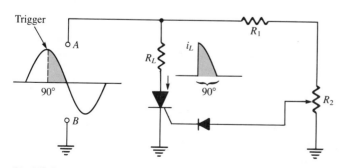

(b)   90° Conduction

FIGURE 18–16
Operation of phase-controlled circuit.

FIGURE 18–16
(continued)

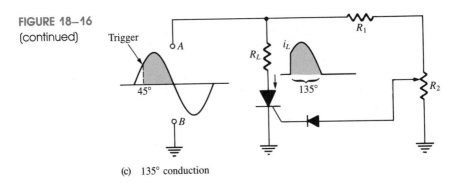

(c) 135° conduction

When the SCR triggers near the beginning of the cycle (approximately 0°), as in Figure 18–16(a), it conducts for approximately 180° and maximum power is delivered to the load. When it fires near the peak of the positive half-cycle (90°), as in part (b), the SCR conducts for approximately 90° and less power is delivered to the load. By adjusting R_2, triggering can be made to occur anywhere between these two extremes, and therefore, a variable amount of power can be delivered to the load. Figure 18–16(c) shows triggering at the 45° point as an example. When the ac input goes negative, the SCR turns off and does not conduct again until the trigger point on the next positive half-cycle. The diode prevents the negative ac voltage from being applied to the gate of the SCR.

Lighting System for Power Interruptions

As another example of SCR applications, we will examine a circuit that will maintain lighting by using a backup battery when there is an ac power failure. Figure 18–17 shows a center-tapped full-wave rectifier used for providing ac power to a low-voltage lamp. As long as the ac power is available, the battery charges through diode D_3 and R_1.

The SCR's cathode voltage is established when the capacitor charges to the peak value of the full-wave rectified ac (6.3 V rms less the drops across R_2 and D_1). The anode is at the 6 V battery voltage, making it less positive than the cathode, thus preventing conduction. The SCR's gate is at a voltage established by the voltage divider made up of R_2 and R_3. Under these conditions the lamp is illuminated by the ac input power and the SCR is off, as shown in Figure 18–17(a).

When there is an interruption of ac power, the capacitor discharges through the closed path D_3, R_1, and R_3, making the cathode less positive than the anode or the gate. This action establishes a triggering condition, and the SCR begins to conduct. Current from the battery is through the SCR and the lamp, thus maintaining illumination, as shown in part (b). When ac power is restored, the capacitor recharges and the SCR turns off. The battery begins recharging.

**SECTION
REVIEW
18–3**

1. If the potentiometer in Figure 18–16 is set at its midpoint, the SCR will conduct for more than _____ degrees but less than _____ degrees of the input cycle.

2. In Figure 18–17, what is the purpose of diode D_3?

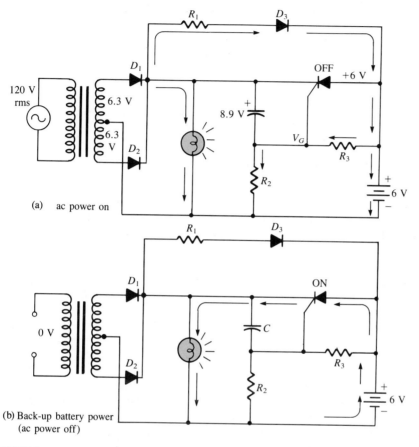

FIGURE 18–17
Automatic back-up lighting circuit.

SILICON-CONTROLLED SWITCH (SCS)

18–4

The SCS is similar in construction to the SCR. The SCS, however, has two gate terminals, as shown in Figure 18–18, the *cathode gate* and the *anode gate*. The SCS can be turned *on* and *off* using either gate terminal. Remember that the SCR can be turned on only using its gate terminal. Normally, the SCS is available only in lower power ratings than those of the SCR.

As with the previous four-layer devices, the basic operation of the SCS can be understood by referring to the transistor equivalent, shown in Figure 18–19. To start, assume that both Q_1 and Q_2 are off, and therefore that the SCS is not conducting. A positive pulse on the cathode gate drives Q_2 into conduction and thus provides a path for Q_1 base current. When Q_1 turns on, its collector current provides base drive to Q_2, thus sustaining the *on* state of the device. This regenerative action is the same as in the turn-on process of the SCR and the Shockley diode, and is illustrated in Figure 18–19(a).

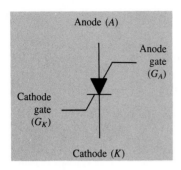

FIGURE 18–18
Silicon-controlled switch (SCS).

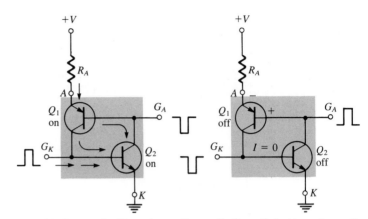

(a) *Turn-on:* Positive pulse on G_K or negative pulse on G_A

(b) *Turn-off:* Positive pulse on G_A or negative pulse on G_K

FIGURE 18–19
SCS operation.

The SCS can also be turned on with a negative pulse on the anode gate, as indicated in part (a). This drives Q_1 into conduction which, in turn, provides base current for Q_2. Once Q_2 is on, it provides a path for Q_1 base current, thus sustaining the on state.

To turn the SCS off, a positive pulse is applied to the anode gate. This reverse-biases the base-emitter junction of Q_1 and turns it off. Q_2, in turn, cuts off and the SCS ceases conduction, as shown in Figure 18–19(b). The device can also be turned off with a negative pulse on the cathode gate, as indicated in part (b). The SCS typically has a faster turn-off time than the SCR.

In addition to the positive pulse on the anode gate or the negative pulse on the cathode gate, there are other methods for turning off an SCS. Figure 18–20(a) and (b) shows two switching methods to reduce the anode current below the holding value. In each case the transistor acts as a switch.

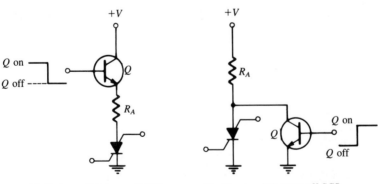

(a) Series switch turns off SCS

(b) Shunt switch turns off SCS

FIGURE 18–20
Transistor switch reduces I_A below I_H and turns off SCS.

Applications

The SCS and SCR are used in similar applications. The SCS has the advantage of faster turn-off with pulses on either gate terminal; however, it is more limited in terms of maximum current and voltage ratings. Also, the SCS is sometimes used in digital applications such as counters, registers, and timing circuits. A typical SCS is shown in Figure 18–21.

FIGURE 18–21
A typical SCS.

1. Explain the difference between an SCS and an SCR.
2. Sketch the schematic symbol for an SCS.

SECTION
REVIEW
18–4

THE DIAC AND TRIAC

Diac

18–5

Both the diac and triac are four-layer devices that can conduct in *both* directions (bilateral). The diac basic construction and schematic symbol are shown in Figure 18–22. Notice that there are two terminals, labelled A_1 and A_2.

Conduction occurs in the diac when the *breakover voltage* is reached with either polarity across the two terminals. The curve in Figure 18–23 illustrates this characteris-

FIGURE 18–22
Diac.

(a) Basic construction

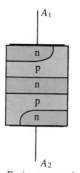

(b) Symbol

FIGURE 18–23
Diac characteristic curve.

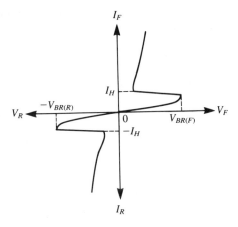

tic. Once breakover occurs, current is in a direction depending on the polarity of the volt-age across the terminals. The device turns off when the current drops below the holding value.

The equivalent circuit of a diac consists of four transistors arranged as shown in Figure 18–24(a). When the diac is biased as in part (b), the pnpn structure from A_1 to A_2 provides the four-layer device operation as was described for the Shockley diode. In the equivalent circuit, Q_1 and Q_2 are forward-biased, and Q_3 and Q_4 are reverse-biased. The device operates on the upper right portion of the characteristic curve in Figure 18–23 under this bias condition. When the diac is biased as shown in Figure 18–24(c), the pnpn structure from A_2 to A_1 is used. In the equivalent circuit, Q_3 and Q_4 are forward-biased,

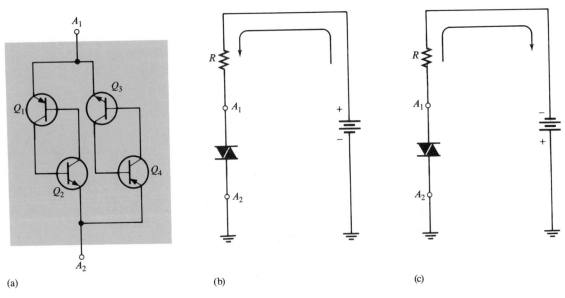

(a)   (b)   (c)

FIGURE 18–24
Diac equivalent circuit and bias conditions.

and Q_1 and Q_2 are reverse-biased. The device operates on the lower left portion of the characteristic curve, as shown in Figure 18–23.

Triac

The triac is like a diac with a gate terminal. The triac can be turned on by a pulse of gate current and does not require the breakover voltage to initiate conduction, as does the diac. Basically, the triac can be thought of simply as two SCRs connected in parallel and in opposite directions with a common gate terminal. Unlike the SCR, the triac can conduct current in either direction when it is triggered on, depending on the polarity of the voltage across its A_1 and A_2 terminals. Figure 18–25(a) and (b) shows the basic construction and schematic symbol for the triac. Typical package configurations are shown in part (c). The characteristic curve is shown in Figure 18–26. Notice that the breakover potential decreases as the gate current increases, just as with the SCR.

As with other four-layer devices, the triac ceases to conduct when the anode current drops below the specified value of the holding current I_H. The only way to turn off the triac is to reduce the current to a sufficiently low level.

Figure 18–27 shows the triac being triggered into both directions of conduction. In part (a), terminal A_1 is biased positive with respect to A_2, so the triac conducts as shown when triggered by a positive pulse at the gate terminal. The transistor equivalent circuit in

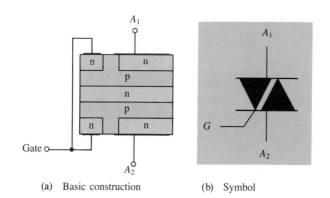

(a)   Basic construction   (b)   Symbol

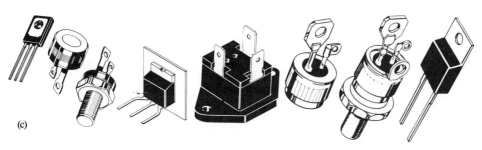

(c)

FIGURE 18–25
Triac.

FIGURE 18–26
Triac characteristic curves.

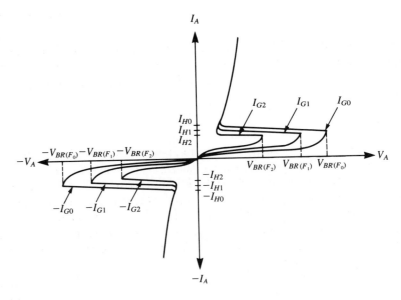

FIGURE 18–27
Bilateral operation of a triac.

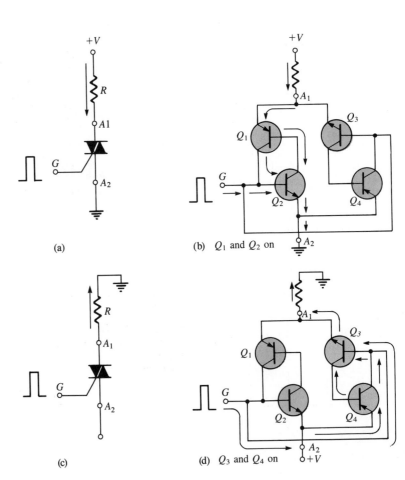

(a)

(b) Q_1 and Q_2 on

(c)

(d) Q_3 and Q_4 on

part (b) shows that Q_1 and Q_2 conduct when a positive trigger pulse is applied. In Figure 18–27(c), terminal A_2 is biased positive with respect to A_1, so the triac conducts as shown. In this case, Q_3 and Q_4 conduct as indicated in part (d) upon application of a positive trigger pulse.

Applications

Like the SCR, triacs are also used to control average power to a load by the method of phase control. The triac can be triggered such that the ac power is supplied to the load for a controlled portion of each half-cycle. During each positive half-cycle of the ac, the triac is off for a certain interval, called the *delay angle* (measured in degrees), and then it is triggered on and conducts current through the load for the remaining portion of the positive half-cycle, called the *conduction angle*. Similar action occurs on the negative half-cycle except that, of course, current is conducted in the opposite direction through the load. Figure 18–28 illustrates this action.

One example of phase control using a triac is illustrated in Figure 18–29(a). Diodes are used to provide trigger pulses to the gate of the triac. Diode D_1 conducts during the

FIGURE 18–28
Basic triac phase control.

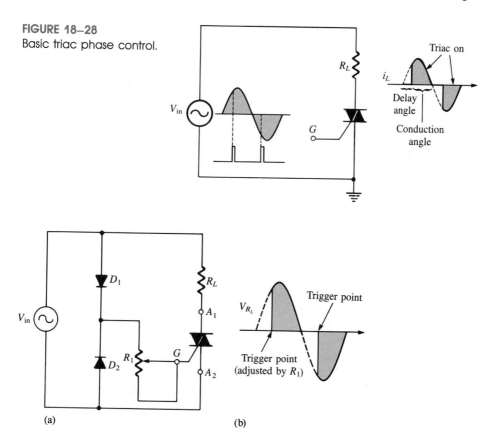

(a)

(b)

FIGURE 18–29
Triac phase-control circuit.

positive half-cycle. The value of R_1 sets the point on the positive half-cycle at which the triac triggers. Notice that during this portion of the ac cycle, A_1 and G are positive with respect to A_2.

Diode D_2 conducts during the negative half-cycle, and R_1 sets the trigger point. Notice that during this portion of the ac cycle, A_2 and G are positive with respect to A_1. The resulting waveform across R_L is shown in Figure 18–29(b).

In the control circuit application, it is necessary that the triac turn off at the end of each positive and each negative alternation of the ac. Figure 18–30 illustrates that there is an interval near each 0 crossing where the triac current drops below the holding value, thus turning the device off.

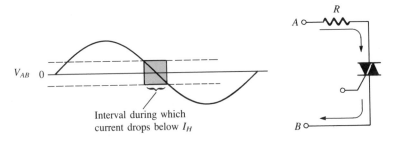

FIGURE 18–30
Triac turn-off interval.

---

1. Compare the triac with the SCR in terms of basic operation.
2. How does a triac differ from a diac?

THE UNIJUNCTION TRANSISTOR (UJT)

18–6

The unijunction transistor is a three-terminal device whose basic construction is shown in Figure 18–31; the schematic symbol appears in part (b). Notice the terminals are labelled *emitter, base 1 (B1)*, and *base 2 (B2)*. Do not confuse this symbol with that of a JFET; the difference is that *the arrow is at an angle for the UJT*. The UJT has only one pn junction, and therefore, the characteristics of this device are very different from those of either the bipolar junction transistor or the FET, as you will see.

Equivalent Circuit

The equivalent circuit for the UJT, shown in Figure 18–32(a), will aid in understanding the basic operation. The diode shown in the figure represents the pn junction, r_{B1} represents the internal bulk resistance of the silicon bar between the emitter and base 1, and r_{B2} represents the bulk resistance between the emitter and base 2. The sum $r_{B1} + r_{B2}$ is the total resistance between the base terminals and is called the *interbase resistance, r_{BB}*.

$$r_{BB} = r_{B1} + r_{B2} \qquad \text{(18–4)}$$

FIGURE 18–31
Unijunction transistor (UJT).

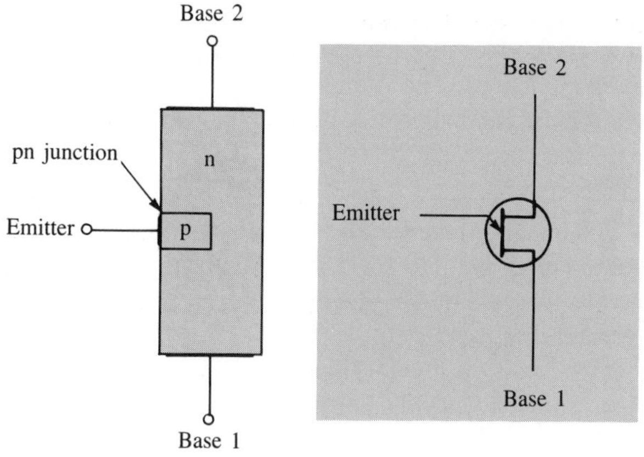

(a)   Basic construction   (b)   Symbol

FIGURE 18–32
UJT equivalent circuit.

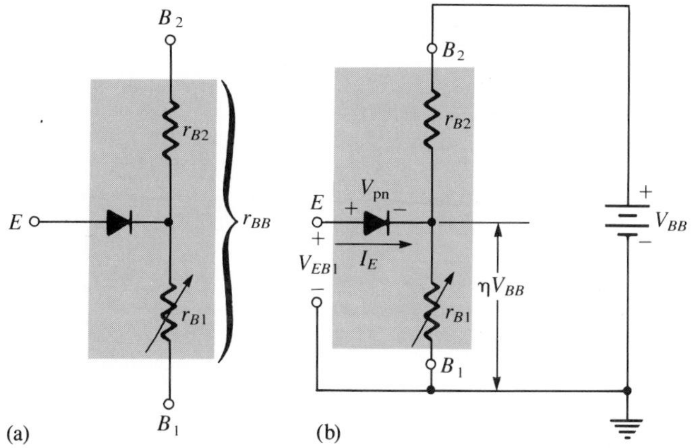

(a)   (b)

The value of r_{B1} varies inversely with emitter current I_E, and therefore, it is shown as a variable resistor. Depending on I_E, the value of r_{B1} can vary from several thousand ohms down to tens of ohms. r_{B1} and r_{B2} form a voltage divider when the device is biased as shown in Figure 18–32(b). The voltage across the resistance r_{B1} can be expressed as

$$V_{r_{B1}} = \left(\frac{r_{B1}}{r_{BB}} \right) V_{BB} \qquad (18-5)$$

Standoff Ratio

The ratio r_{B1}/r_{BB} is a UJT characteristic called the *intrinsic standoff ratio* and designated by η (Greek *eta*).

$$\eta = \frac{r_{B1}}{r_{BB}} \qquad (18-6)$$

As long as the applied emitter voltage V_{EB_1} is less than $V_{r_{B1}} + V_{pn}$, there is no emitter current because the pn junction is not forward-biased (V_{pn} is the barrier potential of the pn junction). The value of emitter voltage which causes the pn junction to become forward-biased is called V_P (peak-point voltage) and is expressed as

$$V_P = \eta V_{BB} + V_{pn} \tag{18-7}$$

When V_{EB_1} reaches V_P, the pn junction becomes forward-biased and I_E begins. Holes are injected into the n-type bar from the p-type emitter. This increase in holes causes an increase in free electrons, thus increasing the conductivity between emitter and B1 (decreasing r_{B1}).

After turn-on, the UJT operates in a *negative resistance* region up to a certain value of I_E, as shown by the characteristic curve in Figure 18-33. As you can see, after the *peak point* ($V_E = V_P$ and $I_E = I_P$), V_E decreases as I_E continues to *increase*, thus producing the *negative resistance* characteristic. Beyond the *valley point* ($V_E = V_V$ and $I_E = I_V$), the device is in saturation, and V_E increases very little with an increasing I_E.

FIGURE 18-33
UJT characteristic curve for a fixed value of V_{BB}.

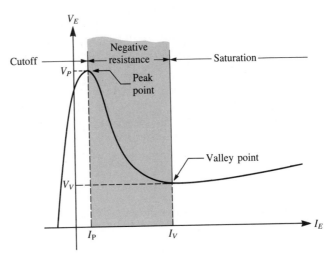

EXAMPLE 18-3
The data sheet of a certain UJT gives $\eta = 0.6$. Determine the peak-point emitter voltage V_P if $V_{BB} = 20$ V.

Solution

$$V_P = \eta V_{BB} + V_{pn}$$
$$= 0.6(20 \text{ V}) + 0.7 \text{ V}$$
$$= 12.7 \text{ V}$$

UJT Applications

The UJT is often used as a trigger device for SCRs and triacs. Other applications include nonsinusoidal oscillators, sawtooth generators, phase control, and timing circuits. Figure

FIGURE 18–34
Relaxation oscillator.

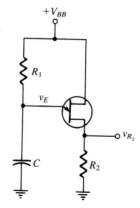

18–34 shows a UJT relaxation oscillator as an example of one application. Also, this type of circuit is basic to other timing and trigger circuits.

The operation is as follows. When dc power is applied, the capacitor C charges exponentially through R_1 until it reaches the peak-point voltage V_P. At this point, the pn junction becomes forward-biased, and the emitter characteristic goes into the negative resistance region (V_E decreases and I_E increases). The capacitor then quickly discharges through the forward-biased junction, r_B, and R_2. When the capacitor voltage decreases to the valley-point voltage V_V, the UJT turns off, the capacitor begins to charge again, and the cycle is repeated, as shown in the emitter voltage waveform in Figure 18–34. During the discharge time of the capacitor, the UJT is conducting. Therefore a voltage is developed across R_2, as shown in the waveform diagram in Figure 18–35.

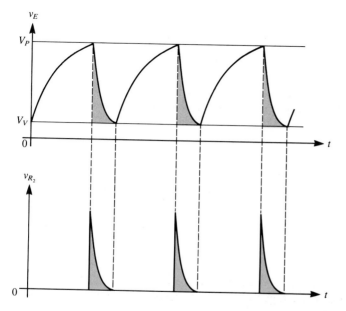

FIGURE 18–35
Waveforms for UJT relaxation oscillator.

Conditions for Turn-On and Turn-Off

In the relaxation oscillator of Figure 18–35, certain conditions must be met for the UJT to reliably turn on and turn off. First, to insure turn-on, R_1 must not limit I_E at the peak point to less than I_P. To insure this, the voltage drop across R_1 at the peak point should be greater than $I_P R_1$. Thus, the condition for turn-on is

$$V_{BB} - V_P > I_P R_1$$

or

$$R_1 < \frac{V_{BB} - V_P}{I_P} \qquad\qquad \textbf{(18–8)}$$

To insure turn-off of the UJT at the valley point, R_1 must be large enough that I_E (at the valley point) can decrease below the specified value of I_V. This means that the voltage across R_1 at the valley point must be less than $I_V R_1$. Thus, the condition for turn-off is

$$V_{BB} - V_V < I_V R_1$$

or

$$R_1 > \frac{V_{BB} - V_V}{I_V} \qquad\qquad \textbf{(18–9)}$$

So, for a proper turn-on and turn-off, R_1 must be in the range

$$\frac{V_{BB} - V_P}{I_P} > R_1 > \frac{V_{BB} - V_V}{I_V}$$

EXAMPLE 18–4

Determine a value of R_1 in Figure 18–36 that will insure proper turn-on and turn-off of the UJT. The characteristic of the UJT exhibits the following values: $\eta = 0.5$, $V_V = 1$ V, $I_V = 10$ mA, $I_P = 20$ μA, and $V_P = 14$ V.

FIGURE 18–36

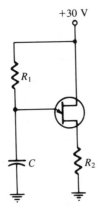

+30 V

R_1

C R_2

Solution

$$\frac{V_{BB} - V_P}{I_P} > R_1 > \frac{V_{BB} - V_V}{I_V}$$

$$\frac{30 \text{ V} - 14 \text{ V}}{20 \text{ } \mu A} > R_1 > \frac{30 \text{ V} - 1 \text{ V}}{10 \text{ mA}}$$

$$800 \text{ k}\Omega > R_1 > 2.9 \text{ k}\Omega$$

As you can see, R_1 has quite a wide range of possible values.

Practice Exercise 18–4
Determine a value of R_1 in Figure 18–36 that will insure proper turn-on and turn-off for the following values: $\eta = 0.33$, $V_V = 0.8$ V, $I_V = 15$ mA, $I_P = 35$ μA, and $V_P = 18$ V.

Another interesting application that includes a UJT, triac, bipolar transistor, and diodes is the temperature-sensitive heater control circuit shown in Figure 18–37. The basic operation is as follows. The ac line voltage is full-wave rectified by the bridge formed by D_1 through D_4. The output of the bridge is applied to the control circuit through R_1 and clamped to a fixed voltage by the zener diode. The thermistor (temperature-sensitive resistor) R_T and the variable resistor R_2 control the bias for Q_1. R_2 is adjusted so that transistor Q_1 is off at a predetermined temperature. When Q_1 is off, the capacitor C_1 is uncharged, and therefore both the UJT (Q_2) and the triac are off.

When the temperature decreases below the desired temperature, the resistance of R_T increases, thus causing Q_1 to conduct. This allows C_1 to charge up to a voltage sufficient

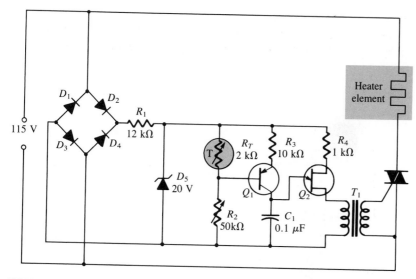

FIGURE 18–37
Temperature-sensitive heater control.

to trigger the UJT. The resulting UJT output pulse is coupled through the transformer and triggers the triac, which conducts current through the heater element (load). As the element heats up, its resistance increases until the current decreases below the holding level of the triac, thus turning it off. If the temperature continues to decrease, the resistance of R_T increases more and causes Q_1 to conduct harder, thus charging C_1 faster. This triggers the triac sooner in the ac cycle, so that more power is delivered to the load. When the temperature increases, the resistance of R_T decreases, causing Q_1 to conduct less. As a result, C_1 takes longer to charge and the triac is triggered later in the ac cycle, so that less power is delivered to the load. When the desired temperature is reached, Q_1 is off, the triac is off, and no more power is delivered to the load.

SECTION REVIEW 18–6

1. Name the UJT terminals.
2. What is the *intrinsic standoff ratio*?
3. In a basic UJT relaxation oscillator such as in Figure 18–34, what three factors determine the period of oscillation?

THE PROGRAMMABLE UJT (PUT)

18–7

The structure of the PUT is similar to that of an SCR (four-layer) except that the gate is brought out, as shown in Figure 18–38. Notice that the gate is connected to the n region adjacent to the anode. This pn junction controls the *on* and *off* states of the device. The gate is always biased positive with respect to the cathode. When the anode voltage exceeds the gate voltage by approximately 0.7 V, the pn junction is forward-biased and the PUT turns on. When the anode voltage falls below this level, the PUT turns off.

FIGURE 18–38
The programmable unijunction transistor (PUT).

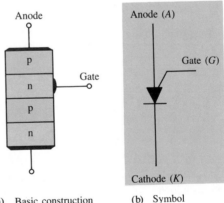

(a) Basic construction   (b) Symbol

Setting the Trigger Voltage

The gate can be biased to a desired voltage with an external voltage divider, as shown in Figure 18–39(a), so that when the anode voltage exceeds this "programmed" level, the PUT turns on.

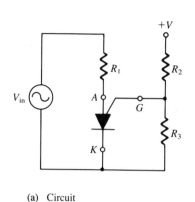

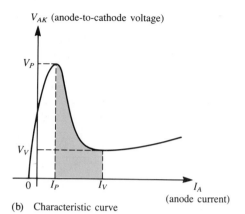

(a) Circuit

(b) Characteristic curve

FIGURE 18–39
PUT biasing.

Application

A plot of the anode-to-cathode voltage V_{AK} versus anode current I_A in Figure 18–39(b) reveals a characteristic curve similar to that of the UJT. Therefore, the PUT replaces the UJT in many applications. One such application is the relaxation oscillator in Figure 18–40(a). Its basic operation is as follows.

The gate is biased at +9 V by the voltage divider consisting of resistors R_2 and R_3. When dc power is applied, the PUT is off and the capacitor charges toward +18 V through R_1. When the capacitor reaches $V_G + 0.7$ V, the PUT turns on and the capacitor rapidly discharges through the low on resistance of the PUT and R_4. A voltage spike is developed across R_4 during the discharge. As soon as the capacitor discharges, the PUT turns off and the charging cycle starts over, as shown by the waveforms in Figure 18–40(b).

---

1. What does the term *programmable* mean as used in *programmable unijunction transistor* (PUT)?

2. Compare the structure and the operation of a PUT to those of other devices.

SECTION
REVIEW
18–7

A SYSTEM APPLICATION

18–8

It is desirable in some applications to protect the system's circuitry from a potential over-voltage condition, which may result from the failure of the regulator in the dc power supply. If the zener opens in a simple zener diode regulator, for example, the output voltage would suddenly increase and perhaps destroy the circuits to which the dc voltage is connected.

Figure 18–41 shows a simple over-voltage protection circuit, sometimes called a "crowbar" circuit, in a dc power supply. The dc output voltage from the regulator is monitored by the zener diode D_1 and the resistive voltage divider (R_1 and R_2). The upper

FIGURE 18–40
PUT relaxation oscillator.

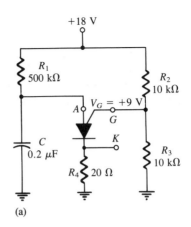

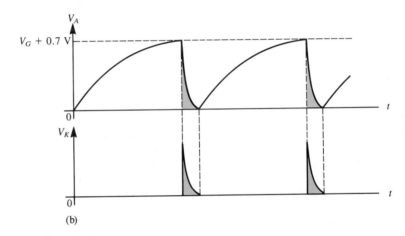

(b)

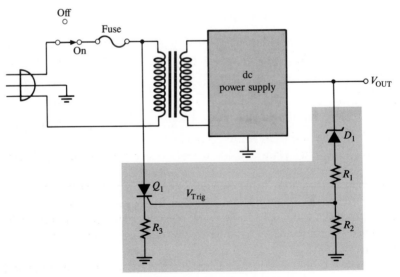

FIGURE 18–41
A basic SCR over-voltage protection circuit.

limit of the output voltage is set by the zener voltage. If this voltage is exceeded, the zener conducts and the voltage divider produces an SCR trigger voltage. The trigger voltage fires the SCR, which is connected across the line voltage. The SCR current causes the fuse to blow, thus disconnecting the line voltage from the power supply.

Shockley diode

FORMULAS

(18–1) $I_{B_1} = (1 - \alpha_{dc_1})I_A - I_{CBO_1}$ Shockley diode, Q_1 base current

(18–2) $I_{C_2} = \alpha_{dc_2}I_K + I_{CBO_2}$ Shockley diode, Q_2 base current

(18–3) $I_A = \dfrac{I_{CBO_1} + I_{CBO_2}}{1 - (\alpha_{dc_1} + \alpha_{dc_2})}$ Shockley diode, anode current (or cathode current)

Unijunction transistor

(18–4) $r_{BB} = r_{B1} + r_{B2}$ UJT interbase resistance

(18–5) $V_{r_{B1}} = \left(\dfrac{r_{B1}}{r_{BB}}\right)V_{BB}$ UJT gate-to-$B1$ voltage

(18–6) $\eta = \dfrac{r_{B1}}{r_{BB}}$ UJT intrinsic standoff ratio

(18–7) $V_P = \eta V_{BB} + V_{pn}$ UJT peak-point voltage

(18–8) $R_1 < \dfrac{V_{BB} - V_P}{I_P}$ Turn-on condition, relaxation oscillator

(18–9) $R_1 > \dfrac{V_{BB} - V_V}{I_V}$ Turn-off condition, relaxation oscillator

SUMMARY

☐ The Shockley diode is a pnpn device which conducts when the voltage across its terminals exceeds the breakover potential.

☐ Thyristors are devices constructed with four semiconductor layers.

☐ Thyristors include Shockley diodes, SCRs, SCSs, diacs, triacs, and PUTs.

☐ The SCR can be triggered *on* by a pulse at the gate and turned *off* by reducing the anode current below the specified *holding* value.

☐ The SCS has two gate terminals and can be turned *on* by a pulse at the cathode gate and turned *off* by a pulse at the anode gate.

☐ The diac can conduct current in either direction and is turned on when a *breakover* voltage is exceeded. It turns off when the current drops below the *holding* value.

☐ The triac, like the diac, is a bidirectional device. It can be turned on by a pulse at the gate and conducts in a direction depending on the voltage polarity across the two anode terminals.

☐ The intrinsic standoff ratio of a UJT determines the voltage at which the device will trigger on.

☐ The PUT can be externally programmed to turn on at a desired voltage level.

SELF-TEST

1. Identify each symbol in Figure 18–42.

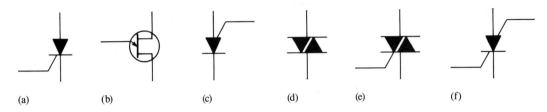

(a)   (b)   (c)   (d)   (e)   (f)

FIGURE 18–42

2. Label the terminals for each device in Figure 18–42.
3. Explain how an SCR is turned on and turned off.
4. Sketch the V_R waveform for the circuit in Figure 18–43, given the indicated relationship of the input waveforms.

FIGURE 18–43

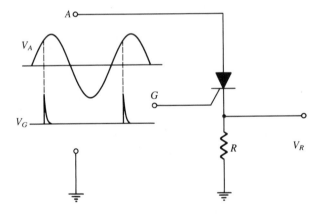

5. Repeat problem 4 with the SCR replaced by a triac.
6. A certain UJT has an $r_{B1} = 1$ kΩ and an $r_{B2} = 1.5$ kΩ. What is the intrinsic standoff ratio?
7. What unique property of a UJT allows it to be used as a relaxation oscillator?
8. Sketch the voltage waveform across R_1 in Figure 18–44 in relation to the input voltage waveform.

PROBLEMS

Section 18–1

18–1 The Shockley diode in Figure 18–45 is biased such that it is in the forward-blocking region. For this particular bias condition, $\alpha_{dc} = 0.38$ for both internal transistor structures. $I_{CBO_1} = 75$ nA and $I_{CBO_2} = 80$ nA. Determine the anode current and cathode current.

FIGURE 18–44

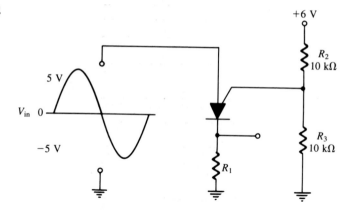

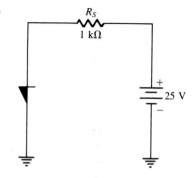

FIGURE 18–45

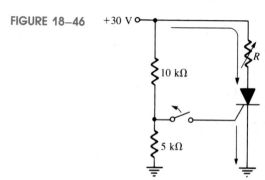

18–2 (a) Determine the forward resistance of the Shockley diode in Figure 18–45 in the forward-blocking region.

(b) If the forward-breakover voltage is 50 V, how much must V_{AK} be increased to switch the diode into the forward-conduction region?

Section 18–2

18–3 Explain the operation of an SCR in terms of its transistor equivalent.

18–4 To what value must the variable resistor be adjusted in Figure 18–46 in order to turn the SCR off? Assume $I_H = 10$ mA.

FIGURE 18–46

Section 18–3

18–5 Show how you would modify the circuit in Figure 18–15 so that the SCR triggers and conducts on the negative half-cycle of the input.

18–6 What is the purpose of diodes D_1 and D_2 in Figure 18–17?

Section 18–4

18–7 Explain the turn-on and turn-off operation of an SCS in terms of its transistor equivalent.

18–8 Name the terminals of an SCS.

Section 18–5

18–9 Sketch the current waveform for the circuit in Figure 18–47. The diac has a breakover potential of 20 V. $I_H = 20$ mA.

FIGURE 18–47

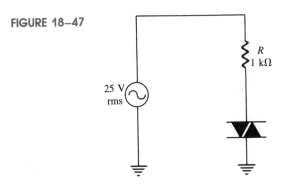

18–10 Repeat problem 18–9 for the triac circuit in Figure 18–48. The breakover potential is 25 V and $I_H = 1$ mA.

FIGURE 18–48

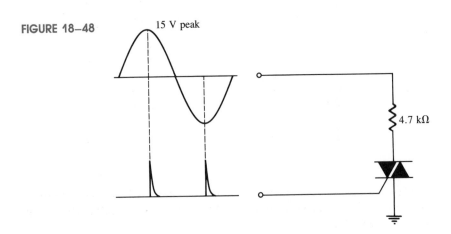

Section 18–6

18–11 In a certain UJT, $r_{B1} = 2.5$ kΩ and $r_{B2} = 4$ kΩ. What is the intrinsic standoff ratio?

18–12 Determine the peak-point voltage for the UJT in problem 18–11 if $V_{BB} = 15$ V.

18–13 Find the value of R_1 in Figure 18–49 that will insure proper turn-on and turn-off of the UJT. $\eta = 0.68$, $V_V = 0.8$ V, $I_V = 15$ mA, $I_P = 10$ μA, and $V_P = 10$ V.

FIGURE 18–49

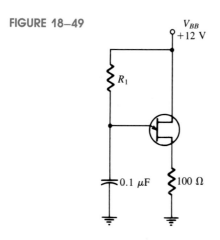

Section 18–7

18–14 At what anode voltage (V_A) will each PUT in Figure 18–50 begin to conduct?

FIGURE 18–50

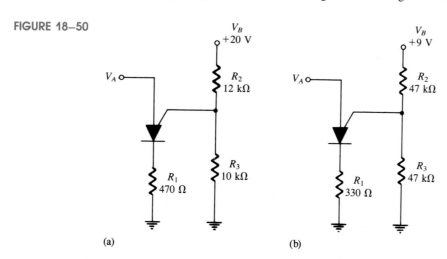

(a)   (b)

18–15 Draw the current waveform for each circuit in Figure 18–50 when the source produces a 10 V peak sine wave voltage at the anode. Neglect the forward voltage of the PUT.

**ANSWERS
TO
SECTION
REVIEWS**

Section 18–1

1. See Figure 18–51.

FIGURE 18–51

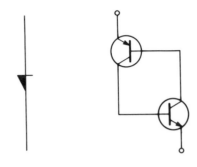

2. Off.

Section 18–2

1. Silicon-controlled rectifier.

2. Anode, cathode, gate.

3. Positive gate pulse, reduce I_A below I_H.

Section 18–3

1. 90°, 180°.

2. Prevents negative voltage on the gate.

Section 18–4

1. An SCS can be turned off with gate; an SCR cannot.

2. See Figure 18–52.

FIGURE 18–52

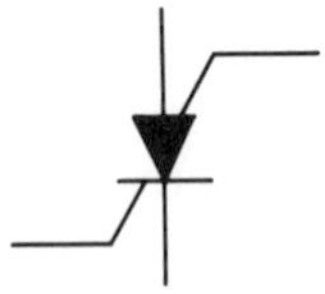

Section 18–5

1. A triac conducts in both directions; an SCR does not.

2. A diac has no gate terminal.

Section 18–6

1. Base 1, base 2, emitter.

2. Ratio of r_{B1} to r_{BB}.

3. R, C, η.

Section 18–7

1. The turn-on voltage can be adjusted or programmed to a desired level.

2. Structure of PUT is similar to that of SCR (four-layer). The anode-to-gate voltage determines when device turns on.

Example 18–4: $343 \text{ k}\Omega > R_1 > 1.95 \text{ k}\Omega$

Optoelectronic Devices

In this chapter you will learn

☐ What light is and how wavelength determines whether light is visible, ultraviolet, or infrared
☐ The characteristics of photoconductive cells
☐ How photodiodes work and how they can be applied
☐ The basic operation and use of phototransistors
☐ The meaning of *dark current*
☐ The construction and operation of solar cells
☐ How light-activated SCRs can be applied
☐ The principle of the light-emitting diode (LED)
☐ What a laser diode is
☐ The principles of optical isolators
☐ The basics of fiber optics
☐ How optical devices are used in a specific system application

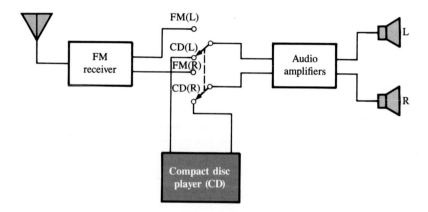

Optoelectronics is an area that combines electronics with optical technology. The emergence of the field of optoelectronics has produced a variety of devices that either produce light under electrical stimulus or respond to light in a way that converts light energy into electrical energy or affects some electrical parameter. These devices are classified as either *light-emitting* or *light-activated*.

This chapter covers several important types of optoelectronic devices and their applications. Two of the devices are the *laser diode* and the *photodiode*. An interesting application of these particular devices in audio systems is with the compact disc (CD) player. As indicated in the block diagram, many modern stereo systems include a CD player as a component.

CHARACTERISTICS OF LIGHT

19–1

The quantum theory states that light consists of discrete packets of energy called *photons*. The energy contained in a photon depends on the frequency of the light and is expressed as

$$W = \hbar f \tag{19–1}$$

where W is the symbol for energy, f is the frequency, and $\hbar$ is *Planck's constant*, equal to 6.624×10^{-34} joule-second.

You can see from equation (19–1) that the light energy is directly related to *frequency*. As the frequency goes up, so does the energy, and vice versa. Light is usually referred to in terms of its *wavelength* rather than its frequency. The wavelength, λ, is expressed as

$$\lambda = \frac{c}{f} \tag{19–2}$$

where c is the speed of light (3×10^8 meters/second). When c is in meters/second and f is in Hz, λ comes out in meters. However, the *angstrom* (Å) is a more commonly used unit for the wavelength; 1 Å equals 1×10^{-10} meter.

The wavelength of light determines its color in the visible range and whether it is ultraviolet or infrared outside the visible range, as Figure 19–1 illustrates.

Any light source, including optoelectronic devices, emits light over a limited range of wavelengths. When the amount of energy versus the wavelength (or frequency) is plotted on a graph, it is called the *emission spectrum*. Any optoelectronic device that is light-activated rather than light-emitting responds to a certain range of wavelengths. This characteristic is known as the *spectral response*.

The amount of light emitted from a source is usually measured in *lumens* (lm). *Light intensity* is a measure of the lumens falling on a specified surface area and is usually expressed in lumens per square meter (lm/m^2), foot-candles (fc), or watts per square meter (W/m^2). One foot-candle is one lumen per square foot and can be expressed in other common units as follows:

$$1 \text{ fc} = 10.764 \text{ lm/m}^2 = 1.609 \times 10^{-12} \text{ W/m}^2 \tag{19–3}$$

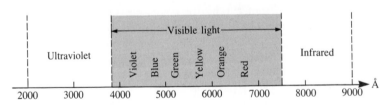

FIGURE 19–1
Wavelength spectrum of light energy.

1. What unit is commonly used to measure the wavelength of light?
2. What is the speed of light?
3. What factor determines the color of light?
4. What are the units of light intensity?

PHOTOCONDUCTIVE CELLS

19–2

The photoconductive cell is a semiconductor device whose resistance varies inversely with the intensity of light that falls upon its photosensitive material. Typical photoconductive cells and the schematic symbol are shown in Figure 19–2. These devices are also known as *photoresistive cells* or *photoresistors*.

A graph of cell resistance versus light intensity is shown in Figure 19–3(a). With no incident light, the cell resistance is maximum; this is called the *dark resistance*. As the light intensity increases, the resistance decreases significantly. Photoconductive cells are typically constructed of cadmium compounds such as cadmium sulfide (CdS) or cadmium selenide (CdSe). Spectral response curves for CdSe and CdS are shown in Figures 19–3(b) and (c). These curves show that the CdSe cell is most responsive to light having a wavelength of slightly less than 7000 Å, and the CdS cell responds best to about 5500 Å.

FIGURE 19–2
Photoconductive cell.

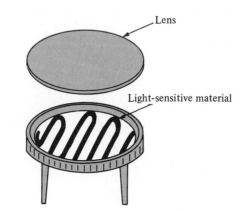

Lens

Light-sensitive material

(a) Typical photoconductive cell

(b) Schematic symbol

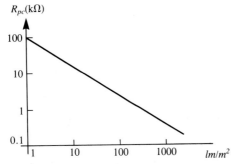

(a) Cell resistance versus light intensity

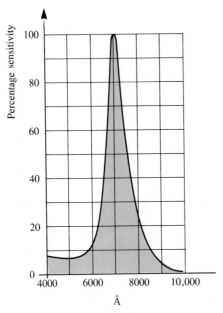

(b) Spectral response for CdSe

(c) Spectral response for CdS

FIGURE 19–3
Typical photoconductive cell characteristics.

EXAMPLE 19–1

Determine the current in the circuit of Figure 19–4 when the light intensity is 10 lm/m². The photoconductive cell has the characteristic shown in Figure 19–3(a).

Solution
From the graph in Figure 19–3(a), $R_{pc} \cong 20$ kΩ at 10 lm/m². Thus

$$I = \frac{V}{R_{pc}} = \frac{10 \text{ V}}{20 \text{ k}\Omega} = 0.5 \text{ mA rms}$$

FIGURE 19–4

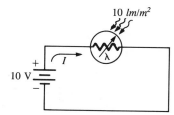

10 *lm/m²*

10 V

I

Photoconductive cells are used in a variety of applications, including smoke detectors, burglar alarms, outdoor lighting controls, photographic instruments, and punched card readers, to name a few.

---

1. The resistance of a photoconductive cell varies (directly, inversely) with light intensity.

2. Name two common materials used in photoconductive cells.

SECTION REVIEW 19–2

PHOTODIODES

19–3

The photodiode is a pn junction device that operates in reverse bias, as shown in Figure 19–5(a). Note the schematic symbol for the photodiode. The photodiode has a small transparent window that allows light to strike the pn junction. Typical photodiodes are shown in part (b).

Recall that a rectifier diode has a very small reverse leakage current when reverse-biased. The same is true for the photodiode. The reverse-biased current is produced by thermally generated electron-hole pairs in the depletion layer, which are swept across the junction by the electric field created by the reverse voltage. In a rectifier diode, the reverse leakage current increases with temperature due to an increase in the number of electron-hole pairs.

A photodiode differs in that the reverse current increases with the light intensity at the pn junction. When there is no incident light, the reverse current I_λ is almost negligible and is called the *dark current*. An increase in the amount of light energy produces an increase in the reverse current, as shown by the graph in Figure 19–6(a). For a given value of reverse-bias voltage, part (b) shows a set of characteristic curves for a typical

FIGURE 19–5
Photodiode.

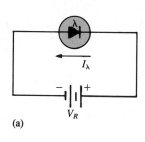

I_λ

V_R

(a)

(b)

FIGURE 19–6
Typical photodiode characteristics.

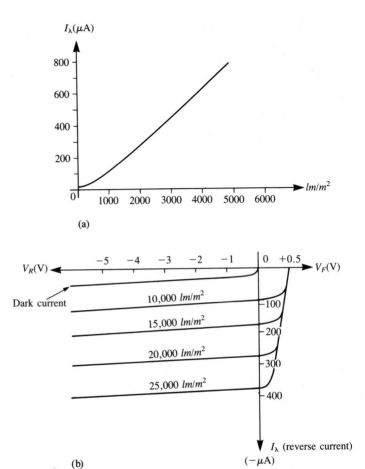

(a)

(b)

photodiode. From the characteristic curve in Figure 19–6(b), the dark current for this particular device is approximately 25 μA at a reverse bias voltage of 3 V. Therefore, the resistance of the device with no incident light is

$$r_R = \frac{V_R}{I_\lambda} = \frac{3 \text{ V}}{25 \ \mu\text{A}} = 120 \text{ k}\Omega$$

At 25,000 lm/m², the current is approximately 375 μA at −3 V. The resistance under this condition is

$$r_R = \frac{V_R}{I_\lambda} = \frac{3 \text{ V}}{375 \ \mu\text{A}} = 8 \text{ k}\Omega$$

    These calculations show that the photodiode can be used as a variable resistance device controlled by light intensity. The photodiode differs, of course, from the photoconductive cell in that it conducts current in only one direction.

SECTION REVIEW 19–3

1. There is a very small reverse current in a photodiode under no-light conditions. What is this current called?

2. What is the resistance of the photodiode in Figure 19–7?

FIGURE 19–7

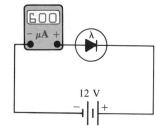

12 V

PHOTOTRANSISTORS

19–4

The phototransistor has a light-sensitive, collector-base pn junction. It is exposed to incident light through a lens opening in the transistor package. When there is no incident light, there is a small thermally generated collector-to-emitter leakage current, I_{CEO}; this is called the *dark current* and is typically in the nA range. When light strikes the collector-base pn junction, a base current I_λ is produced that is directly proportional to the light intensity. This action produces a collector current which increases with I_λ according to the relationship in equation (19–4). Except for the way base current is generated, the phototransistor behaves as a conventional bipolar transistor. In many cases, there is no electrical connection to the base.

$$I_C = \beta_{dc} I_\lambda \qquad (19\text{–}4)$$

The schematic symbol and some typical phototransistors are shown in Figure 19–8. Since the actual photogeneration of base current occurs in the collector-base region, the larger the physical area of this region, the more base current is generated. Thus, a typical phototransistor is designed to offer a large area to the incident light, as the simplified structure diagram in Figure 19–9 illustrates.

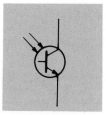

(a) Schematic symbol

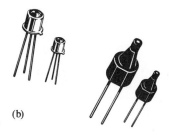

(b)

FIGURE 19–8
Phototransistor.

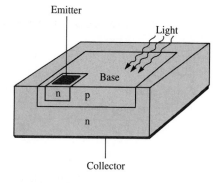

FIGURE 19–9
Typical phototransistor chip structure.

A phototransistor can be either a two-lead or a three-lead device. In the three-lead configuration, the base lead is brought out so that the device can be used as a conventional bipolar transistor with or without the additional light-sensitivity feature. In the two-lead configuration, the base is not electrically available, and the device can be used only with light as the input. In many applications, the phototransistor is used in the two-lead version. Figure 19–10 shows a phototransistor with a biasing circuit and typical collector characteristic curves. Notice that each individual curve on the graph corresponds to a certain value of light intensity (in this case, the units are mW/cm^2) and that the collector current increases with light intensity.

Phototransistors are not sensitive to all light but only to light within a certain range of wavelengths. They are most sensitive to particular wavelengths, as shown by the peak of the spectral response curve in Figure 19–11.

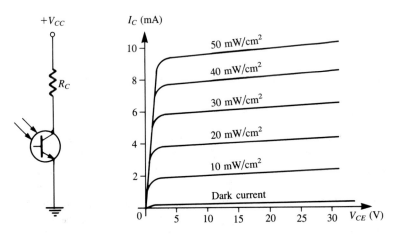

FIGURE 19–10
Phototransistor bias circuit and typical collector characteristic curves.

FIGURE 19–11
Typical phototransistor spectral response.

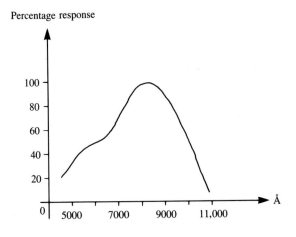

Photodarlington

The photodarlington consists of a phototransistor connected in a darlington arrangement with a conventional transistor, as shown in Figure 19–12. Because of the higher current gain, this device has a much higher collector current and exhibits a greater light sensitivity than does a regular phototransistor.

FIGURE 19–12
Photodarlington.

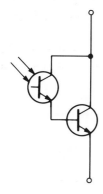

Applications

Phototransistors are used in a wide variety of applications. A few are presented in this section. A light-operated relay circuit is shown in Figure 19–13. The phototransistor Q_1 drives the bipolar transistor Q_2. When there is sufficient incident light on Q_1, transistor Q_2 is driven into saturation, and collector current through the relay coil energizes the relay.

Figure 19–14 shows a circuit in which a relay is de-energized by incident light on the phototransistor. When there is *insufficient* light, transistor Q_2 is biased on, keeping the relay energized. When there is *sufficient* light, phototransistor Q_1 turns on; this pulls the base of Q_2 low, thus turning Q_2 off and de-energizing the relay.

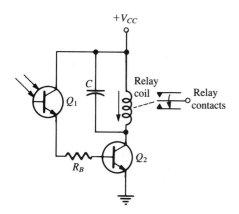

FIGURE 19–13
Light-operated relay circuit.

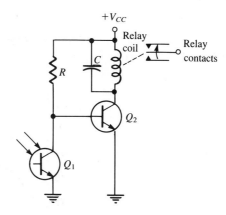

FIGURE 19–14
Darkness-operated relay circuit.

These relay circuits can be used in a variety of applications such as automatic door activators, process counters, and various alarm systems. Another simple application is illustrated in Figure 19–15. The phototransistor is normally *on,* holding the gate of the SCR low. When the light is interrupted, the phototransistor turns off. The high-going transition on the collector triggers the SCR and sets off the alarm mechanism. The momentary contact switch S_1 provides for resetting the alarm. Smoke and intrusion detection are possible uses for this circuit.

FIGURE 19–15
Light-interruption alarm.

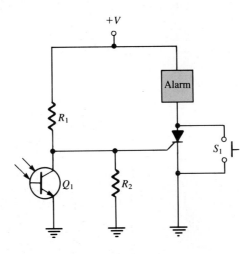

SECTION REVIEW 19–4

1. How does a phototransistor differ from a conventional bipolar transistor?
2. Many, but not all, phototransistors have _____ external leads for electrical connections.
3. The collector current in a phototransistor circuit depends on what two factors?

SOLAR CELLS

19–5

The operation of solar cells is based on the principle of *photovoltaic action,* the process of converting light energy directly to electrical energy. This action occurs in all semiconductors that are constructed to absorb incident light energy. The electrical symbol and a typical solar cell array are shown in Figure 19–16.

A basic solar cell consists of n-type and p-type semiconductor material forming a pn junction. The bottom surface that is always away from the light (dark side) is covered with a continuous conductive contact to which a wire lead is attached. The upper surface has a maximum area exposed to sunlight with a small contact, often along one edge or around the perimeter, as shown in Figure 19–17.

Electrons within the semiconductive material absorb energy from sunlight photons that penetrate the exposed surface. Many of these electrons acquire sufficient energy to break away from the parent atoms, thus creating electron-hole pairs. An electric field is established in the vicinity of the pn junction by the positive and negative ions created when the electron-hole pairs are produced, inducing a potential across the junction. The

FIGURE 19–16
Solar cell array.

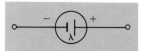

(a) Symbol

(b)

FIGURE 19–17
Basic solar cell construction

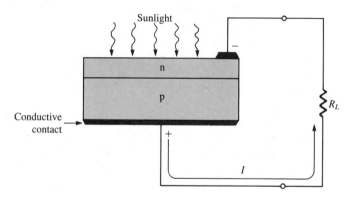

number of electron-hole pairs far exceeds the number needed for thermal equilibrium, and therefore, many of the electrons are pulled across the junction by the force of the electric field. Those that cross the pn junction contribute to the current in the cell and through the external load.

Typical solar cells generate 0.4 V or less with currents ranging from the μA range to the mA range, depending on the external load. Usually, a large number of solar cells are arranged in an *array* in order to achieve higher voltages and currents, as shown in Figure 19–18.

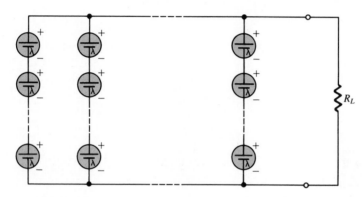

FIGURE 19–18
A solar cell array diagram.

SECTION
REVIEW
19–5

1. What is the principle upon which a solar cell works?
2. In what way is a solar cell similar to a diode?

THE LIGHT-ACTIVATED SCR (LASCR)

19–6

The LASCR operates essentially as does the conventional SCR except that it can also be light-triggered. Most LASCRs have an available gate terminal so that the device can also be triggered by an electrical pulse just as a conventional SCR. Figure 19–19 shows an LASCR schematic symbol and typical packages.

The LASCR responds best to light when the gate terminal is open. If necessary, a resistor from the gate to the cathode can be used to lower the light sensitivity. Figure 19–20 shows an LASCR used to energize a latching relay. The input source turns on the lamp, and the resulting incident light triggers the LASCR. The anode current energizes the relay and closes the contact. Notice that the input source is *electrically isolated* from the rest of the circuit.

SECTION
REVIEW
19–6

1. Can most LASCRs be operated as conventional SCRs?
2. Show what is required in Figure 19–20 to turn off the LASCR to de-energize the relay.

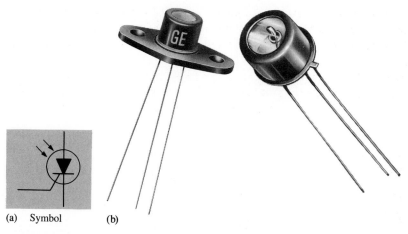

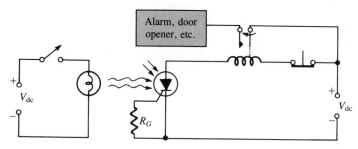

(a)   Symbol   (b)

FIGURE 19–19
Light-activated SCR.

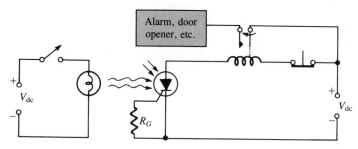

FIGURE 19–20
Light-activated SCR circuit.

THE LIGHT-EMITTING DIODE (LED)

19–7

All the optoelectronic devices covered so far were light-activated. Now we will discuss a type of light-emitting device, the LED. Whereas the photodiode *absorbs* light energy and produces a *reverse* current, the operation of the LED is essentially opposite: It *emits* light in response to a sufficient *forward* current, as shown in Figure 19–21(a). The amount of light output is directly proportional to the forward current, as indicated in Figure 19–21(b). Typical LEDs are shown in part (c).

The basic operation of the LED is as follows. When the device is forward-biased, electrons cross the pn junction from the n-type material and *recombine* with holes in the p-type material. Recall that these free electrons are in the conduction band, whereas the holes are in the valence band and, therefore, at a higher energy level. When recombination takes place, the recombining electrons release energy in the form of *heat* and *light*. A large exposed surface area on one layer of the semiconductor material permits the light photons to be emitted as visible light. Figure 19–22 illustrates this process, which is called *electroluminescence*.

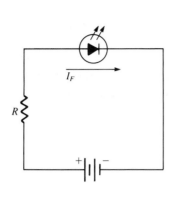

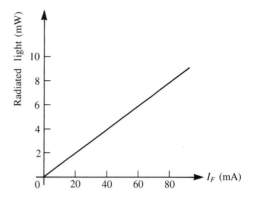

(a)   Forward-biased operation

(b)   Typical light output versus forward current

(c)

FIGURE 19–21
Light-emitting diodes (LEDs).

FIGURE 19–22
Electroluminescence in an LED.

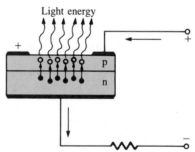

The semiconductor materials used in LEDs are *gallium arsenide* (GaAs), *gallium arsenide phosphide* (GaAsP), or *gallium phosphide* (GaP). Silicon and germanium are not used because they are essentially heat-producing materials and are very poor at producing light. GaAs LEDs emit *infrared* radiation (IR), GaAsP produces either *red* or *yellow* light, and GaP emits *red* or *green* light.

Section 19–2

1. Inversely.

2. Cadmium sulfide (CdS), cadmium selenide (CdSe).

Section 19–3

1. Dark current.

2. 20 kΩ.

Section 19–4

1. The base current is light-induced.

2. Two.

3. β_{dc} and I_λ.

Section 19–5

1. Photovoltaic effect.

2. It is a pn junction device.

Section 19–6

1. Yes.

2. A switch in parallel with the LASCR.

Section 19–7

1. An LED emits light when forward-biased. A photodiode's reverse current changes with incident light.

2. GaAs, GaAsP, and GaP.

Section 19–8

1. Light Amplification by Stimulated Emission of Radiation.

2. Coherent light has one wavelength; incoherent light consists of a wide band of wavelengths.

Section 19–9

1. LED.

2. Phototransistor, photodarlington, LASCR, phototriac, linear amplifier.

APPENDIX A:
Data Sheets

MOTOROLA
Semiconductors
BOX 20912 • PHOENIX, ARIZONA 85036

SILICON CONTROLLED RECTIFIERS

1.6 AMPERE RMS
50 thru 400 VOLTS

REVERSE BLOCKING TRIODE THYRISTORS

These devices are glassivated planar construction designed for gating operation in mA/μA signal or detection circuits.

- Low-Level Gate Characteristics —
  I_{GT} = 10 mA (Max) @ 25°C

- Low Holding Current —
  I_H = 5.0 mA (Typ) @ 25°C

- Glass-to-Metal Bond for Maximum Hermetic Seal

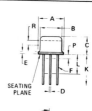

*MAXIMUM RATINGS (T_J = 125°C unless otherwise noted).

Rating	Symbol	Value	Unit
Repetitive Peak Reverse Blocking Voltage	V_{RRM}		Volts
2N1595		50	
2N1596		100	
2N1597		200	
2N1598		300	
2N1599		400	
Repetitive Peak Forward Blocking Voltage	V_{DRM}		Volts
2N1595		50	
2N1596		100	
2N1597		200	
2N1598		300	
2N1599		400	
RMS On-State Current (All Conduction Angles)	$I_{T(RMS)}$	1.6	Amps
Peak Non-Repetitive Surge Current (One Cycle, 60 Hz, T_J = –65 to +125°C)	I_{TSM}	15	Amps
Peak Gate Power	P_{GM}	0.1	Watt
Average Gate Power	$P_{G(AV)}$	0.01	Watt
Peak Gate Current	I_{GM}	0.1	Amp
Peak Gate Voltage — Forward	V_{GFM}	10	Volts
Reverse	V_{GRM}	10	
Operating Junction Temperature Range	T_J	–65 to +125	°C
Storage Temperature Range	T_{stg}	–65 to +150	°C

*Indicates JEDEC Registered Data.

STYLE 3:
PIN 1. CATHODE
2. GATE
3. ANODE (CONNECTED TO CASE)

DIM	MILLIMETERS MIN	MILLIMETERS MAX	INCHES MIN	INCHES MAX
A	8.89	9.40	0.350	0.370
B	8.00	8.51	0.315	0.335
C	6.10	6.60	0.240	0.260
D	0.406	0.533	0.016	0.021
E	0.229	3.18	0.009	0.125
F	0.406	0.483	0.016	0.019
G	4.83	5.33	0.190	0.210
H	0.711	0.864	0.028	0.034
J	0.737	1.02	0.029	0.040
K	12.70	–	0.500	–
L	6.35	–	0.250	–
M	45° NOM		45° NOM	
P	–	1.27	–	0.050
Q	90° NOM		90° NOM	
R	2.54	–	0.100	–

All JEDEC dimensions and notes apply.

CASE 79-02
TO-39

DS 6503 R1

ELECTRICAL CHARACTERISTICS (T_C = 25°C unless otherwise noted).

Characteristic	Symbol	Min	Typ	Max	Unit
*Peak Reverse Blocking Current (Rated V_{RRM}, T_J = 125°C)	I_{RRM}	–	–	1000	μA
*Peak Forward Blocking Current (Rated V_{DRM}, T_J = 125°C)	I_{DRM}	–	–	1000	μA
*Peak On-State Voltage (I_F = 1.0 Adc, Pulsed, 1.0 ms (Max), Duty Cycle ≤ 1%)	V_{TM}	–	1.1	2.0	Volts
*Gate Trigger Current (V_{AK} = 6.0 V, R_L = 12 Ohms)	I_{GT}	–	2.0	10	mA
*Gate Trigger Voltage (V_{AK} = 6.0 V, R_L = 12 Ohms) (V_{AK} = 6.0 V, R_L = 12 Ohms, T_J = 125°C)	V_{GT}	– 0.2	0.7 –	3.0 –	Volts
Reverse Gate Current (V_{GK} = 10 V)	I_{GR}	–	17	–	mA
Holding Current (V_{AK} = 12 V)	I_H	–	5.0	–	mA
Turn-On Time (I_{GT} = 10 mA, I_F = 1.0 A) (I_{GT} = 20 mA, I_F = 1.0 A)	t_{gt}	– –	0.8 0.6	– –	μs
Turn-Off Time (I_F = 1.0 A, I_R = 1.0 A, dv/dt = 20 V/μs, T_J = 125°C)	t_q	–	10	–	μs

*Indicates JEDEC Registered Data.

CURRENT DERATING

FIGURE 1 – CASE TEMPERATURE REFERENCE

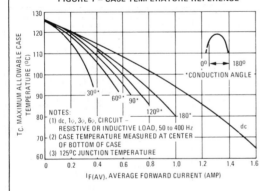

FIGURE 2 – AMBIENT TEMPERATURE REFERENCE

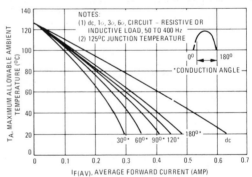

MOTOROLA Semiconductor Products Inc.

BOX 20912 • PHOENIX, ARIZONA 85036 • A SUBSIDIARY OF MOTOROLA INC.

2N4877

MEDIUM-POWER NPN SILICON TRANSISTOR

. . . designed for switching and wide band amplifier applications.

- Low Collector-Emitter Saturation Voltage —
  $V_{CE(sat)}$ = 1.0 Vdc (Max) @ I_C = 4.0 Amp

- DC Current Gain Specified to 4 Amperes

- Excellent Safe Operating Area

- Packaged in the Compact TO-39 Case for Critical Space-Limited Applications.

4 AMPERE POWER TRANSISTOR

**NPN SILICON
60 VOLTS
10 WATTS**

JANUARY 1971 — DS 3189

* MAXIMUM RATINGS

Rating	Symbol	Value	Unit
Collector-Emitter Voltage	V_{CEO}	60	Vdc
Collector-Base Voltage	V_{CB}	70	Vdc
Emitter-Base Voltage	V_{EB}	5.0	Vdc
Collector Current — Continuous	I_C	4.0	Adc
Base Current	I_B	1.0	Adc
Total Device Dissipation @ T_C = 25°C Derate above 25°C	P_D	10 57.2	Watts mW/°C
Operating and Storage Junction Temperature Range	T_J, T_{stg}	–65 to +200	°C

*Indicates JEDEC Registered Data

THERMAL CHARACTERISTICS

Characteristic	Symbol	Max	Unit
Thermal Resistance, Junction to Case	θ_{JC}	17.5	°C/W

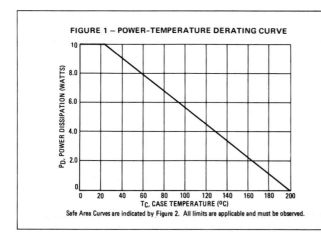

FIGURE 1 – POWER–TEMPERATURE DERATING CURVE

P_D, POWER DISSIPATION (WATTS)

T_C, CASE TEMPERATURE (°C)

Safe Area Curves are indicated by Figure 2. All limits are applicable and must be observed.

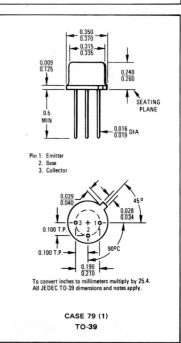

Pin 1. Emitter
2. Base
3. Collector

To convert inches to millimeters multiply by 25.4.
All JEDEC TO-39 dimensions and notes apply.

**CASE 79 (1)
TO-39**

*ELECTRICAL CHARACTERISTICS ($T_C = 25^\circ C$ unless otherwise noted)

Characteristic	Symbol	Min	Max	Unit
OFF CHARACTERISTICS				
Collector-Emitter Sustaining Voltage (1) (I_C = 200 mAdc, I_B = 0)	$V_{CEO(sus)}$	60	–	Vdc
Collector Cutoff Current (V_{CE} = 70 Vdc, $V_{EB(off)}$ = 1.5 Vdc) (V_{CE} = 70 Vdc, $V_{EB(off)}$ = 1.5 Vdc, T_C = 100°C)	I_{CEX}	– –	100 1.0	µAdc mAdc
Collector Cutoff Current (V_{CB} = 70 Vdc, I_E = 0)	I_{CBO}	–	100	µAdc
Emitter Cutoff Current (V_{BE} = 5.0 Vdc, I_C = 0)	I_{EBO}	–	100	µAdc
ON CHARACTERISTICS(1)				
DC Current Gain (I_C = 1.0 Adc, V_{CE} = 2.0 Vdc) (I_C = 4.0 Adc, V_{CE} = 2.0 Vdc)	h_{FE}	30 20	– 100	–
Collector-Emitter Saturation Voltage (I_C = 4.0 Adc, I_B = 0.4 Adc)	$V_{CE(sat)}$	–	1.0	Vdc
Base-Emitter Saturation Voltage (I_C = 4.0 Adc, I_B = 0.4 Adc)	$V_{BE(sat)}$	–	1.8	Vdc
DYNAMIC CHARACTERISTICS				
Current-Gain-Bandwidth Product (I_C = 0.25 Adc, V_{CE} = 10 Vdc, f = 1.0 MHz) (I_C = 0.25 Adc, V_{CE} = 10 Vdc, f = 10 MHz)**	f_T	4.0 30	– –	MHz
SWITCHING CHARACTERISTICS				
Rise Time (V_{CC} = 25 Vdc, I_C = 4.0 Adc, I_{B1} = 0.4 Adc)	t_r	–	100	ns
Storage Time (V_{CC} = 25 Vdc, I_C = 4.0 Adc,	t_s	–	1.5	µs
Fall Time $I_{B1} = I_{B2}$ = 0.4 Adc)	t_f	–	500	ns

*Indicates JEDEC Registered Data.
**Motorola guarantees this value in addition to JEDEC Registered Data.
Note 1: Pulse Test: Pulse Width ≤ 300 µs, Duty Cycle ≤ 2.0%.

FIGURE 2 – ACTIVE-REGION SAFE OPERATING AREA

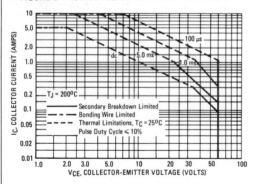

FIGURE 3 – SWITCHING TIME TEST CIRCUIT

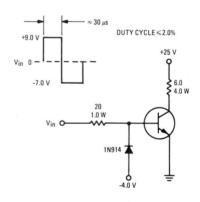

There are two limitations on the power handling ability of a transistor: average junction temperature and second breakdown. Safe operating area curves indicate I_C–V_{CE} limits of the transistor that must be observed for reliable operation; i.e., the transistor must not be subjected to greater dissipation than the curves indicate.

The data of Figure 2 is based on $T_{J(pk)}$ = 200°C; T_C is variable depending on conditions. Second breakdown pulse limits are valid for duty cycles to 10% provided $T_{J(pk)} \leq$ 200°C. At high case temperatures, thermal limitations will reduce the power that can be handled to values less than the limitations imposed by second breakdown. (See AN-415)

MOTOROLA Semiconductor Products Inc.

BOX 20912 • PHOENIX, ARIZONA 85036 • A SUBSIDIARY OF MOTOROLA INC.

MOTOROLA

1N4001
thru
1N4007

Designers' Data Sheet

"SURMETIC"▲ RECTIFIERS

. . . subminiature size, axial lead mounted rectifiers for general-purpose low-power applications.

Designers Data for "Worst Case" Conditions

The Designers▲ Data Sheets permit the design of most circuits entirely from the information presented. Limit curves — representing boundaries on device characteristics — are given to facilitate "worst case" design.

LEAD MOUNTED SILICON RECTIFIERS

50-1000 VOLTS DIFFUSED JUNCTION

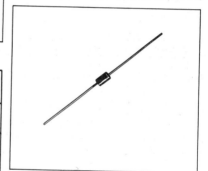

*MAXIMUM RATINGS

Rating	Symbol	1N4001	1N4002	1N4003	1N4004	1N4005	1N4006	1N4007	Unit
Peak Repetitive Reverse Voltage Working Peak Reverse Voltage DC Blocking Voltage	V_{RRM} V_{RWM} V_R	50	100	200	400	600	800	1000	Volts
Non-Repetitive Peak Reverse Voltage (halfwave, single phase, 60 Hz)	V_{RSM}	60	120	240	480	720	1000	1200	Volts
RMS Reverse Voltage	$V_{R(RMS)}$	35	70	140	280	420	560	700	Volts
Average Rectified Forward Current (single phase, resistive load, 60 Hz, see Figure 8, $T_A = 75°C$)	I_O	1.0							Amp
Non-Repetitive Peak Surge Current (surge applied at rated load conditions, see Figure 2)	I_{FSM}	30 (for 1 cycle)							Amp
Operating and Storage Junction Temperature Range	T_J, T_{stg}	-65 to +175							°C

*ELECTRICAL CHARACTERISTICS

Characteristic and Conditions	Symbol	Typ	Max	Unit
Maximum Instantaneous Forward Voltage Drop (i_F = 1.0 Amp, T_J = 25°C) Figure 1	v_F	0.93	1.1	Volts
Maximum Full-Cycle Average Forward Voltage Drop (I_O = 1.0 Amp, T_L = 75°C, 1 inch leads)	$V_{F(AV)}$	—	0.8	Volts
Maximum Reverse Current (rated dc voltage) T_J = 25°C T_J = 100°C	I_R	0.05 1.0	10 50	μA
Maximum Full-Cycle Average Reverse Current (I_O = 1.0 Amp, T_L = 75°C, 1 inch leads	$I_{R(AV)}$	—	30	μA

*Indicates JEDEC Registered Data.

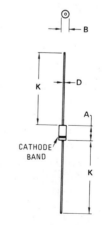

CATHODE BAND

MECHANICAL CHARACTERISTICS

CASE: Transfer Molded Plastic
MAXIMUM LEAD TEMPERATURE FOR SOLDERING PURPOSES: 350°C, 3/8" from case for 10 seconds at 5 lbs. tension
FINISH: All external surfaces are corrosion-resistant, leads are readily solderable
POLARITY: Cathode indicated by color band
WEIGHT: 0.40 Grams (approximately)

DIM	MILLIMETERS		INCHES	
	MIN	MAX	MIN	MAX
A	5.97	6.60	0.235	0.260
B	2.79	3.05	0.110	0.120
D	0.76	0.86	0.030	0.034
K	27.94	—	1.100	—

CASE 59-04
Does Not Conform to DO-41 Outline.

▲Trademark of Motorola Inc.

DS 6015 R3

FIGURE 1 — FORWARD VOLTAGE

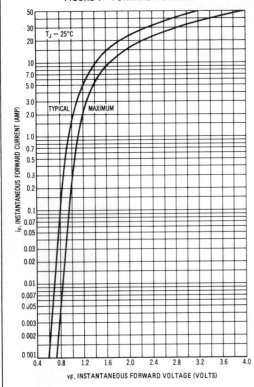

i_F, INSTANTANEOUS FORWARD CURRENT (AMP)

v_F, INSTANTANEOUS FORWARD VOLTAGE (VOLTS)

FIGURE 2 — NON-REPETITIVE SURGE CAPABILITY

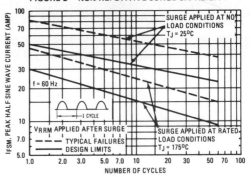

I_{FSM}, PEAK HALF SINE WAVE CURRENT (AMP)

NUMBER OF CYCLES

FIGURE 3 — FORWARD VOLTAGE TEMPERATURE COEFFICIENT

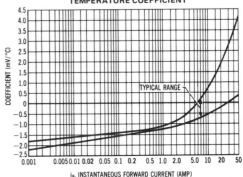

COEFFICIENT (mV/°C)

i_F, INSTANTANEOUS FORWARD CURRENT (AMP)

FIGURE 4 — TYPICAL TRANSIENT THERMAL RESISTANCE

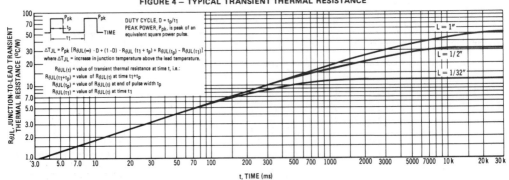

DUTY CYCLE, D = t_p/t_1
PEAK POWER, P_{pk}, is peak of an equivalent square power pulse.

$\Delta T_{JL} = P_{pk} [R_{\theta JL}(\infty) \cdot D + (1-D) \cdot R_{\theta JL}(t_1 + t_p) + R_{\theta JL}(t_p) - R_{\theta JL}(t_1)]$
where ΔT_{JL} = increase in junction temperature above the lead temperature.

$R_{\theta JL}(t)$ = value of transient thermal resistance at time t, i.e.:
$R_{\theta JL}(t_1+t_p)$ = value of $R_{\theta JL}(t)$ at time t_1+t_p
$R_{\theta JL}(t_p)$ = value of $R_{\theta JL}(t)$ at end of pulse width t_p
$R_{\theta JL}(t_1)$ = value of $R_{\theta JL}(t)$ at time t_1

$R_{\theta JL}$, JUNCTION-TO-LEAD TRANSIENT THERMAL RESISTANCE (°C/W)

t, TIME (ms)

The temperature of the lead should be measured using a thermocouple placed on the lead as close as possible to the tie point. The thermal mass connected to the tie point is normally large enough so that it will not significantly respond to heat surges generated in the diode as a result of pulsed operation once steady-state conditions are achieved. Using the measured value of T_L, the junction temperature may be determined by:

$$T_J = T_L + \Delta T_{JL}.$$

 MOTOROLA *Semiconductor Products Inc.*

MOTOROLA

Designers▲ Data Sheet

ONE WATT HERMETICALLY SEALED GLASS SILICON ZENER DIODES

- Complete Voltage Range — 2.4 to 100 Volts
- DO-41 Package — Smaller than Conventional DO-7 Package
- Double Slug Type Construction
- Metallurgically Bonded Construction
- Nitride Passivated Die

Designer's Data for "Worst Case" Conditions

The Designers▲ Data sheets permit the design of most circuits entirely from the information presented. Limit curves — representing boundaries on device characteristics — are given to facilitate "worst case" design.

1N4728, A thru 1N4764, A

1.0 WATT ZENER REGULATOR DIODES

3.3–100 VOLTS

---

*MAXIMUM RATINGS

Rating	Symbol	Value	Unit
DC Power Dissipation @ $T_A = 50^oC$ Derate above 50^oC	P_D	1.0 6.67	Watt mW/^oC
Operating and Storage Junction Temperature Range	T_J, T_{stg}	–65 to +200	^oC

MECHANICAL CHARACTERISTICS

CASE: Double slug type, hermetically sealed glass

MAXIMUM LEAD TEMPERATURE FOR SOLDERING PURPOSES: 230^oC, 1/16'' from case for 10 seconds

FINISH: All external surfaces are corrosion resistant with readily solderable leads.

POLARITY: Cathode indicated by color band. When operated in zener mode, cathode will be positive with respect to anode.

MOUNTING POSITION: Any

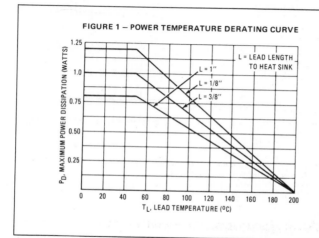

FIGURE 1 — POWER TEMPERATURE DERATING CURVE

L = LEAD LENGTH TO HEAT SINK

L = 1"

L = 1/8"

L = 3/8"

P_D, MAXIMUM POWER DISSIPATION (WATTS)

T_L, LEAD TEMPERATURE (^oC)

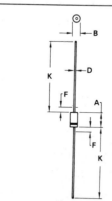

NOTE:
1. POLARITY DENOTED BY CATHODE BAND
2. LEAD DIAMETER NOT CONTROLLED WITHIN "F" DIMENSION.

DIM	MILLIMETERS		INCHES	
	MIN	MAX	MIN	MAX
A	4.07	5.20	0.160	0.205
B	2.04	2.71	0.080	0.107
D	0.71	0.86	0.028	0.034
F	–	1.27	–	0.050
K	27.94	–	1.100	–

All JEDEC dimensions and notes apply.

CASE 59-03 (DO-41)

---

*Indicates JEDEC Registered Data
▲Trademark of Motorola Inc.

DS 7039 R1

*ELECTRICAL CHARACTERISTICS (T_A = 25°C unless otherwise noted) V_F = 1.2 V max, I_F = 200 mA for all types.

JEDEC Type No. (Note 1)	Nominal Zener Voltage V_Z @ I_{ZT} Volts (Notes 2 and 3)	Test Current I_{ZT} mA	Maximum Zener Impedance (Note 4)			Leakage Current		Surge Current @ T_A = 25°C
			Z_{ZT} @ I_{ZT} Ohms	Z_{ZK} @ I_{ZK} Ohms	I_{ZK} mA	I_R µA Max	V_R Volts	i_r – mA (Note 5)
1N4728	3.3	76	10	400	1.0	100	1.0	1380
1N4729	3.6	69	10	400	1.0	100	1.0	1260
1N4730	3.9	64	9.0	400	1.0	50	1.0	1190
1N4731	4.3	58	9.0	400	1.0	10	1.0	1070
1N4732	4.7	53	8.0	500	1.0	10	1.0	970
1N4733	5.1	49	7.0	550	1.0	10	1.0	890
1N4734	5.6	45	5.0	600	1.0	10	2.0	810
1N4735	6.2	41	2.0	700	1.0	10	3.0	730
1N4736	6.8	37	3.5	700	1.0	10	4.0	660
1N4737	7.5	34	4.0	700	0.5	10	5.0	605
1N4738	8.2	31	4.5	700	0.5	10	6.0	550
1N4739	9.1	28	5.0	700	0.5	10	7.0	500
1N4740	10	25	7.0	700	0.25	10	7.6	454
1N4741	11	23	8.0	700	0.25	5.0	8.4	414
1N4742	12	21	9.0	700	0.25	5.0	9.1	380
1N4743	13	19	10	700	0.25	5.0	9.9	344
1N4744	15	17	14	700	0.25	5.0	11.4	304
1N4745	16	15.5	16	700	0.25	5.0	12.2	285
1N4746	18	14	20	750	0.25	5.0	13.7	250
1N4747	20	12.5	22	750	0.25	5.0	15.2	225
1N4748	22	11.5	23	750	0.25	5.0	16.7	205
1N4749	24	10.5	25	750	0.25	5.0	18.2	190
1N4750	27	9.5	35	750	0.25	5.0	20.6	170
1N4751	30	8.5	40	1000	0.25	5.0	22.8	150
1N4752	33	7.5	45	1000	0.25	5.0	25.1	135
1N4753	36	7.0	50	1000	0.25	5.0	27.4	125
1N4754	39	6.5	60	1000	0.25	5.0	29.7	115
1N4755	43	6.0	70	1500	0.25	5.0	32.7	110
1N4756	47	5.5	80	1500	0.25	5.0	35.8	95
1N4757	51	5.0	95	1500	0.25	5.0	38.8	90
1N4758	56	4.5	110	2000	0.25	5.0	42.6	80
1N4759	62	4.0	125	2000	0.25	5.0	47.1	70
1N4760	68	3.7	150	2000	0.25	5.0	51.7	65
1N4761	75	3.3	175	2000	0.25	5.0	56.0	60
1N4762	82	3.0	200	3000	0.25	5.0	62.2	55
1N4763	91	2.8	250	3000	0.25	5.0	69.2	50
1N4764	100	2.5	350	3000	0.25	5.0	76.0	45

NOTE 1 — Tolerance and Type Number Designation. The JEDEC type numbers listed have a standard tolerance on the nominal zener voltage of ±10%. A standard tolerance of ±5% on individual units is also available and is indicated by suffixing "A" to the standard type number.

NOTE 2 — Specials Available Include:

 A. Nominal zener voltages between the voltages shown and tighter voltage tolerances,

 B. Matched sets.

For detailed information on price, availability, and delivery, contact your nearest Motorola representative.

NOTE 3 — Zener Voltage (V_Z) Measurement. Motorola guarantees the zener voltage when measured at 90 seconds while maintaining the lead temperature (T_L) at 30°C ± 1°C, 3/8" from the diode body.

NOTE 4 — Zener Impedance (Z_Z) Derivation. The zener impedance is derived from the 60 cycle ac voltage, which results when an ac current having an rms value equal to 10% of the dc zener current (I_{ZT} or I_{ZK}) is superimposed on I_{ZT} or I_{ZK}.

NOTE 5 — Surge Current (i_r) Non-Repetitive. The rating listed in the electrical characteristics table is maximum peak, non-repetitive, reverse surge current of 1/2 square wave or equivalent sine wave pulse of 1/120 second duration superimposed on the test current, I_{ZT}, per JEDEC registration; however, actual device capability is as described in Figures 4 and 5.

APPLICATION NOTE

Since the actual voltage available from a given zener diode is temperature dependent, it is necessary to determine junction temperature under any set of operating conditions in order to calculate its value. The following procedure is recommended:

Lead Temperature, T_L, should be determined from

$$T_L = \theta_{LA}P_D + T_A$$

θ_{LA} is the lead-to-ambient thermal resistance (°C/W) and P_D is the power dissipation. The value for θ_{LA} will vary and depends on the device mounting method. θ_{LA} is generally 30 to 40°C/W for the various clips and tie points in common use and for printed circuit board wiring.

The temperature of the lead can also be measured using a thermocouple placed on the lead as close as possible to the tie point. The thermal mass connected to the tie point is normally large enough so that it will not significantly respond to heat surges generated in the diode as a result of pulsed operation once steady-state conditions are achieved. Using the measured value of T_L, the junction temperature may be determined by:

$$T_J = T_L + \Delta T_{JL}.$$

ΔT_{JL} is the increase in junction temperature above the lead temperature and may be found as follows:

$$\Delta T_{JL} = \theta_{JL}P_D$$

θ_{JL} may be determined from Figure 3 for dc power conditions. For worst-case design, using expected limits of I_Z, limits of P_D and the extremes of $T_J(\Delta T_J)$ may be estimated. Changes in voltage, V_Z, can then be found from:

$$\Delta V = \theta_{VZ}\Delta T_J$$

θ_{VZ}, the zener voltage temperature coefficient, is found from Figure 2.

Under high power-pulse operation, the zener voltage will vary with time and may also be affected significantly by the zener resistance. For best regulation, keep current excursions as low as possible.

Surge limitations are given in Figure 5. They are lower than would be expected by considering only junction temperature, as current crowding effects cause temperatures to be extremely high in small spots resulting in device degradation should the limits of Figure 5 be exceeded.

 MOTOROLA Semiconductor Products Inc.

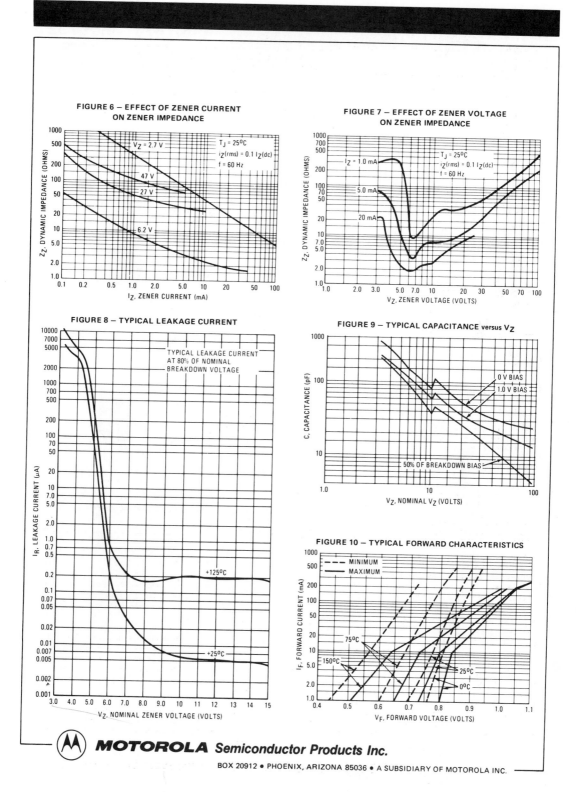

FIGURE 6 — EFFECT OF ZENER CURRENT ON ZENER IMPEDANCE

FIGURE 7 — EFFECT OF ZENER VOLTAGE ON ZENER IMPEDANCE

FIGURE 8 — TYPICAL LEAKAGE CURRENT

FIGURE 9 — TYPICAL CAPACITANCE versus V_Z

FIGURE 10 — TYPICAL FORWARD CHARACTERISTICS

MOTOROLA *Semiconductor Products Inc.*

BOX 20912 • PHOENIX, ARIZONA 85036 • A SUBSIDIARY OF MOTOROLA INC.

MOTOROLA SEMICONDUCTORS

P.O. BOX 20912 • PHOENIX, ARIZONA 85036

MOC7811 MOC7821
MOC7812 MOC7822
MOC7813 MOC7823

OPTO SLOTTED COUPLER/INTERRUPTER MODULES

These devices consist of a gallium arsenide infrared emitting diode facing a silicon NPN phototransistor in a molded plastic housing. A slot in the housing between the emitter and the detector provides a means of interrupting the signal. They are widely used in position and motion indicators, end of tape indicators, paper feed controls and arcless switches.

- 1.0 mm Aperture
- Easy PCB Mounting
- Cost Effective
- Industry Standard Configuration
- Uses Long-Lived LPE IRED

OPTO SLOTTED COUPLER

TRANSISTOR OUTPUT

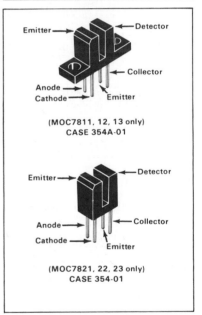

(MOC7811, 12, 13 only)
CASE 354A-01

(MOC7821, 22, 23 only)
CASE 354-01

ABSOLUTE MAXIMUM RATINGS: (25°C)

Rating	Symbol	Value	Unit
TOTAL DEVICE			
Storage Temperature	T_{stg}	–40 to +100	°C
Operating Temperature	T_J	–40 to +100	°C
Lead Soldering Temperature (5 seconds maximum)	T_L	260	°C
INFRARED EMITTING DIODE			
Power Dissipation	P_D	150*	mW
Forward Current (Continuous)	I_F	50	mA
Reverse Voltage	V_R	6	V
PHOTOTRANSISTOR			
Power Dissipation	P_D	150**	mW
Collector-Emitter Voltage	V_{CEO}	30	V

*Derate 2.0 mW/°C above 25°C ambient.
**Derate 2.0 mW/°C above 25° ambient.

INDIVIDUAL ELECTRICAL CHARACTERISTICS: (25°C) (See Note 1)

Characteristic	Symbol	Min	Typ	Max	Unit
EMITTER					
Reverse Breakdown Voltage (I_R = 100 μA)	$V_{(BR)R}$	6	—	—	V
Forward Voltage (I_F = 50 mA)	V_F	—	1.3	1.8	V
Reverse Current (V_R = 6.0 V, R_L = 1.0 MΩ)	I_R	—	50	—	nA
Capacitance (V = 0, f = 1 MHz)	C_i	—	25	—	pF
DETECTOR					
Breakdown Voltage (I_C = 10 mA, H ≈ 0)	$V_{(BR)CEO}$	30	—	—	V
Collector Dark Current (V_{CE} = 10 V, H ≈ 0)	I_{CEO}	—	—	100	nA

Note 1: Stray irradiation can alter values of characteristics. Adequate shielding should be provided. © MOTOROLA INC. 1982 DS2579

COUPLED ELECTRICAL CHARACTERISTICS: (25°C) (See Note 1)

Characteristics	Symbol	MOC7811/7821			MOC7812/7822			MOC7813/7823			Unit
		Min	Typ	Max	Min	Typ	Max	Min	Typ	Max	
$I_F = 5.0$ mA, $V_{CE} = 5.0$ V	$I_{CE(on)}$	0.15	—	—	0.30	—	—	0.60	—	—	mA
$I_F = 20$ mA, $V_{CE} = 5.0$ V	$I_{CE(on)}$	1.0	—	—	2.0	—	—	4.0	—	—	mA
$I_F = 30$ mA, $V_{CE} = 5.0$ V	$I_{CE(on)}$	1.9	—	—	3.0	—	—	5.5	—	—	mA
$I_F = 20$ mA, $I_C = 1.8$ mA	$V_{CE(sat)}$	—	—	—	—	—	0.40	—	—	0.40	V
$I_F = 30$ mA, $I_C = 1.8$ mA	$V_{CE(sat)}$	—	—	0.40	—	—	—	—	—	—	V
$V_{CC} = 5.0$ V, $I_F = 30$ mA, $R_L = 2.5$ kΩ	t_{on}	—	12	—	—	12	—	—	12	—	μs
$V_{CC} = 5.0$ V, $I_F = 30$ mA, $R_L = 2.5$ kΩ	t_{off}	—	60	—	—	60	—	—	60	—	μs

Note 1: Stray irradiation can alter values of characteristics. Adequate shielding should be provided.

FIGURE 1 — OUTPUT CURRENT versus INPUT CURRENT

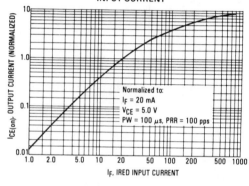

Normalized to:
$I_F = 20$ mA
$V_{CE} = 5.0$ V
PW = 100 μs, PRR = 100 pps

I_F, IRED INPUT CURRENT

$I_{CE(on)}$, OUTPUT CURRENT (NORMALIZED)

FIGURE 2 — t_{on}, t_{off} versus LOAD RESISTANCE

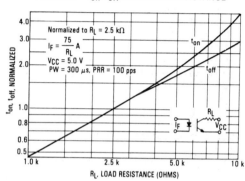

Normalized to $R_L = 2.5$ kΩ
$$I_F = \frac{75}{R_L} \text{ A}$$
$V_{CC} = 5.0$ V
PW = 300 μs, PRR = 100 pps

t_{on}, t_{off}, NORMALIZED

R_L, LOAD RESISTANCE (OHMS)

FIGURE 3 — OUTPUT CURRENT versus POSITION OF SHIELD COVERING APERTURE

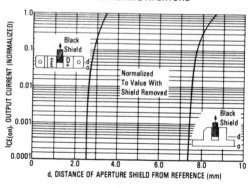

Black Shield

Normalized To Value With Shield Removed

Black Shield

d, DISTANCE OF APERTURE SHIELD FROM REFERENCE (mm)

$I_{CE(on)}$, OUTPUT CURRENT (NORMALIZED)

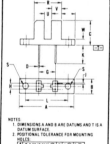

NOTES:
1. DIMENSIONS A AND B ARE DATUMS AND T IS A DATUM SURFACE.
2. POSITIONAL TOLERANCE FOR MOUNTING HOLES:
3. POSITIONAL TOLERANCE FOR LEAD DIMENSION J
4. POSITIONAL TOLERANCE FOR LEAD DIMENSION D
5. DIMENSIONING AND TOLERANCING PER Y14.5, 1973.

DIM	MILLIMETERS		INCHES	
	MIN	MAX	MIN	MAX
A	24.13	25.01	0.950	0.985
B	6.19	6.50	0.244	0.256
C	10.64	10.94	0.419	0.431
D	0.35	0.55	0.014	0.022
E	6.24	6.55	0.246	0.258
G	7.36 BSC		0.290 BSC	
H	2.54 BSC		0.100 BSC	
J	0.43	0.55	0.017	0.022
K	7.36		0.290	
L	19.05 BSC		0.750 BSC	
R	11.98	12.19	0.472	0.480
S	3.24	3.37	0.124	0.133
U	3.07	3.32	0.115	0.129
V	4.36	4.52	0.172	0.178
W	2.38	2.69	0.094	0.106

CASE 354A-01

NOTES:
1. DIMENSIONS R AND B ARE DATUMS AND T IS A DATUM SURFACE.
2. POSITIONAL TOLERANCE FOR LEAD DIMENSION J
3. POSITIONAL TOLERANCE FOR LEAD DIMENSION D
4. DIMENSIONING AND TOLERANCING ARE PER Y14.5, 1973.

DIM	MILLIMETERS		INCHES	
	MIN	MAX	MIN	MAX
B	6.19	6.50	0.244	0.256
C	10.64	10.94	0.419	0.431
D	0.35	0.55	0.014	0.022
G	7.36 BSC		0.290 BSC	
H	2.54 BSC		0.100 BSC	
J	0.43	0.55	0.017	0.022
K	7.36		0.290	
R	11.98	12.19	0.472	0.480
U	3.07	3.32	0.115	0.129
V	4.36	4.52	0.172	0.178
W	2.38	2.69	0.094	0.106

CASE 354-01

 MOTOROLA *Semiconductor Products Inc.*

BOX 20912 • PHOENIX, ARIZONA 85036 • A SUBSIDIARY OF MOTOROLA INC.

MC1741, MC1741C
MC1741N, MC1741NC

OPERATIONAL AMPLIFIER
SILICON MONOLITHIC
INTEGRATED CIRCUIT

INTERNALLY COMPENSATED, HIGH PERFORMANCE OPERATIONAL AMPLIFIERS

. . . designed for use as a summing amplifier, integrator, or amplifier with operating characteristics as a function of the external feedback components.

- No Frequency Compensation Required
- Short-Circuit Protection
- Offset Voltage Null Capability
- Wide Common-Mode and Differential Voltage Ranges
- Low-Power Consumption
- No Latch Up
- Low Noise Selections Offered — N Suffix

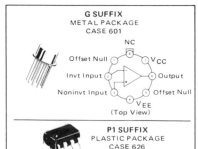

G SUFFIX
METAL PACKAGE
CASE 601

P1 SUFFIX
PLASTIC PACKAGE
CASE 626
(MC1741C, MC1741NC)

U SUFFIX
CERAMIC PACKAGE
CASE 693

MAXIMUM RATINGS (T_A = +25°C unless otherwise noted)

Rating	Symbol	MC1741C	MC1741	Unit
Power Supply Voltage	V_{CC}	+18	+22	Vdc
	V_{EE}	-18	-22	Vdc
Input Differential Voltage	V_{ID}	±30		Volts
Input Common Mode Voltage (Note 1)	V_{ICM}	±15		Volts
Output Short Circuit Duration (Note 2)	t_S	Continuous		
Operating Ambient Temperature Range	T_A	0 to +70	-55 to +125	°C
Storage Temperature Range Metal, Flat and Ceramic Packages Plastic Packages	T_{stg}	-65 to +150 -55 to +125		°C

Note 1. For supply voltages less than ± 15 V, the absolute maximum input voltage is equal to the supply voltage.

Note 2. Supply voltage equal to or less than 15 V.

L SUFFIX
CERAMIC PACKAGE
CASE 632
TO-116

P2 SUFFIX
PLASTIC PACKAGE
CASE 646
(MC1741C, MC1741NC)

EQUIVALENT CIRCUIT SCHEMATIC

F SUFFIX
CERAMIC PACKAGE
CASE 606-04
TO-91

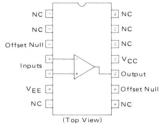

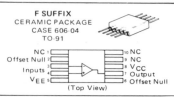

DS 9123 R3

ELECTRICAL CHARACTERISTICS (V_{CC} = 15 V, V_{EE} = 15 V, T_A = 25°C unless otherwise noted).

Characteristic	Symbol	MC1741 Min	MC1741 Typ	MC1741 Max	MC1741C Min	MC1741C Typ	MC1741C Max	Unit
Input Offset Voltage ($R_S \leqslant$ 10 k)	V_{IO}	—	1.0	5.0	—	2.0	6.0	mV
Input Offset Current	I_{IO}	—	20	200	—	20	200	nA
Input Bias Current	I_{IB}	—	80	500	—	80	500	nA
Input Resistance	r_i	0.3	2.0	—	0.3	2.0	—	MΩ
Input Capacitance	C_i	—	1.4	—	—	1.4	—	pF
Offset Voltage Adjustment Range	V_{IOR}	—	±15	—	—	±15	—	mV
Common Mode Input Voltage Range	V_{ICR}	±12	±13	—	±12	±13	—	V
Large Signal Voltage Gain (V_O = ±10 V, $R_L \geqslant$ 2.0 k)	A_v	50	200	—	20	200	—	V/mV
Output Resistance	r_o	—	75	—	—	75	—	Ω
Common Mode Rejection Ratio ($R_S \leqslant$ 10 k)	CMRR	70	90	—	70	90	—	dB
Supply Voltage Rejection Ratio ($R_S \leqslant$ 10 k)	PSRR	—	30	150	—	30	150	μV/V
Output Voltage Swing	V_O							V
($R_L \geqslant$ 10 k)		±12	±14	—	±12	±14	—	
($R_L \geqslant$ 2 k)		±10	±13	—	±10	±13	—	
Output Short-Circuit Current	I_{os}	—	20	—	—	20	—	mA
Supply Current	I_D	—	1.7	2.8	—	1.7	2.8	mA
Power Consumption	P_C	—	50	85	—	50	85	mW
Transient Response (Unity Gain — Non-Inverting)								
(V_I = 20 mV, $R_L \geqslant$ 2 k, $C_L \leqslant$ 100 pF) Rise Time	t_{TLH}	—	0.3	—	—	0.3	—	μs
(V_I = 20 mV, $R_L \geqslant$ 2 k, $C_L \leqslant$ 100 pF) Overshoot	os	—	15	—	—	15	—	%
(V_I = 10 V, $R_L \geqslant$ 2 k, $C_L \leqslant$ 100 pF) Slew Rate	SR	—	0.5	—	—	0.5	—	V/μs

ELECTRICAL CHARACTERISTICS (V_{CC} = 15 V, V_{EE} = 15 V, T_A = *T_{high} to T_{low} unless otherwise noted.)

Characteristic	Symbol	MC1741 Min	MC1741 Typ	MC1741 Max	MC1741C Min	MC1741C Typ	MC1741C Max	Unit
Input Offset Voltage ($R_S \leqslant$ 10 kΩ)	V_{IO}	—	1.0	6.0	—	—	7.5	mV
Input Offset Current	I_{IO}							nA
(T_A = 125°C)		—	7.0	200	—	—	—	
(T_A = -55°C)		—	85	500	—	—	—	
(T_A = 0°C to +70°C)		—	—	—	—	—	300	
Input Bias Current	I_{IB}							nA
(T_A = 125°C)		—	30	500	—	—	—	
(T_A = -55°C)		—	300	1500	—	—	—	
(T_A = 0°C to +70°C)		—	—	—	—	—	800	
Common Mode Input Voltage Range	V_{ICR}	±12	±13	—	—	—	—	V
Common Mode Rejection Ratio ($R_S \leqslant$ 10 k)	CMRR	70	90	—	—	—	—	dB
Supply Voltage Rejection Ratio ($R_S \leqslant$ 10 k)	PSRR	—	30	150	—	—	—	μV/V
Output Voltage Swing	V_O							V
($R_L \geqslant$ 10 k)		±12	±14	—	—	—	—	
($R_L \geqslant$ 2 k)		±10	±13	—	±10	±13	—	
Large Signal Voltage Gain ($R_L \geqslant$ 2 k, V_{out} = ±10 V)	A_v	25	—	—	15	—	—	V/mV
Supply Currents	I_D							mA
(T_A = 125°C)		—	1.5	2.5	—	—	—	
(T_A = -55°C)		—	2.0	3.3	—	—	—	
Power Consumption (T_A = +125°C)	P_C	—	45	75	—	—	—	mW
(T_A = -55°C)		—	60	100	—	—	—	

*T_{high} = 125°C for MC1741 and 70°C for MC1741C
T_{low} = -55°C for MC1741 and 0°C for MC1741C

MOTOROLA *Semiconductor Products Inc.*

MOTOROLA SEMICONDUCTORS

P.O. BOX 20912 • PHOENIX, ARIZONA 85036

MPC100

Advance Information

SMARTpower SERIES
10 AMPERES POSITIVE VOLTAGE REGULATOR

This fixed voltage regulator is a series pass monolithic integrated circuit capable of supplying current up to 10 amperes. SMARTpower technology, utilizing a combination of a high-power bipolar output transistor in conjunction with small-signal CMOS control circuitry offers a unique monolithic chip with the following features:

● Internal Thermal Protection

● Internal Short Circuit Protection

● Low Differential Voltage — Typ 1.5 V @ 10 A

POSITIVE 5.0 VOLT FIXED VOLTAGE REGULATOR

10 AMPERES
80 WATTS

MAXIMUM RATINGS

Rating	Symbol	Value	Unit
Input Voltage	V_{in}	25	Vdc
Output Current	I_o	10	Adc
Total Power Dissipation @ T_C = 25°C Derate above T_C = 25°C	P_D	80 0.8	Watts W/°C
Storage Temperature Range	T_{stg}	0 to 150	°C
Operating Junction Temperature Range	T_J	0 to 125	°C

THERMAL CHARACTERISTICS

Thermal Resistance, Junction-to-Case	$R_{\theta JC}$	1.25	°C/W
Maximum Lead Temperature for Soldering Purposes 1/8" from Case for 5.0 sec	T_L	275	°C

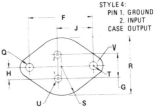

SEATING PLANE

STYLE 4:
PIN 1. GROUND
2. INPUT
CASE OUTPUT

NOTES:
1. DIAMETER V AND SURFACE W ARE DATUMS.
2. POSITIONAL TOLERANCE FOR HOLE Q:
   $\boxed{\oplus \; \phi \; 0.25 \, (0.010) \; \textcircled{M} \; | \; W \; | \; V \; \textcircled{M}}$
3. POSITIONAL TOLERANCE FOR LEADS:
   $\boxed{\oplus \; \phi \; 0.30 \, (0.012) \; \textcircled{M} \; | \; W \; | \; V \; \textcircled{M} \; | \; Q \; \textcircled{M}}$

DIM	MILLIMETERS		INCHES	
	MIN	MAX	MIN	MAX
A	—	39.37	—	1.550
B	—	21.08	—	0.830
C	6.35	7.62	0.250	0.300
D	0.97	1.09	0.038	0.043
E	1.40	1.78	0.055	0.070
F	30.15 BSC		1.187 BSC	
G	10.92 BSC		0.430 BSC	
H	5.46 BSC		0.215 BSC	
J	16.89 BSC		0.665 BSC	
K	11.18	12.19	0.440	0.480
Q	3.81	4.19	0.150	0.165
R	—	26.67	—	1.050
U	2.54	3.05	0.100	0.120
V	3.81	4.19	0.150	0.165

CASE 1-04
TO-204AA (TYPE)

STANDARD APPLICATION

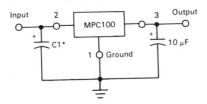

Input   2   MPC100   3   Output

C1*   10 μF

1   Ground

*C1 is required if the regulator is located an appreciable distance from the power supply main filter capacitor.

© MOTOROLA INC., 1982

ADI-711

ELECTRICAL CHARACTERISTICS (1)

Characteristic	Symbol	Min	Typ	Max	Unit
Output Voltage (10 mA $\leq I_O \leq$ 10 A) (10 mA $\leq I_O \leq$ 5.0 A, 6.5 V $\leq V_{in} \leq$ 20 V)	V_O	4.75	—	5.25	Vdc
Line Regulation (2) (6.5 V $\leq V_{in} \leq$ 20 V) (6.5 V $\leq V_{in} \leq$ 10 V)	Regline	— —	— —	100 50	mV
Load Regulation (2) (10 mA $\leq I_O \leq$ 10 A, T_C = 0 to 125°C)	Regload	—	—	50	mV
Quiescent Current (10 mA $\leq I_O \leq$ 10 A)	I_B	—	—	25	mA
Ripple Rejection (V_{in} = 8.0 V, f = 120 Hz)	RR	—	45	—	db
Dropout Voltage (I_O = 10 A, T_C = 0 to 125°C)	$V_{in}-V_O$	—	1.5	2.0	Vdc
Short Circuit Current (3) (V_{in} = 10 Vdc)	I_{SC}	—	—	20	Adc
Averge Temperature Coefficient of Output Voltage (I_O = 10 A)	TCV_O	—	1.6	—	mV/°C
Output Noise Voltage (10 Hz $\leq$ f $\leq$ 100 kHz)	V_N	—	5.0	—	mV
Output Resistance (f = 120 Hz)	R_O	—	2.0	—	mΩ

(1) Unless otherwise specified, test conditions are: T_J = 25°C, V_{in} = 7.0 Vdc, I_O = 5.0 Adc, $P_O \leq P_{max}$ and C_O = 10 μF.
(2) Load and line regulation are specified at constant junction temperature. Changes in V_O due to heating effects must be taken into account separately. Pulse
testing is used with pulse width $\leq$3.0 ms and duty cycle $\leq$ 1.0%. Kelvin contacts must be used for these tests.
(3) Depending on heat sinking and power dissipation, thermal shutdown may occur.

SMARTpower VERTICAL PROFILE

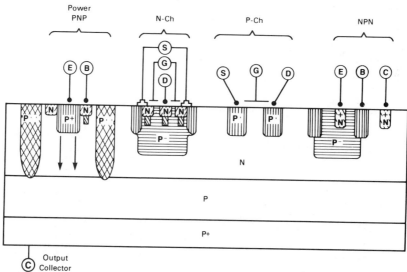

The PNP power transistor uses standard fabrication technology which assures a low saturation voltage and results in a low input-
output differential voltage. The die bond is the output collector contact which results in superior load regulation.

 MOTOROLA *Semiconductor Products Inc.*

APPENDIX B:
Derivations of Selected Equations

EQUATION (3–1)

The average value of a half-wave rectified sine wave is the area under the curve divided by the period (2π). The equation for a sine wave is

$$v = V_p \sin \theta$$

$$V_{\text{AVG}} = \frac{\text{area}}{2\pi}$$

$$= \frac{1}{2\pi} \int_0^{\pi} V_p \sin \theta \, d\theta$$

$$= \frac{V_p}{2\pi}(-\cos \theta)\Big|_0^{\pi}$$

$$= \frac{V_p}{2\pi}[-\cos \pi - (-\cos 0)]$$

$$= \frac{V_p}{2\pi}[-(-1) - (-1)]$$

$$= \frac{V_p}{2\pi}(2)$$

$$= \frac{V_p}{\pi}$$

EQUATION (3–13)

Referring to Figure B–1, when the filter capacitor discharges, the voltage is

$$v_C = V_{p(\text{in})}e^{-t/RC}$$

Since the discharge time of the capacitor is from one peak to approximately the next peak, $t_{\text{dis}} \cong T$ when v_C reaches its minimum value.

$$v_{C(\text{min})} = V_{p(\text{in})}e^{-T/RC}$$

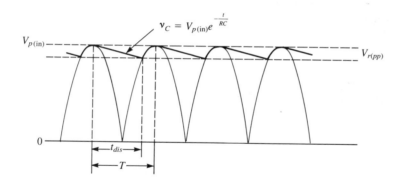

FIGURE B–1

Since $RC >> T$, T/RC becomes much less than 1 (which is usually the case); $e^{-T/RC}$ approaches 1 and can be expressed as

$$e^{-T/RC} \cong 1 - \frac{T}{RC}$$

Therefore,

$$v_{C(\text{min})} = V_{p(\text{in})}\left(1 - \frac{T}{RC}\right)$$

The peak-to-peak ripple voltage is

$$V_{r(pp)} = V_{p(\text{in})} - V_{C(\text{min})}$$

$$= V_{p(\text{in})} - V_{p(\text{in})} + \frac{V_{p(\text{in})}T}{RC}$$

$$= \frac{V_{p(\text{in})}T}{RC}$$

For $f = 120$ Hz (full-wave),

$$V_{r(pp)} = \frac{V_{p(\text{in})}}{RCf} = \frac{0.00833 V_{p(\text{in})}}{RC}$$

To obtain the dc value, one-half of the peak-to-peak ripple is subtracted from the peak value.

$$V_{\text{dc}} = V_{p(\text{in})} - \frac{V_{r(pp)}}{2}$$

$$= V_{p(\text{in})} - \frac{0.00833 V_{p(\text{in})}}{2RC}$$

$$= \left(1 - \frac{0.00417}{RC}\right)V_{p(\text{in})}$$

EQUATION (3–14)

The peak ripple voltage is

$$V_{r(p)} = \frac{0.00833 V_{p(\text{in})}}{2 R_L C}$$

Since the ripple waveform is a sawtooth, we divide by $\sqrt{3}$ to convert peak to rms.

$$V_{r(\text{rms})} = \frac{0.00833 V_{p(\text{in})}}{2 R_L C(\sqrt{3})} = \frac{0.0024 V_{p(\text{in})}}{R_L C}$$

DERIVATION OF THE RIPPLE VOLTAGE FOR A FULL-WAVE RECTIFIED SIGNAL AS USED IN EXAMPLE 3–9

The ac component of a full-wave rectified signal is the total voltage minus the dc value.

$$v = v_t - V_{\text{dc}}$$

The rms value of V_{ac} is

$$V_{r(\text{rms})} = \left(\frac{1}{2\pi} \int_0^{2\pi} V_{\text{ac}}\, d\theta \right)^{1/2}$$

$$= \left(\frac{1}{2\pi} \int_0^{2\pi} (v_t - V_{\text{dc}})^2\, d\theta \right)^{1/2}$$

$$= \left(\frac{1}{2\pi} \int_0^{2\pi} (v_t^2 - 2v_t V_{\text{dc}} + V_{\text{dc}}^2)\, d\theta \right)^{1/2}$$

$$= \left[\frac{1}{2\pi} \left(\int_0^{2\pi} v_t^2\, d\theta - \int_0^{2\pi} 2v_t V_{\text{dc}}\, d\theta + \int_0^{2\pi} V_{\text{dc}}^2\, d\theta \right) \right]^{1/2}$$

$$= \left[\frac{1}{2\pi} \left(\int_0^{2\pi} v_t^2\, d\theta - 2V_{\text{dc}} \int_0^{2\pi} v_t\, d\theta + V_{\text{dc}}^2 \int_0^{2\pi} d\theta \right) \right]^{1/2}$$

$$= (V_{t(\text{rms})}^2 - 2V_{\text{dc}}^2 + V_{\text{dc}}^2)^{1/2}$$

$$= (V_{t(\text{rms})}^2 - V_{\text{dc}}^2)^{1/2}$$

For a full-wave rectified voltage:

$$V_{t(\text{rms})} = \frac{V_p}{1.414}$$

$$V_{\text{dc}} = \frac{2V_p}{\pi}$$

$$V_{r(\text{rms})} = \sqrt{\left(\frac{V_p}{1.414} \right)^2 - \left(\frac{2V_p}{\pi} \right)^2}$$

$$= V_p \sqrt{\left(\frac{1}{1.414} \right)^2 - \left(\frac{2}{\pi} \right)^2}$$

$$= 0.308 V_p$$

The Shockley equation for the base-emitter pn junction is

$$I_E = I_R(e^{VQ/kT} - 1)$$

EQUATION
(8–10)

where I_E = the total forward current across the base-emitter junction.
   I_R = the reverse saturation current.
   V = the voltage across the depletion layer.
   Q = the charge on an electron.
   k = a number known as Boltzmann's constant.
   T = the absolute temperature.

At ambient temperature, $Q/kT \cong 40$, so

$$I_E = I_R(e^{40V} - 1)$$

Differentiating, we get

$$\frac{dI_E}{dV} = 40I_R e^{40V}$$

Since $I_R e^{40V} = I_E + I_R$,

$$\frac{dI_E}{dV} = 40(I_E + I_R)$$

Assuming $I_R \ll I_E$,

$$\frac{dI_E}{dV} \cong 40I_E$$

The ac resistance r_e of the base-emitter junction can be expressed as dV/dI_E.

$$r_e = \frac{dV}{dI_E} \cong \frac{1}{40I_E} \cong \frac{25 \text{ mV}}{I_E}$$

The emitter-follower is represented by the r parameter ac equivalent circuit in Figure B–2(a). By Thevenizing from the base back to the source, the circuit is simplified to the form shown in Figure B–2(b).

EQUATION
(8–31)

$$V_{\text{out}} = V_e, \ I_{\text{out}} = I_e, \text{ and } I_{\text{in}} = I_b$$

$$R_{\text{out}} = \frac{V_e}{I_e}$$

$$I_e \cong \beta I_b$$

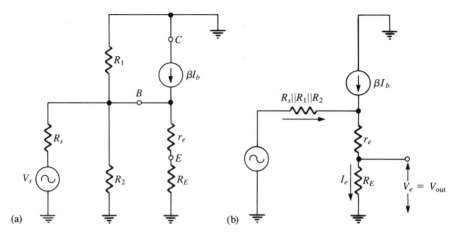

FIGURE B–2

With $V_s = 0$ and with I_b produced by V_{out}, and neglecting the base-to-emitter voltage drop (and therefore r_e):

$$I_b \cong \frac{V_e}{R_1 \| R_2 \| R_s}$$

Assuming that $R_1 >> R_s$ and $R_2 >> R_s$,

$$I_b \cong \frac{V_e}{R_s}$$

$$I_{out} = I_e = \frac{\beta V_e}{R_s}$$

$$\frac{V_{out}}{I_{out}} = \frac{V_e}{I_e} = \frac{V_e}{\beta V_e / R_s} = \frac{R_s}{\beta}$$

Looking from the emitter, R_E appears in parallel with R_s/β. Therefore,

$$R_{out} = \left(\frac{R_s}{\beta}\right) \| R_E$$

EQUATION (9–7)

$$I_D = I_{DSS}\left(1 - \frac{I_D R_S}{V_{GS(off)}}\right)^2$$

$$= I_{DSS}\left(1 - \frac{I_D R_S}{V_{GS(off)}}\right)\left(1 - \frac{I_D R_S}{V_{GS(off)}}\right)$$

$$= I_{DSS}\left(1 - \frac{2I_D R_S}{V_{GS(off)}} + \frac{I_D^2 R_S^2}{V_{GS(off)}^2}\right)$$

$$= I_{DSS} - \frac{2I_{DSS} R_S}{V_{GS(off)}} I_D + \frac{I_{DSS} R_S^2}{V_{GS(off)}^2} I_D^2$$

Rearranging into a standard quadratic equation form:

$$\left(\frac{I_{DSS}R_S^2}{V_{GS(off)}^2}\right)I_D^2 - \left(1 + \frac{2I_{DSS}R_S}{V_{GS(off)}}\right)I_D + I_{DSS} = 0$$

The coefficients and constant are

$$A = \frac{R_S I_{DSS}^2}{V_{GS(off)}^2}$$

$$B = -\left(1 + \frac{2R_S I_{DSS}}{V_{GS(off)}}\right)$$

$$C = I_{DSS}$$

In simplified notation, the equation is

$$A I_D^2 + B I_D + C = 0$$

The solutions to this quadratic equation are

$$I_D = \frac{-B \pm \sqrt{B^2 - 4AC}}{2A}$$

An inverting amplifier with feedback capacitance is shown in Figure B–3.

For the input:

$$I_1 = \frac{V_1 - V_2}{X_C}$$

Factoring V_1 out:

$$I_1 = \frac{V_1(1 - V_2/V_1)}{X_C}$$

The ratio V_2/V_1 is the voltage gain, $-A_v$.

$$I_1 = \frac{V_1(1 + A_v)}{X_C} = \frac{V_1}{X_C/(1 + A_v)}$$

**EQUATION
(11–1)**

FIGURE B–3

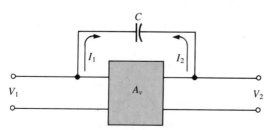

The effective reactance as seen from the input terminals is

$$X_{C_{\text{in(Miller)}}} = \frac{X_C}{1 + A_v}$$

or

$$\frac{1}{2\pi f C_{\text{in(Miller)}}} = \frac{1}{2\pi f C(1 + A_v)}$$

Cancelling and inverting, we get

$$C_{\text{in(Miller)}} = C(A_v + 1)$$

EQUATION (11–2)

For the output in Figure B–3,

$$I_2 = \frac{V_2 - V_1}{X_C} = \frac{V_2(1 - V_1/V_2)}{X_C}$$

Since $V_1/V_2 = -1/A_v$,

$$I_2 = \frac{V_2(1 + 1/A_v)}{X_C} = \frac{V_2}{X_C/(1 + 1/A_v)} = \frac{V_2}{X_C/[(A_v + 1)/A_v]}$$

The effective reactance as seen from the output is

$$X_{C_{\text{out(Miller)}}} = \frac{X_C}{(A_v + 1)/A_v}$$

$$\frac{1}{2\pi f C_{\text{out(Miller)}}} = \frac{1}{2\pi f C[(A_v + 1)/A_v]}$$

Cancelling and inverting, we get

$$C_{\text{out(Miller)}} = C\left(\frac{A_v + 1}{A_v}\right)$$

EQUATIONS (11–35) AND (11–36)

The *rise time* is defined as the time required for the voltage to increase from 10 percent of its final value to 90 percent of its final value, as indicated in Figure B–4. Expressing the curve in its exponential form gives

$$v = V_{\text{final}}(1 - e^{-t/RC})$$

Answers to Odd-Numbered Problems

Chapter 3

3–1 (a) Reverse   **(b)** Forward
     (c) Forward   **(d)** Forward

3–3 (a) Open   **(b)** Open   **(c)** Shorted
     (d) Functioning

3–5 See Figure ANS–1.

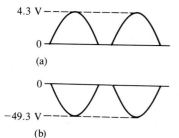

FIGURE ANS–1

3–7 23 V rms

3–9 (a) 1.59 V   **(b)** 63.66 V
     (c) 16.37 V   **(d)** 0.92 V

3–11 172.79 V

3–13 78.54 V

3–15 $V_r = 2.4$ V, $V_{dc} = 25.3$ V

3–17 160 μF

3–19 $V_r = 1.46$ V, $V_{dc} = 30.49$ V

3–21 Coil is open. Capacitor is shorted.

3–23 The circuit should not fail because the diode ratings exceed the actual PIV and maximum current.

3–25 See Figure ANS–2.

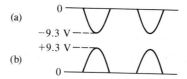

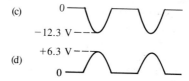

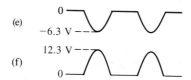

FIGURE ANS–2

3–27 (a) A sine wave with a positive peak at +0.7 V, a negative peak at −7.3 V, and a dc value of −3.3 V.

     (b) A sine wave with a positive peak at +29.3 V, a negative peak at −0.7 V, and a dc value of +14.3 V.

(c) A square wave varying from +0.7 V down to −15.3 V, with a dc value of −7.3 V.

(d) A square wave varying from +1.3 V down to −0.7 V, with a dc value of +0.3 V.

3–29 56.56 V

Chapter 4

4–1 See Figure ANS–3.

7.5 V   0.5 Ω

FIGURE ANS–3

4–3 5 Ω

4–5 6.92 V

4–7 14.93 V

4–9 See Figure ANS–4.

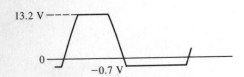

13.2 V

0

−0.7 V

FIGURE ANS–4

4–11 15.1%

4–13 3.13%

4–15 3 V

4–17 2.2 V

4–19 Oscillator circuits

4–21 Dark current

Chapter 5

5–1 Holes

5–3 The base is thin and lightly doped so that a small recombination (base) current is generated compared to the collector current.

5–5 8.98 mA

5–7 0.99

5–9 **(a)** $V_{BE} = 0.7$ V, $V_{CE} = 6.9$ V, $V_{BC} = -6.2$ V

(b) $V_{BE} = -0.7$ V, $V_{CE} = -3.74$ V, $V_{BC} = 3.04$ V

5–11 $I_B = 26$ μA, $I_E = 1.3$ mA, $I_C = 1.274$ mA

5–13 0.003 mA

5–15 0.425 W

5–17 33.33

5–19 0.5 mA, 3.33 μA, 4.03 V

5–21 **(a)** Good **(b)** Good **(c)** Bad

(d) Bad

Chapter 6

6–1 (a) and (c)

6–3 $V_{CE} = 20$ V, $I_{C(sat)} = 2$ mA

6–5 See Figure ANS–5.

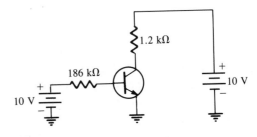

1.2 kΩ

186 kΩ

10 V

10 V

FIGURE ANS–5

6–7 I_C changes in the circuit using a common V_{CC} and V_{BB} supply, because a change in V_{CC} causes I_B to change, which in turn changes I_C.

6–9 59.57 mA, 5.96 V

6–11 754 Ω

6–13 $R_E >> \dfrac{R_B}{\beta_{dc}}$

6–15 69.12

6–17 $I_C \cong 0.81$ mA, $V_{CE} = 13.24$ V

6–19 See Figure ANS–6.

6–21 2.53 kΩ

6–23 **(a)** Open collector or R_E open

(b) No problems

(c) Transistor shorted collector-to-emitter

(d) Open emitter

Chapter 7

7–1 **(a)** Narrows **(b)** Increases

7–3 −5 V

7–5 **(a)** 4 V **(b)** 10 mA

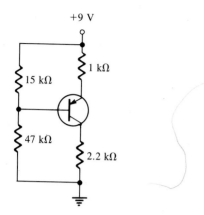

FIGURE ANS–6

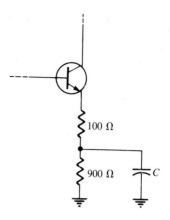

FIGURE ANS–7

7–7   4 V

7–9 -2.63 V

7–11 $g_m = 1429$ μS, $y_{fs} = 1429$ μS

7–13   **(a)** Depletion   **(b)** Enhancement
      (c) 0 bias   **(d)** Depletion

7–15   **(a)** n-channel
      (b) 0 mA, 0.32 mA, 1.28 mA, 2.88 mA,
         5.12 mA, 8 mA, 11.52 mA, 15.68 mA,
         20.48 mA, 25.92 mA, 32 mA

7–17   1.08 mA

7–19   105.3 Ω

7–21   83.33 MΩ

7–23   **(a)** $V_{GS} = 3.33$ V, $V_{DS} = 4$ V
      (b) $V_{GS} = 2.5$ V, $I_D = 2$ mA, $V_{DS} = 2$ V

7–25   6.799 V

Chapter 8

8–1   Approximately 0.75 mA

8–3   **(a)** $h_{ie} = 134$ Ω **(b)** $h_{re} = 0.0001$
      (c) $h_{fe} = 146.67$ **(d)** $h_{oe} = 3.33$ mS

8–5 $r_e \cong 19.08$ Ω

8–7   **(a)** $V_B = 3.25$ V   **(b)** $V_E = 2.55$ V
      (c) $I_E = 2.55$ mA   **(d)** $I_C \cong 2.55$ mA
      (e) $V_C = 9.59$ V   **(f)** $V_{CE} = 7.04$ V

8–9 $A'_v = 173.7$, $\phi = 180°$

8–11 $A_{v(max)} = 61.54$, $A_{v(min)} = 1.94$

8–13   See Figure ANS–7.

8–15 $R_{in} = 3.22$ kΩ, $V_{OUT} = 1.13$ V

8–17   248 Ω

8–19   8.3

8–21 $R_{in(emitter)} = 2.36$ Ω, $A_v = 508.5$, $A_i = 1$,
      $A_p = 508.5$

8–23   400

8–25   **(a)** $A_{v1} = 93.57$, $A_{v2} = 302.2$
      (b) $A'_v = 28{,}276.9$
      (c) A_{v1} (dB) $= 39.42$ dB,
         A_{v2} (dB) $= 49.61$ dB,
         A'_v (dB) $= 89.03$ dB

8–27 $V_{B1} = 2.16$ V, $V_{E1} = 1.46$ V, $V_{C1} \cong 6.16$ V,
      $V_{B2} = 6.16$ V, $V_{E2} = 5.46$ V, $V_{C2} \cong 6.54$ V,
      $A_{v1} = 42.5$, $A_{v2} = 218.4$, $A'_v = 9282$

8–29

Test point	dc volts	ac volts (rms)
Input	0 V	25 μV
Q_1 base	2.99 V	20.79 μV
Q_1 emitter	2.29 V	0 V
Q_1 collector	7.44 V	1.95 mV
Q_2 base	2.99 V	1.95 mV
Q_2 emitter	2.29 V	0 V
Q_2 collector	7.44 V	589 mV
Output	0 V	589 mV

8–31   **(a)** 1.41   **(b)** 2   **(c)** 3.16
      (d) 10   **(e)** 100

8–33   Cutoff, 10 V

Chapter 9

9–1   **(a)** n-channel DE MOSFET with zero-bias;
         $V_{GS} = 0$

(b) p-channel JFET with self-bias; $V_{GS} = -0.99$ V

(c) n-channel E MOSFET with voltage-divider bias; $V_{GS} = 4$ V

9–3 **(a)** n-channel DE MOSFET
    (b) n-channel JFET
    (c) p-channel E MOSFET

9–5 5.71 kΩ

9–7 2.85

9–9 1.06 V

9–11 85.71 MΩ

9–13 $V_{GS} = 9$ V, $I_D = 3.125$ mA, $V_{DS} = 13.31$ V, $V_{ds} = 675$ mV

9–15 $R_{in} = 99.97$ MΩ, $A_v = 0.7826$

9–17 $R_{in(source)} = 250$ Ω

9–19 **(a)** $V_{D_1} = V_{DD}$; no output signal
    (b) $V_{D_1} = 0$ V; no output signal
    (c) $V_{GS_1} = 0$ V; $V_S = 0$ V; clipped output signal
    (d) Positive bias voltage on Q_2 gate ($V_{G_2} = V_{D_1}$); distorted output signal.
    (e) $V_{D_2} = V_{DD}$; no output signal.

Chapter 10

10–1 2.11 mA, 12 V

10–3 8.47 V

10–5 **(a)** A 50 percent increase in β_{dc} will have little effect on the bias values because $R_{IN(base)} \gg R_2$ already.
    (b) Since $R_{IN(base)}$ is not ten times R_2, a 50 percent increase in β_{dc} will result in a reduction in V_B which, in turn, causes an increase in I_E and I_C.

10–7 **(a)** 1.01 V rms   **(b)** 13.56 mV rms

10–9 **(a)** $P_{OUT} = 1.35$ mW, $\eta = 0.038$
    (b) $P_{OUT} = 14.8$ mW, $\eta = 0.164$

10–11 $V_{B_1} = 10.7$ V, $V_{E_1} = 10$ V, $V_{E_2} = 10$ V, $V_{CEQ_1} = 10$ V, $V_{CEQ_2} = 10$ V

10–13 $P_{OUT} = 3.13$ W, $P_{dc} = 3.97$ W

10–15 $I_{CC} = 477.46$ mA, $P_{dc} = 11.46$ W, $P_{OUT} = 9$ W

10–17 See Figure ANS–8.

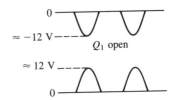

FIGURE ANS–8

10–19 50.33 kHz

10–21 $\eta = 0.998$

10–23 One of the transistors is open between collector and emitter.

Chapter 11

11–1 If $C_1 = C_2$, the critical frequencies are equal, and they will both cause the gain to drop at 40 dB/decade below f_c.

11–3 Bipolar: C_{be}, C_{bc}
    FET: C_{gs}, C_{gd}

11–5 See Figure ANS–9.

11–7 10 dB

11–9 −8.3 dB

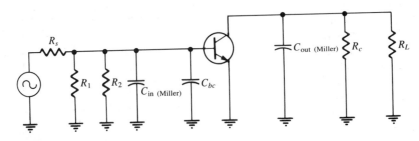

FIGURE ANS–9

11–11 (a) 318.31 Hz   (b) 1.59 kHz

11–13 At f_c: A_v = 35.23 dB
At $0.1f_c$: A_v = 18.23 dB
At $10f_c$: A_v = 38.23 dB

11–15 Input network: f_c = 4.61 MHz
Output network: f_c = 106.10 MHz
Input f_c is dominant.

11–17 f_{cl} = 136 Hz, f_{ch} = 8 kHz

11–19 BW = 5.26 MHz, f_{ch} = 5.26 MHz

11–21 Input RC network: f_c = 0.48 Hz
Output RC network: f_c = 2.89 kHz
Output f_c is dominant.

11–23 Input network: f_c = 12.69 MHz
Output network: f_c = 32 MHz
Input f_c is dominant.

11–25 f_{cl} = 350 Hz, f_{ch} = 17.5 MHz

Chapter 12

12–1 *Practical op-amp:* High open-loop gain, high input impedance, low output impedance, large bandwidth, high CMRR.
Ideal op-amp: Infinite open-loop gain, infinite input impedance, 0 output impedance, infinite bandwidth, infinite CMRR.

12–3 (a) Single-ended input; differential output
(b) Single-ended input; single-ended output
(c) Differential input; single-ended output
(d) Differential input; differential output

12–5 8.1 μA

12–7 107.96 dB

12–9 0.3

12–11 40 μs

12–13 (a) 11   (b) 101   (c) 47.81   (d) 23

12–15 (a) 1   (b) −1   (c) 21   (d) −10

12–17 (a) 0.5 mA   (b) 0.5 mA
(c) −10 V   (d) −10

12–19 (a) $Z_{in(VF)}$ = 1.32 × 10^{12} Ω
$Z_{out(VF)}$ = 0.455 mΩ
(b) $Z_{in(VF)}$ = 5 × 10^{11} Ω
$Z_{out(VF)}$ = 0.6 mΩ
(c) $Z_{in(VF)}$ = 40,000 MΩ
$Z_{out(VF)}$ = 1.5 mΩ

12–21 (a) 75 Ω placed in feedback path
(b) 150 μV

12–23 200 μV

Chapter 13

13–1 70 dB

13–3 1.67 kΩ

13–5 (a) 79,603   (b) 56,569
(c) 7960   (d) 80

13–7 (a) −0.67°   (b) −2.69°
(c) −5.71°   (d) −45°
(e) −71.22°   (f) −84.29°

13–9 (a) 0   (b) −20 dB/decade
(c) −40 dB/decade   (d) −60 dB/decade

13–11 4.05 MHz

13–13 21.14 MHz

13–15 (a) 2.4 kHz
(b) 97.5 kHz
(c) Circuit A has smaller BW.

13–17 (a) 150°   (b) 120°   (c) 60°
(d) 0°   (e) −30°

13–19 (a) Unstable   (b) Stable
(c) Marginally stable

13–21 25 Hz

Chapter 14

14–1 24 V

14–3 See Figure ANS–10.

14–5 See Figure ANS–11.

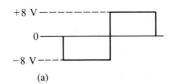

(a)

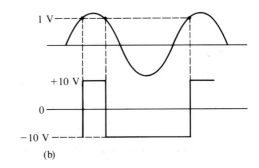

(b)

FIGURE ANS–10

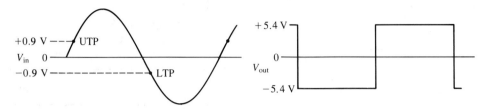

FIGURE ANS–11

14–7   110 kΩ

14–9 $V_{OUT} = -3.66$ V, $I_f = 0.366$ mA

14–11 -5 mV/μs

14–13   1 mA

14–15   2.51 V rms

14–17   (a)   6.8 mA   (b)   60 mA

14–19 D_2 shorted

Chapter 15

15–1   dc power supply

15–3 $^1/_{75}$

15–5   733.33 mV

15–7   50 kΩ

15–9   7.5 V, 4

15–11   87 kΩ

15–13   9.4

15–15   Change R_1 to 14.14 kΩ

15–17 $R_4 = 70$ kΩ

15–19   3.33 V, 6.67 V

15–21   0.007 μF

15–23 $f_{min} = 42.5$ kHz, $f_{max} = 57.5$ kHz

15–25   25 kHz

Chapter 16

16–1   800 Hz

16–3   20 dB/decade

16–5   15 kHz

16–7   (a) $R_1 = 680$ Ω
           (c) $R_1 = 1$ kΩ, $R_3 = 1$ kΩ

16–9   189.8 Hz

16–11   Add another identical stage.

16–13   Double R or C.

16–15   (a) $f_0 = 4949$ Hz, $BW = 3848$ Hz
           (b) $f_0 = 448.6$ Hz, $BW = 96.5$ Hz
           (c) $f_0 = 15.92$ kHz, $BW = 838$ Hz

16–17   Sum the low-pass and high-pass outputs with a two-input adder (see Figure 16–24).

Chapter 17

17–1   0.00417%

17–3   1.01%

17–5   7 V

17–7   9.45 V

17–9   500 mA

17–11   10 mA

17–13 $I_{L(max)} = 250$ mA, $P_{R_1} = 6.25$ W

17–15   40%

17–17   Increases

17–19   pin 1—NC, pin 2—Current limit, pin 3—Current sense, pin 4—Inverting input, pin 5—Noninverting input, pin 6—V_{REF}, pin 7—$V-$, pin 8—NC, pin 9—V_Z, pin 10—V_{OUT}, pin 11—V_C, pin 12—$V+$, pin 13—FREQ COMP, pin 14—NC

17–21   20 kΩ

17–23   (a)   Fold-back current-limiting regulator.
           (b)   Basic current-limiting positive regulator.
           (c)   Basic negative regulator.

Chapter 18

18–1 $I_A = I_K = 645.82$ nA

18–3   See pages 716–18.

18–5   Add a transistor to provide inversion of negative half-cycle in order to obtain a positive gate trigger.

18–7   See pages 723–25.

18–9   See Figure ANS–12.

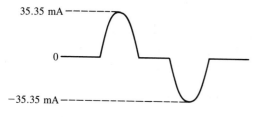

FIGURE ANS–12

18–11 0.385

18–13 746.67 $\Omega < R_1 < 200$ kΩ

18–15 See Figure ANS–13.

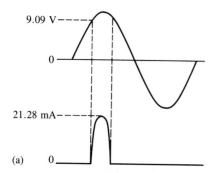

(a)

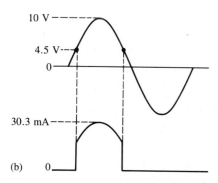

(b)

FIGURE ANS–13

Chapter 19

19–1 3.312×10^{-20} J

19–3 4.65 fc

19–5 3.09 mA

19–7 30 kΩ, 8.57 kΩ, 5.88 kΩ

19–9 **(a)** 12 V   **(b)** 0 V

19–11 1.40 V, 1.5 mA

19–13 0.133

19–15 See Figure ANS–14.

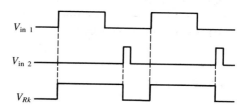

FIGURE ANS–14

19–17 LEDs emit light energy when forward-biased; regular diodes do not.

19–19 True

19–21 16.67 mA

Solutions to Self-Tests

Chapter 1

1. A rectifier diode allows current in only one direction so that ac can be converted to dc.

2. Zener diodes are used for voltage regulation.

3. An amplifier increases the amplitude of a signal voltage.

4. A rectifier converts ac to dc.

5. The two basic types of transistor are bipolar junction and field-effect.

6. Three types of thyristors are the silicon-controlled rectifier (SCR), the diac, and the triac.

7. An important type of linear integrated circuit is the operational amplifier (op-amp).

8. A voltage regulator maintains a constant value of dc voltage for variations of the input voltage and the load.

Chapter 2

1. The atom is the smallest particle of an element that retains the characteristics of that element.

2. According to the Bohr model, atoms have a planetary type of structure with electrons orbiting a nucleus that contains protons and neutrons.

3. Electrons have negative charge, protons have positive charge, and neutrons have no charge.

4. Atomic number of silicon is 14.

5. Atomic weight of germanium is 64.

6. Silicon and germanium have a valence of 4.

7. The valence shell in silicon is M.

8. A neutral atom has the same number of electrons and protons; a positive ion has more protons than electrons; a negative ion has more electrons than protons.

9. Covalent bonding is created by the sharing of valence electrons by two or more atoms.

10. Free electrons exist in the conduction band.

11. When a valence electron acquires sufficient energy to jump into the conduction band, a vacancy (hole) is left in the valence band.

12. Recombination occurs when a free electron falls into a hole in the valence band.

13. Electron current and hole current.

14. The energy gap between the valence band and conduction band of an insulator is much greater than that of a semiconductor.

15. A trivalent impurity adds holes.

16. A pentavalent impurity adds electrons.

17. Electrons are majority carriers in an n-type semiconductor.

18. Approximately 0.7 V.

19. $I = \dfrac{2 \text{ V}}{110 \text{ }\Omega} = 18.18 \text{ mA}$

Chapter 3

1. (a) 0.7 V   (b) 0.3 V

2. $I = \dfrac{5 \text{ V} - 0.7 \text{ V}}{1 \text{ k}\Omega} = 4.7 \text{ mA}$

   $V_R = (4.3 \text{ mA})(1 \text{ k}\Omega) = 4.3 \text{ V}$

   Therefore, $V_K = +4.3$ V, referenced to the negative terminal of the source. (See Figure S–1.)

3. (a) Bad   (b) Good

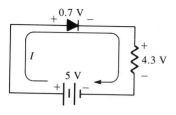

FIGURE S–1

4. An ideal forward-biased diode is a short. An ideal reverse-biased diode is an open.

5. **(a)** $I_F = \dfrac{10 \text{ V}}{75 \text{ Ω} + 25 \text{ Ω}} = 100 \text{ mA}$

(b) $I_R = \dfrac{500 \text{ V}}{100 \text{ MΩ} + 1 \text{ MΩ}} = 4.95 \text{ μA}$

(c) $I_R = \dfrac{12 \text{ V}}{100 \text{ MΩ} + 1 \text{ kΩ}} = 0.12 \text{ μA}$

(d) $I_F = \dfrac{5 \text{ V}}{50 \text{ Ω} + 25 \text{ Ω}} = 66.67 \text{ mA}$

6. Rectification is the process of converting ac to pulsating dc.

7. $V_{AVG} = \dfrac{V_p}{\pi} = \dfrac{200 \text{ V}}{\pi} = 63.66 \text{ V}$

8. $f_{out} = 60 \text{ Hz}$

9. $I_{L(p)} = \dfrac{4.3 \text{ V}}{100 \text{ Ω}} = 43 \text{ mA}$

$V_{L(p)} = 5 \text{ V} - 0.7 \text{ V} = 4.3 \text{ V}$

(See Figure S–2.)

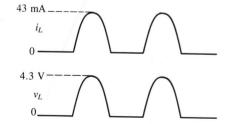

FIGURE S–2

10. Yes

11. $V_{AVG} = \dfrac{2V_p}{\pi} = \dfrac{2(75 \text{ V})}{\pi} = 47.75 \text{ V}$

12. $f_{out} = 120 \text{ Hz}$

13. $V_{out} = \dfrac{V_s}{2} = \dfrac{125 \text{ V}}{2} = 62.5 \text{ V rms}$

14. $PIV = 2V_{p(out)} = 2(100 \text{ V}) = 200 \text{ V}$

15. $PIV = V_{p(out)} = 1.414(20 \text{ V}) = 28.28 \text{ V}$

16. Peak

17. $r = \dfrac{V_r}{V_{dc}} = \dfrac{100 \text{ mV}}{20 \text{ V}} = 0.005$

18. $V_r = \dfrac{0.0024V_{p(in)}}{R_L C} = \dfrac{0.0024(60 \text{ V})}{(10 \text{ kΩ})(10 \text{ μF})} = 1.44 \text{ V}$

19. $C = \dfrac{0.0024}{R_L r} = \dfrac{0.0024}{(22 \text{ kΩ})(0.001)} = 109 \text{ μF}$

20. $V_{dc(out)} = \left(\dfrac{R_L}{R_W + R_L}\right)V_{dc(in)} = \left(\dfrac{1.5 \text{ kΩ}}{1.53 \text{ kΩ}}\right)15 \text{ V}$

$= 14.7 \text{ V}$

21. The output becomes half-wave with increased ripple.

22. See Figure S–3.

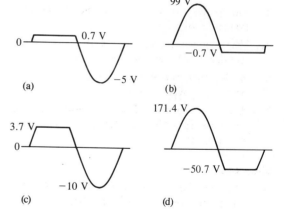

FIGURE S–3

23. 15 V

24. $V_{OUT} = 1.414(12 \text{ V})(3) = 50.9 \text{ V}$

Chapter 4

1. Zener breakdown

2. $r_Z = \dfrac{\Delta V_Z}{\Delta I_Z} = \dfrac{0.1 \text{ V}}{10 \text{ mA}} = 10 \text{ Ω}$

3. $V_{Z(min)} = V_Z + I_{Zr}r_Z = 12\text{ V} + (0.5\text{ mA})(10\text{ }\Omega)$

   $= 12.005\text{ V}$

   $V_{Z(max)} = V_Z + I_{Zr}r_Z = 12\text{ V} + (35\text{ mA})(10\text{ }\Omega)$

   $= 12.35\text{ V}$

4. % input regulation $= \left(\dfrac{\Delta V_{out}}{\Delta V_{in}}\right)100$

   $= \left(\dfrac{0.2\text{ V}}{5\text{ V}}\right)100$

   $= 4\%$

5. % load regulation $= \left(\dfrac{V_{NL} - V_{FL}}{V_{FL}}\right)100$

   $= \left(\dfrac{3.6\text{ V} - 3.4\text{ V}}{3.4\text{ V}}\right)100$

   $= 5.88\%$

6. $\Delta V = 6\text{ V} - 2\text{ V} = 4\text{ V}$

   $\Delta C = (5\text{ pF/V})(4\text{ V}) = 20\text{ pF}$

   At 8 V, $C = 30\text{ pF} - \Delta C = 10\text{ pF}$

7. $R = \dfrac{0.1\text{ V}}{-0.5\text{ mA}} = -0.2\text{ k}\Omega = -200\text{ }\Omega$

8. (a) Reverse-biased, no light
   (b) Reverse-biased, no light
   (c) Forward-biased, emits light
   (d) Forward-biased, emits light

9. Decrease

10. PIN diode is constructed of three semiconductor regions.

Chapter 5

1. Collector, emitter

2. Negative, positive

3. $I_C = I_E - I_B = 5.34\text{ mA} - 475\text{ }\mu\text{A} = 4.865\text{ mA}$

4. $\alpha = \dfrac{I_C}{I_E} = \dfrac{8.23\text{ mA}}{8.69\text{ mA}} = 0.947$

5. $\beta_{dc} = \dfrac{I_C}{I_B} = \dfrac{25\text{ mA}}{200\text{ }\mu\text{A}} = 125$

6. $\beta_{dc} = \dfrac{\alpha_{dc}}{1 - \alpha_{dc}} = \dfrac{0.96}{1 - 0.96} = 24$

7. $P_{max} = I_{C(max)}V_{CE}$

   $I_{C(max)} = \dfrac{P_{max}}{V_{CE}} = \dfrac{0.5\text{ W}}{8\text{ V}} = 62.5\text{ mA}$

8. $A_v = \dfrac{1.5\text{ V}}{10\text{ mV}} = 150$

9. Saturation

10. Yes, the base-emitter junction is reverse-biased.

11. Leakage current (I_{CBO}, I_{CEO}) and β_{dc}.

Chapter 6

1. Too close to saturation.

2. $I_C = \beta_{dc}I_B = (75)(150\text{ }\mu\text{A}) = 11.25\text{ mA}$

   $V_{CE} = V_{CC} - I_C R_C = 18\text{ V} - (11.25\text{ mA})(1\text{ k}\Omega)$

   $= 6.75\text{ V}$

3. $I_{C(sat)} \cong \dfrac{V_{CC}}{R_C} = \dfrac{18\text{ V}}{1\text{ k}\Omega} = 18\text{ mA}$

4. $V_{CE(cutoff)} = V_{CC} = 18\text{ V}$

5. $V_{BB} = V_{CC}, V_E = 0\text{ V}$

   $I_B = \dfrac{V_{CC} - 0.7\text{ V}}{R_B} = \dfrac{12\text{ V} - 0.7\text{ V}}{22\text{ k}\Omega} = \dfrac{11.3\text{ V}}{22\text{ k}\Omega}$

   $= 0.514\text{ mA}$

   $I_C = \beta_{dc}I_B = (90)(0.514\text{ mA}) = 46.26\text{ mA}$

   $V_{CE} = V_{CC} - I_C R_C$

   $= 12\text{ V} - (46.26\text{ mA})(100\text{ }\Omega) = 7.37\text{ V}$

6. $I_{CQ} = (180)(0.514\text{ mA}) = 92.52\text{ mA}$

   $V_{CEQ} = 12\text{ V} - (92.52\text{ mA})(100\text{ }\Omega) = 2.75\text{ V}$

7. See Figure S–4.

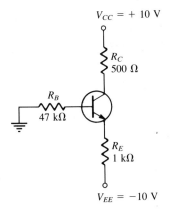

$V_{CC} = +10\text{ V}$

R_C
$500\text{ }\Omega$

R_B
$47\text{ k}\Omega$

R_E
$1\text{ k}\Omega$

$V_{EE} = -10\text{ V}$

FIGURE S–4

8. $V_B \cong 0, V_E \cong -0.7\text{ V}$

   $I_E = \dfrac{-0.7\text{ V} - (-10\text{ V})}{1\text{ k}\Omega} = \dfrac{9.3\text{ V}}{1\text{ k}\Omega} = 9.3\text{ mA}$

$I_d \cong I_E = 9.3$ mA

$V_C = V_{CC} - I_C R_C = 10$ V $- (9.3$ mA$)(500 \ \Omega)$

$\quad = 5.35$ V

$V_{CE} = 5.35$ V $- (-0.7$ V$) = 6.05$ V

9. $R_{\text{IN(base)}} = \beta_{dc} R_E = (125)(300 \ \Omega) = 37.5$ kΩ

10. (a) $R_{\text{IN(base)}} = \beta_{dc} R_E = (50)(600 \ \Omega) = 30$ kΩ

$V_B = \dfrac{5 \text{ k}\Omega \| 30 \text{ k}\Omega}{33 \text{ k}\Omega + (5 \text{ k}\Omega \| 30 \text{ k}\Omega)} (12 \text{ V})$

$\quad = 1.38$ V

(b) $R_{\text{IN(base)}} = (50)(1200 \ \Omega) = 60$ kΩ

Since $R_{\text{IN(base)}} > 10R_2$, it can be neglected.

$V_B = \left(\dfrac{5 \text{ k}\Omega}{33 \text{ k}\Omega + 5 \text{ k}\Omega} \right) 12 \text{ V} = 1.58$ V

11. (a) $V_{EQ} = V_B - 0.7$ V $= 1.38$ V $- 0.7$ V

$\quad = 0.68$ V

$I_{CQ} \cong I_E = \dfrac{V_E}{R_E} = \dfrac{0.68 \text{ V}}{600 \ \Omega} = 1.13$ mA

$V_{CQ} = V_{CC} - I_C R_C$

$\quad = 12$ V $- (1.13$ mA$)(1.8$ k$\Omega) = 9.97$ V

$V_{CEQ} = 9.97$ V $- 0.68$ V $= 9.29$ V

(b) $P_{D(\text{min})} = I_{CQ} V_{CEQ} = (1.13$ mA$)(9.29$ V$)$

$\quad = 10.5$ mW

12. Yes. If V_{CC} is increased, V_B will increase. If V_B increases, V_E will increase. When V_E increases, I_E and (therefore) I_C increase.

No. When I_C increases, V_C decreases. A decrease in V_C and an increase in V_E result in a decrease in V_{CE}, since $V_{CE} = V_C - V_E$.

13. See Figure S–5.

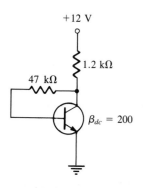

FIGURE S–5

$I_C = \dfrac{V_{CC} - V_{BE}}{R_C + R_B/\beta_{dc}} = \dfrac{12 \text{ V} - 0.7 \text{ V}}{1.2 \text{ k}\Omega + 47 \text{ k}\Omega/200}$

$\quad = 7.87$ mA

$V_C = V_{CC} - I_C R_C = 12$ V $- (7.87$ mA$)(1.2$ k$\Omega)$

$\quad = 2.56$ V

14. $V_C = 8$ V indicates that the transistor is cut off. $V_B = 0$ V indicates a shorted base-emitter junction or an open upper voltage-divider resistor.

15. If the emitter is open, there is no emitter or collector current. The collector voltage will be 10 V, and the emitter voltage 0 V.

Chapter 7

1. See Figure S–6.

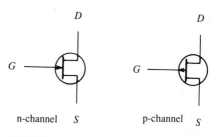

n-channel   S   p-channel   S

FIGURE S–6

2. See Figure S–7.

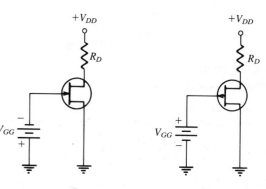

FIGURE S–7

3. $V_{DS(p)} = V_p + V_{GS} = 5$ V $+ (-2$ V$) = 3$ V

4. (a) $I_D = I_{DSS} = 20$ mA

(b) $I_D = 0$ A

(c) I_D increases

5. $I_D = I_{DSS}\left(1 - \dfrac{V_{GS}}{V_{GS(off)}}\right)^2$

$= 12\ \text{mA}\left(1 - \dfrac{-2.5\ \text{V}}{-8\ \text{V}}\right)^2$

$= 12\ \text{mA}(1 - 0.3125)^2 = 5.67\ \text{mA}$

6. $g_m = g_{m_0}\left(1 - \dfrac{V_{GS}}{V_{GS(off)}}\right) = 2000\ \mu\text{S}\left(1 - \dfrac{2\ \text{V}}{6\ \text{V}}\right)$

$= 1333\ \mu\text{S}$

This is a p-channel because V_{GS} is positive.

7. $R_{IN} = \dfrac{V_{GS}}{I_{GSS}} = \dfrac{10\ \text{V}}{10\ \text{nA}} = 1000\ \text{M}\Omega$

8. A MOSFET has no pn junction, and its gate is insulated from the channel.

9. See Figure S–8.

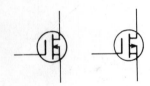

n-channel *DE*   p-channel *DE*

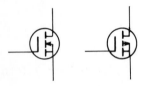

n-channel *E*   p-channel *E*

FIGURE S–8

10. $I_D = I_{DSS}\left(1 - \dfrac{V_{GS}}{V_{GS(off)}}\right)^2 = 20\ \text{mA}(1 - 0)^2$

$= 20\ \text{mA}$

11. Enhancement

12. An E MOSFET has no physical channel or depletion mode.

13. $K = \dfrac{I_{D(on)}}{[V_{GS} - V_{GS(th)}]^2} = \dfrac{4\ \text{mA}}{(8\ \text{V} - 2\ \text{V})^2}$

$= 0.111\ \text{mA/V}^2$

$I_D = K[V_{GS} - V_{GS(th)}]^2$

$= 0.111\ \text{mA/V}^2(6\ \text{V} - 2\ \text{V})^2 = 1.78\ \text{mA}$

14. The gate is insulated from the channel by an SiO_2 layer.

15. $V_{GS} = -I_D R_S = -(12\ \text{mA})(100\ \Omega) = -1.2\ \text{V}$

16. $R_S = \left|\dfrac{V_{GS}}{I_D}\right| = \left|\dfrac{-4\ \text{V}}{5\ \text{mA}}\right| = 800\ \Omega$

17. Since $V_{GS} = 0$, $I_D = I_{DSS} = 15\ \text{mA}$

$V_{DS} = V_{DD} - I_D R_D = 15\ \text{V} - (15\ \text{mA})(500\ \Omega)$

$= 7.5\ \text{V}$

18. $V_{GS(min)} = V_{GS(th)} = 3\ \text{V}$

Chapter 8

1. From the graph in Figure 8–3, the approximate values are:

$I_{c(pp)} \cong 6.4\ \text{mA} - 3.5\ \text{mA} = 2.9\ \text{mA}$

$V_{ce(pp)} \cong 2.25\ \text{V} - 0.75\ \text{V} = 1.5\ \text{V}$

2. $h_{ie} = \dfrac{V_b}{I_b} = \dfrac{100\ \mu\text{V}}{5\ \mu\text{A}} = 20\ \Omega$

$h_{fe} = \dfrac{I_c}{I_b} = \dfrac{800\ \mu\text{A}}{5\ \mu\text{A}} = 160$

3. $\beta = h_{fe} = 100$

$r_e = \dfrac{h_{re}}{h_{oe}} = \dfrac{0.001}{50\ \mu\text{S}} = 20\ \Omega$

$r_c = \dfrac{h_{re} + 1}{h_{oe}} = \dfrac{0.001 + 1}{50\ \mu\text{S}} = 20.02\ \text{k}\Omega$

4. $r_e \cong \dfrac{25\ \text{mV}}{I_E} = \dfrac{25\ \text{mV}}{3\ \text{mA}} = 8.33\ \Omega$

5. See Figure S–9.

6. $R_{IN(base)} = \beta_{dc} R_E = 90\ \text{k}\Omega$

Neglecting $R_{IN(base)}$:

$V_B = \left(\dfrac{R_2}{R_1 + R_2}\right)V_{CC} = \left(\dfrac{5\ \text{k}\Omega}{22\ \text{k}\Omega + 5\ \text{k}\Omega}\right)15\ \text{V}$

$= 2.78\ \text{V}$

$V_E = V_B - 0.7\ \text{V} = 2.78\ \text{V} - 0.7\ \text{V} = 2.08\ \text{V}$

$I_E = \dfrac{V_E}{R_E} = \dfrac{2.08\ \text{V}}{1\ \text{k}\Omega} = 2.08\ \text{mA}$

$V_C = V_{CC} - I_C R_C = 15\ \text{V} - (2.08\ \text{mA})(2\ \text{k}\Omega)$

$= 10.84\ \text{V}$

7. (a) $r_e = \dfrac{25\ \text{mV}}{I_E} = \dfrac{25\ \text{mA}}{2.08\ \text{mA}} = 12.02\ \Omega$

$R_{in(base)} = \beta(r_e + R_E) = 100(1012.02\ \Omega)$

$= 101.202\ \text{k}\Omega$

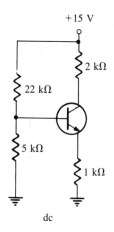

+15 V

2 kΩ

22 kΩ

5 kΩ

1 kΩ

dc

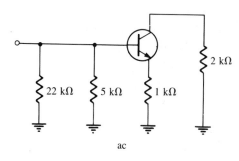

2 kΩ

22 kΩ   5 kΩ   1 kΩ

ac

FIGURE S–9

(b) $R_{in} = R_{in(base)}\|R_1\|R_2$
$= 101.202\ k\Omega\|22\ k\Omega\|5\ k\Omega = 3.92\ k\Omega$

(c) $A_v = \dfrac{R_C}{R_E + r_e} = \dfrac{2\ k\Omega}{1012.202\ \Omega} = 1.98$

8. (a) $R_{in(base)} = \beta r_e = (100)(12.02\ \Omega) = 1202\ \Omega$

(b) $R_{in} = 1.202\ k\Omega\|22\ k\Omega\|5\ k\Omega = 928\ \Omega$

(c) $A_v = \dfrac{R_C}{r_e} = \dfrac{2\ k\Omega}{12.02\ \Omega} = 166.39$

9. (a) $R_{in(base)} = 1.202\ k\Omega$

(b) $R_{in} = 928\ \Omega$

(c) $A_v = \dfrac{R_c}{r_e} = \dfrac{R_C\|R_L}{r_e} = \dfrac{2\ k\Omega\|10\ k\Omega}{12.02\ \Omega} = 138.66$

10. $R_{in(base)} = \beta R_E = (125)(1\ k\Omega) = 125\ k\Omega$
$R_{in} = R_1\|R_2\|R_{in(base)} = 100\ k\Omega\|50\ k\Omega\|125\ k\Omega$
$= 26.32\ k\Omega$

11. $R_{in(base)} = \beta\beta R_E = (125)^2(1\ k\Omega) = 15.625\ M\Omega$
$R_{in} = 100\ k\Omega\|50\ k\Omega\|15.625\ M\Omega = 33.26\ k\Omega$

12. $V_{B_1} = \left(\dfrac{R_2}{R_1 + R_2}\right)V_{CC} = \left(\dfrac{50\ k\Omega}{150\ k\Omega}\right)9\ V = 3\ V$

$V_{E_1} = 2.3\ V$

$V_{E_2} = 1.6\ V$

$I_{E_2} = \dfrac{V_E}{R_E} = \dfrac{1.6\ V}{1\ k\Omega} = 1.6\ mA$

$r_{e_2} = \dfrac{25\ mV}{I_{E_2}} = \dfrac{25\ mV}{1.6\ mA} = 15.63\ \Omega$

$I_{E_1} = \dfrac{I_{E_2}}{\beta_{dc}} = \dfrac{1.6\ mA}{125} = 12.8\ \mu A$

$r_{e_1} = \dfrac{25\ mV}{I_{E_1}} = \dfrac{25\ mV}{12.8\ \mu A} = 1.95\ k\Omega$

13. $V_B = \left(\dfrac{R_2}{R_1 + R_2}\right)V_{CC} = \left(\dfrac{10\ k\Omega}{34\ k\Omega}\right)(-10\ V)$
$= -2.94\ V$

$V_E = V_B + 0.7 = -2.94\ V + 0.7\ V = -2.24\ V$

$I_E = \dfrac{V_E}{R_E} = \dfrac{2.24\ V}{1\ k\Omega} = 2.24\ mA$

$R_{in} = r_e \cong \dfrac{25\ mV}{I_E} = \dfrac{25\ mV}{2.24\ mA} = 11.16\ \Omega$

$A_v = \dfrac{R_C}{r_e} = \dfrac{1.5\ k\Omega}{11.16\ \Omega} = 134.4$

$A_i \cong 1$

$A_p \cong A_v = 134.4$

14. $A'_v = (15)(15)(15)(15) = 50,625$
$A_v\ (dB) = 20\ \log A'_v = 20\ \log(50,625) = 94.09\ dB$

15. Using data from Problem 8:

First stage

$R_c = R_C\|R_{in(2)} = 2\ k\Omega\|928\ \Omega = 633.88\ \Omega$

$A_{v_1} = \dfrac{633.88\ \Omega}{12.02\ \Omega} = 52.74$

Second stage

$A_{v_2} = 166.39$

Overall gain

$A'_v = A_v A_v = (52.74)(166.39) = 8775.4$

$A_v\ (dB) = 20\ \log(8775.4) = 78.87\ dB$

Chapter 9

1. From the curve in Figure 9–7(a):

$I_{D(pp)} \cong 3.9\ mA - 1.1\ mA = 2.8\ mA$

2. From the curve in Figure 9–7(b):

$I_{D(pp)} \cong 6 \text{ mA} - 2 \text{ mA} = 4 \text{ mA}$

From the curve in Figure 9–7(c):

$I_{D(pp)} \cong 4.5 \text{ mA} - 1.3 \text{ mA} = 3.2 \text{ mA}$

3. $A_v = \dfrac{V_{ds}}{V_{gs}} = \dfrac{3.2 \text{ V}}{0.28 \text{ V}} = 11.43$

4. $A_v = g_m R_d = (3.8 \text{ mS})(1.2 \text{ k}\Omega) = 4.56$

5. See Figure S–10.

+15 V

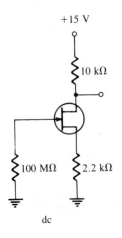

dc

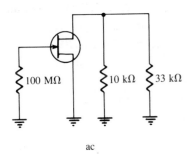

ac

FIGURE S–10

6. $I_D = \dfrac{I_{DSS}}{2} = \dfrac{15 \text{ mA}}{2} = 7.5 \text{ mA}$

7. $A_v = \dfrac{g_m R_d}{1 + g_m R_s}$

$= \dfrac{(5000 \ \mu\text{S})(10 \text{ k}\Omega\|33 \text{ k}\Omega)}{1 + (5000 \ \mu\text{S})(22 \text{ k}\Omega)}$

$= \dfrac{(5000 \ \mu\text{S})(7.67 \text{ k}\Omega)}{1 + (5000 \ \mu\text{S})(22 \text{ k}\Omega)} = 3.2$

8. $A_v = g_m R_d = (5000 \ \mu\text{S})(10 \text{ k}\Omega\|33 \text{ k}\Omega\|4.7 \text{ k}\Omega)$

$= (5000 \ \mu\text{S})(2.92 \text{ k}\Omega) = 14.6$

9. $I_D = \dfrac{I_{DSS}}{2} = \dfrac{9 \text{ mA}}{2} = 4.5 \text{ mA}$

$V_{GS} = -I_D R_S = -(4.5 \text{ mA})(330 \ \Omega) = -1.49 \text{ V}$

$V_{DS} = V_{DD} - I_D(R_D + R_S)$

$= 9 \text{ V} - (4.5 \text{ mA})(1.33 \text{ k}\Omega) = 3 \text{ V}$

10. $A_v = g_m R_d = (2500 \ \mu\text{S})(1 \text{ k}\Omega) = 2.5$

$V_{out} = A_v V_{in} = (2.5)(10 \text{ mV}) = 25 \text{ mV}$

11. $V_{GS} = \left(\dfrac{R_2}{R_1 + R_2}\right) V_{DD} = \left(\dfrac{6.8 \text{ k}\Omega}{24.8 \text{ k}\Omega}\right) 20 \text{ V}$

$= 5.48 \text{ V}$

$K = \dfrac{I_{D(on)}}{[V_{GS} - V_{GS(th)}]^2} = \dfrac{18 \text{ mA}}{(10 \text{ V} - 2.5 \text{ V})^2}$

$= 0.32 \text{ mA/V}^2$

$I_D = K[V_{GS} - V_{GS(th)}]^2$

$= 0.32 \text{ mA/V}^2 (5.48 \text{ V} - 2.5 \text{ V})^2 = 2.84 \text{ mA}$

$V_{DS} = V_{DD} - I_D R_D = 20 \text{ V} - (2.84 \text{ mA})(1 \text{ k}\Omega)$

$= 17.16 \text{ V}$

12. **(a)** $A_v = \dfrac{g_m R_s}{1 + g_m R_s} = \dfrac{(3000 \ \mu\text{S})(5 \text{ k}\Omega)}{1 + (3000 \ \mu\text{S})(R_s)}$

$= 0.938$

(b) $A_v = \dfrac{g_m R_s}{1 + g_m R_s} = \dfrac{(4300 \ \mu\text{S})(100 \ \Omega)}{1 + (4300 \ \mu\text{S})(100 \ \Omega)}$

$= 0.301$

13. **(a)** $A_v = \dfrac{g_m R_s}{1 + g_m R_s}$

$= \dfrac{(3000 \ \mu\text{S})(5 \text{ k}\Omega\|10 \text{ k}\Omega)}{1 + (3000 \ \mu\text{S})(5 \text{ k}\Omega\|10 \text{ k}\Omega)}$

$= \dfrac{(3000 \ \mu\text{S})(3.33 \text{ k}\Omega)}{1 + (3000 \ \mu\text{S})(3.33 \text{ k}\Omega)} = 0.909$

(b) $A_v = \dfrac{g_m R_s}{1 + g_m R_s}$

$= \dfrac{(4300 \ \mu\text{S})(100 \ \Omega\|10 \text{ k}\Omega)}{1 + (4300 \ \mu\text{S})(100 \ \Omega\|10 \text{ k}\Omega)}$

$= 0.299$

14. $R_{in(gate)} = \left|\dfrac{V_{GS}}{I_{GSS}}\right| = \left|\dfrac{15 \text{ V}}{30 \text{ pA}}\right| = 500 \times 10^9 \ \Omega$

15. $A_v = g_m R_d = (3500 \ \mu\text{S})(10 \text{ k}\Omega) = 35$

$$R_{in} = \left(\frac{1}{g_m}\right)\|2\ k\Omega = \left(\frac{1}{3500\ \mu S}\right)\|2\ k\Omega$$

$$= 250\ \Omega$$

16. Using A_v of each amplifier found in Problem 12:

$$A'_v = A_{v_1}A_{v_2} = (0.938)(0.301) = 0.282$$

Chapter 10

1. Bottom peak.

2. $I_{c(sat)} = I_{CQ} + \dfrac{V_{CEQ}}{R_c} = 2\ mA + \dfrac{3\ V}{3\ k\Omega}$

$$= 3\ mA$$

3. $V_{ce(cutoff)} = V_{CEQ} + I_{CQ}R_c$

$$= 3\ V + (2\ mA)(3\ k\Omega) = 9\ V$$

4. From the Q-point of 3 V, V_{ce} can swing down to 0. This limits the peak swing to 3 V. The peak-to-peak value is 6 V.

5. $A_v = \dfrac{R_c}{r'_e} = \dfrac{500\ \Omega}{18\ \Omega} = 27.78$

6. $A_p = A_v A_i = (50)(75) = 3750$

7. $P_{out} = 0.5V_{CEQ}I_{CQ} = 0.5(5\ V)(10\ mA)$

$$= 25\ mW$$

8. $I_C \cong 0\ A,\ V_C \cong 20\ V,\ V_B = 0.7\ V$

9. $I_{c(sat)} = \dfrac{V_{CEQ}}{R_e} = \dfrac{10\ V}{50\ \Omega} = 0.2\ A$

$$V_{CC} = 20\ V$$

10. $P_{out} = 0.5V_{CEQ}I_{c(sat)} = 0.5(10\ V)(0.2\ A) = 1\ W$

$$P_{dc} = V_{CC}I_{CC} = \frac{V_{CC}I_{c(sat)}}{\pi} = 20\ V\left(\frac{0.2\ A}{\pi}\right)$$

$$= 1.2732\ W$$

$$\eta = \frac{P_{out}}{P_{dc}} = \frac{1\ W}{1.2632\ W} = 0.785$$

11. $I_{CC} = \dfrac{I_{c(sat)}}{\pi} = \dfrac{20\ mA}{\pi} = 6.37\ mA$

12. Because the transistor is on for a small part of the input cycle. Therefore, the average power dissipation is small.

13. $P_{D(avg)} = \left(\dfrac{t_{on}}{T}\right)V_{CE(sat)}I_{C(sat)}$

$$= \left(\frac{0.5\ \mu s}{1\ \mu s}\right)(0.3\ V)(200\ mA) = 30\ mW$$

14. $P_{out} = \dfrac{0.5V_{CC}^2}{R_c} = \dfrac{0.5(20\ V)^2}{100} = 2\ W$

15. $\eta = \dfrac{P_{out}}{P_{out} + P_{D(avg)}} = \dfrac{2\ W}{2\ W + 30\ mW} = 0.985$

Chapter 11

1. Low-frequency response: C_1, C_2, and C_3
High-frequency response: C_{bc} and C_{be}

2. $V_E \cong \left(\dfrac{R_2}{R_1 + R_2}\right)V_{CC} - 0.7\ V$

$$= \left(\frac{5\ k\Omega}{38\ k\Omega}\right)20\ V - 0.7\ V = 1.93\ V$$

$$I_E = \frac{V_E}{R_E} = \frac{1.93\ V}{500\ \Omega} = 3.86\ mA$$

$$r_e = \frac{25\ mV}{3.86\ mA} = 6.48\ \Omega$$

$$A_v = \frac{R_c}{r_e} = \frac{2\ k\Omega\|5\ k\Omega}{6.48\ \Omega} = 220.5$$

$$C_{in(Miller)} = C_{bc}(A_v + 1) = 4\ pF(220.5 + 1)$$

$$= 886\ pF$$

3. $C_{out(Miller)} = C_{bc}\left(\dfrac{A_v + 1}{A_v}\right) = 4\ pF\left(\dfrac{221.5}{220.5}\right)$

$$\cong 4\ pF$$

4. $A_v\ (dB) = 20\ log\ A_v = 20\ log(220.5) = 46.87\ dB$

5. *Input* $(r_e = 7\ \Omega$; assume R_E is bypassed)

$$R_{in} = R_1\|R_2\|R_{in(base)} = 33\ k\Omega\|5\ k\Omega\|\beta r_e$$

$$= 33\ k\Omega\|5\ k\Omega\|(150)(7\ \Omega) = 846\ \Omega$$

$$f_c = \frac{1}{2\pi R_{in}C_1} = \frac{1}{2\pi(846\ \Omega)(0.1\ \mu F)}$$

$$= 1.88\ kHz$$

Output

$$f_c = \frac{1}{2\pi(R_C + R_L)C_2}$$

$$= \frac{1}{2\pi(2\ k\Omega + 5\ k\Omega)(0.1\ \mu F)} = 227.4\ Hz$$

Bypass

$$R_{TH} = R_1\|R_2\|R_s = 33\ k\Omega\|5\ k\Omega\|50\ \Omega \cong 50\ \Omega$$

$$f_c = \frac{1}{2\pi[(r_e + R_{TH}/\beta)\|R_E]C_3}$$

$$= \frac{1}{2\pi[(6.48\ \Omega + 50\ \Omega/150)\|500\ \Omega]10\ \mu F}$$

$$= 2.37\ kHz$$

6. *Input*

$$C_T = C_{in(Miller)} + C_{be} = 886 \text{ pF} + 10 \text{ pF} = 896 \text{ pF}$$

$$f_c = \frac{1}{2\pi(R_s\|R_1\|R_2\|\beta r_e)C_T}$$

$$= \frac{1}{2\pi(50\ \Omega\|33\ k\Omega\|5\ k\Omega\|972\ \Omega)896\ pF}$$

$$= \frac{1}{2\pi(47\ \Omega)(896\ \mu F)} = 3.78 \text{ MHz}$$

Output

$$f_c = \frac{1}{2\pi R_c C_{out(Miller)}} = \frac{1}{2\pi(2\ k\Omega\|5\ k\Omega)4\ pF}$$

$$= 27.85 \text{ MHz}$$

7. $f_{ch} = 3.78 \text{ MHz}, f_{cl} = 2.37 \text{ kHz}$

8. $BW = f_{ch} - f_{cl} = 3.78 \text{ MHz} - 2.37 \text{ kHz}$

$$\cong 3.78 \text{ MHz}$$

9. $f_T = A_{v(mid)}BW$

$$A_{v(mid)} = \frac{f_T}{BW} = \frac{75 \text{ MHz}}{10 \text{ MHz}} = 7.5$$

10. $f_{ch} = \dfrac{0.35}{t_f} = \dfrac{0.35}{10 \text{ ns}} = 35 \text{ MHz}$

11. $BW = f_{ch} = 35 \text{ MHz}$

Chapter 12

1. Inverting input, noninverting input, output, positive dc supply, and negative dc supply.

2. **(b)** Low power   **(f)** dc isolation

3. V_1:   Differential output voltage
V_2:   Noninverting input voltage
V_3:   Single-ended output voltage
V_4:   Differential input voltage
I_1:   Bias current

4. An op-amp is typically made up of more than one diff-amp.

5. $V_{out(diff)} = V_{C_1} - V_{C_2}$

$$V_{C_1} = V_{CC} - I_{C_1}R_{C_1}$$

$$V_{C_2} = V_{CC} - I_{C_2}R_{C_2}$$

$$V_{out(diff)} = (I_{C_1} - I_{C_2})5 \text{ k}\Omega$$

$$= (1.35 \text{ mA} - 1.29 \text{ mA})5 \text{ k}\Omega$$

$$= 0.3 \text{ V}$$

6. $I_{OS} = |I_1 - I_2| = |50\ \mu A - 49.3\ \mu A| = 0.7\ \mu A$

7. Differential input impedance is the total impedance between the two input terminals. Common-

mode input impedance is the impedance from each input to ground.

8. **(a)** Infinite
     (b) The one with a CMRR of 100 dB.

9. Slew rate $= \dfrac{\Delta V_{out}}{\Delta t} = \dfrac{8 \text{ V}}{12\ \mu s} = 0.667 \text{ V}/\mu s$

10. **(a)** Voltage follower   **(b)** Noninverting
      (c) Inverting

11. $B = \dfrac{R_i}{R_i + R_f} = \dfrac{1 \text{ k}\Omega}{101 \text{ k}\Omega} = 0.0099$

$$V_f = BV_{out} = (0.0099)5 \text{ V} = 0.0495 \text{ V} = 49.5 \text{ mV}$$

12. $A_{cl(NI)} = \dfrac{1}{B} = \dfrac{1}{0.0099} = 101$

13. **(c)** Noninverting   **(d)** $A_{cl} \cong 1$
      (e) High Z_{in}

14. $A_{cl(I)} = -\dfrac{R_f}{R_i} = -\dfrac{220 \text{ k}\Omega}{2.2 \text{ k}\Omega} = -100$

15. No, except for the voltage-follower configuration, which is always approximately 1.

16. In a voltage follower, the feedback attenuation is unity. In a noninverting amplifier, the feedback attenuation is determined by a voltage divider and is usually much less than unity.

17. Voltage follower.

18. $A_{c(NI)} = \dfrac{1}{B} = \dfrac{1}{0.025} = 40$

19. Inverting

20. R_1:   Input resistor establishes input impedance and closed-loop gain.
R_2:   Feedback resistor establishes closed-loop gain.
R_3:   Used for bias current compensation when equal to $R_i\|R_f$.
R_4:   Used for input offset voltage compensation.

Chapter 13

1. Open-loop gain is the voltage gain of an op-amp without external feedback. Closed-loop gain is the voltage gain of an op-amp with negative feedback.

2. $BW = f_{ch} - f_{cl} = 10 \text{ kHz} - 1 \text{ kHz} = 9 \text{ kHz}$

3. $BW = f_{ch} = 100 \text{ kHz}$

4. **(a)** $X_C = \dfrac{1}{2\pi fC} = \dfrac{1}{2\pi(10 \text{ kHz})(0.07\ \mu F)}$

$$= 1.59 \text{ k}\Omega$$

$$\frac{V_{out}}{V_{in}} = \frac{X_C}{\sqrt{R^2 + X_C^2}}$$

$$= \frac{1.59 \text{ k}\Omega}{\sqrt{(1 \text{ k}\Omega)^2 + (1.59 \text{ k}\Omega)^2}}$$

$$= 0.846$$

(b) $X_C = \dfrac{1}{2\pi(10 \text{ kHz})(0.002 \text{ }\mu\text{F})} = 7.96 \text{ k}\Omega$

$$\frac{V_{out}}{V_{in}} = \frac{7.96 \text{ k}\Omega}{\sqrt{(12 \text{ k}\Omega)^2 + (7.96 \text{ k}\Omega)^2}} = 0.553$$

(c) $X_C = \dfrac{1}{2\pi(10 \text{ kHz})(50 \text{ pF})} = 318 \text{ k}\Omega$

$$\frac{V_{out}}{V_{in}} = \frac{318 \text{ k}\Omega}{\sqrt{(100 \text{ k}\Omega)^2 + (318 \text{ k}\Omega)^2}} = 0.954$$

(d) $X_C = \dfrac{1}{2\pi(10 \text{ kHz})(0.1 \text{ }\mu\text{F})} = 159.15 \text{ }\Omega$

$$\frac{V_{out}}{V_{in}} = \frac{159.15 \text{ }\Omega}{\sqrt{(33 \text{ k}\Omega)^2 + (159.15 \text{ }\Omega)^2}}$$

$$= 0.00482$$

5. **(a)** $A_{ol} = \dfrac{A_{o(mid)}}{\sqrt{1 + f^2/f_{c(ol)}^2}}$

$$= \frac{80,000}{\sqrt{1 + (100 \text{ Hz})^2/(500 \text{ Hz})^2}}$$

$$= 78,446$$

(b) $A_{ol} = \dfrac{80,000}{\sqrt{1 + (1 \text{ kHz})^2/(500 \text{ Hz})^2}} = 35,777$

(c) $A_{ol} = \dfrac{80,000}{\sqrt{1 + (5 \text{ kHz})^2/(500 \text{ Hz})^2}} = 7960$

(d) $A_{ol} = \dfrac{80,000}{\sqrt{1 + (20 \text{ kHz})^2/(500 \text{ Hz})^2}} = 1999$

6. **(a)** A_{ol} (dB) $= 20 \log(78,446) = 97.89$ dB
(b) A_{ol} (dB) $= 20 \log(35,777) = 91.07$ dB
(c) A_{ol} (dB) $= 20 \log(7960) = 78.02$ dB
(d) A_{ol} (dB) $= 20 \log(1999) = 66.02$ dB

7. **(a)** $\phi = -\arctan\left(\dfrac{R}{X_C}\right) = -\arctan\left(\dfrac{1 \text{ k}\Omega}{1.59 \text{ k}\Omega}\right)$

$$= -32.17°$$

(b) $\phi = -\arctan\left(\dfrac{12 \text{ k}\Omega}{7.96 \text{ k}\Omega}\right) = -56.44°$

(c) $\phi = -\arctan\left(\dfrac{100 \text{ k}\Omega}{318 \text{ k}\Omega}\right) = -17.46°$

(d) $\phi = -\arctan\left(\dfrac{33 \text{ k}\Omega}{159.15 \text{ }\Omega}\right) = -89.72°$

8. $\phi_{tot} = -90° - 90° - 90° = -270°$

9. $BW_{cl} = BW_{ol}(1 + BA_{ol(mid)})$

$$= 150 \text{ Hz}[1 + (0.004)(100,000)]$$

$$= 60.15 \text{ kHz}$$

10. Gain-bandwidth product $= A_{cl}f_{c(cl)}$

$$= 1(800 \text{ kHz})$$

$$= 800 \text{ kHz}$$

11. ≈ 50 dB (lowest point on the -20 dB/decade slope)

12. $\theta_{pm} = 180° - \phi_{tot} = 180° - 120° = 60°$

The amplifier is stable as long as $\phi_{tot} < 180°$.

13. When $\theta_{pm} = 0°$

14. Less

15. The op-amp can be compensated only to the degree necessary, therefore optimizing the bandwidth.

16. 20 dB

Chapter 14

1. **(a)** Maximum negative
(b) Maximum positive
(c) Maximum negative

2. $V_{UTP} = \left(\dfrac{R_2}{R_1 + R_2}\right)(+10 \text{ V}) = \left(\dfrac{20 \text{ k}\Omega}{70 \text{ k}\Omega}\right)10 \text{ V}$

$$= 2.86 \text{ V}$$

$$V_{LTP} = \left(\dfrac{20 \text{ k}\Omega}{70 \text{ k}\Omega}\right)(-10 \text{ V}) = -2.86 \text{ V}$$

3. $V_{HYS} = V_{UTP} - V_{LTP} = 2.86 \text{ V} - (-2.86 \text{ V})$

$$= 5.72 \text{ V}$$

4. $+6.2$ V and -0.7 V

5. **(a)** $V_{OUT} = -(+1 \text{ V} + 1.5 \text{ V}) = -2.5 \text{ V}$

(b) $V_{OUT} = -\dfrac{R_f}{R}(0.1 \text{ V} + 1 \text{ V} + 0.5 \text{ V})$

$$= -2(1.6 \text{ V}) = -3.2 \text{ V}$$

6. **(a)** $A_{cl(1)} = \dfrac{47 \text{ k}\Omega}{5 \text{ k}\Omega} = 9.4$

$$A_{cl(2)} = \frac{47 \text{ k}\Omega}{12 \text{ k}\Omega} = 3.92$$

(b) $A_{cl(1)} = \dfrac{100 \text{ k}\Omega}{10 \text{ k}\Omega} = 10$

$A_{cl(2)} = \dfrac{100 \text{ k}\Omega}{20 \text{ k}\Omega} = 5$

$A_{cl(3)} = \dfrac{100 \text{ k}\Omega}{40 \text{ k}\Omega} = 2.5$

7. See Figure S–11.

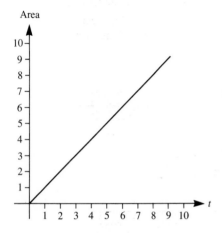

FIGURE S–11
Graph represents increase in area under curve with time.

8. $\Delta V/\Delta t = 2.5$ V/s. The derivative is the rate of change at a point.

9. (a) Differentiator (b) Integrator

10. $V_{out} = \pm RC\left(\dfrac{V_{PP}}{T/2}\right)$

$\qquad = \pm(15 \text{ k}\Omega)(0.05 \ \mu\text{F})\left(\dfrac{2 \text{ V}}{0.5 \text{ ms}}\right) = \pm 3 \text{ V}$

(See Figure S–12.)

11. $\dfrac{\Delta V_{out}}{\Delta t} = -\dfrac{V_{IN}}{RC} = -\dfrac{5 \text{ V}}{(100 \text{ k}\Omega)(10 \ \mu\text{F})}$

$\qquad = -5$ V/ms

To reach the saturated output level, it takes

$\dfrac{-12 \text{ V}}{-5 \text{ V/ms}} = 2.4$ ms. (See Figure S–13.)

Chapter 15

1. An oscillator produces an output waveform without an input signal. An amplifier requires an input signal.

2. A circuit will oscillate when the phase shift around the closed positive feedback loop is 0° and the closed-loop gain is unity.

3. Start-up means that oscillation begins when the dc power is turned on.

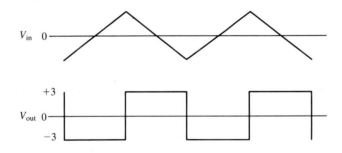

FIGURE S–12

FIGURE S–13

4. For start-up, $A_{cl} > 1$. For sustained oscillation, $A_{cl} = 1$.

5. The Wien-bridge is formed by a voltage divider and a lead-lag network.

6. 1 MHz

7. **(a)** Colpitts   **(b)** Clapp   **(c)** Hartley

8. The RC time constant.

9. There is no feedback. Oscillation is produced by the charging and discharging of a capacitor.

10. The frequency of a 555 astable is determined by the external RC network. A duty cycle of less than 50 percent can be achieved by bypassing one of the external resistors with a diode.

11. A 555 VCO can be implemented by adding a variable-control voltage to the CONT input (pin 5) of the astable configuration.

12. A voltage-controlled oscillator (VCO) is used in a PLL.

Chapter 16

1. **(a)** Band-pass   **(b)** High-pass
    (c) Low-pass   **(d)** Band-stop

2. **(a)** High-pass, first order
    (b) Low-pass, second order
    (c) Band-pass, second order
    (d) High-pass, second order

3. **(a)** $f_c = \dfrac{1}{2\pi RC} = \dfrac{1}{2\pi(680\ \Omega)(0.01\ \mu F)}$

$= 23.4$ kHz

(b) $f_c = \dfrac{1}{2\pi\sqrt{R_A R_B C_A C_B}} = \dfrac{1}{2\pi\sqrt{R^2 C^2}}$

$= \dfrac{1}{2\pi RC} = \dfrac{1}{2\pi(3.3\ k\Omega)(0.22\ \mu F)} = 219$ Hz

(c) $f_c = \dfrac{1}{2\pi C}\sqrt{\dfrac{R_1 + R_3}{R_1 R_2 R_3}}$

$= \dfrac{1}{2\pi(0.05\ \mu F)}\sqrt{\dfrac{33\ k\Omega + 2.2\ k\Omega}{(33\ k\Omega)(150\ k\Omega)(2.2\ k\Omega)}}$

$= \dfrac{1}{2\pi(0.05\ \mu F)}(0.0000569) = 18.1$ kHz

(d) $f_c = \dfrac{1}{2\pi\sqrt{R_A R_B C_A C_B}} = \dfrac{1}{2\pi RC}$

$= \dfrac{1}{2\pi(1\ k\Omega)(0.05\ \mu F)} = 3.18$ kHz

4. **(a)** 20 dB/decade
    (b) 40 dB/decade
    (c) 40 dB/decade
    (d) 40 dB/decade

5. **(a)** $DF = 2 - \dfrac{R_1}{R_2} = 2 - \dfrac{1\ k\Omega}{1.8\ k\Omega}$

$= 1.44$ (approx. Butterworth)

(b) $DF = 1.44$ (approx. Butterworth)

6. **(a)** Chebyshev   **(b)** Butterworth
    (c) Bessel   **(d)** Butterworth

7. See Figure S–14.

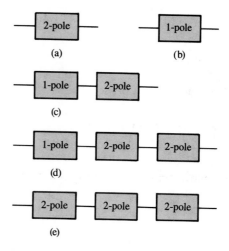

FIGURE S–14

8. **(a)** Decrease R or C.
    (b) Increase the 1.5 kΩ resistor or decrease the 2.7 kΩ.

Chapter 17

1. See Figure S–15.

2. Line regulation maintains a constant output voltage with changes in input voltage. Load regulation maintains a constant output voltage with changes in load current.

3. **(a)** Series   **(b)** Shunt
    (c) Switching (step-down)

4. $V_{OUT} = \left(1 + \dfrac{R_2}{R_3}\right)V_{REF} = \left(1 + \dfrac{33\ k\Omega}{10\ k\Omega}\right)2$ V

$= 8.6$ V

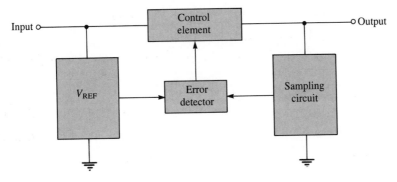

FIGURE S–15

5.   See Figure S–16.

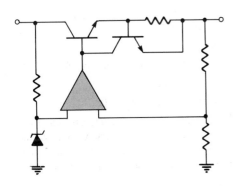

FIGURE S–16

6. $I_{L(max)} = \dfrac{V_{IN}}{R_1} = \dfrac{12\ V}{10\ \Omega} = 1.2\ A$

7.   Transistor Q_1 is used to switch the input voltage at a varying duty cycle based on the regulator's load requirements.

8. $V_{OUT} = \left(\dfrac{t_{on}}{T}\right)V_{IN} = \left(\dfrac{1\ ms}{1\ ms + 3\ ms}\right)20\ V$

$\qquad = 5\ V$

9.   **(a)**   Precision voltage regulator
     (b)   Three-terminal voltage regulators

10. $V_{OUT} = \left(1 + \dfrac{R_1}{R_2}\right)V_{REF} = \left(1 + \dfrac{12\ k\Omega}{4\ k\Omega}\right)1.6\ V$

$\qquad = 6.4\ V$

11. With constant current limiting, the load current is restricted to a specified maximum value. With fold-back current limiting, the load current drops below its maximum value under overload conditions.

12. $P_{max} = \dfrac{V_{IN} - V_{OUT}}{I_{L(max)}} = \dfrac{30\ V - 15\ V}{1\ A} = 15\ W$

13. The negative supply tracks changes in the positive supply.

Chapter 18

1.   **(a)** SCR   **(b)** UJT   **(c)** PUT
     (d) Diac   **(e)** Triac   **(f)** SCS

2.   See Figure S–17.

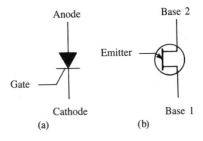

(a)   (b)

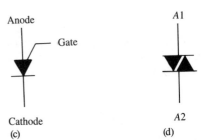

(c)   (d)

FIGURE S–17

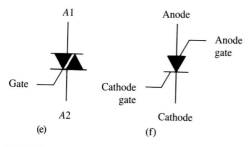

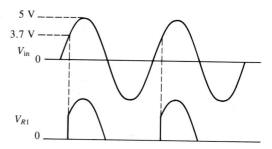

FIGURE S–17 (continued)

FIGURE S–20

3. The SCR is turned on by a positive pulse of current at the gate. The SCR is turned off by reducing the anode current to less than the holding current.

4. See Figure S–18.

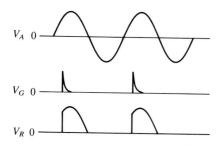

FIGURE S–18

5. See Figure S–19.

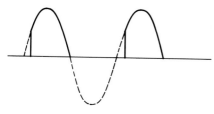

FIGURE S–19

Once triggered on, the triac may or may not turn off at the 0 crossing, depending on the rate of change of voltage. If it turns off, it will not turn back on until the next trigger.

6. $\eta = \dfrac{r_{B_1}}{r_{BB}} = \dfrac{r_{B_1}}{r_{B_1} + r_{B_2}} = \dfrac{1\ k\Omega}{2.5\ k\Omega} = 0.4$

7. The negative resistance characteristic.

8. See Figure S–20.

Chapter 19

1. (a) Photodiode   (b) Photoconductive cell
   (c) Phototransistor   (d) LED
   (e) LASCR   (f) Photodarlington

2. $\lambda = \dfrac{c}{f} = \dfrac{3 \times 10^8\ \text{m/s}}{6 \times 10^{14}\ \text{Hz}} = 0.5 \times 10^{-6}\ \text{m}$

3. $1\ \text{Å} = 1 \times 10^{-10}\ \text{m}$

   $(350 \times 10^{-9}\ \text{m})(1 \times 10^{-10}\ \text{Å/m}) = 350 \times 10^{-19}\ \text{Å}$

4. $1\ \text{fc} = 10.764\ \text{lm/m}^2$

   $(3.5\ \text{fc})(10.764\ \text{lm/m}^2/\text{fc}) = 36.674\ \text{lm/m}^2$

5. The resistance of both the photoconductive cell and the photodiode varies inversely with light intensity. The photodiode is a unidirectional device, whereas the photoconductive cell is bidirectional.

6. Dark current

7. $r_R = \dfrac{5\ \text{V}}{18\ \mu\text{A}} = 277.78\ k\Omega$

8. The collector current increases with light intensity.

9. The spectral response of a device indicates how it responds to various wavelengths of light.

10. The photodarlington has a much greater sensitivity to light than the phototransistor.

11. Photovoltaic action is the process of converting light energy directly to electrical energy with a semiconductor pn junction device called the *solar cell*.

12. Light-activated SCR

13. Forward-biased

14. LEDs emit incoherent light; laser diodes emit coherent light (one wavelength).

15. The length of the pn junction

16. Transfer gain is the ratio of output voltage to input current.

Index

AGC, 14, 92, 200, 292
Alpha, 141, 142
Amplification, 7, 137, 153–155, 313–316
Amplifier
  audio, 14, 16, 200, 330, 379
  averaging, 549, 550
  capacitive-coupled, 285
  class A, 346–361
  class B, 361–370
  class C, 370–378
  common-base, 258, 281–283, 284
  common-collector, 258, 276–281, 284
  common-drain, 325–327, 329
  common-emitter, 258, 264–276, 284
  common-gate, 328–329
  common-source, 316–325, 329
  differential, 450–459
  direct-coupled, 287, 412
  IF, 14, 16, 200
  instrumentation, 558–561
  large-signal, 345, 346
  multistage, 283–289
  power, 361–370
  push-pull, 361–370
  RF, 13, 16, 243, 333
  small-signal, 256–258, 308–313
  summing, 547–549, 567
  transformer-coupled, 288
Amplitude, 12
Amplitude modulation (AM), 12–14, 90–92, 200
Analog, 1
Angstrom, 748
Anode, 46, 47, 715, 724
Antenna, 13
Armstrong oscillator, 600

Astable multivibrator, 610–613
Attenuation, 269, 270
Atom, 22–25
Atomic
  bond, 25
  number, 22
  weight, 22
Automatic gain control (AGC), 14, 92, 200, 292
Avalanche breakdown, 38, 108
Average value, 52

Bandwidth, 422, 499, 506, 634
Barrier potential, 32, 35, 48, 52
Base, 138, 730
Base bias, 178–181
Beta, 141, 142, 149, 179, 263, 280
Bias
  base, 178–181
  collector-feedback, 194–197
  emitter, 181–185
  forward, 34, 35, 46, 139, 157
  reverse, 35, 46, 139, 157
  self, 227, 308
  voltage-divider, 185–194
  zero, 310
BIFET, 480–481
Bipolar junction transistor (BJT), 6, 138–161, 172–201, 256–292
Bode plot, 408–412, 420–422, 499, 503–504, 507, 510–513, 515–522
Breakdown
  avalanche, 38, 108
  reverse, 38, 720
  zener, 108
Bridge rectifier, 61–63

WE VALUE YOUR OPINION—PLEASE SHARE IT WITH US

Merrill Publishing and our authors are most interested in your reactions to this textbook. Did it serve you well in the course? If it did, what aspects of the text were most helpful? If not, what didn't you like about it? Your comments will help us to write and develop better textbooks. We value your opinions and thank you for your help.

Text Title _____ Edition _____

Author(s) _____

Your Name (optional) _____

Address _____

City _____ State _____ Zip _____

School _____

Course Title _____

Instructor's Name _____

Your Major _____

Your Class Rank _____ Freshman _____ Sophomore _____ Junior _____ Senior

_____ Graduate Student

Were you required to take this course? _____ Required _____ Elective

Length of Course? _____ Quarter _____ Semester

1. Overall, how does this text compare to other texts you've used?

   _____ Superior _____ Better Than Most _____ Average _____ Poor

2. Please rate the text in the following areas:

	Superior	Better Than Most	Average	Poor
Author's Writing Style	_____	_____	_____	_____
Readability	_____	_____	_____	_____
Organization	_____	_____	_____	_____
Accuracy	_____	_____	_____	_____
Layout and Design	_____	_____	_____	_____
Illustrations/Photos/Tables	_____	_____	_____	_____
Examples	_____	_____	_____	_____
Problems/Exercises	_____	_____	_____	_____
Topic Selection	_____	_____	_____	_____
Currentness of Coverage	_____	_____	_____	_____
Explanation of Difficult Concepts	_____	_____	_____	_____
Match-up with Course Coverage	_____	_____	_____	_____
Applications to Real Life	_____	_____	_____	_____

3. Circle those chapters you especially liked:
1 2 3 4 5 6 7 8 9 10 11 12 13 14 15 16 17 18 19 20
What was your favorite chapter? _____
Comments:

4. Circle those chapters you liked least:
1 2 3 4 5 6 7 8 9 10 11 12 13 14 15 16 17 18 19 20
What was your least favorite chapter? _____
Comments:

5. List any chapters your instructor did not assign. _____

6. What topics did your instructor discuss that were not covered in the text?_____

_____

7. Were you required to buy this book? _____ Yes _____ No

   Did you buy this book new or used? _____ New _____ Used

   If used, how much did you pay? _____

   Do you plan to keep or sell this book? _____ Keep _____ Sell

   If you plan to sell the book, how much do you expect to receive? _____

   Should the instructor continue to assign this book? _____ Yes _____ No

8. Please list any other learning materials you purchased to help you in this course (e.g., study guide, lab manual).

   _____

9. What did you like most about this text? _____

   _____

10. What did you like least about this text? _____

    _____

11. General comments:

    May we quote you in our advertising? _____ Yes _____ No

    Please mail to: Boyd Lane
                     College Division, Research Department
                     Box 508
                     1300 Alum Creek Drive
                     Columbus, Ohio 43216

    Thank you!